Thinking through
Science and Technology

Thinking through Science and Technology

Philosophy, Religion, and Politics in an Engineered World

Edited by

Glen Miller, Helena Mateus Jerónimo, and Qin Zhu

ROWMAN & LITTLEFIELD
Lanham • Boulder • New York • London

Published by Rowman & Littlefield
An imprint of The Rowman & Littlefield Publishing Group, Inc.
4501 Forbes Boulevard, Suite 200, Lanham, Maryland 20706
www.rowman.com

86-90 Paul Street, London EC2A 4NE

Copyright © 2023 by The Rowman & Littlefield Publishing Group, Inc.

All rights reserved. No part of this book may be reproduced in any form or by any electronic or mechanical means, including information storage and retrieval systems, without written permission from the publisher, except by a reviewer who may quote passages in a review.

British Library Cataloguing in Publication Information Available

Library of Congress Cataloging-in-Publication Data

Names: Miller, Glen, 1975- editor. | Jerónimo, Helena Mateus, editor. | Zhu, Qin, 1983- editor.
Title: Thinking through science and technology : philosophy, religion, and politics in an engineered world / edited by Glen Miller, Helena Mateus Jerónimo, and Qin Zhu.
Description: Lanham : Rowman & Littlefield Publishers, [2023] | Includes bibliographical references and index.
Subjects: LCSH: Technology--Philosophy. | Technology and state. | Technology--Social aspects.
Classification: LCC T14 .T446 2023 (print) | LCC T14 (ebook) | DDC 601–dc23/eng/20230202
LC record available at https://lccn.loc.gov/2022055691
LC ebook record available at https://lccn.loc.gov/2022055692

ISBN: 978-1-5381-7650-4 (cloth : alk. paper)
ISBN: 978-1-5381-7651-1 (pbk. : alk. paper)
ISBN: 978-1-5381-7652-8 (ebook)

∞™ The paper used in this publication meets the minimum requirements of American National Standard for Information Sciences—Permanence of Paper for Printed Library Materials, ANSI/NISO Z39.48-1992.

For Carl,

whose work has forged so many intellectual spaces and connections in which scholars begin to see each other as bound together through shared conversation and thinking, an existence that is truly human (cf. Nicomachean Ethics, 1170b10–14),

who quickly learned and taught so many others that "to study and at due times practice what one has studied, is this not a pleasure? When friends come from distant places, is this not joy? To remain unsoured when his talents are unrecognized, is this not a junzi [君子]?" (Analects, 1.1),[1]

whose being-in-the-world is "to realize in oneself all humanity in all moments" (Passagem das Horas, Álvaro de Campos/Fernando Pessoa).[2]

---

1. From Dr. Robert Eno's translation of the *Analects*, available at https://chinatxt.sitehost.iu.edu/Analects_of_Confucius_(Eno-2015).pdf. We appreciate that Dr. Eno granted us permission to use this translation.

2. The English translation of *Time's Passage* is from *Pessoa: An Experimental Life*, translated by Richard Zenith (New York: Grove Press, 1998). The original excerpt of the poem by Álvaro de Campos (one of Pessoa's heteronyms) is *"Realizar em si toda a humanidade de todos os momentos."*

Contents

Foreword: Possessed by Engineering and Technology   xi

Preface   xv
   Glen Miller, Helena Mateus Jerónimo, and Qin Zhu

Chapter 1: Editors' Introduction   1
   Glen Miller, Helena Mateus Jerónimo, and Qin Zhu

PART I: PHILOSOPHY AND TECHNOLOGY   11

Chapter 2: The Enigma of Technology   13
   Andrew Feenberg

Chapter 3: Organization as Technique: A Blind Spot in the Philosophy of Technology   29
   Daniel Cérézuelle

Chapter 4: Technology as Process   55
   Mark Coeckelbergh

Chapter 5: Political Philosophy of Technology: After Leo Strauss   69
   Carl Mitcham

Chapter 6: The Nuclear Menace and the Prophecy of Doom   93
   Jean-Pierre Dupuy

Chapter 7: The End of Technology and the Renewal of Reality   113
   Albert Borgmann

PART II: PHILOSOPHY AND ENGINEERING — 117

Chapter 8: An Engineer Considers Technological (Non)Neutrality: "But Where Are the Values?" — 119
Byron Newberry

Chapter 9: How Engineers Can Care from a Distance: Promoting Moral Sensitivity in Engineering Ethics Education — 141
Janna van Grunsven, Lavinia Marin, Taylor Stone, Sabine Roeser, and Neelke Doorn

Chapter 10: Parallel Steps toward Philosophy of Engineering in China and the West — 165
Nan WANG and LI Bocong

Chapter 11: The Development of the Philosophy of Engineering in China: Engaging the Scholarship of Carl Mitcham — 191
Tong LI and Yongmou LIU

PART III: RELIGION, SCIENCE, AND TECHNOLOGY — 207

Chapter 12: Christianity, Power, and Technological Domination: A Typological Approach to the Church — 209
José Antonio Ullate

Chapter 13: Technology in Cosmic Terms: The World Council of Churches in Amsterdam, 1948 — 225
Jennifer Karns Alexander

Chapter 14: Beyond Tools, Means, and Ends: Explorations into the Post-Instrumental Erewhon — 243
Jean Robert

Chapter 15: Understanding Bureaucratic Order: The Theological Paradigms of Modern Hierarchy — 255
Sajay Samuel

Chapter 16: What Religion, What Technology?: A Wittgensteinian Approach — 277
Andoni Alonso

Chapter 17: Bioethics, Philosophy, and Religious Wisdom: A Critical Assessment of Leon Kass's Thought — 297
Larry Arnhart

PART IV: SCIENCE AND TECHNOLOGY STUDIES — 317

Chapter 18: Ethics and the Search for Scientific Knowledge: The Whole Truth and Nothing but the Truth? — 319
Carlos Verdugo-Serna

Chapter 19: A Short History of Science, Truth, and Politics in the United States, 1945–2021 — 335
Daniel Sarewitz

Chapter 20: Moral Narratives of Technological Change in the Early Green Revolution — 345
Suzanne Moon

Chapter 21: Momentum, Interrupted: Developing Habits of Discernment in Engineering and Beyond — 367
Jen Schneider

Chapter 22: Innovation Policy Driven by the Market: The Second Great Disembeddedness — 391
José Luís Garcia

PART V: SCIENCE AND TECHNOLOGY POLICY — 415

Chapter 23: Irrational Energy Ethics — 417
Adam Briggle

Chapter 24: Paradoxical Policy in Sub-Saharan Africa: Women's Farming, Oil, and Sustainable Development — 435
Tricia Glazebrook and Gordon Akon-Yamga

Chapter 25: The Pandemic and Clamor for Vaccines: Ethical-Legal Considerations **for** Intellectual Property Rights and Technology Sharing — 461
Pamela Andanda

Chapter 26: An Effective History of the Basic-Applied Distinction in "Science" Policy — 483
J. Britt Holbrook

Chapter 27: Technological Risks, Institutional Wariness, and the Dynamics of Trust — 499
José A. López Cerezo

Index 517
About the Contributors 557
About the Editors 565

Foreword

Possessed by Engineering and Technology

> Depending on how close you stand, there can be a lot more than you in the mirror.
>
> —P Hans Sun, "The Sound of One Mirror Clapping"

This book is possessed by engineering, science, and technology in manifold forms, from professional practice to political and religious expressions. As one similarly possessed, I've been invited and then encouraged by the editors to add a personal note.

It's hard to know how most appropriately to do this—except by expressing surprise and thanks to the Glen Miller, Helena Jerónimo, and Qin Zhu editorial team for their broad interdisciplinary, cross-continental collective engagement. Independent of any relationship it has to me, this volume surely provides a unique, compelling invitation to think broadly and deeply about life in an increasingly engineered and engineering world.

In their preface the editors mention that their project was somehow motivated by key characteristics found in the work of a guy called "Carl Mitcham." Although I know there is a person with that name somewhere in the world, I have a hard time recognizing him. I'm not sure who he really is. It may be a phenomenon paradoxically promoted by a culture of high self-consciousness. He's seldom in the mirror when I look. There are only pictures from an exhibition. As Ivan Illich once described the present, we are like crabs pushing ourselves forward while looking backward into the mirror of the past, stumbling into our next backward step into the future with stories of how we got here.

There are two stories I've told myself about how engineering and technology took hold of my thinking. My father was a mechanical engineer who occasionally took me to work. I was taken by the contrast between his work and that of an uncle who owned a farm in central Texas. When spending summers at the farm, there was always work to do: helping to milk the cows, feed and water chickens, to pen them up at night, to collect eggs in the morning, to ride the combine with my cousin. With my father I could make no contribution; at his office there was nothing for me to do. While he sat at his drafting table solving design problems, I could only sit at another desk and play, creating interesting shapes with straight edge, triangle, and compass. Even at the age of five or six the contrast registered at some level, to return again and again as if in a dream.

A second dream takes place as a dialogue with Donald Davidson. I was for a short period an undergraduate at Stanford University when Davidson was working through "Actions, Reasons, and Causes." I signed up for a course on philosophy of action, but very shortly became disconcerted. One day after class I approached Professor Davidson and asked, why in a course on action, we were not considering more than the difference between me raising my arm and my arm going up. What about real actions such as dropping atomic bombs and polluting the natural environment? (Recall that 1962 witnessed both the Cuban missile crisis and publication of Rachel Carson's *Silent Spring*.)

The first story points toward the need to reach out from engineering into philosophy; the second further out from philosophy to what emerged in the late 1960s as science, technology, and society (STS) studies. But because of the great good fortune, after dropping out of Stanford, of encountering at the University of Colorado one mentor who had studied with Étienne Gilson and another with Leo Strauss, I became committed to an STS-ism that retains a strong philosophical tinge.

Reading this collection now, it's possible to detect another story. According to Aristotle, the first name of the good is that which is our own, and the practice of philosophy is the gradual expansion of our sense of the whole. What happens when in the initial moment a world fails to make such an appearance, or insufficiently appears as good, or when the good appears as other? Are strange attractors no more than orientalisms? Can't crossing boundaries, cultural and disciplinary, become more than superficial, establishing their own distinctive depths—especially when we try to comprehend something as daunting as our scientific, engineering, technological experience?

As a necessarily biased reader, I'd submit that this collection offers some of the best evidence available in the affirmative. On the surface, one finds a spectrum of articles reflecting multiple interactions between philosophy, technology, engineering, religion, science, STS, and science policy by contributors from Europe, the Americas, Asia, and Africa. Particularly notable, along with representatives from two non-American centers for philosophy of technology—the Netherlands and China—are multiple Ibero-American contributors who deserve better recognition in the Anglophone world. Collectively, they are philosophy all the way across and down, exhibiting a consistent effort to reflect on the familiar, that which is closest to us in our engineering world, as nevertheless uncanny, if not mysterious. Socrates (in Xenophon's remembrance) "never wearied of discussing human things: What is piety? Impiety? The beautiful? The ugly? The noble? The base? What is meant by just and unjust? Sobriety and madness? Courage and cowardice? What is the *polis*? Politics? Kingship? Character?" In this Miller, Jerónimo, and Zhu creation we find these questions compelling brought to bear on human things under conditions of a historically profound mutation.

Carl Mitcham (Alamo, Colorado)

Preface

Glen Miller, Helena Mateus Jerónimo, and Qin Zhu

Readers who are even somewhat familiar with philosophy of technology or philosophy of engineering literature have likely noticed that our title plays off of Carl Mitcham's *Thinking through Technology: The Path between Engineering and Philosophy* (Chicago: University of Chicago Press, 1994). Those who are familiar with his body of work will note that each part roughly coincides with one part of his broad scholarly career, which spans philosophy, religion and theology, history, education, policy, science, technology, and engineering, and more: he is interdisciplinarity personified! Some of you may recognize that many of the contributors are members of a not-at-all exclusive club of individuals who have collaborated with him at some point over the last fifty years. And some of you may, like us, hear echoes of the voices of some of his early collaborators, such as Robert Mackey and Jim Grote, in its chapters.

Carl was indeed the motivating force behind this collection, but what motivated us was not a desire to draw attention to his ideas, as worthy as such an effort would have been. Instead, we were motivated by two of his character traits. One is his insatiable interest in hearing the ideas of others, especially junior scholars, and his tireless efforts assisting them in developing their ideas, many times by connecting them with other ideas, scholars, or institutions that will help. With this recognition, we immediately realized that the overarching goal for the volume should be to assemble a collection of papers that he would like to read, which is the mandate we gave the contributors.

His second character trait that motivated our work is his commitment to editorial excellence. While we are certain we fall short of the gold standard, we aimed to imitate the painstaking care—and seemingly endless enthusiasm that he somehow always summons—that he has given so many other texts over the course of his career. We are grateful for our contributors, not only for the original ideas they have developed, but also for their patience and amicability in response to comments on early drafts of the papers: editing an interdisciplinary book is difficult, and we have learned a great deal from you!

We hope our readers, those well-schooled in philosophy of technology and those who are not yet, appreciate the innovative and thoughtful arguments in the collection. If we hit the perfect pitch, the book will be a physical manifestation of Carl's warm intellectual embrace of those who, regardless of prior expertise, disciplinary background, and cultural context, want to think philosophically about science, technology, and engineering, in order to better understand their world. Following Carl's spirit, we hope that students of all ages, at all ranks, from all disciplines find this book provocative, that it serves as an additional spur toward discourse, in conversations as well as in texts.

We would be remiss not to mention our gratitude to the team at Rowman & Littlefield, especially to our acquisitions editors Frankie Mace, who helped initiate the project, and Natalie Mandziuk, whose assistance, patience, and support helped bring it to fruition. We are grateful to the proposal reviewers for their advice, which helped round out the volume with several chapters that were not part of our original vision. Thanks to the Colorado School of Mines Daniels Fund Program in Professional Ethics Education that funded the translation of one of the chapters, and Virginia Tech Department of Engineering Education for providing funding for indexing. The index is the fine work of Laura Shelley.

And lastly, thank you, Carl, for your support of this project and through the years, and for all you have done to build a global intellectual community committed to thinking through what is good, beautiful, and true in the world.

Chapter 1

Editors' Introduction

Glen Miller, Helena Mateus Jerónimo, and Qin Zhu

Thinking through science and technology is paradoxical. One could say that it has never been easier. Science now plays an important part of the K–12 education system, with a never-ending chorus of STEM apologists reinforcing its importance and power. Never before have more people been educated in scientific methods, which are usually held as the "gold standard" for understanding, not just in the West, where it originated, but in cultures around the world. Governments and policymakers seem to worry about scientific literacy more than they worry about civic, historical, or moral literacy, as the latter has less influence on the economic and military competitiveness of a people. In an interesting twist, as what is considered science has shrunk from its original sense of knowledge understood broadly ("archaic" or "obsolete," according to the *Oxford English Dictionary*) to now only refer to what is obtained according to methods that use careful observation, usually instrumentally measured, and stepwise reasoning, these more limited methods have been applied to an increasing range of subjects: from the natural sciences that described the movement of bodies of Copernicus, Galileo, and Isaac Newton to ecology, transformed from natural history to ecosystem analysis; psychology, now a data-driven, standardized field with "proven" best practices; and the workplace, now understood and optimized by industrial engineering and industrial and organizational psychology.

Technology directly contributes to the experimental processes of modern science when it is employed to precisely measure various attributes of scientific subjects, so thinking through science is also thinking through technology. It is also used to construct laboratories where scientists are able to control the environment so that they can measure what they desire while holding everything else constant, and it is used in interventions that attempt to test the accuracy of the predictions of hypotheses. Similar to science, technology has also had a modern transformation: one sees its etymological origin in the Greek concept of *techné*, which in one sense makes a fundamental distinction between artificial objects and those that are natural (*physis*), but now it commonly refers only to artifacts developed using insights from modern science and modern production techniques—that is, those that are engineered.

Technology has become more affordable, more available, and more powerful: it increasingly satisfies the related goals of "convenience and use." Channeling media theorist Marshall McLuhan, technology is an extension of man: human capabilities to observe, analyze, plan, remember, and act are supplemented, replaced, and transformed by the artifacts that humans have created. Instruments capture the movement and activities of most individuals in developed countries and offer views into distant galaxies. Computing resources are used to analyze this data, to find correlations and trends that went previously undetected. These same resources can be used to coordinate natural, human, and technological resources for projects of a scale previously barely imaginable that depend on complex supply chains with outposts around the world. Technologies provide a historical record of what we say and do, part of a collective memory that differs in important ways from the way human memory has evolved and that takes a different form from writing technologies—which already drew Socrates's concern in the *Phaedrus*—as they have become increasingly pervasive. And modern technology permits humans to act at scales (micro and macro) and with an intensity and lasting duration that have no historical precedent. When we think today, science and technology are often means and subject.

But one could also say that thinking through science and technology has never been harder. Their very successes have led to their adoption in nearly all modes of human activity in most cultures: humans in developed countries use science and technology not just in the workplace but throughout the day, from waking to sleep (which itself is increasingly monitored); it is part of a conditioning that leads them to interpret the world in terms of modern science and technology. While we can remember a world with limited mediation—where high technology and intensive science was dominant in only certain delimited domains—we struggle to imagine a future world in which that is the case: the size, complexity, and characteristics of the sociotechnical networks (or, perhaps better, associations) that have been developed are path dependent. Many skills, bodies of knowledge, and attitudes that were essential for human existence and social life for millennia and had developed over long time frames have been lost in a matter of decades. This impermanence stands in contrast to the material products of our sci tech initiatives, some of which last for longer than a human lifetime when they receive the resources and maintenance they need to function; even when they cease to function or are no longer of use they persist as decaying objects. In a world so transformed by technology and engineering that it has been called the Anthropocene, the material and social world have been so transformed that it seems one must think through science and technology.

Science and technology provide a lens through which the world is interpreted and structure the space in which one acts. The lens provides focus, but also, at least for a moment, it limits peripheral vision, and making sense of what one sees requires a new complex synthesis—one that is always incomplete and perhaps inadequate to what has been constructed. The conception of truth utilized by technology and modern science emphasizes practical utility, which relegates questions about how one should live and the ultimate foundations and ends of human existence to the sidelines: most of philosophy and religion is seen as distracting from solving problems at hand that can be solved by dedicating more resources to science and technology initiatives. But thinking of humans as intelligent machines is a fallacy of reduction even when intelligence is understood more broadly than economists' "rational agents" who seek to optimally use resources to lead to their advantage, which is often imagined as a primarily hedonistic calculus.

At a time when science and technology is pervasive, its values (especially convenience, efficiency, and utility) have in many areas been adopted as the primary set of human values. Human ingenuity, uniqueness, and creativity become vulnerabilities when systems thinking is dominant: systems are easiest to maintain when they are composed of easily replaceable cogs. This shift in values has rarely taken place through reasoned democratic discourse; rather, it has come about through economic and political machinations that have coincided with, and often contributed to, the weakening of prior institutions, arrangements, and value systems. One unsurprising result of this shift, one that is rather ironic given its emphasis on what is instrumentally measurable and functional, is that STEM has obtained an aura that exaggerates its benefits while camouflaging its limitations and failures, both in what has been experienced and what is promised. It has received de facto protection from critical assessment, even as it increasingly consumes resources (most importantly human attention, but also increasing amounts of energy and metals, including rare earth minerals) and structures intellectual and social existence. When we think today, it often seems impossible to transcend—or even understand—the distortions, limitations, and power of science and technology.

The present volume seeks to plumb this paradox with a critical and timely reflection on science and technology that counters trends toward technological optimism, on the one hand, and disciplinary and cultural regionalization, on the other. While science and modern technology have some universalizing characteristics, their implementations, concerns, consequences, and meaning differ across cultures. Similarly, those attempting to understand and respond to science and technology will benefit from the expertise of a diverse group of scholars working from a variety of disciplines that, put together, provide a global view.

Part I, on "Philosophy and Technology," consists of research that opens a comprehensive understanding of technology itself and its inscription in a wider web of social and political meanings, values, and civilizational change. It begins with Andrew Feenberg's problematization of the enigma of technology—that it is somehow determined by humans and also somehow determines them. As he puts it, "Technology lies at the intersection of causality and teleology." Further developing Martin Heidegger's concept of "world," Feenberg argues that an ontology of technical domain that starts in the constant interaction between "ought" and "is" is needed to restore the notion of potentiality, eliminated by modern science, which reveals the contingency of the "real."

Two chapters offer insights on technology that have been obscured since the "empirical turn" or "thing turn" in philosophy of technology in the 1990s. Mark Coeckelbergh sketches the outlines of a process-oriented phenomenology and hermeneutics of technology grounded on becoming (rather than being). In this framework, technology is a relational, performative, historical, and narrative process that has both bodily and social-cultural dimensions. Seen in this way, engineers and designers of technology are also, as he puts it, "designers of processes and performances, of subjects, of time, of forms of life, and of narratives." Daniel Cérézuelle argues that the "thing-turn" has led to a "blind spot" in philosophy of technology regarding intangible technologies, even as techniques directed at humans increasingly organize and determine the spatial, temporal, and relational dimensions of daily life through management, education, planning, propaganda, and so on. These techniques were criticized by personalist philosophers such as Nikolai Berdyaev, Bernard Charbonneau, and Jacques Ellul, who were sensitive to their depersonalizing forces and sought to preserve freedom in the face of those pursuing the total technoscientific organization of society.

While critiques of technology are often based on social consequences, there has been little done to develop a political dimension of philosophy of technology that draws substantively on the resources developed in political philosophy. As a response to this shortcoming, Carl Mitcham shows how ideas from Leo Strauss can be used to chart a course of inquiry that considers the historical quarrel between revelation (Jerusalem) and reason (Athens) in light of the dominance of technoscience characteristic of modernity. Both biblical and philosophical wisdom seem impotent, yet without such wisdom, engineering and technology will become ever more vicious and destabilizing. Jean-Pierre Dupuy analyzes the surprising effectiveness of the mutually assured destruction (MAD) approach to nuclear deterrence, explaining it by way of a "projected time" metaphysics in which one can question whether an event that will never take place must be taken as impossible. He argues that the prophet announcing doom oscillates between catastrophism, in stating what is inscribed in the future, and optimism, in that what he says will prevent these events. The chapter by Albert Borgmann provides a coda to the first part, arguing that, by virtue of its very successes, technology has lost its motivating force, as it has made so many things commodities, conveniently available but only at the cost of decontextualization. He argues that a "renewal of reality" requires re-embedding things and practices in a time, place, and community, and sees the new urbanism, the artisan economy, and organic farming as movements that succeed in drawing focus to things that matter.

Part II, "Philosophy and Engineering," begins with Byron Newberry's critical reflection on his engagement, as an engineer and engineering professor, with the technological neutrality debate. To explore the tension between the idea held by many engineers—including many of his students—that technology is value-neutral with the consensus view in philosophy of technology that it is value-laden, Newberry provides a comprehensive analysis of the flurry of recent literature on technological (non-)neutrality and explains why it matters for engineers and engineering students. The development of engineering students is also at the core of Janna van Grunsven, Lavinia Marin, Taylor Stone, Sabine Roeser, and Neelke Doorn's paper. They propose a novel framework to inject care ethics into engineering education by adapting a theoretical framework on moral sensitivity developed in nursing ethics. They argue that care ethics arises differently in nursing than in engineering, as the former is characterized by proximity and the latter by distance. Nurses offer *particularized care*, whereas care at a distance demands *generalized care* attuned to representatives affected by the engineer's work, as well as *universalized care*, which recognizes the technology-dependent vulnerability of everyone affected by the individual's engineering.

The last two chapters in Part II analyze the development of philosophy of engineering. Nan WANG and LI Bocong's comparative historical study of the content and institutionalization of the philosophy of engineering in China and the West shows near simultaneous progress. The embryonic stage (1990s) was marked by a few Chinese and Western scholars working independently but in parallel; from its emerging stage (2000s) and the developmental stage (2010s), conferences and publications became increasingly regular, and more and more scholarly exchanges occurred as the growth of philosophy of engineering became a global phenomenon. Tong LI and Yongmou LIU explore the development of philosophy of engineering in China by showing how its scholars have been "in negotiation" with ideas that arose in the Western context, especially those of Carl Mitcham. While there is much in common, they argue that the Chinese tradition is distinct because Chinese engineering has always had a political character—in part because it arose in response to fears of Western colonialism—and because the milieu in which it developed was shaped by Marxism, Confucianism, and other Chinese intellectual traditions, which results in the view that engineering should be transformative while also finding and preserving harmony.

Chapters in part III investigate the complex relationship of "Religion, Science, and Technology" as they have evolved in Western culture. José Antonio Ullate proposes that an understanding of the relationship between Christianity and technology based primarily on biblical exegesis and the historical record of Europe and North America, especially during the modern period, is incomplete. In a novel application of the typographical approach, he argues that insights into the way that technological power (and power in general) is understood by Christians and nation-states influenced by it can be gained with a nuanced review of the formation of the Jewish people. Jennifer Karns Alexander explores a more recent historical event, the 1948 World Council of Churches, that analyzed and responded to the technological context that existed in the wake of World War II. This global ecumenical conference, animated in part by the ideas of Ellul, developed a consensus critique of the technological hubris of the time that required a radical conversion of how Christians should respond to God and relate to each other.

Jean Robert draws from Ellul and Ivan Illich, in particular the latter's historical analysis of the medieval theological innovation of *causa instrumentalis*, to explain the transition from technological artifacts closely connected to the ends of those who use them, proportionate in their means, and limited in some way, to their present manifestation as part of technological systems in which means and ends are bureaucratically determined and to which individuals become subsystems. Sajay Samuel directs his focus on the theological paradigms that help to understand the rise of the bureaucratic order, an "organizational technique" that mediates human action in countries around the world, regardless of their economic or political orientation. With an analysis grounded on the work of Illich, Michel Foucault, and Giorgio Agamben, he illuminates the theological antecedents of the hierarchical arrangement, instrumental logic, and universal scope found in modern bureaucracy.

Inspired by insights of Ludwig Wittgenstein as well as Ellul and Illich, among others, Andoni Alonso plumbs how religion, which speaks of things beyond sense, relates to science and technology. Rather than finding contradiction, his investigation reveals multifaceted and disputed relationships that are sometimes seemingly inverted in the modern period, as science and technology are treated as though they are religions, which is especially apparent in cases of technological idolatry. Larry Arnhart submits Leon Kass's historical progression of philosophical, religious, and scientific beliefs to the test, assessing their coherence and how they influenced his views on human nature, human dignity, and the appropriate limits of biotechnology. While Kass advocated for "a richer bioethics" informed by classics in the Western tradition to discern those limits, Arnhart instead argues for a Darwinian science of human nature as part of liberal education.

The first two contributions in part IV, "Science and Technology Studies," explore the ethics and politics of science. Carlos Verdugo-Serna proposes that an adequate ethics of science, which in most accounts collapses into research ethics, needs to include a moral fundamental inquiry into the kinds of research that should be prohibited or discouraged, in contrast to an absolute right to inquiry that is usually assumed. Even basic science or pure research, he argues, which often are assumed to "immunize" scientists from ethical inquiry, needs to be examined. In response to concerns that we are now living a post-truth era, Daniel Sarewitz identifies some key moments in United States history, starting with the 1945 bombing of Hiroshima through the present, that serve as indications that the connection between science and truth has been shown to be—and needs to be considered as—more complicated, problematic, and politicized than is commonly accepted.

The other chapters in the part are interdisciplinary inquiries into innovation, technology adoption, and engineering. Whereas most historians of technology (and STS practitioners in general) tend to ignore ethics, which is often reduced to an anecdotal analysis of power when it is considered, Suzanne Moon draws off of the work of philosopher Margaret Urban Walker to develop a richer, truly interdisciplinary approach to technomoral history through her investigation of the moral questions, assumptions, and narratives made by farmers, the governments, and international entities during Indonesia's Green Revolution, and how and why they changed over its course. The way we think and talk about engineering is investigated by Jen Schneider. She employs a metacritical analysis to analyze Carl Mitcham's body of work, which she uses as a proxy for Western philosophy of engineering, to discern its key "terministic screens" that emphasize or diminish certain concerns and themes, and, in doing so, outlines its past trajectory and forecasts its future. José Luís Garcia reveals the key assumptions and forces behind the development and adoption of innovation policy, now largely considered a fait accompli and interpreted as a manifestation of what he dubs "technological historicism." He argues that over the last forty years the pursuit of technological and "marketological" innovation under neoliberal governance has resulted in a form of disembeddedness in which the relationship between society and the market is inverted, where the former becomes almost completely subservient to the latter.

Chapters in part V, titled "Science and Technology Policy," address questions of whether and how institutions (especially nations) should support and fund initiatives that are often markedly politicized and ethically contentious and have global implications. The first two focus on energy and its socio-environmental impacts. Adam Briggle denounces the irrationality—and inevitability—of the economic and technological models that shape current and future energy consumption. He invites the reader to imagine being an energy czar who, on the one hand, must deal with existing energy consumption patterns built into developed society to obtain popular consent; on the other hand, a rational czar should also aim to reduce overall energy usage, distribute its benefits more fairly, and reduce its ecological impacts. Climate change is already affecting agricultural practices in Africa, and Tricia Glazebrook and Gordon Akon-Yamga expose the policy paradoxes encountered as sub-Saharan Africa seeks to reduce poverty through oil and gas exploration and development. While many would accept such development to reduce hunger, its capital- and market-based systems overlook the critical contributions of women subsistence farmers, who are "invisible" in economic assessments and policy design. Changes in rainfall patterns have already affected the crops they grow and the yields they obtain, resulting in marked decreases in food security that have worsened in the last five years. The COVID-19 pandemic and the uneven distribution of vaccines, tests, and related technologies, especially between developed and developing countries, provide the departure point for Pamela Andanda's work on intellectual property rights (IPRs). Her critical and detailed analysis of the key policies that govern international arrangements and the decisions made by various political entities and corporations shows shortcomings of the current system. She outlines a new IP paradigm, a rights-based approach grounded on the principles of justice, fairness, and global solidarity.

The other texts in this part explore more general concerns regarding scientific and technological policy. On the seventy-fifth anniversary of Vannevar Bush's *Science, the Endless Frontier* and the seventieth anniversary of the US National Science Foundation, J. Britt Holbrook argues that a policy for "science" that is wholly dedicated to applied science runs the risk of becoming a totalizing instrument of the state, especially if it is seen as a means to technological supremacy. Science policy should be broadly understood to include technology policy, but it also needs to be positioned within a broad spectrum of knowledge and knowledge production that includes philosophy and the humanities. José Antonio López Cerezo argues that the significant crises in the food, health, and industrial sectors that occurred in the past decades have eroded trust in institutions and technology, leading to what has been called a "post-trust society." Rather than seeing this turn of events as catastrophic, he sees the development of critical awareness and science-informed distrust as essential in responding to inescapable risks created by new technologies and critical for democratic governance in contemporary societies.

While some paradoxes are only apparent, the present volume does not claim to resolve what was described at the beginning of this introduction. Its aim is far more modest: to help readers develop a critical and multi-perspectival understanding of science and technology; to be more aware of their power and potential, of their effects on individuals, society, and the natural environment, as well as their limitations; to consider and reconsider what they believe, what they value, how they act, what they build, and what they seek; and to spur other thinkers to do the same.

PART I

Philosophy and Technology

Chapter 2

The Enigma of Technology

Andrew Feenberg

Technology is instrumental to our goals, but our goals grow in large measure out of our technologically situated existence. This is the strange loop of society and technology. The enigma of technology is this undecidable alternative between determined and determining. Philosophy of technology has oscillated between the alternatives since its origins. Either technology is an extension of human powers or it is a power in its own right that dominates its human creators. The enigma is deepened by the role of rationality in the organization of modern societies and the technology that supports them. Is rationality in this sense an expression of human agency or an autonomous power operating according to its own logic? In this article I will address this enigma through a discussion of Heidegger's philosophy of technology.

Heidegger is rarely discussed in philosophy of technology today, still less in Science and Technology Studies. He is among the predecessors against whom the new disciplines constitute themselves. Disciplinary boundary-drawing has precluded serious confrontation with his thought, a situation no doubt aggravated by his infamous obscurity. There are insights but also problems with Heidegger's several accounts of technology. His most important insight is his concept of "world." What we have learned about technology in recent years, both from academic research and political mobilization, points toward a specific revision of this concept that will help us address the enigma of technology.

TECHNOLOGY AND WORLD

By "world" Heidegger means a system of meaningful entities that refers back to an agent, which he calls *Dasein,* capable of interpreting its environment and entertaining purposes. This phenomenological concept of world is difficult to separate from the usual commonsense and naturalistic concepts. Because it presupposes meaning and intention, "world" is not identical with the totality of entities, as common sense would have it, nor with the cosmos studied by natural science. They treat what Heidegger calls "world" as a system of subjective attributions with no ontological significance. But Heidegger regards world in his sense as ontologically fundamental and claims that our ordinary common sense and natural science are founded on it.

Heidegger develops the concept of world as an *existentiale*, that is, as a universal feature of being in its relation to human being. World is a "category" in the Aristotelian sense, but a category dependent on *Dasein*. The universality of such categories overleaps any particular cultural limitation to define the human as such in its relation to being. Thus in his early work Heidegger seeks a philosophical equivalent of the universality achieved by traditional philosophy and the natural sciences. But he seeks it *within* the substance of everyday experience rather than through speculative construction or the methods of quantification and experimentation by which the sciences transcend that experience.

Heidegger defines world as "beings in their accessibility" (1995a, 199). By "accessible" he means understandable *as*, taken *as*, enacted *as*. Thus the chair on which I sit is not simply there as an object but is treated by me as a chair, that is, as intended for sitting. No such relation to the chair is possible for the papers I stack on it temporarily in my preparations for leaving the office. Those papers are supported by the chair, but not *as* a chair. *Dasein* establishes a different type of relation, a relation of meaning. In this sense, then, worlds are existential situations, not collections of things. Perhaps the closest we come to Heidegger's own usage is in expressions such as "the world of the theatre," "the medieval world." Such worlds are not merely subjective but nor are they reducible to the natural objects they encompass.[1]

In his important lecture course *Fundamental Concepts of Metaphysics*, Heidegger (1995a) distinguishes three relations to world in his sense of the term. Natural objects have no world. Their relations to other things are purely causal. Animals are "poor in world." They enact meanings and have purposes, but they cannot represent their objects in language and consider them outside the context of their needs and reflexes. Only human beings have a world in the full sense. They can relate freely to their environment and represent the nature of their objects.

Being and Time explains the concept of world on the model of the workshop and its tools.[2] Given this initial focus it is understandable that Heidegger treats tools in functional terms: artifacts serve. He writes,

Now in the production of equipment the plan is determined in advance by the serviceability [*Dienlichkeit*] of the equipment. This serviceability is regulated by anticipating what purposes the piece of equipment or indeed the machine are to serve. All equipment is what it is and the way it is only within a particular context. This context is determined by a totality of involvements [*Bewandtnisganzheit*] in each case. (1995a, 215)

This totality is a system of references between entities in *Dasein's* world. Heidegger calls this system "significance" and treats it as an open space of meaning within which particular usages are enabled. Technical artifacts thus belong to a world without themselves having a world of their own. Their specific difference from other natural objects is owing to the plan that presides over their production and use.

Departing now from Heidegger's vocabulary, we can reformulate his point in terms of the relation of causal and symbolic dimensions of artifacts. Their parts and their environment stand in causal relations that are more or less coherent. But these relations are *designed*, and their design affects their internal form and configuration. As observers we make sense of this in terms of a symbolic language of function, a teleological attribute not found in nature as natural science understands it. That attribute is not merely extrinsic as might be the attribution of function to a piece of green paper called "money." Rather it is practically effective in the construction of the artifact. It is this materialization of function that makes it possible to activate artifacts for our purposes. Thus a reductive causal account of technical artifacts without regard for their purposes cannot explain them. Their essential characteristic is the parallelism between causal and symbolic levels. Heidegger's analysis of world operates at the symbolic level while presupposing the causal coherence of the objects it contains. In this Heideggerian world, Spinoza's famous dictum prevails: "The order and connection of ideas" is identical to the "order and connection of things."

There is something anachronistic about the workshop as a model. Heidegger would have had more difficulty had he discussed the technologies that surrounded him in the early twentieth century. He notes that the autonomous functioning of machines distinguishes them from tools. That suggests a deceptive similarity between machines and animals (Heidegger 1995a, 215). But Heidegger dismisses the comparison on the grounds that machines must be designed and activated, whereas animals have no designer and "activate" themselves. Machines are thus just a species of equipment [*Zeug*], like hammers and nails. If this is so, one wonders why they go unmentioned in the explanation of world.

Can technology be simply classified with ordinary tools? Doesn't its autonomy have significant implications lost in this amalgam? Heidegger fails to note the specific rationality of modern technology which incorporates an understanding of causality in the relation of its parts. This rationality extends to the whole society not just in machines but in social institutions such as bureaucracies and markets that operate more or less mechanically, that is, independent of individual intentions and in accordance with specific causal principles and relations.

In his later reflections on technology, Heidegger recognizes the specificity of the machine and these social systems in emphatic terms, but there he denies that they are autonomous. Instead, he claims, they are entirely under the sway of an anonymous ordering power he identifies with the "technological revealing." Autonomy has now moved up a notch, to the level of the "revealing." This is a principle of intelligibility defining for the modern era. It has a clear relation to the idea of reification in Lukács and instrumental rationality in the Frankfurt School. But the technological revealing undoes worlds, reducing everything it touches to "objectlessness." Machines simply play their part in the process (Heidegger 1977, 17).

Something has gone wrong in Heidegger's early evasion and later apocalyptic understanding of modern technology. The autonomy of technology implies a different relation to the "totality of involvements" than that of a tool. Technology functions as a kind of intermediary in *Dasein's* world, implementing references "mechanically," that is, establishing relations of meaning through its operations even as it obeys a causal logic.

Consider the mobile phone. It has no function outside a network in which it is connected to many other telephones. But the existence of this network has effects that are not intended by any particular user. The network has the kind of autonomy Heidegger feared, but the invocation of enframing is not a satisfactory substitute for concrete analysis. Among these effects, the mobile phone has eroded the separation of work and private life. It re-signifies these basic categories in establishing physical connections with its wires and transmissions, where formerly spatial separations demarcated social spheres. And it is the telephone that does this, not the users, or rather it is the hybrid of users and telephone network that operates this effect invisibly and anonymously (Latour 1999, ch. 6). This is one of many world-constituting effects of the telephone network. Every important technology effects changes of comparable scope.

The simplest account of these changes would be a causal one: the mobile phone has various material effects. But just as the design of the technical artifact cannot be explained without reference to its symbolic dimension, so too the effects only make sense as meanings. The case is similar to that of the hand gestures that punctuate speech. They can be exhaustively analyzed in terms of the physical structures they activate, the muscles, nerves, and bones, but no matter how far the causal account is pursued, it cannot explain gestures *qua* gestures, as communicating a meaning. Only a hermeneutic approach succeeds at that level. Similarly, the effects of mobile phones, like every other technical artifact that impinges on our world, must be explained in the same dual accounting required to understand design.

In this respect technology differs from tools that play their role in the system of references in which worlds consist without *operating* those references. This is significant because the world is now mediated by technical devices down to its finest details beyond ordinary instrumental control. Technologies belong to a specific environment that they modify both causally and symbolically. It is not that technologically mediated worlds have become meaningless, as the later Heidegger claims, but rather they have become meaningful in new ways that escape foresight and understanding. The structures of ordering and control he identifies coexist with these meanings encountered in a lifeworld that resists reduction to pure instrumentality.

TECHNÉ AND TECHNOLOGY

The ambiguities in Heidegger's overextended concept of enframing can be traced back to his ambivalent understanding of *techné*. The theory of world in *Being and Time* is foreshadowed in Heidegger's early discussions of Aristotle. He writes that for Aristotle "Being means *being-produced*" (2002, 128). Aristotle understood the central role of the making of artifacts for human existence, but he interpreted it objectivistically. In *Being and Time* Heidegger restored the existential dimension to this early Greek conception. Technical work opens up the realm of possibility he calls *world*. He does not limit this "projection" to a menu of specific items. It is an open domain in which beings exhibit meaning and within which human beings are free to entertain projects. The process of world-constitution is thus essentially technical, not in the pragmatic sense but as the emergence of meanings in the course of coping with the environment.[3] But a few years later in "The Origin of the Work of Art" it is art that reveals worlds while technology is dismissed as merely useful (1971a, 64).

The difference in perspective seems to reflect an evolution in Heidegger's views. His *Introduction to Metaphysics* reflects the transition. There Heidegger engages in some questionable hermeneutic moves in order to substitute artistic creation for technical making as the practice of world-constitution. Interpreting an ode in Sophocles's *Antigone* that praises man for his technical skills, Heidegger argues that art is the essential *techné* of the Greeks. Why? Because crafted objects withdraw into inconspicuousness in use, whereas art reveals meanings, enables interpretative acts in which a world appears.

"Through the artwork as Being that *is* [*das Seinde Sein*], everything else that appears and that we can find around us first becomes confirmed and accessible, interpretable and understandable as a *being*, or else as a non-being" (Heidegger 2000, 170).

"The Origin of the Work of Art" goes further, affirming that "*techné* never signifies the action of making" (1971a, 59). Craft is no longer relevant; what matters is only the world-constituting power of the artwork. This interpretation of art is developed around several examples, including Vincent van Gogh's painting of boots and the Parthenon. The example of the Parthenon shows how an artifact can lay out the framework of a city and the life it supports. The Parthenon is of course a great work of art and not a mundane technological artifact. It imposes the structure of the city through its visible form and location.

Eventually, in a much later essay entitled "The Thing," even technical artifacts are granted a weak world-constituting power, but only insofar as they are *not* considered as "object[s] of making" (1971b, 168). Instead, they are said to "gather" human beings, "gods" and nature. *Techné* in the usual sense is definitely banished at the very point where Heidegger grants artifacts a role in the constitution of meaning. Heidegger's main example throughout his essay on the thing is a jug, not a modern technology. Technology is barely mentioned. The nearly contemporaneous essay, "The Question Concerning Technology," explains it as world-destroying, not world-constituting.

Albert Borgmann usefully develops Heidegger's contrast between things and technologies in his book *Technology and the Character of Contemporary Life* (Borgmann 1984). "A thing," he writes, "is inseparable from its context, namely, its world, and from our commerce with the thing and its world, namely engagement" (41). This is close to Heidegger's late notion of the thing as a gathering. "In a device," by contrast, "the relatedness of the world is replaced by a machinery, but the machinery is concealed, and the commodities, which are made available by a device, are enjoyed without the encumbrance of or engagement with a context" (47). The device paradigm flattens the technologically mediated environment into a one-dimensional surface without depth and potential. Borgmann's book is an analysis of the triumph of the "device paradigm" over the relation to things in modern times, the transformation of things into devices.

There is one brief seemingly positive mention of technology in the essay on "The Thing." Heidegger describes the gathering of a harvest wagon and lumber cart on the old stone bridge in his usual sentimental terms, and then suddenly remarks that "The highway bridge is tied into the network of long-distance traffic, as calculated for maximum yield. Always and ever differently the bridge escorts the lingering and hastening of men so that they may get to other banks" (Heidegger 1971a, 44). This curious passage has given encouragement to sympathetic interpreters of Heidegger who defend him against the charge of romantic anti-modernism.[4] Indeed, this treatment of a technological artifact as a gathering suggests that Heidegger might have accepted a richer account of the modern lifeworld than his most critical remarks on technology imply.

Such an account might require a fourth level beyond the natural object, the animal, and the human. The machine cannot be reduced to the indifference of a natural object since it operates autonomously. Nor can all its symbolic effects be traced back to *Dasein*. Of course its operation is not based on consciousness of its environment, perceived meanings, motivations, and so on. It must be activated to operate, but once in action it exhibits some of the characteristics of a meaningful relation to its environment, weaving a world out of the references it establishes. Its activation thus grants it a kind of quasi-world. Depending on the extent of technical mediation, that quasi-world either plays a role in *Dasein's* world or, on the contrary, defines it. In the latter case, which is our case, *Dasein* could be said to live within a technological world, laid out in its essential characteristics not by human intentions but by the artifacts humans create. However, such a revisionary account of enframing would call into question Heidegger's original turn away from technical artifacts toward art.

As noted above, what distinguishes technical making from art, according to Heidegger, is the withdrawal of the technical object in use versus the signifying role of art in bringing a world to presence. As Heidegger points out, it is characteristic of art to command attention and bring other objects to presence in this way. But modern technology belongs to a society stripped of most meaningful ritual and in which art has been reduced to a private experience where it is not simply a market good. In this context the wide reach of technology as it shapes society becomes the principal medium through which meaning is made accessible. But isn't the withdrawal of the technological object precisely the most powerful making present? How else are meanings revealed in modern societies except through the often unnoticed web of enactments and consequences of the technological system? This dispensation is as though crystallized in objects that by their obtrusiveness symbolize the system as a whole, objects such as cars, smart phones, skyscrapers, nuclear reactors, airplanes, and so on.

TECHNOLOGICAL MYTHOLOGY

There is a curious passage in *The Fundamental Concepts of Metaphysics* which gives a hint of how these privileged objects could be understood on

Heideggerian terms. He asks whether a human being can entertain an empathetic relation to a natural object and answers "yes."

> There are two fundamental ways in which this can happen: first when human Dasein is determined in its existence by *myth,* and second in the case of *art.* But it would be a fundamental mistake to try and dismiss such animation as an exception or even as a purely metaphorical procedure which does not really correspond to the facts, as something fantastical based upon the imagination, or as mere illusion. What is at issue here is not the opposition between actual reality and illusory appearance, but the distinction between quite different *kinds* of possible *truth.* (1995b, 204)

What Heidegger misses in this account of the varieties of truth is the specifically modern kind, the kind that is transmitted by the technological system. Instead, he claims that we persist in a purely instrumental understanding of technology even as it reveals an un-world of destroyed meanings. This interpretation of the everyday relation to technology precludes not only empathy with things, but an understanding of them as charged with meaning in their technical context. That we have such an understanding, and that it is not a "mere illusion," is confirmed by the particular way in which technological artifacts are present, as possessing significance within a more or less coherent shared world. In sum, the automobile shapes the city as effectively as the Parthenon, if with considerably less beauty and grace. It thus participates in the creation of what Roland Barthes calls a modern "mythology." At that level, its design must be analyzed in relation to users' perceptions and appropriations independent of any naturalistic account of function. This approach brings into focus the cultural dimension of the interpretative flexibility of technology.

At another level, the automobile belongs to a system of urban transportation that dictates routes, relations, and lifestyles. This second level is structured by social forces that cannot be explained at the first level. Yet the two levels are imbricated; they fit together to shape a world. The systemic level appears immediately as a second nature as, quite simply, "the way things are," the "facts of life." It leaves no room for doubt or contestation under normal circumstances. This aspect of the world-constituting power of technology is well captured by the concept of enframing. Its political implications are developed in Marcuse's analysis of one-dimensionality.

Heidegger's concept of world can be made fruitful for the study of technology if it is reinterpreted to encompass not only the reductive and leveling system effects with which he identifies technology in his late work but also the meaning-granting functions it plays in his early work. The technological world is shaped by a variety of forces, not by a singular technological rationality or *Ge-Stell*. As Heidegger understands it, it is denuded of meaning, or rather, it has a hidden meaning not yet revealed to us. In practical terms, this future prospect offers little more than meaninglessness. But technology materializes values and meanings through the design process and through the networks that establish a way of life for its users. In sum, technology in modern societies does the work Heidegger attributes to art in bringing meaning to presence. But unlike art, technology is a means for the society it shapes. The population is active in assigning technology its tasks and designing it to serve them. The further implications of any particular choice of technical means gradually emerge from practice and experience, provoking new interventions, reinventions, changes, and abandonments. It is this process, the politics of technology, which is occluded by Heidegger's conception in which Being grants meaning.

Coincidentally, around the time Heidegger was writing on technology, Barthes and Jean Baudrillard were exploring hermeneutic approaches to consumer culture. Their investigations show how enframing establishes new meanings in which functionality itself is mythologized. These interpretations concern what in my terminology I call the "domain codes" of consumer goods. Domain codes define progress for whole technological regions. They govern the general *type* of a range of related artifacts, essential qualities that identify those artifacts as appropriately designed. These qualities include not only functional attributes but also the way in which they are presented to users. The surplus of meaning, beyond an abstract conception of function, requires cultural contextualization to be understood.

Barthes and Baudrillard study an important transformation that took place in the twentieth century in the domain codes under which consumer goods were experienced. Their work reveals the variety of processes involved in what appears in Heidegger as a single unique destiny. These processes include effects of meaning on design on which I will focus here.

These French philosophers take a semiotic approach to the study of discourses and technical objects. In Baudrillard, function is transcended toward an imaginary meaning that is not explicit but lived as a background assumption or quality of experience taken for natural and self-evident. Barthes calls this naturalization of the object in bourgeois society "myth." The structure of myth as Barthes explains it resembles an allegory in which a primary signifying act takes on a second-order meaning expressing an ideological content. "The myth operates a prestidigitation which transforms the real, empties it of history and fills it with nature, which removes their human meaning from things" (Barthes 1957, 230). This is one aspect of reification as Lukács describes it.

Barthes's book *Mythologies* analyzes popular images and everyday clichés, with the exception of a chapter on the automobile which suggests how his semiotics might be applied more generally to objects of consumption. Writing in the 1950s, Barthes observed a significant change in the meaning of the automobile, as reflected in the design of the new Citroën DS. From a heavy and clumsy machine, the automobile becomes a goddess ("*Déesse*"). Its extensive glass surfaces reduce the steel of which it is constructed to a mere support. The arrow-like symbol on its grill now speaks of wings "as though one passed from an order of propulsion to an order of movement, from an order of motor vehicle to an organic order" (1957, 152). Finally, Barthes turns to the dashboard, which he finds transformed as well. It no longer resembles the controls of an industrial machine but has become smooth and modern. It reminds him of a nice new kitchen, signifying "a sort of control exercised over movement, conceived henceforth as comfort rather than as performance" (230).

In *Le Système des Objets,* Baudrillard argues that technical artifacts must be understood not only at the functional level, but also as they are experienced. He introduces the linguistic distinction between denotation and connotation to describe these double aspects. What might be called the functional proposition of the artifact, its "denotation," is interpreted and ultimately modified through the "connotations" it acquires in lived experience. Unlike language, which is fairly stable in the face of usage, technology changes rapidly as usage changes. "The connotations of the object visibly mark and alter the technical structures. . . . From all this it results that the system of objects, unlike that of language, can only be described *scientifically* insofar as one considers it as the *simultaneous* result of the continual impact of a system of practices on a technical system" (Baudrillard 1968, 16–17).[5]

Like Barthes, Baudrillard applies a semiotic approach to the automobile. He discusses the fishtail, all the rage at the time when he was writing his book. The fishtail symbolizes the conquest of space, the triumph of speed over distance. Of course it has no real function beyond this symbolic one, but this in no way diminishes its importance for the consumer. On the contrary, the fishtail is the signifier of the functional quality that makes the automobile useful. "The fishtail is our modern allegory. . . . And it is in the allegory that the discourse of the unconscious speaks" (85). An element of nature here stands for the functional quality of the machine. In the transition from quality to allegorical representation something is hidden, censured in the Freudian sense of the term, namely, the fantasies surrounding the automobile as an expression of narcissism and phallic supremacy.

Baudrillard pursues the analysis at a more abstract level as it appears in the general enthusiasm for specific types of stylization of consumer products. His discussion of what he calls handiness (*maniabilité*) shows how the elimination of effort constructs a new relation between hand and technical artifact. Now the hand does not labor, it simply touches a button or a switch that controls the hidden workings of the machine. The ideal design is the perfect accord of control panel and hand, the first neatly adapted to the second. Baudrillard finds this type of design proliferating in the home where the inhabitant relates to the surroundings through a smooth accommodation, like that of hand and glove: we no longer sit upright on the chair but are inserted into its form-fitting curves. "Functionality is thus no longer imposed by real labor, but has become the adaptation of one form to another (the handle to the hand), and, thereby, the elision, the omission of the real labor processes" (75).

Something important has changed in the relation of human beings to artifacts. A world in Heidegger's sense emerges from the change, but it is a world constructed by technology itself in conjunction with *Dasein*. This is a paradoxical world, combining rational order and fantasy. On the one hand it is the product of the most extreme rationalization of production and consumption. On the other hand the functional relationships it establishes are overlaid with fantasies incorporated into design. In a sense different from that intended by Theodor Adorno and Max Horkheimer (1972, xiii), one can truly say that Enlightenment turns into myth.

"OUGHT" AND "IS"

Philosophers from David Hume, Immanuel Kant, and Jeremy Bentham to G. E. Moore assure us that values and facts belong to different ontological registers: values are not real in the way in which facts are real, and no value can be attributed to facts simply because they exist. This is an essential aspect of a modern scientific worldview in contrast with the Aristotelian science of essences. The arguments for this modern view are persuasive so long as the facts are conceived in the framework of a Kantian or a naturalistic ontology. Science and ethics pretend to divide the world and no generally convincing connection between them can be forged. But it is difficult to fit the social "fact" of technology into such a framework. Heidegger's merit is to have broken with the fact/value dilemma and raised the question of technology at the ontological level. The limitation of his breakthrough is his failure to engage with the role of meaning in a technological world.

As Barthes and Baudrillard show, technologies realize the values that preside over their creation and shape the daily lives of those who live them. New values emerge as excluded voices demand changes in existing technology. Something like essences reappears in the environmental movement as concepts such as sustainability mobilize populations for the protection of "nature," not of course the disenchanted nature of physics or biology, but some other nature that was supposed to have disappeared in the enframing. Are we dealing here with mere confusion and, worse yet, regression to a magical worldview? Or is the value-fact conundrum undergoing a fundamental sea-change now that technology intrudes on every aspect of human life?

With rare exceptions, modern philosophy lacks the means to account for the technological world in which values and facts routinely communicate and exchange places. Perhaps the best resource in the tradition for such an account is Hegel's notion of "objective spirit" in which ethical substance is realized concretely. He defines it as follows:

> But the purposive action of this will is to realise its concept, Liberty, in these externally objective aspects, making the latter a world moulded by the former, which in it is thus at home with itself, locked together with it: the concept accordingly perfected to the Idea. Liberty, shaped into the actuality of a world, receives the form of Necessity, the deeper substantial nexus of which is the system or organisation of the principles of liberty, whilst its phenomenal nexus is power or authority, and the sentiment of obedience awakened in consciousness. (1971, ¶ 484)

Hegel is thinking here of custom and law, the forms in which he understood the realization of values in the world. Artifacts do not appear in his work except as property. Today, surrounded as we are by technology, we might consider the artifactual world also to realize the will in precisely Hegel's sense. Certainly we have molded a technological world in accordance with our purposes. But technology's causal efficacy has always been understood to place it in the domain of the sciences. This is the source of much confusion. How can technology realize values when it is an application of scientific knowledge?

Technology lies at the intersection of causality and teleology. This puzzling locus is the one we must explore if we are to correctly address the enigma of technology. Heidegger's apocalyptic vision misses this locus, as does the currently popular study of ethical dilemmas arising from the use of technology.

What would an ontology look like that started out not from the absolute divide between value and fact required by modern science, but rather from the fluid relationship between them evident in the technical domain? Such an ontology would restore the notion of potentiality eliminated by modern science in its turn from quality to quantity. Heidegger acknowledged the role of potentiality in his considerations on Greek *techné*. The materials of craft stand in need of the craftsman to become what they can be. He gives the example of the potter. The clay he works is not simply *there* to be formed into a jug; insofar as it is part of the process of production, it *demands* the achievement of form. "With the transformation of the clay into the bowl, the lump also loses its form, but fundamentally it loses its formlessness; it gives up a lack, and hence the tolerating here is at once a positive contribution to the development of something higher" (1995b, 74). It is the disappearance of potentiality that transforms the materials into mere resources, *Bestand,* in modern technology.

But in reality, the technical domain is fraught with unrealized possibilities that are only slightly more contingent than the actually existing technologies. What is properly called a potentiality depends on the logic of social development. Potentialities do not appear out of the blue but arise from the relation between existing technology and human needs. Some of these needs are scientifically determinable and provide the basis on which scientific knowledge enters the lifeworld as appreciation of risk or opportunity. The environmental movement has given us many examples of such needs. Other needs belong to what can only be called the "meaning" of life for the members of the society. Meaning in this sense is a matter of imaginative investments that have general appeal. It may be understood in terms of a generalized concept of the aesthetic as what appears "good" in the way of taste. Meanings shape technology to fit the way of life they imply. Potentialities reflect these two types of needs that eventually are realized technically, only to again project a new horizon of potentialities into the future.

The status of the "real" is thus uncertain and insecure. A current reality may appear inadequate and be replaced—become unreal—as it is confronted with its own unrealized potentialities. A value proposed today may be a fact tomorrow. Indeed, values are simply potentialities excluded from the networks of design and so not essentially different from the existing facts. Since such exclusion may well be temporary, a sliding ontological scale joins given technical realizations to future designs represented discursively at present but potentially capable of realization in the future. The notion that "ought" and "is" belong to different worlds thus collapses in the face of their constant interaction in the technical domain.

The role of time in a technical ontology is Bergsonian rather than Newtonian. From the point of view of modern science, values appear to be a merely subjective feature of individual consciousness, hence without significance for understanding the real. But potentiality is precisely the trace of the future in the present and as such it is more than an idle thought. It is alive in the present as the horizon of development under which the real reveals its impermanence. This horizon is invisible to modern science, but it appears to the creative subjects of technique, capable of articulating unfulfilled needs and demanding change. Such subjects include the users and victims of technology along with the engineers, government, and corporate officials who are the privileged actors in technocratic representations of the technical domain.

Bruno Latour has made the useful distinction between the Janus face of science that looks to the past of science—science that has been made—and opposing Janus face that looks to the future of science in the making. The parallel distinction in the technical domain replaces the traditional contrast of fact and value. The solid world of current technology appears fixed and finished to the backward glance that takes present existence as the touchstone of reality. The forward glance into the future sees more. It sees the contingency of all that is and the potentiality of a more fulfilling future.

Latour's Janus is purely theoretical, but in the case of technology the two perspectives are imbricated in reality. This is the dynamic aspect of the enigma of technology. Technology is neither determining nor determined, but both at the same time. It has the paradoxical form of a "strange loop," a figure in which the top and bottom of a logical hierarchy exchange places. Technology both determines the world and is determined by it through the changes in taste and values it provokes and that ultimately shape and reshape its design. Heidegger's attempt to capture this complexity was short-circuited twice: first, in his early work, by his relegation of technology to mere utility, and second, in his later work, by his casting the technological revealing as a destiny beyond human powers. We are able to better face the enigma, if not to resolve it, as a consequence of the exercise of those powers in so many cultural and political movements transforming the technologies that transform us.

NOTES

1. The sense of "meaning" in this Heideggerian context is complex because he does not consider it to be simply an idea in the mind of a thinker, but rather attributes it to the entity of which it is the meaning. For an account of Heidegger's concept of meaning in relation to subjectivity, see Crowell (2001).

2. The principal discussion of tools is in Heidegger, *Being and Time*, part 1, section III; The distinction in world relations is developed at length in Heidegger (1995a, part 2, ch. 3).

3. I defend this argument against the many commentators who project Heidegger's later anti-productivist stance back to his early work in *Heidegger and Marcuse: The Catastrophe and Redemption of History* (Feenberg 2005, ch. 2). See also Feenberg (2006, 81–93).

4. For an interesting example, see Dreyfus (1995).

5. This argument is similar to that of Pinch and Bijker (1987) in their discussion of the parallel between symbolic and causal levels of design.

REFERENCES

Adorno, Theodor, and Max Horkheimer. 1972. *Dialectic of Enlightenment*. Translated by J. Cummings. New York: Herder and Herder.
Barthes, Roland. 1957. *Mythologies*. Paris: Seuil.
Baudrillard, Jean. 1968. *Le système des objets*. Paris: Gallimard.
Borgmann, Albert. 1984. *Technology and the Character of Contemporary Life: A Philosophical Inquiry*. Chicago: University of Chicago Press.
Crowell, Steven Galt. 2001. *Husserl, Heidegger, and the Space of Meaning: Paths toward Transcendental Phenomenology*. Evanston, IL: Northwestern University Press.
Dreyfus, Hubert L. 1995. "Heidegger on Gaining a Free Relation to Technology." In *Technology and the Politics of Knowledge*, edited by Andrew Feenberg and Alastair Hannay, 97–107. Bloomington: Indiana University Press.
Feenberg, Andrew. 2005. *Heidegger and Marcuse: The Catastrophe and Redemption of History*. New York: Routledge.
———. 2006. "Reply to Dahlstrom and Scharff." *Techné: Research in Philosophy and Technology* 9 (3): 81–93.
Hegel, G. W. F. 1971. *Philosophy of Mind: Part Three of the Encyclopaedia of the Philosophical Sciences*. Translated by William Wallace. Oxford: Clarendon Press.
Heidegger, Martin. 1971a. "The Origin of the Work of Art." In *Poetry, Language, Thought*. Translated by Albert Hofstadter. New York: Harper & Row.
———. 1971b. "The Thing," In *Poetry, Language, Thought*. Translated by Albert Hofstadter. New York: Harper & Row.
———. 1977. *The Question Concerning Technology and Other Essays*. Translated by William Lovitt. New York: Harper & Row.
———. 1995a. *The Fundamental Concepts of Metaphysics: World, Finitude, Solitude*. Translated by William McNeill and Nicholas Walker. Bloomington: Indiana University Press.
———. 1995b. *Aristotle's Metaphysics Θ 1–3: On the Essence and Actuality of Force*. Translated by Walter Brogan and Peter Warnek. Bloomington: Indiana University Press.
———. 2000. *Introduction to Metaphysics*. Translated by Gregory Fried and Richard Polt. London; New Haven, CT: Yale University Press.
———. 2002. "Phenomenological Interpretations in Connection with Aristotle." In *Supplements: From the Earliest Essays to Being and Time and Beyond*, edited by John Van Buren, 111–46. Albany: State University of New York Press.
Latour, Bruno. 1999. *Politiques de la nature*. Paris: La Découverte.
Pinch, Trevor J., and Wiebe E. Bijker. 1987. "The Social Construction of Facts and Artefacts." In *The Social Construction of Technological Systems*, edited by Wiebe E. Bijker, Thomas P. Hughes, and Trevor Pinch, 17–50. Cambridge, MA: MIT Press.

Chapter 3

Organization as Technique
A Blind Spot in the Philosophy of Technology

Daniel Cérézuelle

Translated by Christian Roy

That "organization is a technique" (Ellul 1954, 19) has receded from the contemporary understanding. Especially since the 1990s, philosophy of technology has taken an "empirical turn," refocusing on the study of technical objects and social practices organized around their origins or use. Philosophers such as Peter Kroes and Anthonie W. M. Meijers (2000), Hans Achterhuis (2001), and Philip Brey (2010) explained that to clearly understand the sense and place of technology in our world, we must eschew global approaches, such as those suggested by thinkers like Martin Heidegger, Jacques Ellul, Ivan Illich, and Lewis Mumford. These approaches are too critical and too pessimistic. They are too focused on the difficulties raised by technical progress and do not allow for constructive suggestions for the future. Only approaches that are much more empirical and concretely describe how technical artifacts operate and are applied allow us to constructively understand the place of technical objects and the role of engineering in community life.

This return to things (*thing turn*) is also a feature of the French approach to the philosophy of technology which, as expressed by Gilbert Simondon, assigns a more significant role to the study of technical objects. The introduction to the collective work *French Philosophy of Technology* (Loeve, Guchet, and Bensaude-Vincent 2018) points out that, from now on, most contemporary thought leaders "construct their philosophical analysis based on a careful empirical study of technical objects" and the way in which they exist in the world. Bruno Latour, who insists on a technical approach that is decidedly empirical, goes even further: according to him, one must repudiate global approaches like Ellul's "technical system." As such, "the idea of a technical system, for example, is a philosophical perspective that, once again, is not based on any empirical study" (Latour 1994, 176). Latour is of the opinion that, to carefully consider the role of technology, we must now focus on technical *objects* when they are being created and on the *agents* involved in their creation either directly or indirectly.

This methodological approach fundamentally sees technology as an instrumental reality effecting material action on the physical world. This focus on material objects and their practical use, coupled with a desire to come up with constructive solutions, leads most contemporary philosophers and sociologists to focus their investigation on technical objects and related processes and practices. It is thus the materially operative dimension of humans' technical activity within the world that provides the dominant paradigm. It is clear that this construction of the theoretical field of contemporary philosophy of technology reduces the risk of straying into metaphysical generalities. And yet, there is a cost to this purported realism: it leaves aside a whole side of contemporary technological reality—that of practices and processes which are at once technical and intangible.

These practices and processes are, in a sense, the invisible continent, ignored by many. But some serious thinkers have recognized this hidden realm. Karl Popper argues that a "social technology is needed whose results can be tested by piecemeal social engineering" (1966, 2:222, italics removed), which is "the application of the critical and rational methods of science to the problems of the open society" (1966, 1:1). His recommendation of a case-by-case approach (piecemeal engineering) instead of a global approach reinforces his belief that social engineering technology has an operative capacity that has not been tapped. Similarly, epistemologist Mario Bunge noted that there are intangible technologies focused on organization. In "Technology as Applied Science," aside from *physical techniques* such as mechanical engineering and *biological techniques* such as pharmacology, Bunge distinguishes a field of *social techniques* such as operational research and one of *thought techniques* such as computer science (we shall briefly return to these techniques later in this chapter). Social techniques are based on knowledge, so "psychology can be used by the industrial psychologist in the interest of production" (Bunge 1966, 331). Forty years later, Bunge takes up these ideas again in *Political*

Philosophy, Facts, Fiction and Vision (Bunge 2009), where he affirms that "a social technology is a science-based discipline capable of tackling social problems in a rational and efficient fashion. Its aim, in other words, is to alter human behavior so as to either eliminate the problem or mitigate the sufferings they cause" (313). There are many social techniques from "education science, social work, and marketing to law, management science, normative macroeconomics, military strategy and diplomacy. All of them have been invented to tackle social problems of various kinds and sizes, from distribution of resources to conflict resolution. They help meeting one of the conditions of good governance—efficiency" (313–14). Dominique Raynaud, in *Qu'est-ce que la technologie? (What Is Technology?)* (2016), holds that the application of scientific knowledge does not necessarily produce a material object: "I call any operation based on scientific knowledge a technological process, whether it be physical or intellectual" (31). He distinguishes engineering, which produces material artifacts, from technologies that produce intangible effects. "We must distinguish between engineering and technology, so as not to restrict the field of technology solely to productive technologies. There is thus a fiscal technology whose purpose is to set tax bases and define tax brackets using the previous year's income. But this technology is not used to produce any artifacts. It is not engineering proper" (31).

Most French proponents of the thing turn, who place the technical object at the center of their reflection, like to lay claim to Gilbert Simondon's foundational analyses, but a paradox should be noted: the author of *Du mode d'existence des objets techniques (On the Mode of Existence of Technical Objects)* (Simondon 1989) devoted several pages of his long treatise to establishing the existence of "techniques of the human world," intangible technologies that are related to material technologies through an order of succession: "Man's techniques arose as separate technologies at the moment when natural world development techniques, by their sudden expansion, altered social and political systems" (216). Furthermore, "after the development of the natural world, technical thought turned to that of the human world, which it analyzes and deconstructs into elementary processes, and then reconstructs using operative models, preserving figural structures, and leaving out background qualities and forces" (215). But, having pointed out the existence of a whole world of intangible human techniques, and recognized their use to be very problematic, he does not elaborate. The same is true for most contemporary philosophers of technology who promote an "object" approach to technique. To be sure, they have read Simondon and are aware of the existence of an important area of techniques of the human world, but, like Simondon, they are not very interested in techniques that do not produce objects or sets of material techniques. In this way, under the pretext of objectivity and a respect for experience, a whole dimension of modern technicity has been neglected by philosophical models of contemporary technique and its relationship with social life and, more generally, with the human mode of existence within a technological society.

Now this field of intangible technologies should not be neglected on the pretext that they are a special area, separate from physical techniques. Not only is its development a result of the former's progress, as Simondon mentions, but, in addition, it has in turn become one of the conditions for the development of techniques producing material effects. We should not forget that techno-scientific innovation has now become dependent on Research and Development (R&D) management techniques. One is condemned to a merely limited understanding of the unfolding of material techniques if one does not also take into consideration the unfolding of intangible techniques along with the interactions between the two. The "realism" that motivates the thing turn in philosophy of technology springs from a one-sided bias, because it excludes a whole side of the technical world that continues to increase in importance in our concrete lives. Strategy, management, logistics, administration, planning, and propaganda are among the many intangible techniques of organization that increasingly narrowly frame the spatial, temporal, and relational dimensions of our daily lives. Reflecting on the growing importance of management technologies, Baptiste Rappin suggests that "management appears to be the condition of ontological possibility of our contemporary lives: we come into the world and are thrown into organizations, we spend our entire life as part of them, or connected to them, and even our dead bodies have the hardest time extricating themselves from them. We must reconcile ourselves to looking at management as an existential condition. Throwness-into organizations is an ontological structure within human existence in the era of planetarization" (Rappin 2015, 60).

INSIGHTS FROM THE PERSONALIST CRITIQUE

While philosophers from a variety of backgrounds have recognized the existence of a realm of intangible techniques aimed at organizing the human world, the dominant currents of the philosophy and sociology of technology have hardly drawn any consequences from it. And yet, the question of the technical functionalization of personal and collective life was raised by a few thinkers as early as the aftermath of the First World War. The Jünger brothers and Romano Guardini raised the question but without much theoretical elaboration.[1] The same is true of Heidegger who, early on, understood the risk of humankind being enframed by technology (*Gestell*). But, as Rappin notes, despite a brief, perceptive remark on the fundamental nature of organizations,[2] he did not delve further into this question.

One striking exception is found in the work of a group of personalist philosophers. The Bordeaux School—especially Bernard Charbonneau (1910–1996) and Jacques Ellul (1912–1994)—assigned great importance to the development of intangible techniques of organization and attempted to characterize the role they play in establishing a technological society, especially as they affect freedom and depersonalization. Let me start, though, with Nikolai Berdyaev, another personalist philosopher, who also focused on freedom and stressed the organizational more than the mechanical dimension of modern technique. Berdyaev's texts were known to and discussed among French personalist thinkers, and it is certainly possible that they influenced Charbonneau and Ellul.

Berdyaev and the Concept of Technical Organization

In 1933, in *Man and Machine*, Berdyaev, advocating for an existential, socialist, and spiritualist personalism, explained that humankind has entered the "technical era" ([1933] 2019, 40). The advent of mechanization and the sudden increase in material and economic power are only the most visible aspect of a deeper phenomenon—the installation of a new anthropological system which is that of *organization* and which affects every dimension of life. The technical world thus extends much further than material machines. "We are not only talking about industrial, military technique, a technique having to do with locomotion and the comforts of life, we are also talking about the technique of thought and versification, of dance and law, even that of spiritual life and mystical development. . . . Any technology teaches how to obtain the best result with the least amount of effort" (19). As such, the phenomenon of technique profoundly affects our relationship with the world, our culture. According to Berdyaev, technique brings about an anthropological transformation that goes undetected. It is no longer about having at our disposal the most efficient tools to act upon things: we are entering a different world. Technique, he writes, "makes man a cosmiurge" (43).

For to a given world endowed with symbolic organic unity, both social and natural, man strives to substitute a world constructed in the mode of "organization." For Berdyaev, what is organized (by man) is the opposite of what is organic (created by nature). This is why he thinks that, "from the perspective of organic life, technology corresponds to a disembodiment, to a rupture that takes place inside historical bodies, to a schism between the flesh and the spirit. Technology creates a new order; henceforth, it produces organized bodies" (27). From an ontological perspective, this is what is totally new, because "organized elements do not appear before man does, they come into being after man's arrival and come to us from him" (32). Technique creates a world by organizing in a very specific way. Through the organization of a set of material elements or intangible processes, the organizer sets its goal, from the outside, as it were in some cases. The behavior of elements does not come from inside them: it is imposed on them by force. Man thus organizes a new world, a "second nature" (28), on the sole basis of what his rational mind knows. For "technique, for its part, remains foreign to symbols, it is realistic, it is pragmatic, reflecting nothing, it creates a new reality, everything within it is present. It takes man away from nature as well as from the beyond" (24). It modifies man's relationship with the world. What results is a desymbolized world, stripped of any organic inner cohesiveness. For example, "technology leads man to conceive of the earth as a planet, giving him a completely different sense of it than the one he used to have" (35).[3] The result is a transformation by technology of culture that is so profound that the task of organization no longer knows any limit. "The old organic order collapses and a new form of organization, created by technology, inevitably prevails" (38). In the process, what began as a liberating exercise of transitive action on things ends up becoming a reflexive action of man on himself and makes organization by man himself necessary: "the organization linked to technology assumes an organizing subject who cannot be transformed into a machine; however, such organization precisely tends to turn him into a mechanism" (28), so that "the organizational capacity manifested by man disorganizes him from inside," in such a way that "technicization of the spirit and reason can lead to their annihilation" (34),[4] which calls for yet more organization. This is why Berdyaev, wanting to break the vicious circle, writes: "It is impossible to tolerate the machine's autonomy, to leave it a full freedom of action" (46).

It seems that Berdyaev borrowed this notion of organization from Ernest Renan who, in *L'avenir de la science* (*The Future of Science*), wrote: "to scientifically organize mankind is therefore the last word of modern science, such is its bold but legitimate claim" ([1890] 1995, 36–37). As for Renan, it seems that he borrowed this notion of organization from the Saint-Simonians, for whom it played a central role in reestablishing the unity of the social body being dissociated by nascent industrialization. These approaches are just as political as the analyses of Charbonneau and Ellul, but they are oriented in an opposite direction. The Saint-Simonians, concerned with Unity, start from considering the social whole and the necessity to weave back together the torn social fabric, while our two Gascons, concerned with freedom, ground their reflection in the condition of the individual in industrial and technological society and seek, on the basis of this experience, to identify and characterize the social and technical factors of this condition.

Charbonneau and the Idea of Social Totalization

From his youth, Charbonneau was convinced that his century would be that of the plunder of nature and of totalitarianism. Charbonneau's entire work is a call to become aware that techno-industrial and scientific development, what he calls the "Great Moulting" of mankind, may well deprive man of nature and freedom. If the issue of organization becomes central in his thought, it is because Charbonneau has a very original understanding, social and not political, of the totalitarian phenomenon. For him, the essence of totalitarianism is not to be sought on the side of political ideologies, but instead on the side of the deeper social transformations driven by technical and industrial progress. This position is stated as early as 1935: "What characterizes the world in which we live is the symbiosis of the political and the technical" (Charbonneau 1945, 206)—that is to say that the progresses of the State as much as those of technique tend toward the same type of organization of the whole of social life. At the end of the Second World War, thinking about the use of the atom bomb by Protestant and liberal America, he writes that "the main thing is no longer ideological superstructures but the unleashing of power techniques and the mental attitude it spawns" (206–7). Then, contrary to Hannah Arendt's analyses, he states in *L'état* that "the novelty of the totalitarian spirit does not lie in a theory but in an absence of theory" (1987, 303), since the first manifestations of the total (rather than totalitarian) organization of social life occur before the appearance of regimes leading to totalitarian politics.

For Charbonneau, the experience of the First World War, which was carried out by liberal nation-states, was crucial for understanding the essence of the totalitarian fact: the unprecedented explosion of industrialized violence is but one of the dimensions, to be sure a revolting one, of the "total war" nations have waged against each other. We must always consider another dimension of the conflict, much less spectacular but crucial and novel, namely the emergence of a new type of social organization that has made possible, among other things, large-scale slaughters. For it is over the course of this war that we see falling into place—for the first time in history—the main features of a total organization of social life. Economic life is strictly harnessed to the war effort. Everything becomes raw material to be mobilized for industrial production, including women's hair. Science is drafted to design new weapons, intellectual, cultural, and artistic life is controlled by propaganda services, school is enrolled to the service of intellectual and moral mobilization of children (Audoin-Rouzeau 2004). War becomes a total social fact: the whole of social life submits to the logic of military power—and all of that follows from the simple necessity of the war effort, without this being motivated by any kind of totalitarian project. From the 1920s onward, this total mobilization of society for war serves as a model for revolutionary projects of the right and the left to which vast human masses will again be sacrificed.

According to Charbonneau, to understand the totalitarian phenomenon, one must not start from the ideologies of these regimes that appeared after the war, but instead ask ourselves why, as a result of war, Western liberal societies came to adopt a totalitarian functioning (and not a "regime"). The military motive for warlike totalization, just like the ideological motive of political totalitarianisms, is not the cause but rather an indicator of the gravity of a deeper social transformation which, within liberal society, had already put in place the conditions for this totalization. Totalitarian regimes were not the ones that tore peoples out of the countryside to concentrate in urban masses, uprooted and susceptible to propaganda, furthering the depersonalization of work and the dehumanization of life settings in the name of economic efficiency, creating the press industry and methods for controlling opinions and behavior, as well as a vast weapons industry, designing administrative techniques to anonymously and painstakingly manage all aspects of the daily life of said masses, and so on. The reign of bureaucratic depersonalization and the loss of mastery over one's own life are not an invention of totalitarian ideologues. It was first under liberal regimes that people fell into the habit of no longer deciding for themselves about any of the conditions of their personal and collective life. There is nothing totalitarian regimes have done which was not first prepared by liberal society.

To those not blinded by the violent character of political totalitarianisms, it is clear that it is the progress of the State and Technique under a liberal regime that makes the totalitarian State first possible, and then hard to avoid. According to Charbonneau and Ellul, there exists a totalizing dynamic inherent to the State as much as to Technique, so that both of these "impersonal structures" tend to first develop, and then converge, in an autonomous manner. Furthermore, they have such a kinship that "the totalitarian regime may be defined as a sudden fulfillment of Technique's incipient social content. The same spirit animates both: the means understood independently of any end" (Charbonneau 1987, 350). This then raises a risk that is more difficult to think through, that of a future total—and not political—organization of industrial and technological society under cover of scientific rationality and technical efficiency. And to think through this risk, the concept of organization is a central one.

Charbonneau and Organization

Charbonneau is highly sensitive to the experience of the powerlessness of individuals before the impersonal and depersonalizing logic of modern social organization. While the concept of *organization* does not appear in Charbonneau's first writings as one of the intellectual tools needed to describe the technical civilization then taking shape, the germ of this idea already exists in the *Directives pour un manifeste personnaliste*, written in 1935 with Ellul, to explain the powerlessness and loss of responsibility of individuals as a result of concentration, anonymousness, and massification: "Man was absolutely powerless before the Banks, the Stock Exchange, insurance contracts, Hygiene, the wireless, Production, etc." (Charbonneau and Ellul 1935, 49–50). "The means of concentration is technique" (56), which requires, as we can see in the development of industrial machinism, that human behaviors passively conform to impersonal rules: there is inevitably a contradiction between personal autonomy and the technicization of a domain of action that submits each element of a whole to a predetermined routine of acts and procedures.

Recall that in the 1930s, debates about technique mainly dealt with industrial machinism, understood as a perfecting of the tool that allows us to act upon matter, for example in Henri Bergson's *Les deux sources de la morale et de la religion* (*The Two Sources of Morality and Religion*), published in 1932, three years before the *Directives*. Now, extending Berdyaev's analyses, the two Gascon Personalists Charbonneau and Ellul put forward a startling approach of technique, very different in being non-mechanicist ("not an industrial procedure but a general procedure" (56)), and the examples they give are mostly characterized by the intangibility of the processes involved: intellectual Technique, economic Technique, political Technique ("one of the first areas affected by technique"), juridical Technique, mechanical Technique. It is significant that mechanical technique is mentioned last and intellectual technique ("determination of an official intelligence by immutable principles"[5]) first.

The technical phenomenon is interpreted first as inseparable from a state of mind that can apply the same rules and principles in all areas of human action, not just action upon matter. As a result of this technicization, social life is structured by power relations of a new type that favor the power of technicians before which there hardly stands any counterforce. "In capitalist society, the powerful guys are not the capitalists but the managers" (Charbonneau and Ellul 1935, 56). This statement anticipates analyses by Bruno Rizzi ([1939] 1976) and James Burnham (1941) about the role of managers. Thirty years later, these ideas will issue in the development of the concept of technostructure by J. K. Galbraith in *The New Industrial State* (1968). It is in 1937, in "Le sentiment de la nature, force révolutionnaire," that the term *organization* appears to refer to the forces that subjugate social life under an impersonal and abstract order that leaves nothing untouched. The word *organization* seems here synonymous with "the social armature"— the framework and control—that must accompany "progress" (178). Thus, for instance, Charbonneau explains that in a world that technique makes more and more artificial, "the last areas of free nature appear condemned and if there still remain some wild countries, it is by a refinement of organization" (172). As for reason, it is reduced to "mere organizational power" (179).

In 1949, Charbonneau self-publishes *L'état*. In this big book, a solitary meditation on the experience of the Second World War leads him to further develop the concept of organization and to give it a central importance for understanding the social transformations of his time. He comes back to modern reason understood as "the extreme *rationalization* of all activities" ([1949] 1987, 9, emphasis added) driven by the progress of technique and that of the State as they reinforce each other. As for the technical advances favored by liberal society in the nineteenth century, he points out that

> wherever technique penetrates, freedom is driven back, for technique, like the law, imposes to all the same disciplines, and everywhere it takes hold, that law takes hold which alone can make its applications possible: totalitarian discipline in its seemingly legitimate aspects is merely industrial discipline openly expressed. Thus, under cover of liberalism, economic evolution realizes in the daily life of individuals the basic condition of the totalitarian regime: man's abdication, whether it be the blank indifference of the many to determinations that go over their head, or the frenzied participation of a few. (275)

As for describing the advances of the State, the term *organization* now recurs: "to turn society into an efficient *organization*, the State conquers it, . . . replacing the diversity of the natural [i.e., spontaneous] order with the unity of an *organization* in which everything starts from a center, where an apparatus of outer determinations replaces inner bonds" (52, emphasis added); "to a world of conflicts and powerlessness, it imposes the peace of an *organization* befitting the reasons of his will to power" (53, emphasis added)—and commenting upon the gains of the French Revolution and the liberal State: "the function of the State is not limited to ensuring a minimum of indispensable conditions for life; it must realize the perfect society by a total *organization*" (73, emphasis added). Thus, the world of factories is also that of the reign of offices, of regulation and organization whose progress is demanded by those of the State and of technique in a kind of circular causality. On the one hand, organization is only the extension of material techniques that from now on are aimed at the world of human behaviors; on the other hand, it is the progress of the State that has enabled the accelerated technical development of the industrial age: "it is within the framework of the State that the methodical and realistic spirit of modern civilization has arisen: by its ruthless automatism, management foreshadows the machine" (53). And this new techno-industrial world is one in which new modes of domination take shape. Taking up anew the thesis of the 1935 *Directives*, Charbonneau writes: "The managerial class can be viewed, even in the USSR, as a genuine ruling class" (364). He generalizes by stating that the revolutions of the twentieth century, be they communist or fascist, have the common feature of bringing to power the "manager" class.

In 1973, in *Le système et le chaos*, Charbonneau puts the issue of organization at the heart of his reflection on the costs of progress and on its freedom-destroying consequences.

> People have long wrongly reduced industrial civilization to the machine; it is its most visible aspect, but also its most superficial one. Our true machines are factories and offices. . . . Our cities, our nations, strictly subject to norms called regulations or laws, tied together in a network of rails, pipes and lines, are enormous apparatuses, ever more run following technical rules. And if the machine can be viewed as a concrete organization, political organization: the State, must be viewed as an abstract machine. It is *organization* and not the machine that characterizes our time." ([1973] 2012, 49, emphasis added)

Thus, organization appears as the intangible fulfillment of the logic of technique. Furthermore, what makes the development of organization almost irresistible is the fact that it feeds off the very failures of the technical enterprise.

An important part of Charbonneau's work is dedicated to the description and the analysis of various "costs of progress": the loss of balanced and enriching contact with nature, the destruction of the countryside by agrobusiness, the standardization of food, grave environmental imbalances due to many hubristic actions upon nature, the destruction of local cultures disintegrated by the shock of progress too rapid to be integrated, the generalization of the bureaucratic technocracy of States and industrial corporations (which generates in turn the rise of nationalisms, identity-driven neuroses, and terrorism), and so on. Thus, at the same time as industrial civilization is becoming generalized and the strictures of technical organization become tighter, the sudden rise in power of humankind's capacity for action has as a counterpart a dangerous environmental, cultural, social, and political disorganization—what Charbonneau calls the risk of chaos. As more and more human groups have access to nuclear power, the catastrophic consequences of this rise of chaos become unacceptable. This is why the rise of disorganization calls in turn for the development of organization and of ever more rigorous techniques for the monitoring and control of the various dimensions of social life.

> For every progress of organization surrounds itself with a halo of disorganization, . . . which makes organization all the more necessary. Indeed, beyond a certain point, it breaks the balance of nature and from now on increases on its own. . . . Perhaps we have already reached this point; technical enterprise can

no longer stop halfway, it will need to artificially reconstruct the natural totality broken by the intervention of human freedom. . . . When man becomes the master of action over man and over society, technique must replace . . . family, the people, God himself: 'Science will organize society, and after having organized society, will organize God himself' (Renan, *L'avenir de la science*). Past a certain level of organization, the only choice left is between chaos and the system, which reconstructs from outside this universe being destroyed from the inside. (53)

From 1975 onward, Charbonneau worries about the risks involved in generalized computerization, understanding that "the computer will make it possible to order Earth as we wish [and] above all allow cybernetics to at last govern man mechanically" (1975). In 1980, it is already clear that "the translation of nature and of man into figures by science, thus making them digestible by the computer . . . turns them into interchangeable data or commodities, . . . into a statistical or juridical element to be stockpiled and managed" (2018, 119). *Because* it is "the totalizer," the computer enables "the reduction of qualities to quantity that the calculation of countless factors used to make impossible. It is the march towards a de facto totalitarianism without 'ism' or dictator" (2010, 164)—"until the great Automaton is devised whenever there will only be one State, computers operate for corporations and governments" ([1973] 2012, 230), for, even now that they have become personal and miniaturized, they are made to "order, integrate, mechanize . . . society as a whole . . . into an enormous apparatus, cybernetic itself, . . . with people—or rather elements—acting and speaking as the machine dictates" (2012, 136). Thus, "to overcome the disorder of the urban and industrial explosion, and avoid apoplexy, we are witnessing a process of organization that information technology apparently makes it possible to realize down to the smallest detail" (328).

Techniques of Organization According to Jacques Ellul

Charbonneau's association, and even the identification, of the concepts of technique and organization in his identification of the risk of social totalization in technicist and industrial, and hence of necessity statist, liberal society would later be systematized by his friend Jacques Ellul using the concept of *techniques of organization* that plays such an important role in his two books *La technique ou l'enjeu du siècle* (*The Technological Society*, 1954) and *Le système technicien* (*The Technological System*, 1977).

From the standpoint of the philosophy of technique, Ellul's originality consists in three theses:

1. "Modern" technique unfolds in an autonomous manner (an often-misunderstood thesis).
2. Techniques of organization are the necessary extension of material techniques.
3. Modern technique tends to organize as a *totalizing* system.

As he writes in *L'illusion politique*, echoing Charbonneau's theses, "If government multiplies techniques of organization, psychological action techniques, public relations techniques, mobilizes all forces for productivity, planifies the economy and social life, bureaucratizes all activities, reduces law to a technique for social control, socializes daily life . . . it is a totalitarian government" (Ellul [1965] 2004, 318–19).

Ellul devotes the first twenty pages of *La technique ou l'enjeu du siècle* (1954) to explaining how the role of modern technique cannot be understood without factoring in organization (with reference to James Burnham's *The Managerial Revolution*, 1941) as it represents a higher stage of technical progress: "It is technique applied to social, economic and administrative life." It makes it possible to integrate collectives or masses in the world created by the progress of material techniques. To specify what he means by "organization," Ellul puts forward a classification of the various types of techniques (1954, 19–20).

Under the heading of traditional techniques and of the beginnings of industrial society, he includes *mechanical production techniques* and *intellectual techniques* (files, libraries, etc.). Under the heading of modern techniques, which correspond with Charbonneau's idea of organization, he distinguishes three families. *Economic technique* has to do with the area of production and ranges from the organization of work to planification. *Organizational technique* has to do with great masses, and the action collectives take to make their action more efficient by building an impersonal action framework. Technique of organization applies just as much to business and industrial affairs (and therefore belongs to the area of economics) as to States, to administrative or police life and to war. At this point in time, it also covers all things juridical. Economic technique and techniques of organization correspond to what today is called management technique. To these two families of techniques, Ellul adds a third, *technique of man*, which, contrary to the previous ones, applies to individuals: "here man himself is the object of technique" (20).

In the rest of this book, Ellul returns to the theme of organization without which it is impossible to understand the social and human consequences of technical progress. Deepening the thesis of the 1935 *Directives* according to which technique must be viewed as a "general process," he states that "organization is precisely technique itself" (83). A little further in a development on "the ripple effect of techniques," he shows how these techniques of organization generate each other in various areas: urban planning, economics, work, public administration, and so on (103). It is therefore essential to take into account the fact that technique tends to become interiorized, to dematerialize: "The more precise material techniques become, the more they make intellectual and psychic techniques more necessary" (106). This is why Ellul devotes a whole chapter to "techniques of man" that are the necessary complements of techniques of organization. "Without them, man will no longer be at the same level as organizations and machines; without them, technique cannot be absolutely safe." Ellul mentions several: schooling technique, work technique, professional counseling, propaganda, entertainment, sport, medicine.

In *Le système technicien,* Ellul ([1977] 2004) also emphasizes organization. He notes that the development of material techniques calls forth that of organization and that we have come to a stage where "productive forces are no longer the infrastructure; they have become a superstructure—that is, they cannot develop, make new progress if there is a social organization infrastructure that may at once perform indispensable research for such progress, and host this progress within the social body" (76). He devotes long passages to showing how much the progress of techniques of organization owes to the advances of information technology and the development of the computer: "With it, knowledge becomes an organizational force" (85). These monitoring and control techniques are henceforth needed by large organizations in order to promote social and professional integration of personnel: "The byword of techniques of man is adaptation" (315). In the final analysis, taken together, each of the pragmatic and limited techniques of organization contribute to a totalizing movement: "We are dealing with total technicization when every aspect of human life is subject to control and manipulation, experimentation and observation so that demonstrable efficiency is obtained everywhere" (94). And this process does not seem to know any limit. Even if the creation of material techniques is somehow limited, the development of the technical system will shift toward techniques of organization or techniques of the human.

Ellul's warnings about the risk that technical super-organization represents for freedom have not been taken seriously. His theory has been portrayed as an exaggeration, misleading, and ideological in extending technique beyond the material, and that impact of techniques of organization and of the human are neither measurable nor demonstrable. As a counterpoint, in 2013, the world market for commercial advertising was $518 billion USD, roughly the equivalent of a quarter of France's GNP—not taking into account States' "communication" budgets—spent because sponsors believe that their investment will produce measurable effects. Likewise, consider the fact that businesses compete to hire management professionals coming out of institutions such as Harvard Business School. The intimate link between material and intangible techniques of organization is illustrated by considering agricultural techniques. The advances of mechanization, of "chemicalization," of the cold chain (i.e., refrigeration), of genetic engineering, and the other material techniques would not be enough to ensure the establishment of the industrial agribusiness system on which our physical survival now depends; intangible techniques would also be necessary: the full employment of the possibilities opened by agricultural mechanization called for a willful reconfiguration of rural space and the agricultural world. Administrative techniques, land management techniques, land consolidation, training, and professional supervision techniques, and so on were devised to realize the potential of material techniques. Finally, a great variety of techniques of organization are implemented by the professionals who occupy these offices. The fact that the appearance of our cities has been transformed by the multiplication of these offices and these buildings in which they are housed gives a palpable indication of the importance of this technical field. Thus, on the social and political plane, the progress of material power is everywhere accompanied by a technocratization of political life and a growing bureaucratization of our lives that are ever more subjected to centralized, hierarchical, and opaque management and technical patterns over which we have little control.

FROM ORGANIZATION TO TOTAL MANAGEMENT: A PROJECT LONG IN THE MAKING

Charbonneau and Ellul were among the first to critically highlight how technical action, at first transitive, becomes reflective: turned toward things, it turns inward and is then methodically applied to humans, down to the most intimate dimensions of their existence. However, the idea of techniques of organization appears from the outset of what is known as the Industrial Revolution, as insightful thinkers understood that the technical will to power could not stop at technical mastery of the things of nature. It also had to tackle the mastery of human things. Realizing the socially disorganizing power of industrialization, in 1819 already, in *L'organisateur*, Saint-Simon (1760–1825) and his disciples called for the advent of a reign of organization and laid the foundations for what would later become the science of management. The scientific-industrial organization of society, that nightmare that Charbonneau and Ellul wanted to avoid to save freedom, is precisely what Saint-Simon wanted to bring about to rescue mankind from the divisions tearing it apart. In spite of this radical opposition, Charbonneau, Ellul, and Saint-Simon share a conviction that organization is the fulfilment of technique.

Saint-Simon borrows the concept of organization from biology where it started being used in the eighteenth century. An organism is a living whole, functionally unified, in which every part, or *organ*, acts in concert with all the others. Organization is the ability to enable the various parts of this whole to hold together. After the French Revolution, this concept would be extended to the political realm. Any risk of social disintegration and of conflict between opposing forces demanded organizing, reestablishing, or even building up social unity on the model of the living individual (hence the spread of the term "body politic"). The concept of organization corresponds not only to the search for solidarity between various social groups, but also to the search for a coherent development of their various activities.

Saint-Simonism was driven by the religion of industry—for them, industrialization was a salvific process—and the ideal of a *scientific organization of society*. Saint-Simon viewed the "industrial system" as a new future state of society to be built through *organization*, made at once possible and necessary by the evolution of techniques. The responsibility for the regulation of the industrial system would be entrusted to an élite of scientists, engineers, and technicians of finance. The aim was *to replace the governance of people with the administration of things*. Organization is thus meant to bring about the withering away of the political for the benefit that comes from technical management of the whole of social life.

Saint-Simon emphasizes the crucial role of technical transportation and communication networks such as roads, canals, and telegraph—the line of the Chappe telegraph connecting Paris and Lille had been in operation since 1794—that represented the first step toward a universal association of producers. In doing so, he anticipated many aspects of the unifying and dynamizing role of technical and informational networks in contemporary society, as well as the ideology of networks that accompanies their development.

As Pierre Musso notes,[6] Saint-Simon's work has inspired managerial propaganda's universal claims and positivist prophecies of economic and technological progress:

> All bureaucracies and technocracies have grabbed hold of this science of the administration and the organization of men, invented in the second half of the XIXth century by civil engineers. Management brings industrial religion to completion by determining behavioural norms, moral rules about how to live in the factory and within a firm. Engineers, as machine experts, extend their competence beyond mechanized production with "scientific" methods for the organization of work and production, and later with a science of organization, cybernetics; it is enough for them to look upon society as a "vast workshop" and upon the workshop "according to a technological model." (2017, 631)

French Saint-Simonism had a great influence on the networks of American engineers who would organize engineering societies to develop management science and prepare the managerial revolution. The American Association of Industrial Management was founded in 1899. Even earlier, in 1885, a report by Captain Metcalfe, who headed federal arsenals in the United States, explains that "the two great issues organization must deal with are coordination and control." According to him, the administration of arsenals and other workshops is to a large extent an art and depends on the application to a wide variety of cases of certain principles which, taken together, constitute what may be called "the science of administration." Perhaps most importantly, one must optimize the cost/efficiency ratio that undergirds a manufacture's management principles. The vocabulary of scientific management thus crystallizes around the trinity, to direct, control, and evaluate on the basis of daily activity dashboards.

Musso details how the professionalization and the institutionalization of management were on the march during the run-up to the First World War. The Wharton Business School, first of its kind, was created in Philadelphia in 1881, and Harvard Business School in 1909, when the *Engineering Magazine* celebrated the "gospel of efficiency" as the movement's highest value. In 1911, its founder John R. Dunlap stated that "efficiency is subject to the great laws of the universe which are always and everywhere the same. . . . These laws are sovereign, inherent and eternal, like the laws of gravitation or of chemical affinity" (cited in Musso 2017, 644). In 1912, the same magazine published Harrington Emmerson's *Twelve Principles of Efficiency*.[7]

The concept of a "managerial revolution" is used in 1932 in a book by A. Berle and G. Means, *The Modern Corporation and Private Property*. It is the product of a gestation that took forty years. In 1895, Frederick Taylor published his first memoir on scientific work organization based on the breakdown of operations and the timing of gestures. His *Principles of Scientific Management* follows in 1911. This technique is based on a "state of mind" excluding conflicts, hostile to labour unions, aimed at creating a spirit of cooperation within the business and mutualizing the interests of workers and the boss. This approach to management opens the way to a managerial way of handling things that will go beyond the world of the enterprise and extend to the entire planet. Thus, as early as 1918, Lenin writes in "Immediate Tasks of the Power of the Soviets: "We will be able to realize socialism precisely to the extent that we will have succeeded in combining the power of the Soviets and the Soviet management system with the most recent advances of capitalism. We must organize in Russia the study and teaching of the Taylor system, systematically testing and adapting it" (1918). Management thus appears as a global social technology. In 1931, during a US–Soviet meeting, H. S. Person, president of the Taylor Society, gave a paper entitled "Scientific Management as a Philosophy and a Technique of Progressive Industrial Stabilization." The machine model is then extended to "efficient" techniques for the organization and rationalization of production. It is necessary to go toward a directed social economy applying to the entire world. In view of this, Person underlines the crucial role of directors, managers, and organization technicians. Burnham's *The Managerial Revolution* (1941) theorizes the role of these new technicians. According to him, capitalism and socialism are both superseded by the emergence of a new society dominated by managers, a thesis already put forward by the Italian Bruno Rizzi in *La bureaucratisation du monde* (1939). Burnham foresees the emergence of a new class of technicians: the Directors. They are characterized by their ability to organize work and production, but they are also propaganda specialists.

As industrial society became so complex that it was difficult to organize, management was the solution. From this perspective, the distinction between private enterprise and the State tends to be erased: "Government directors have almost the same training, the same functions, the same mental habits as the directors of industry. Differences setting them apart are going to fade away" (Burnham 1947, 191). In practice, what we find is an interchangeability between technicians coming from the public and the private sectors, since they use the same managerial organization and communication techniques aimed at securing adherence and participation.

After the Second World War, the convergence of management techniques with new information technologies yielded a cybernetic model of social organization. In 1948, Norbert Wiener published *Cybernetics or Control and Communication in the Animal and the Machine*. Cybernetics means "the art of governing." Thanks to the automation of communications and information processing and the establishment of feedbacks, Wiener considers the possibility of a scientific-industrial steering as much of industry as of society in rigorously articulating three parameters: goals/means/results in terms of an efficiency standard. Cybernetics aims at a governance of men by automatic steering, numbers, and algorithms, as already was the case in financial markets. Political regulation having failed, cybermanagement claims to at last bring rationality to decision-making. As Rappin (2014, 2015) notes, managerial dogma and the cybernetic paradigm now aim at the scientific administration of men and the governance of things (an inversion of Saint-Simon's project, aimed at *replacing the governance of persons with the administration of things*).

THE CARNAL AND THE IMMATERIAL

Two centuries after Saint-Simon's *L'organisateur*, it is clear that the worries of Charbonneau and Ellul were not paranoia! Since they published their prophetic books, the field of immaterial techniques of organization and control has never stopped developing and extending to new areas of social life. This growth is made necessary by the development of material techniques and, in circular causality, it is also one of the conditions for the development of material technical objects, sets, or systems. Thus, the advances of information technology that allow the collection, storage, and treatment of data in real time made it possible to perfect business management techniques which, as Rappin has shown, foster the depersonalization of work and the emergence of servitudes of a new type, thus confirming the diagnosis of technical depersonalization put forward by Ellul and Charbonneau in the 1930s. But things did not stay at this stage and, after the enterprise, it was administration and then, among others, the whole universe of medical and social institutions which, after having been technicized, and as a result of having been technicized, were later easily taken over by management techniques and had to submit to a process of organizational rationalization and generalized computerized protocolization, resulting in new forms of alienated work and individual and collective malaise. This is seen in the area of maladjusted children, as the process of progressive technicization of educational action, driven at the beginning by a "humanistic" project, centered on the person, has involuntarily favored the progressive enslavement of professionals, whose interventions must follow organizational and managerial constraints that tend to empty educational work of its meaning in subservience to a technicist formalism (Cérézuelle 1996). Likewise, more physicians who, subjected to the same type of computerized management, start "hating their computers," which they experience as the instrument of subjection of their relations with individual patients to depersonalizing managerial dynamics.[8] These results are not outliers.

In addition, aside from management techniques, land use planning techniques were also developed after the war in France that reconfigured space and its use: zoning, urban, and other planning, and so on. The same goes for the propaganda techniques indispensable to ensure the acceptability and social integration of new techniques: all extensions of thought and technical action became necessary as much to enable new advances in material techniques as to adapt to their effects. Although it was not the result of a clearly conceived project, a "complete functionalization of whole swathes of our life" thus takes place (Rappin 2014, 36). Moreover, it is likely that the worsening environmental crises and their social and political consequences are going to call for a reinforcement and an extension of these techniques of organization and control, even as they remain largely overlooked by philosophers and sociologists of technique, who remain focused on "technical objects."

In *Du mode d'existence des objets techniques*, Simondon identified the potential of depersonalization of the "techniques of the human world." Considering "man as a technical material" (1989, 214), their reductionist formalism sets aside important dimensions of social and human reality and can only implement a closed and impoverished technicity. As Rappin notices sixty years later about management techniques, it is the worker's flesh that disappears (2015, 270–71). But aside from a very vague call to "reflective technology," Simondon provides almost no analysis of these intangible techniques of the human world and of the consequence of their development. Seeking to reconcile human culture and techniques, he always comes back to the tool and the man/machine couple.

Since he had given pride of place to cybernetics in his understanding of the man/machine coupling, Simondon's philosophy of Technique purports to be a "mechanology," and he does not seem to be concerned about the striking development that, under his eyes—he died in 1989—the advances of information technology are about to give to management techniques, and more generally, to the intangible techniques of the human world. Driven by a keen desire for reconciliation and unity, as a thinking of generalized analogy, as Gilbert Hottois has underlined (1993), Simondon's philosophy is no doubt sensitive to individuation in process and the sense of freedom that characterize mechanical invention. But it is mostly the freedom of the technician who creates machines as bearers of an open *future* that interests Simondon; apparently, for him, the present is negligible, meant to be superseded. More interested in the movement of individuation than in the individual who actually exists here and now, he hardly dwells on investigating in any depth the issue of how the ordinary individual deals *in real time* in daily life with the new field of intangible techniques.

Simondon's approach ignores the *flesh*, the loss of freedom and meaning that is carnally experienced *on a daily basis*. Thus, even though it purports to be guided solely by experience and a concern for objectivity, the *thing turn* offers no guarantee of realism (Cérézuelle 2019).[9] Conversely, it would seem that a certain ethical sensitivity, a *sense* of carnally experienced freedom, very different from the sense of totality, led certain thinkers to take into consideration the existence of a whole side of contemporary technical reality and to highlight the problematic character of its expanse.

TO CONCLUDE: ANGELISM AND THE SENSE FOR THE FLESH

The price of the "empirical turn" and the new focus on technical *objects* may well be a heavy one, both intellectually and socially. By choosing to practice the *thing turn*, the philosophy and sociology of contemporary techniques run the risk of an all-too-partial understanding, myopic and anodyne when it comes to the relationship between technique and society. Evidently, to throw light on this relationship, it is useful to be able to rely on empirical studies dealing with this or that technical object. But in order to properly understand the various stakes and consequences of the use of a technique, it would often be useful to situate the genesis, development, and functioning over time in the wider field of interactions and interdependencies with the entire set of other techniques, be they material or immaterial. The advances of both types of techniques are inseparable, bound in a relation of circular causality: the progress of material power brings about the progress of techniques of organization and adaptation, while the progress of management techniques in R&D ever more strictly conditions the progress of material techniques. It is the expansion of the whole in close interaction, what Ellul calls *the technical system*, which constitutes from now on our technical *world* that cannot be reduced to a juxtaposition of specific techniques. And we have to understand the functioning of this whole in order to properly situate it and master the role of specific techniques. This is why the decision to stick to the study of the "object" dimension of technique promotes a myopic and one-sided understanding of the technical world within which we live, causing us to overlook some very important issues.

Moreover, this methodological approach that purports to be "empirical" is not as objective and true to concrete experience as it claims. In practice, there is no gaze upon the world that can claim to be axiologically neutral. Quite often, what we take to be the values to which we are attached is a certain *view* of the world that allows us to open our eyes on some dimensions of the world of techniques but blinds us to other dimensions. As Hottois rightly pointed out, Simondon's inability to sufficiently take into account the objectifying power of technique, that can from now on turn man into "technical material," causes his thought to err on the side of a conciliatory irenicism prone to emptying philosophical thought of its polemical force and of its critical power: "A certain angelism constitutes the specific temptation of Simondon's philosophy" (Cérézuelle 2019, 125). And such angelism is not neutral. If Simondon prefers not to see the worrisome side of a movement of generalized technicization of existence that he yet senses, it is because his interest in technique is rooted in technophile *faith*. And this faith, when it is largely shared, becomes a social force that prevents people from asking embarrassing questions, thereby contributing to the acceleration of this objectifying and depersonalizing dynamic.

Similarly, Latour's conciliatory and soothing view of modern technical dynamics is not so much due to being "an empiricist" and attentive to what is given in experience: his thought extends a Jesuitic technicist theology in the line of Teilhard de Chardin, who saw in the progress of material techniques the driver of a progressive spiritualization of the material world (Cérézuelle 2019). From that standpoint, technique is seen as no more than a mediation fostering the progressive incarnation of the spiritual in the world and the unification of minds in a "planetary nervous system" *to come*; technical power therefore could not be laden with a negativity specific to it as such: "everything is negotiable." This prior outlook directs Latour's gaze. Within the framework of this very particular *Weltanschauung*, the issue of the depersonalizing potential that goes along with the increasing power of intangible techniques of the human has no place and remains a blind spot.

By contrast, the attention that Berdyaev, Charbonneau, Ellul, and today Rappin (as well as many others in all likelihood) give to the *flesh*, to the loss of freedom and meaning that is carnally experienced *on a daily basis*, made them sensitive to the depersonalizing role of intangible techniques of organization and to the risk of the total functionalization of existence as a result of their proliferation, which accompanies their rising power (see Cérézuelle 2005).

Thus, even though it purports to be guided solely by experience and a concern for objectivity, the *thing turn* offers no guarantee of realism. Conversely, it would seem that it is a certain ethical sensitivity, a *sense* of carnally experienced freedom, very different from the sense of totality, that led certain thinkers to take into consideration the existence of a whole side of contemporary technical reality and to highlight the problematic character of its expansion.

NOTES

1. Two examples are "An advanced state of technology is accompanied by mechanical theories of the nature of man" and by "efforts to subject man to technical rationality, to a purposeful, all-embracing functionalism" (Jünger [1949] 1956, 155), and "*Ainsi se développe une technique de la sujétion de l'être vivant*" (Guardini 2021).

2. "It was clearly evident to me that the 'organization' is part of an invisible center, not, to be sure, technology, but part of what exists in the history of the being" (Letter from Martin Heidegger to Hannah Arendt, February 15, 1950, in Arendt and Heidegger 2001, 83).

3. This observation anticipates Edmund Husserl's idea that "the earth does not move itself," in manuscript D17 (1934, in Farber 1940, 310).

4. This idea is developed in Charbonneau's *L'état* (1987) and also in *Le système et le chaos* ([1973] 2012).

5. Charbonneau read *L'avenir de la science*, where Ernest Renan maintains that science must be and must only be a "patient study of things," a "pragmatic study of what is." "By all these paths, we thus come to proclaim the right that reason has to reform society through rational science and theoretical knowledge of what is. . . . Organizing mankind scientifically, this then is the last word of modern science, this is his bold but legitimate claim" (Renan [1890] 1995, 151).

6. See also Musso's other works on Saint-Simon (esp. 2004) and the philosophy of networks.

7. Reminder: "In an age of advanced technology, inefficiency is the sin against the Holy Ghost" (Huxley [1932] 1946, xvi).

8. Atul Gawande, "Why Doctors Hate Their Computer," *The New Yorker*, November 5, 2018. Furthermore, a note from the Vantage Technology Consulting Group of December 4, 2018, adds that, even though information systems should make health care "greener, faster, and more productive," the promises have not been realized:

> A 2016 study found that physicians spend 2 hours of computer work for every hour spent with a patient. The University of Wisconsin found that the average workday for family physicians has grown to 11½ hours. The article posits that one unplanned result of going digital is that there is a growing epidemic of burnout among doctors, with 40% of them screening positive for depression and 7% reporting suicidal thoughts; this is double the rate of the general working population. (https://www.newyorker.com/magazine/2018/11/12/why-doctors-hate-their-computers)

9. It should be noted that Gilbert Hottois has pointed out before I did the surprising failure to consider techniques of the human in the philosophy of Gilbert Simondon who, being "hypersensitive to conflict, to separation . . . , dreams of pacification and universal conciliation," so that it labors under "a questionable philosophical irenicism" (1993, 123).

REFERENCES

Achterhuis, Hans. 2001. *American Philosophy and Technology: The Empirical Turn*. Translated by Robert P. Crease. Bloomington: Indiana University Press.

Arendt, Hannah, and Martin Heidegger. 2001. *Lettres et autres documents, 1925–1975*. Translated by Pascal David. Paris: Gallimard.

Audoin-Rouzeau, Stéphane. 2004. *La guerre des enfants, 1914–1918*. Paris: Armand Colin.

Berdyaev, Nicolas. (1933) 2019. *L'homme et la machine*. Paris: R&N Éditions.

Brey, Philip. 2010. "Philosophy of Technology after the Empirical Turn." *Techné: Research in Philosophy and Technology* 14, no. 1.

Bunge, Mario. 1966. "Technology as Applied Science." *Technology and Culture* 7, no. 3 (Summer): 329–47.

———. 2009. *Political Philosophy, Facts, Fiction and Vision*. New Brunswick, NJ: Transaction Publishers.

Burnham, James. 1941. *The Managerial Revolution. What Is Happening in the World*. New York: John Day.

———. 1947. *L'ere des organisateurs* [*The Managerial Revolution*]. Paris: Calmann-Lévy.

Cérézuelle, Daniel. 1996. *Pour un autre développement social: Au-delà des formalismes techniques et économiques*. Toulouse: ERES.

———. 2005. "La technique et la chair. De l'*ensarkosis logou* à la critique de la société technicienne chez Bernard Charbonneau, Jacques Ellul et Ivan Illich." In *Revue Européenne des Sciences Sociales* 43, no. 2: 5–30.

———. 2019. "Une nouvelle théodicée? Remarques sur la sociologie des techniques de Bruno Latour." *Revue du MAUSS semestrielle*, no. 54, 2e semestre, 228–54.

Charbonneau, Bernard. 1937. "Le sentiment de la nature, force révolutionnaire." In Charbonneau and Ellul 2014, 116–92.

———. 1945. "An deux mille." In Charbonneau and Ellul 2014, 193–215.

———. [1949] 1987. *L'état*. Paris: Economica.

———. [1973] 2012. *Le système et le chaos: Critique du développement exponentiel*. Paris: Sang de la Terre.

———. 1975. *La gueule ouverte*, no. 77, October 29, 1975. Quoted in *Le totalitarisme industriel*, 80. Paris, L'Échappée, 2019.

———. 2010. *Finis terrae*. La Bache: À plus d'un titre éditions.

———. 2012. *Une seconde nature*. Paris: Sang de la Terre.

———. 2018. *The Green Light. A Self-Critique of the Ecological Movement*. Translated by Christian Roy. London: Bloomsbury Academic.

Charbonneau, Bernard, and Jacques Ellul. 2014. *Nous sommes des révolutionnaires malgré nous*. Paris: Seuil.

———. 1935. "Directives pour un manifeste personnaliste." In Charbonneau and Ellul 2014, 47–80.

Ellul, Jacques. 1954. *La technique ou l'enjeu du siècle* [*The Technological Society*]. Paris: Armand Colin.

———. [1965] 2004. *L'illusion politique*. Paris: La Table Ronde.

———. [1977] 2004. *Le système technicien*. Paris: Le Cherche Midi.

Husserl, Edmund. 1934. "Grundlegende Untersuchungen zum Phänomenologischen Ursprung der Räumlichkeit der Natur." In *Philosophical Essays in Memory of Edmund Husserl*, edited by Marvin Farber, 307–26. Cambridge, MA: Harvard University Press.

Galbraith, J. K. 1968. *The New Industrial State*. Boston: Houghton Mifflin.

Guardini, Romano. 2021. *Lettres du lac de Come: Sur la technique et l'humanité*. Paris: R&N.

Hottois, Gilbert. 1993. *Simondon et la philosophie de la culture technique*. Brussels: Deboeck Université.

Huxley, Aldous. [1932] 1946. New Foreword to *Brave New World*. New York: Harper & Row.
Kroes, Peter A., and Meijers, Anthony W. M., eds. 2000. *The Empirical Turn in the Philosophy of Technology*. Amsterdam: JAI.
Jünger, Friedrich Georg. [1949] 1956. *The Failure of Technology*. Chicago: H. Regnery.
Latour, Bruno. 1994. "De l'humain dans les techniques, entretien avec Bruno Latour." Interview by Ruth Scheps, in *L'empire des techniques*, 167–79. Paris: Seuil.
Lenin, Vladimir I. 1918. "Immediate Tasks of the Power of the Soviets." *Pravda*, no. 83, April 28.
Loeve, Sacha, Xavier Guchet, and Bernadette Bensaude-Vincent. 2018. *French Philosophy of Technology: Classical Readings and Contemporary Approaches*. New York: Springer.
Musso, Pierre. 2004. "Le présent dans la philosophie politique de Saint-Simon." In *L'actualité du Saint-Simonisme*: *Colloque de Cerisy*, directed by Pierre Musso, 15–34. Paris: Presses Universitaires de France.
———. 2017. *La religion industrielle*. Paris: Fayard.
Popper, Karl. 1966. *The Open Society and Its Enemies*. Vols. 1–2. London: Routledge and Kegan Paul.
Rappin, Baptiste. 2014. *Au fondement du Management*. Nice: Ovadia.
———. 2015. *Heidegger et la question du management*. Nice: Ovadia.
Raynaud, Dominique. 2016. *Qu'est-ce que la technologie?* Foreword by Mario Bunge. Paris: Editions Matériologiques.
Renan, Ernest. [1890] 1995. *L'avenir de la science—pensées de 1848*. Paris: Flammarion.
Rizzi, Bruno [1939] 1976. *L'URSS, collectivisme bureaucratique: La bureaucratisation du monde*. Paris: Champ Libre.
Simondon, Gilbert. 1989. *Du mode d'existence des objets techniques*. Paris: Aubier.

Chapter 4

Technology as Process

Mark Coeckelbergh

Contemporary philosophy of technology after the empirical turn (Achterhuis 2001) predominantly conceptualizes technology as artifact, object, or thing. This is the case for both "engineering" philosophy of technology and "humanities" versions, to use Mitcham's famous distinction (Mitcham 1994). A good example is the analysis of human-technology relations by the (by now) tradition of postphenomenology (Ihde 1990), which Peter-Paul Verbeek (2005) influentially framed in terms of analyzing "what things do." Another example is Luciano Floridi's (2014) information philosophy and information ethics, which sees the world as the totality of informational objects. While these conceptualizations of technology do not reject, and sometimes even include, dynamic aspects, they stand in a philosophical tradition that sees the world as a collection of objects or substances (which *then* may interact).

This being-oriented tradition contrasts with another one, process philosophy, which—finding inspiration in Heraclitus's thesis that everything flows (*panta rhei*)—sees the world as a process of becoming, rather than being. The starting point is not things but processes. If there are (what we call) things, they are dynamic and are the result of processes. They become. Henri Bergson (1944) argued that individual intelligence emerged in the process of evolution that expressed an *élan vital* (life force) and questioned the dichotomy between objective time and experience time. Similarly, Alfred North Whitehead (1978) argued that reality is a continuous process, that existence is a project of becoming, and that entities and experience are both part of becoming; we should not split "objective" reality and "subjective" experience. If there are still subjects and objects at all, they exist only as continually changing. As Tim Barker puts this stress on change and the non-dualism of process philosophy: "This experience is just as real as the physical object of the human body or the hardware of the digital system. In fact, it is the experience of objects, the processes that they go through, and their capacity to respond to these processes that actually makes them what they are" (2012, 146). Humans, too, exist in this way. Our experience and thinking are made in the process. We are made in the process. For Whitehead and Gilles Deleuze (2005), subjects are made by time and we live in time, which exists before we measure and experience it; we become (and perish and make room for others).

What does this process thinking mean for thinking about technology, and in particular for contemporary philosophy of technology? What does it mean to see technology in terms of process(es) and event(s), and to question the subject/object dualisms inherent in the philosophical tradition and perhaps even at work in contemporary philosophy of technology? Responding to Don Ihde's postphenomenology, picking up seminal work by Carl Mitcham, and learning from (digital) media studies, especially Barker's (2012) work, I will distinguish between, and further explore, a number of possible directions in which this journey into process philosophy may take us. In particular, I will explore what it means to conceptualize technology as process by proposing the following process-oriented conceptualizations:

1. processes of technical development, use, and maintenance (reference to Mitcham)
2. processes of mediation through which subjects and objects become (response to Ihde)
3. processes of making time and being made by time, events, and multiple temporalities (learning from Barker)
4. performance processes (references to Barker and my recent work)
5. games, narratives, and cultural processes (drawing on Wittgenstein and Ricoeur)

Taken together, these conceptualizations are meant to highlight different aspects of the process character of technology, and thus may serve as stepping stones for a more comprehensive and more developed process-oriented philosophy of technology.

PROCESSES OF TECHNICAL DEVELOPMENT

Technological artifacts have a history. Technology is not only about what is at a given moment, but also about processes of technical development and change during use. For example, a robot is the result of a process of hardware and software development, and during its use, technical modifications may be made to it. For example, it may be programmed differently. Technological processes are also connected to other processes, for example biological ones. Paul Thompson (2020) has shown that one can use biology and Dewey's pragmatism to approach this from a naturalist point of view: one that connects organisms to their environment, and that, for example, Ihde (often implicitly) endorses. Moreover, as Mark Thomas Young (2021) has reminded us, technology is not only about use and development but also about maintenance. Consider mechanical technologies such as cars, but also the curation of datasets and other "digital" maintenance work such as software updates. Seeing technology as process could help to avoid the bias toward design in contemporary philosophy of technology and engineering. Empirically oriented philosophers may then consider the entire lifecycle of technologies, including maintenance and, for example, recycling, and ask ethical and other questions about each of these stages. For instance, next to existing value-sensitive design and responsible innovation approaches to digital technologies, one could also develop evaluations of how these technologies are maintained and ask what should happen when these technologies are no longer used.

This meaning of technology as process connects to what Mitcham calls "technology as activity": technology is not just about material objects but also about bringing objects into existence (1994, 209). He sums up some relevant activities: crafting, inventing, designing, manufacturing, working, operating, maintaining. He especially discusses engineering action, using, and maintaining as an "intermediary between making and using" (233). These activities highlight the involvement of humans in technological processes. What appears to us as a thing is an artifact, that is, it is made and it is made by people, who in turn use technology in their processes of making.

However, the relations between humans and technology are not just instrumental. Humans are changed and (re-)shaped through their use of technology. Meaning and culture are constituted and changed by technology. Hence technological processes are not just technical. One way of conceptualizing such more-than-instrumental human-technologies has been offered by Ihde, who has combined an implicit pragmatist orientation (in particular a focus on technology use) with what he calls a *postphenomenology* of technology (Ihde 1990, 1993).

PROCESSES OF MEDIATION THROUGH WHICH SUBJECTS AND OBJECTS BECOME

According to Ihde, technological artifacts are not just tools but mediate our relationship to the world, influencing how we experience, perceive, and interpret the world. Through different uses, technology can be meaningful in multiple ways, albeit constrained and shaped by its concrete materiality.

Inspired by phenomenological variations, Ihde calls this "multistability." There is not just one way in which a technology can mediate. And as it is situated in between the user and the world, it can also mediate in different ways; there are different human-technology relations (Ihde 1990). For example, my glasses become embodied: they become part of me and how I see the world. Moreover, as Verbeek has argued, in this relational ontology subjects and objects "mutually constitute each other" (2005, 129). The things we use also shape us. Different uses create different experiences and different people.

However, Ihde and Verbeek conceptualize these mediations and constitutions in a way that leaves their process character implicit and untheorized. Subject and object are seen as part of a rather static picture, with images and spatial metaphors dominating. If we attempt a process philosophy interpretation of postphenomenological mediation theory, however, then postphenomenological mediations can be understood as *becomings* of subjects and objects in human-technology relations. Mediation is then not a kind of thing that sits "in between" subject and object, but a dynamic and historical process from which the entities that are mediated emerge. To use one of Verbeek's examples, a medical imaging technology is not so much a *thing* but rather a *process* through which patients (as subjects) and the technologies (as objects)—that is, the scan and the scanner—come into being.

This interpretation is in line with Barker's process philosophy approach to digital media: "Mediation is not a flow between two preexistent entities; rather, it is a process that re-presents or reconstitutes entities. In short, it is a generative process, setting the conditions for the becoming of entities. This is a temporal process, with technological processes generating particular conditions for becoming" (Barker 2012, 12). It is also compatible with Gilbert Simondon's philosophy of technology (2017), which understands both individuals and technology as outcomes of processes of individuation. A process philosophy approach invites us to reveal these processes of becoming and—combined with postphenomenology and other contemporary approaches in philosophy of technology—show how in this way technologies are so much more than just their instrumental function. Finally, the proposed approach echoes Lenore Langsdorf's (2015) claims that postphenomenology needs a process metaphysics and that Whitehead's process thinking enables us to turn away from substance metaphysics and offers a relational view that rejects "Husserl's predominant focus on consciousness" (Langsdorf 2020, 128).

PROCESSES OF MAKING TIME AND BEING MADE BY TIME, EVENTS, AND MULTIPLE TEMPORALITIES

Process philosophy also helps to conceptualize the relation between technologies and time. And once again it helps to arrive at a better understanding of the non-instrumental role of technology. We can place technologies in time but also talk about the technological making of time and—in the end—the making of experiences and the making of subjects. The so-called "objective" or "scientific" time we know is produced by technological instruments such as clocks. At the same time (!), we are shaped and constrained by that kind of time making. Time itself is produced and subjects are produced. Consider artificial intelligence (AI). As I have recently argued, AI is not neutral toward time and experience, but works as a "time machine," linking and shaping past, present, and future (Coeckelbergh 2021). For example, it may bind me to my past when it makes a biased decision about the present on the basis of what I (or others) have already done. Social media also have this time machine function when they reduce me to past images from my post history. All this happens through use of the technology. But it is not so much the technology as "thing" that does this. It is my use and interaction with the technology. What matters is that interaction and that technological event. Humans are participants in those events. As Barker puts it: "interactive events cannot be reduced to ideas of a subject or user *using* a technological system. Instead, I would like to understand the event as a process of interaction in its fullest sense, as an *interpenetration* of a human with technology" (2012, 7).

Moreover, just as there are many meanings in technology use and interaction (multistability, a concept that is still rather spatial in Ihde's elaboration, as he uses the duck-rabbit and similar images, which refer to spatial forms), there are also many forms and layers of time, which interact and come together in particular events and processes. For example, AI connects different times (Coeckelbergh 2021) as it relates the past (data) to the present (prediction or recommendation), and in social media, digital information about the past comes in contact with present experience; they make that past present (and influence the future). What Barker writes about an art work is true for digital social media and many other digital technologies: "Histories, both personal and collective, are in constant contact with our experiences of the present moment" (21). Social media in the context of the internet thus present us with "thick" forms of temporality in which multiple times come together and are synthesized (29). On social media, we are not just living in the here and now but here and everywhere and in the past; we are involved in many "nows" and pasts, which creates "a turbulence of temporal events" (190) and draws them together. One must understand that there is not only a relationality of subjects and objects, as postphenomenology claims, but also a relationality of events, as conceptualized by process philosophy. We should understand our interaction with digital technology "as a set of relational events, where each present moment draws into itself aspects of the past and future" (191). These events co-constitute us and shape our experience. They also shape what the technology is for us. In a sense, *there is no technological object* before these processes and events take place. For example, Facebook "is" (for me) what it becomes in the process of interacting with it. The technological object cannot be defined a priori but emerges from processes and in events, which shape my perception and experience of the object.

Despite this potential to offer an interesting and original approach to thinking about technology and time, process philosophy has not had much influence on contemporary philosophy of technology. An exception is Simondon, and later Bernard Stiegler, who in turn have influenced Yuk Hui. In *On the Mode of Existence of Technical Objects* (2017), Simondon takes an ontogentic view of technical objects and describes processes of individuation—in other words, he takes a process view of technology and of the world. Stiegler's organological approach (1998–2011)—an account of life that is not just biological but includes the technical (Stiegler 2020)—and Hui's (2016) characterization of digital objects in terms of individuation and evolution within a milieu continue this tradition. These process-oriented and relational views have also influenced reflections on time and technology in terms of how technology exteriorizes memory (Stiegler's "hypomnesis") *and* shapes our imagination of the future (Hui 2016). Stiegler argued that memory became technical, for example in the form of writing or machines. According to Hui, computational relations do not only structure meaning but also temporality. Through mobilizing data within a system of anticipation and production, digital technologies organize time (247) and do not only retain the past but also shape normative futures (see also McKim 2017). This is congruent with what I say about the time machine function of artificial intelligence and deserves more attention, both as a philosophy of technology and as a philosophy of time, although Hui's and Stiegler's Simondonian and Heideggerian language does not render this body of work very accessible for philosophers of technology outside those traditions.

Once we attend to the temporal and process dimensions of technology, we can also reveal and study other phenomena, such as time lags between different phases of technology lifecycles or between different phases of societal development. There are also narratives about technological time and related narratives about modernity and human culture, for example narratives about technological acceleration, technological singularity, and progress. For example, transhumanists such as Ray Kurzweil (2005) and Nick Bostrom (2014) argue that there will be an intelligence explosion, through which increasingly intelligent agents are created, which eventually results in superintelligence that surpasses and perhaps replaces human intelligence. I will say more about the concept of narrative below.

PERFORMANCE PROCESSES

Another relevant term that helps us move toward more process-oriented thinking about technology is performance. I have argued that performance metaphors can help us to think about technology (Coeckelbergh 2019). For example, the concept of choreography can help us to analyze how we move with technologies and how technologies also move us, for example when we have to use our hands and fingers in particular ways in order to use mobile phones and their apps. Another example: theater metaphors help us to understand what goes on when we use digital social media, in which we present ourselves on various stages and take on different roles. Put in the context of the theme of this paper, such metaphors show that what we do to technology are processes and events, that these processes involve the body and bodily movement, and that these processes are deeply social. In contrast to more abstract process philosophy, and also partly in contrast to artifact-oriented concepts of technology, this brings in the concrete, bodily dimension of human-technological processes and interactions. When we use AI or are on social media, we do not leave our body at home. When we are shaped by our technologies and our times, it is not disembodied mind or an abstract thinking that is shaped; we are constituted as embodied, moving, and social beings who are situated in time and space and in a social and political context. For example, we move a computer mouse while sitting at a particular desk, making also other (micro) movements, and these movements are choreographed by the designers of these technologies. While Ihde certainly managed to take into account the bodily dimension in his postphenomenology, more emphasis is needed on movement and on the socially situated temporality of what we do with technology and what technology does with us. Technology use is not only embodied but also involves movement, *kinesis*. This kinetic dimension requires further study and also raises normative questions. There is already some interdisciplinary work in this area from which philosophers of technology can learn. For example, Jaana Parviainen (2021) has discussed mobility inequalities in smart urban environments.

Process philosophy thus helps us to further develop the epistemology of technology in a performative direction. Knowledge is gained *in the process* of interacting with technology—that is, through what we do. If things do things, as Verbeek says, then it should be added that things only can do their things through what we humans do with technology and that this is a temporal process. It is the process and event of our performances with technology that produce experiences, knowledge, meaning, and so on. Barker puts the point as follows: "The knowledge gained from this interaction cannot be separated from the performative action that provided the condition for this knowledge to emerge. In other words, the ways in which we act by means of tools or technologies are linked to, and sometimes productive of, the knowledge of the reality that our actions produce" (2012, 108). And this is also a temporal and historical process. Barker writes:

> The knowledge of how to hit a nail with a hammer, to use a seemingly more rudimentary example, comes from a history of watching your father hammer nails into wood, feeling the weight of the hammer, the contours of the handle, and a history of bruised thumbs. As such, and as pointed out by Whitehead, Deleuze, and Bergson, knowledge is not solely the product of the human mind, but is rather produced through a process, which may involve many human and nonhuman agents. (109)

This participative but also reiterative process can also be described by using the terms *skills* and *habit*. In order to learn to hammer, we have to do it again and again, which itself involves histories: we re-present again and again, connecting different temporalities, and making meaning and knowledge.

GAMES, NARRATIVES, AND CULTURAL PROCESSES

Performance metaphors already help to bring in the social aspect of human-technological processes: we perform with technology, and this may for example involve playing a particular social role. Performance is also an activity and an event that typically involves more than one person: several actors, dancers, musicians, and so on. Moreover, it takes place in a social context. In contrast to (post)phenomenology, what we do and what happens with technology is not mainly conceptualized at the individual level (alone) but is also about *social* processes and dynamics. For example, on social media we may play a particular professional role and our interaction with others, through technology, also constitutes us. Again technology is not just about things; it is also about people and what people do with one another, through technology. This insight is in line with work in STS and history of technology: the history of technology is a technical one but also a history of people and their (techno)performances, involving, for instance, interests and power.

Furthermore, the social does not just take place at the "micro" level, say the level of interaction that Barker talks about, but technology use is also and always connected to wider social and cultural structures. In previous work

I have called those structures games, grammars, and forms of life, taking inspiration from Wittgenstein. Wittgenstein argued that the meaning of words is a matter of language use, which is connected to what he called *language games* and *forms of life* (Wittgenstein 2009). I have argued that, similarly, technology is embedded in what I called *technology games* and *forms of life* (Coeckelbergh 2018). These wider social and cultural structures shape what we do with technology but are also at the same time constituted by what we do. Taking into account the temporal aspect, one could say that there are *iterations* of technology games, which sustain particular games and constitute a particular form of life. For example, digital social media would not change our culture if people used them occasionally. They have their influence on *how we do things here* only because they are used frequently. This constitutes social media use as technology games and eventually co-shapes forms of life. We start to live in and through social media; it is part of how we live and how we do things. Again, making meaning and making culture through technology is not just a one-off act: we develop habits and re-iterate games.

This approach goes further than the linguistic philosophical approach to engineering explored by Mitcham and Robert Mackey (2009): the point is not to develop a linguistic philosophy of technology but to use (in this case Wittgenstein's philosophy of) language as a metaphor for understanding technology. While it is right that technology is linked to language games, my use of the term *technology games* goes further than that: it puts technology in a wider context that includes but is not limited to the ways we use language. Not only does language play a role in technology use—although this is also an interesting area of research—but in addition *technology use is like language use*: it co-constitutes, and is ruled and shaped by, larger cultural structures. And in the spirit of Mitcham's interest in cross-cultural work: we may live in one culture, but there are also dynamics between cultures. Our form of life is neither homogeneous nor isolated from other forms of life. This is a thought that deserves further development, for example in interaction with existing work in intercultural philosophy of technology.

Finally, taking a more temporal perspective with the help of process philosophy also may lead to considering other social and cultural structures, for example narratives or myths. Ricoeur (1983) has inspired recent work on technology and narrativity (Reijers and Coeckelbergh 2020; Reijers, Romele, and Coeckelbergh 2021), myth and engineering (Miller, Portal, and Xin XU 2022), digital hermeneutics (Romele 2020), and, earlier, expanding hermeneutics (Ihde 1998). Building on work Wessel Reijers and I did (2020), one could claim that understanding technology in terms of processes can also be framed in terms of narratives, which have their own temporalities and which shape us. As Ihde already argued (1998), hermeneutics is not just about text: it is also about technology. But in contrast to Ihde, we can directly connect with Ricoeur's work and still talk about a (material) hermeneutics in terms of narratives. Technology is then the object of narratives, but also contribute to our narratives. Technologies are coauthors of our performances and stories. For example, AI "narrates" (Coeckelbergh 2021). Humans are participants, co-narrators.

This amounts to a kind of posthumanist view, according to which humans are not necessarily the center of the performative and hermeneutical processes. Barker (2012) writes about "userness": agency is shared among the part of the assemblage (15). However, the term "assemblage," borrowed from (other) social sciences and in particular from posthumanism, new materialism, Latour, Deleuze, and so on, is a much too spatial metaphor. Process metaphors such as performance and narrative are more suitable to conceptualize the dynamic ways in which humans and technology relate and make meaning.

CONCLUSION: TOWARD A PROCESS-ORIENTED PHENOMENOLOGY AND HERMENEUTICS OF TECHNOLOGY

Inspired by process philosophy, I have presented a number of concepts that promise to give us a more complete phenomenology of what *happens* in and with technology than postphenomenology has done so far. It is more complete because it shows how the relational, mutual constitution of subject and object through technology is a process of *becoming* and a relational, performative, historical, and narrative process that has both bodily and social-cultural dimensions. More work is needed to develop this framework and to further discuss its relations to (post)phenomenology, pragmatism (see, for example, Hickman 1990; Thompson 2020), critical theory, and existing approaches in media theory. But the conceptual framework presented here offers a promising approach to understanding technologies in a more relational, dynamic, and performative way, develops Mitcham's and Ihde's view that philosophy of technology must be situated in its cultural context, and invites us to further conceptualize the relation between technology and time.

Yet this theoretical direction also raises normative considerations: narratives about technology are not neutral, technological processes (for example data processes) are not neutral, and if technology shapes our time and experience, and in the end our games, narratives, and life forms, then that is also normatively relevant. Asking what technologies we want is asking: What performances does a particular technology enable? What time and experience with and through technologies do we want? What are good times with technology? What form of life do we want? What narratives do we want to tell? What do we, as persons and as societies, want to *become*? A process-oriented philosophy of technology therefore offers us an interesting perspective on ethics of technology.

Moreover, in line of the heritage of pioneers such as Mitcham and Ihde, it is also important to ask: what does the proposed approach mean for engineers and designers, for technology development, for scientists? One potential outcome is that when it comes to improving and evaluating technologies, one could pay more attention to entire technological processes rather than single technological artifacts, and for example consider the ethical aspects of maintenance. But technological processes are never merely technical: a further and perhaps more fundamental insight I helped to conceptualize in this chapter by using a process approach is that technology is deeply connected to humans and to culture. What does it mean for designing digital technologies, for instance, that we see the technology not only as a mere instrument for a particular purpose (say writing a text or communicating with a friend) but also as a key participant in histories, narratives, and performances of becoming?

The question how engineers can think more critically about their role in making our world, as Mitcham (2012, 19–20) asked, can now be put in process terms, which more radically question the gap between humans and technology, between culture and artifacts, between subjects and objects, between experience and the things that become. What does it mean to say, for example, that the engineer is not only a designer of technology but also a designer of processes and performances, of subjects, of time, of a form of life, and of a narrative? What are the limits to those attempts to design and engineer these processes and form of life? And how can we, as persons and as a society, engineer intercultural processes? A process-oriented phenomenology and hermeneutics of technology enables us to ask these questions concerning becoming. Technologies, like us, flow. Let's talk about where we want to go and where we are going.

REFERENCES

Achterhuis, Hans. 2001. *American Philosophy of Technology: The Empirical Turn*. Bloomington: Indiana University Press.

Barker, Tim S. 2012. *Time and the Digital: Connecting Technology, Aesthetics, and a Process Philosophy of Time*. Hanover, NH: Dartmouth College Press.

Bergson, Henri. 1944. *Creative Evolution*. Translated by Arthur Mitchell. New York: Random House Modern Library.

Bostrom, Nick. 2014. *Superintelligence: Paths, Dangers, Strategies*. Oxford: Oxford University Press.

Coeckelbergh, Mark. 2018. "Technology Games: Using Wittgenstein for Understanding and Evaluating Technology." *Science and Engineering Ethics* 24: 1503–19. https://doi.org/10.1007/s11948-017-9953-8.

———. 2019. *Moved by Machines: Performance Metaphors and Philosophy of Technology*. New York: Routledge.

———. 2021. "Time Machines: Artificial Intelligence, Process, and Narrative." *Philosophy & Technology* 34: 1623–38. https://doi.org/10.1007/s13347-021-00479-y.

Deleuze, Gilles. 2005. *Cinema 2: The Time Image*. Translated by Hugh Tomlinson and Robert Galeta. London: Continuum.

Floridi, Luciano. 2014. "Informational Realism." Available at SSRN: https://doi.org/10.2139/ssrn.3839564.

Hickman, Larry A. 1990. *John Dewey's Pragmatic Technology*. Bloomington: Indiana University Press.

Hui, Yuk. 2016. *On the Existence of Digital Objects*. Minneapolis: Minnesota University Press.

Ihde, Don. 1990. *Technology and the Lifeworld: From Garden to Earth*. Bloomington: Indiana University Press.

———. 1993. *Postphenomenology*. Evanston, IL: Northwestern University Press.

———. 1998. *Expanding Hermeneutics: Visualism in Science*. Evanston, IL: Northwestern University Press.

Kurzweil, Ray. 2005. *The Singularity Is Near: When Humans Transcend Biology*. New York: Penguin.

Langsdorf, Lenore. 2015. "Why Phenomenology Needs a Metaphysics." In *Postphenomenological Investigations: Essays on Human-Technology Relations*, edited by Robert Rosenberger and Peter-Paul Verbeek, 45–54. Lanham, MD: Lexington.

———. 2020. "Relational Ethics: The Primacy of Experience." In *Reimagining Philosophy and Technology, Reinventing Ihde*, edited by Glen Miller and Ashley Shew, 123–40. Dordrecht: Springer.

McKim, Joel. 2017. "Envisioning a Technological Humanism." Review of Yuk Hui, *On the Existence of Digital Objects*, Minneapolis, University of Minnesota Press, 2016. *Computational Culture* 6 (November 28, 2017). http://computationalculture.net/envisioning-a-technological-humanism-a-review-of-yuk-huis-on-the-existence-of-digital-objects/.

Miller, Glen, Michael Portal, and Xin XU. 2022. "Engineering Myth in China and the United States." In *Engineering, Social Sciences, and the Humanities: Have Their Conversations Come of Age?*, edited by Steen Hyldgaard Christensen, Anders Buch, Eddie Conlon, Christelle Didier, Carl Mitcham, and Mike Murphy, 341–60. Dordrecht: Springer.

Mitcham, Carl. 1994. *Thinking through Technology: The Path between Engineering and Philosophy*. Chicago: University of Chicago Press.

———. 2012. "The True Grand Challenge for Engineering: Self-awareness." *Issues in Science and Technology* 31, no. 1: 19–22.

Mitcham, Carl, and Robert Mackey. 2009. "Comparing Approaches to the Philosophy of Engineering: Including the Linguistic Philosophical Approach." In *Philosophy and Engineering: An Emerging Agenda*, edited by Ibo van de Poel and David Goldberg, 49–59. Springer, Dordrecht. https://doi.org/10.1007/978-90-481-2804-4_5.

Parviainen, Jaana. 2021. "Kinetic Values, Mobility (In)Equalities, and Ageing in Smart Urban Environments." *Ethical Theory and Moral Practice* 24: 1139–53.

Reijers, Wessel, and Mark Coeckelbergh. 2020. *Narrative and Technology Ethics*. Cham, Switzerland: Palgrave.

Reijers, Wessel, Alberto Romele, and Mark Coeckelbergh. 2021. *Interpreting Technology: Ricoeur on Questions concerning Ethics and Philosophy of Technology*. London: Rowman & Littlefield.

Ricoeur, Paul. 1983. *Time and Narrative*. Volume 1. Translated by Kathleen McLaughlin and David Pellauer. Chicago: University of Chicago Press.

Romele, Alberto. 2020. *Digital Hermeneutics: Philosophical Investigations in New Media and Technologies*. New York: Routledge.

Simondon, Gilbert. 2017. *On the Mode of Existence of Technical Objects*. Translated by C. Malaspina with J. Rogove. Minneapolis: Univocal Press.

Stiegler, Bernard. 1998–2011. *Technics and Time*. 3 vols. Stanford: Stanford University Press.

———. 2020. "Elements for a General Organology." *Derrida Today* 13, no. 1: 72–94.

Thompson, Paul. 2020. "Ihde's Pragmatism." In *Reimagining Philosophy and Technology, Reinventing Ihde*, edited by Glen Miller and Ashley Shew, 43–61. Dordrecht: Springer.

Verbeek, Peter-Paul. 2005. *What Things Do: Philosophical Reflections on Technology, Agency, and Design*. University Park: Pennsylvania State University Press.

Whitehead, Alfred North. 1978. *Process and Reality*. New York: Free Press.

Wittgenstein, Ludwig. 2009. *Philosophical Investigations*. Revised Fourth edition. Translated by G. E. M. Anscombe, P. M. S. Hacker, and Joachim Schulte. Malden, MA: Wiley.

Young, Mark T. 2021. "Maintenance." In *The Routledge Handbook of the Philosophy of Engineering*, edited by Diane P. Michelfelder and Neelke Doorn, 356–68. New York: Routledge.

Chapter 5

Political Philosophy of Technology
After Leo Strauss

Carl Mitcham

As the philosophy of engineering and technology has evolved since the mid-twentieth century it has increasingly focused on ethics and social issues, with healthy doses of conceptual and epistemological analysis, but little engagement with political philosophy. The present essay sketches a political philosophy of technology provoked by the thought of Leo Strauss, one of the more consequential if controversial political philosophers of the twentieth century. It is stimulated by selective aspects of Strauss's subtly seductive corpus, without aiming to expand the extensive body of literature explicating, interpreting, and debating his ideas and without too much concern about whether my readings are completely faithful to his intentions. My hypothesis is that Strauss can help us catch sight of something otherwise missing in philosophical discourse on engineering and technology.

The exploration of this hypothesis takes place in two movements. First, I make a case for technology as a persistent reference in Strauss and interrogate a key study from the 1950s that maps the terrain of political thought, identifying different incisions of technology, and then block out Strauss's distinctive historical-philosophical interpretation of political philosophy. Second, I take up with the posthumously published version of an informal talk from the same arguably pivotal decade in the development of Strauss's mature thought. Precisely because of its informality, this text adds a special perspective to Strauss's scholarly studies. It also anticipates a theme that becomes increasingly central in later work, the conflict between reason and revelation, in a way relevant to political philosophy under conditions of engineering and technology.

1

Even though Strauss does not have a political philosophy of technology in the emphatic sense, many of his works treat technology as an important topic. Early in his career, in a 1930s review on Carl Schmitt, Strauss already noted how the modern attempt to establish a new political order entailed creation of a new "faith in technology" (Strauss [1932] 1965, 348). In a 1940s exchange with Alexandre Kojève on tyranny he argued that prospects for a "universal and homogeneous state" depend on modern technology (Strauss 1948). In major sets of lectures at the University of Chicago on natural right and history (1953) and Machiavelli (1958) Strauss more broadly questioned the wisdom of a modern commitment to technology. Consider this lengthy quotation from the final pages of the Machiavelli study:

> The classics were for almost all practical purposes what now are called conservatives. In contradistinction to many present day conservatives, however, they knew that one cannot be distrustful of political or social change without being distrustful of technological change. Therefore they did not favor the encouragement of inventions, except perhaps in tyrannies, i.e., in regimes the change of which is manifestly desirable. They demanded the strict moral-political supervision of inventions; the good and wise city will determine which inventions are to be made use of and which are to be suppressed. Yet they were forced to make one crucial exception. They had to admit the necessity of encouraging inventions pertaining to the art of war. They had to bow to the necessity of defense or of resistance. This means however that they had to admit that the moral-political supervision of inventions by the good and wise city is necessarily limited by the need of adaptation to the practices of morally inferior cities which scorn such supervision because their end is acquisition and ease. They had to admit in other words that in an important respect the good city has to take its bearings by the practice of bad cities or that the bad impose their law on the good. . . . One could say however that it is not inventions as such but the use of science for such inventions which renders impossible the good city in the classical sense. From the point of view of the classics, such use of science is excluded by the nature of science as a theoretical pursuit. Besides, the opinion that there occur periodic cataclysms in fact took care of any apprehension regarding an excessive development of technology or regarding the danger that man's inventions might become his masters and destroyers. Viewed in this light, the natural cataclysms appear as a manifestation of the beneficence of nature. . . . It would seem that the notion of the beneficence of nature or of the primacy of the Good must be restored by being rethought through a return to the fundamental experiences from which it is derived. (Strauss 1958, 298–99)

This sobering assessment of the signal roles of violence and catastrophe in human affairs was complemented the following decade, when introducing the publication of another set of lectures, these given at the University of Virginia, Strauss (1964a) called particular attention to the threat of cataclysms by nuclear weapons. Since Strauss wrote, of course, other existential threats have appeared on the horizon, not the least of which is an anthropogenically mutated (and mutating) climate. Civilizational catastrophe, Strauss suggests, is more than a remote possibility to be set aside in favor of immediate issues, nor should it be dismissed with appeals to hope for salvation through technological innovation. It deserves a serious place in any political philosophy, especially I would argue, in political philosophical reflections on technology.

Aware of such texts when coediting a theme issue of the journal *Philosophy Today* on the philosophy of technology, I wrote Strauss inviting him to contribute. He replied that he was unable to do so (I was unaware of his poor health, which led to his death two years later) while recommending I contact his colleague Joseph Cropsey, who would become his literary executor. Cropsey in turn suggested Leon Kass, who had recently published an article on "The New Biology: What Price Relieving Man's Estate?" (1971). (In a subsequent 1985 book elaborating his argument regarding the ethics of biomedical technologies, Kass included a brief account of Strauss's death.)

So many references to technology, however scattered, could be taken to justify construction of a genuinely Straussian political philosophy of engineering and technology. It might even be claimed that Kass, along with a few others (e.g., William Galston), have already done so, albeit without engaging philosophy of technology as it was emerging during Strauss's lifetime. As noted, however, this is not my intent. Although one of my teachers studied under Strauss, and for some years one of Strauss's most argumentative students was a colleague, I am not even close to being sufficiently Straussian. Instead, I am more concerned to invite those who think of themselves as philosophers of technology to pay attention to Strauss, to be challenged if not inspired, to think political philosophy of technology after him. To do so it is necessary to appreciate what Strauss means by political philosophy.

2

Strauss's most succinct demarcation between political philosophy and its rivals occurs in the title study of *What Is Political Philosophy?* (1959). Part one (the longest) directly addresses "The Problem of Political Philosophy" as (especially in the 1950s) a more or less moribund inquiry. To begin, he notes that political philosophy arises whenever people "make it their explicit goal to acquire knowledge of the good life and of the good society."

Political philosophy deals with political matters in a manner that is meant to be relevant for political life; therefore its subject must be identical with the goal, the ultimate goal of political action. The theme of political philosophy is mankind's great objectives, freedom and government or empire—objectives which are capable of lifting all men beyond their poor selves. Political philosophy is that branch of philosophy which is closest to political life, to nonphilosophic life, to human life. (1959, 10)

"In former epochs," Strauss observes, intelligent people could acquire political knowledge by simply "looking around and by devoting themselves to public affairs." But this "is no longer sufficient because we live in 'dynamic mass societies,' i.e., in societies which are characterized by both immense complexity and rapid change. Political knowledge is more difficult to come by and it becomes obsolete more rapidly than in former times" (15). The lifeworld is continually transformed and refracted by science, engineering, and technology, so attempts to acquire knowledge of the good life or good society—that is, to practice "political philosophy"—should take these activities into account as the branch of philosophy of technology closest to this world. From an Aristotelian framework, in which political philosophy is a prolongation of ethics, it would also be complementary to much current discussion on the ethics of technology.

Almost immediately Strauss offers a second description of political philosophy as "the attempt to replace opinion about the nature of political things by knowledge of the nature of political things" (11–12). This definition assumes an idea of philosophy itself as "the attempt to replace opinions about the whole by knowledge of the whole," that is, of "all things." "Quest for knowledge of 'all things' means . . . quest for knowledge of the natures of all things: the natures in their totality are 'the whole,'" with an emphasis that "Philosophy is essentially not possession of the truth, but quest for truth" (11). Philosophers qua philosophers are deeply aware of the limitations of their knowledge—and indeed of all knowledge.

In the first definition, political philosophy emerges, as it were, from the ground up, within human affairs. This is *political* political philosophy and a characteristic possession of the πολιτικός (*politicos*, politician, commonly translated as "statesman" or "statesperson"). In the second, it takes form top-down, as a branch of philosophy, as *philosophical* political philosophy and is characteristic of the φιλόσοφος (*philosophos* philosopher). The former has a more direct or positive relationship to political life than the latter. In fact, political philosophy in the philosophical sense, as well as philosophy *tout court*, involves a kind of "elitist," disciplined withdrawal from and simultaneously astringent criticism of unexamined human affairs. Such a stance cannot help but take on a fraught relationship with any social order in which it exists, even while it is existentially dependent on this order. Although thinking is a higher activity than eating, philosophers do not live by thought alone.

Taking a cue from Socrates and Aristophanes, as a persistent theme in his political philosophy, Strauss repeatedly spells out and explores alternative ways of managing this inherently conflictual relationship in the Western philosophical tradition before, during, and after Socrates. Political philosophy requires practicing philosophy politically, Strauss argues, in self-protection and out of respect for ways of life that are less than philosophical.

The problem of political philosophy in Strauss is thus not simply the problem of acquiring knowledge of the good life, but also navigating between both versions of political philosophy and, even more crucially, attending to the distance demanded by the second definition and the political order in which one perforce must live. Insofar as a political order becomes ever more infused with science and engineering, these problems, especially the third, necessarily implicate political philosophy of technology. However, the primary focus in "The Problem of Political Philosophy" is the effort to revive political philosophy as something of more than mere historical interest. In this Strauss was of course anticipating work by, among others, Jürgen Habermas and John Rawls, although along a path different than theirs.

3

Following his characterization of political philosophy, and before turning to a fourth problem, Strauss further distinguishes political philosophy from political thought, political theory, political theology, and political science—all of which have philosophy of technology–related faces. Political thought includes any "reflection on, or the exposition of, . . . any politically significant 'phantasm, notion, species, or whatever it is about which the mind can be employed in thinking' concerning the political fundamentals." Political philosophy is always political thought, but seldom is political thought political philosophy. "By political theory, people frequently understand today comprehensive reflections on the political situation which leads up to the suggestion of a broad policy. Such reflections appeal in the last resort to principles accepted by public opinion or a considerable part of it; i.e., they dogmatically assume principles which can well be questioned" (Strauss 1959, 13).

Strauss might well consider the political theories of Habermas and Rawls as cases in point. Strauss would also probably question whether analyses of politics and technology relationships by Hannah Arendt, Langdon Winner, or Andrew Feenberg would count as political philosophy in this sense; indeed, these three thinkers more commonly style their work as political theory.

Strauss further distinguishes political theology, political philosophy, and social philosophy:

> By political theology we understand political teachings which are based on divine revelation. Political philosophy is limited to what is accessible to the unassisted human mind. As regards social philosophy, it has the same subject matter as political philosophy, but it regards it from a different point of view. Political philosophy rests on the premise that the political association—one's country or one's nation—is the most comprehensive or most authoritative association, whereas social philosophy conceives of the political association as a part of a larger whole which it designates by the term "society." (13)

The difference between teachings concerning politics based on divine revelation and teachings concerning politics based in philosophy refers to what Strauss argues throughout his corpus is a fundamental and inescapable tension in Western civilization: the opposition between reason and revelation, Athens and Jerusalem. In Strauss, this tension assumes a political philosophical dimension distinct from that common in the history of ideas.

Just as there can be political teachings based on divine revelation, there are also political studies of technology based on revelation. The work of Jacques Ellul is an obvious case. Although Ellul's primary interest was not politics but society (e.g., Ellul 1954), he analyzed how technology informs contemporary politics and offered in contrast a Christian theological appraisal (especially Ellul 2014). Simplifying, Ellul argues that modern technology constitutes an effort by human beings to deny their dependency on God and to assert self-determinacy, to practice self-assertion. Note that Strauss's definition implies there is no political theology outside the Abrahamic religious traditions; his thinking appears restricted, as was Ellul's, to one of the historically intertwined Jewish, Christian, and Islamic traditions.

Strauss's remark about social philosophy deserves notice too. The modern city disaggregates family, education, economy, religion, government, science, and so on. Galileo promoted the disentanglement of science and religion; Adam Smith, the separation of economics and politics; the American constitution, government and religion. These elements are then re-entangled in society, where they compete for influence if not dominance. Liberal philosophies advocate enhancing structural differentiation in order to grant each social institution (and its individual members) as much autonomy as possible, in principle not allowing any one to exercise sovereignty, thereby de-centering the political (and everything else). Political philosophy becomes just another companion to philosophy of science, of religion, of economics, of education, of language, of art, and more. This diminishing of the authoritative character of the political was formalized in modern social contract theory, which conceives of the state as a social creation rather than (as in classical philosophy) social formation as enabled by the state. Additionally it went hand in hand, whether essentially or accidentally, with creation of the bourgeois state and culture, in which engineering and technology become subtly dominant as apparently neutral (and neutering) means available to be used for the advancement of any and all institutional interests. What Walter Benjamin calls the loss of aura in a world of technological reproduction of art applies to politics as well—until, that is, war erupts to reaffirm the uniquely existential seriousness of the political.

Finally, Strauss notes how "political science" has become "an ambiguous term." For Aristotle, political science was a synonym for political philosophy in a way that combines contributions from statesmen and philosophers in an effort to acquire knowledge of the good life and for the good regime. Today it "designates such investigations of political things as are guided by the model of natural science" (Strauss 1959, 13) in order to provide means for use by any and all regimes. Contemporary political science could thus be, and on occasion has been, described as a kind of social or political engineering.

4

When political science becomes guided by the model of natural science, insofar as this model distinguishes between facts and values and marginalizes if not dismisses a quest for knowledge of the good, it sets a stage for the fourth problem. A situation is created in which political philosophy needs to be reconstructed or defended as something more than a fool's errand or waste of time. The challenge for Strauss, as he experienced it originally during the intellectual collapse of the Weimar Republic, took two forms: positivism and historicism. Positivism is a belief that only modern science produces knowledge; historicism considers values, including the value of modern science, more historical than rational. In such a *Weltanschauung*, efforts to reason about the good are replaced by cultivation of the willed, nihilistic affirmation of historical fate, absent any appeal to guidance by the good. "The crucial issue concerns the status of those permanent characteristics of humanity, such as the distinction between the noble and the base [and whether] these permanencies [can] be used as criteria for distinguishing between good and bad dispensations of fate" (Strauss 1959, 26).

In their classical form philosophy or science (two words for the same way of life) was a search for knowledge of the whole that might offer insight into the nature of the good. Yet over the course of Western history, from philosophy or science as quest for knowledge of all things, specialized sciences gradually peeled off, leaving (analytic and phenomenological) philosophy as a kind of rump. The specializations in turn have been in the modern period reconceived not so much as quests but as methods for producing positive results (as in physics, chemistry, and biology)—results that are able, much more than premodern sciences, to be deployed to enhance human practice (most obviously through engineering). These methods together with their positive results have then been positioned as the best place to begin to construct knowledge of the whole, including knowledge of human things.

Not being gods, we cannot approach the whole immediately; we must approach the whole from or through one of its parts. There is a truth to standpoint theory. Modern political science differs from classical political science in the part it privileges as the departure point from which to seek access to the nature of the whole. In this disagreement it may even be said to agree with the pre-Socratics, who focused attention on being rather than human being.

According to Aristotle, philosophers were φυσιολόγοι (those who discourse about nature) as opposed to θεολόγοι (those who discourse on the gods). The central theme of philosophy is thus nature. The Greek word φύσις (*physis*, nature) refers primarily to growth and then to the stable being that plants and animals manifest in maturity, until decaying and passing away. Things that do not grow but are constructed, such as beds and houses, are not "by nature" but "by τέχνη (*techné*, technics)"; such things nevertheless remain dependent on nature. Above nature, as it were, and on which nature depends, are the ρχαί *archai* (first things or principles). Still another distinction from *physis* is found in νόμος (*nomos*, convention or law): "things which are only by virtue of being held in reverence or, more generally stated and perhaps more precisely stated, by virtue of being held—using 'holding' here in the sense in which it is used of judges, the holding of a judge [—that is], of being believed in" (Strauss 2018, 207). Such things include the gods and *nomoi* or laws of a city. Whether approaching the whole through cosmology or through human affairs, philosophy is liable to question gods and laws, creating problems for both the city and philosophy. Socrates's questioning of law in relation to nature, technics, and the city was "decisive for the emergence of political philosophy" (207).

> Nature, however understood, is not known by nature. Nature had to be discovered. The Hebrew Bible, for example, does not have a word for nature. The equivalent in biblical Hebrew of "nature" is something like "way" or "custom." Prior to the discovery of nature, men knew each thing or kind of things has its "way" or its "custom"—its form of "regular behavior." (Strauss and Cropsey 1963, 3)

As soon as nature as something transcending way or custom was discovered, it was possible to ask whether any political things were by nature and, if so, in what way. According to Cicero, who echoes the views of both Aristophanes and Xenophon, the person to first ask this question was Socrates. In Cicero's words, Socrates was "the first to call philosophy down from the heavens and set her in the cities . . . and compel her to ask questions about life and morality and things good and bad" (*Tusculan Disputations* V, iv, 11).

For Socrates, the nature of justice is more important to us than the nature of the cosmos, even though we are part of the cosmos; indeed, because it is more immediate, inquiries into justice offer the best entrée into understanding the cosmos. For Socrates, political philosophy replaces cosmology as first philosophy, just as for Strauss natural right and its occlusion in history replace Being and its *Ereignis* in time.

More than two millennia later, the relationship between things independent of humans, dependent on human *techne*, and dependent on human reverence has emerged in an even more consequential form. Things which are by virtue of human making are now so engineerable as to become like things by nature and things by nature are increasingly dependent on being protected by law; first principles are considered quaint if not obsolete. The political philosophical consequences of these enhancements in human power and in the centrality of human belief can scarcely be overstated. Belief dependency absent reverence opens the door to nihilistic engineering without end.

5

Strauss's response to the fourth problem of political philosophy implicates classical responses to the first and second—that is, to the quest for an understanding of the good life and of the good society and relationships between top-down and bottom-up political philosophy. "The Classical Solution" ("What Is Political Philosophy?" part two, the shortest of its three sections) describes political philosophy as practiced prior to the attacks of positivism and historicism. The central paragraph explores how what is central in classical political philosophy is the *politeia* or regime and the question, given the plurality of possible regimes, of which is best by nature.

The central appeal in the classical tradition was to a sense of the natural, in contrast to what is merely "human, all too human." As Strauss presents it, during the dual (bottom-up and top-down) emergence of political philosophy, between a pre-philosophical all too human power politics and the establishment of a tradition of political philosophy, the philosophical engagement with politics waxed with a luminosity that has since waned. An erotic encounter occurred that subsequently settled into a tradition.

> Classical political philosophy is non-traditional, because it belongs to the fertile moment when all political traditions were shaken, and there was not yet in existence a tradition of political philosophy. In all later epochs, the philosophers' study of political things was mediated by a tradition of political philosophy which acted like a screen between the philosopher and political things, regardless of whether the individual philosopher cherished or rejected that tradition. From this it follows that the classical philosophers see the political things with a freshness and directness which have never been equaled. (Strauss 1959, 27)

The Socratic reorientation of the study of the whole, of first philosophy, from cosmological ontology to human anthropology is most fully developed in Plato's *The Laws*, a dialogue where Socrates is conspicuous by his absence. The political philosophical tradition catalyzed by *The Laws* extended from the Greeks to the late medieval period in Europe.

The Socratic orientation was, it is true, significantly altered if not distorted by the Christian notion of divine revelation and the assertion of primacy for the ecclesial over the political order. But it was most radically challenged by a new or modern political philosophy grounded in the primacy of the nation-state, a regime characterized by two affirmations absent in classical political philosophy: democracy (the people) and technology (engineered mass wealth).

For democracy to be viable it "must become rule by the educated, and this goal will be achieved by universal education. But universal education presupposes that the economy of scarcity has given way to an economy of plenty. And the economy of plenty presupposes the emancipation of technology from moral and political control" (37)—or, perhaps more accurately, technology is subjected to a new utilitarian morality that shifts the standards of good from virtue for the few to material convenience for the many.

> The difference between the classics and us with regard to democracy consists exclusively in a different estimate of the virtues of technology. But we are not entitled to say that the classical view has been refuted. Their implicit prophecy that the emancipation of technology ... from moral and political control would lead to disaster or the dehumanization of man has not yet been refuted. (37)

A complementary study in *What Is Political Philosophy?* reiterates how ancient or classical political philosophy was directly related to political life, whereas modern political science is "related to political life through the medium of modern natural science, or of the reaction to modern natural science" (60). As Strauss remarks in a pivotal paragraph, "The attitude of classical political philosophy toward political things was always akin to that of the enlightened statesman" rather than of "the social 'engineer'" (90).

6

Among the four problems of political philosophy that can be explicated from Strauss it is the third—concerning relationships between philosophical political philosophy and the city—that most insistently points toward a political philosophy of technology in our time, in a city now at fever pitch with engineering and technology. Strauss approached this issue in a manner complementary to his scholarly studies in a 1952 set of talks at the University of Chicago Hillel House that were posthumously edited as "Progress or Return? The Contemporary Crisis in Western Civilization."[1] Offered to a Jewish audience in that post–World War II moment, these informal reflections display an appropriately existential seriousness.

Strauss begins by contrasting the traditional Jewish call to *t'shuvah*, repentance or return to observance of the law, and the typically modern commitment to two forms of progress: intellectual and social. In the case of *t'shuvah* there is return not only within Judaism but to Judaism by assimilated or secularized Jews, a return from modern non-observance to traditional observance that takes place on the basis of a reassessment of the modern project. Strauss offers intellectual support for such a return, although he practiced it only to the extent of affirming his Jewishness: he did not attend temple or synagogue.

With regard to progress, Strauss distinguished modern and premodern conceptions. For moderns, progress is conceived as at once intellectually open ended and bringing about social progress. Intellectual progress is exemplified by science and technology, social progress by increased wealth and personal freedom. The classical conception admitted "that infinite intellectual progress in secondary matters is theoretically possible," but it qualified practical possibilities. According to the classics, the visible cosmos is either of finite duration or eternal but subject to periodic civilizational destructive cataclysms. Additionally, "the classical conception [lacks any] guaranteed parallelism between intellectual and social progress" (Strauss 1989, 236).

In positivist social science terms, an absence of parallelism has been termed "cultural lag" (Ogburn 1922) and described simply as a social problem waiting to be solved and signaling technology as the primary mechanism of progressive social change leading to progress. For Strauss, however, the modern belief in a correspondence or conjunction between intellectual and social progress is fundamentally questionable.

> To mention only one point, perhaps the most massive one, the idea of progress was bound up with the notion of the conquest of nature, of man making himself the master and owner of nature for the purpose of relieving man's estate. The means for that goal was the new science. We all know the enormous successes of the new science and technology which is based on it, and we all can witness

the enormous increase of man's power. Modern man is a giant in comparison to earlier man. But we have also to note that there is no corresponding increase in wisdom or goodness. Modern man is a giant of whom we do not know whether he is better or worse than earlier man. (238–39)

This idea of an expanding disproportion between technological power and wisdom is a theme in early twentieth-century political thought. In the 1930s, for example, Winston Churchill already observed that, "while men are gathering knowledge and power with ever-increasing and measureless speed, their virtues and their wisdom have not shown any notable improvement as the centuries have rolled" (Churchill 1932, 202). Their "one supreme task" is thus to take control of this new knowledge and powers (Churchill 1937).

Henri Bergson—whom Strauss ranked one of "the four greatest philosophers" of the first half of the century (Strauss 1959, 17)—framed the issue as a divide between mechanism and mysticism that called for a *supplément d'âme* to bridge it:

> So let us not merely say ... that the mystical summons up the mechanical. We must add that the body, now larger, calls for a bigger soul, and that mechanism should mean mysticism. The origins of the process of mechanization are indeed more mystical than we might imagine. Machinery will find its true vocation again, it will render services in pro-portion to its power, only if mankind, which it has bowed still lower to the earth, can succeed, through it, in standing erect and looking heavenwards. (Bergson 1935, 310)

It was left to Strauss, however, to interpret the gap as a stimulus to fundamentally reconsider the wisdom of the modern technological project.

The issue is theoretically acute for Strauss because of the gap between facts and values created by the modern cultivation and commitment to science cut free from philosophy or more accurately from political philosophy—a science that repositions human beings in a merciless cosmos at the mercy of their passions. Ten years later, in another public talk on "The Crisis of Our Time" Strauss phrased it this way: "While the study of [science] succeeded ever more in increasing man's power, the ensuing discredit of reason precluded distinction between the wise, or right, and the foolish, or wrong, use of power. Science, separated from philosophy, cannot teach wisdom" (Strauss 1964b, 49).[2]

The same decade Strauss called attention to the problem as manifested in the lives of progressive promoting philosopher-scientists:

> Originally the philosopher-scientist was thought to be in control of the progressive enterprise. Since he had no power, he had to work through the princes. The

control was then in fact in the hands of the princes, if of enlightened princes. But with the progress of enlightenment, the tutelage of the princes was no longer needed. Power could be entrusted to the people. It is true that the people did not always listen to the philosopher-scientists. But apart from the fact that the same was true of princes, society came to take on such a character that it was more and more compelled to listen to the philosopher-scientists if it desired to survive. Still there remained a lag between the enlightenment coming from above and the way in which the people exercised its freedom. One may even speak of a race: will the people come into full possession of its freedom before it has become enlightened, and if so, what will it do with its freedom and even with the imperfect enlightenment which it will already have received? (Strauss 1968, 21)

Under conditions created by such a progressive race between freedom and power the alternative of return returns, but "Return to what?" (Strauss 1989, 245). With this question Strauss implicitly questions the adequacy of any political effort to take control or philosophical appeal to an amorphous spiritual will.

7

The main body of "Progress or Return?" shifts from "contrasting the life characterized by the idea of return with the life characterized by the idea of progress" (Strauss 1989, 229) to reflecting on substantive options for return. When the modern project is shaken in its foundations, to what should we return? "Obviously," says Strauss, adopting a context-dependent option,

> to Western civilization in its premodern integrity, to the principles of Western civilization. Yet there is a difficulty here, because Western civilization consists of two elements, or has two roots, which are in radical disagreement with each other. We may call these elements . . . Jerusalem and Athens, or, to speak in nonmetaphorical language, the Bible and Greek philosophy. (245)

According to Strauss this most fundamental of disagreements is obscured by two thousand years of effort to harmonize the Bible and Greek philosophy. To state the fundamental disagreement

> simply and therefore somewhat crudely, the one thing needful according to Greek philosophy is the life of autonomous understanding. The one thing needful as spoken by the Bible is the life of obedient love. The harmonizations and synthesizations are possible because Greek philosophy can use obedient love in a subservient function, and the Bible can use philosophy as a handmaid; but what is so used in each case rebels against such use, and therefore the conflict is really a radical one. (246)

Yet the disagreement, as with all disagreements, rests in some agreement. In this case, both the Bible and Greek philosophy agree in criticizing what Strauss described as three key features of modernity: its anthropocentrism (as opposed to biblical theocentrism and Greek cosmocentrism), its emphasis on rights and freedom over duty and discipline, and the valorization of history over nature. There is also a fundamental agreement about morality: "regarding the importance of morality, regarding the content of morality, and regarding its ultimate insufficiency" (246).

> The Bible and Greek philosophy agree in assigning the highest place among the virtues, not to courage or manliness, but to justice. And by justice both understand, primarily, obedience to the law. (247)

In biblical language, the guidance of the law of the Torah "is your life" and "is the tree of life for those who cling to it" (Proverbs 3:18). "In the words of Plato, 'The law effects the blessedness of those who obey it' (see *Laws* 718b)" (247).

The Bible and Greek philosophy unite in affirming the importance of law, that law demands restraint public and private, including restraint in technological production and consumption—but that law is incomplete as the final good. They disagree deeply about what it is that completes the law and on the source of the law. For the Bible, what completes the law is charity or love; for the Greeks, it is contemplation, philosophy. For the Bible, the source of law is divine revelation through a prophet chosen by God; for Greek philosophy, it is reason, as manifested through a human lawgiver if not a philosopher-king, the founder or founders of a state (who may subsequently be divinized). On Strauss's interpretation, the inherent tensions between these two roots of revelation and reason are at the core of what constituted Western civilization prior to its modernization. The paradox of this split was distinctively creative insofar as the two roots became symbiotically intertwined with revelation restraining reason and reason moderating revelation, without ever achieving complete harmony or one ever fully subordinated to the other.

Modernization undertook not so much to harmonize revelation and reason as to marginalize if not destroy the biblical root, thereby opening a door to the unbridled development of technology—that is, to the engineering of production and consumption. Reason undertook to decisively demonstrate the irrationality of revelation and thus remove its superstitions and restraints from political affairs if not from culture generally. Along this path, an argument initiated by Machiavelli was brought to completion by Spinoza. In a parallel, synergistic initiative, Bacon and Descartes created a new science for the conquest of nature dedicated to unlimited human use and convenience. The combined transformed powers, political and technoscientific, would outstrip law, not only biblical but also philosophical, producing transcontinental wars and a world now under the threat of nuclear annihilation.

In the new Athens, a globalized technoscientific Athens, which has reduced revelation to a personal subjective experience, Jerusalem, Strauss seems to suggest, is more necessary than ever. Indeed, an attempt to defend revelation against Enlightenment criticism occupies a significant portion of "Progress or Return?" and makes major appearances in any number of other Strauss texts. As Strauss reiterates the problem in another (posthumously published) talk from 1940,

> The inability of modern science or philosophy to give man an evident teaching as regards the fundamental question, the question of the right life, led people to turn from science or reason to authority, to the authority of the State or the authority of Revelation. Politics and theology, as distinguished from science of all kind[s], appeared to be much more closely connected with the basic interests of man as man than science *and all culture*: the political community and the word of the living God are basic; compared with them everything else is derived and relative. *"Culture" is superseded by politics and theology, by "political theology."* (Strauss 2006, 129, emphasis in original)

Political theology, recall, is politics based on or justified by supernatural revelation.

8

Strauss repeatedly describes his core question as the theological-political problem. The problem has been stated in more than one way by Strauss, and interpreters of Strauss have debated its precise features. That Strauss does not easily divulge his deepest thinking is an understatement. As Strauss's core question it must nevertheless be broached by any attempt to think political philosophy of technology after Strauss.

Consider what appears on the surface to be a basic question concerning political philosophy of technology: The political community is constituted as such by law, which in the first instance tames or moderates those passions that lead to political violence and that naturally emerge among people congregated in cities. The theological-political problem concerns how much religion (in its most emphatic form as revelation, represented by Jerusalem) and/or reason (represented by Athens) is necessary for this law or for the community sufficiently to revere the law—and then, particularly under conditions of modernity and, for present purposes, under modernity infused with and transformed by engineering and technology. Again, in what ways if any must religion and politics collaborate in order to tame the passions and enable us to manage, create, and use our enhanced technological powers wisely? For Strauss, the answer is decisively not political theology.

For Strauss, the theological-political problem emerged with his interpretation of Spinoza's *Tractatus Theologico-Politicus* (1670) and its companion *Tractatus Politicus* (1677), to which he dedicated his first book. Spinoza was the first to outline a comprehensive theory of the liberal state: that is, one in which religion is radically divorced from politics and the state is constructed with the single goal of providing peace, material welfare, and personal freedom as opposed to the traditional state, which aimed at what Spinoza, along with his companion founders of modernity, considered impossible if not fictitious human virtues.

In historically preceding states the founding law was popularly accepted on the basis of belief, belief either that it was a revelation from God or a gift from a founder so superior in nobility as to be for all practical purposes divine. The revelation of the *Torah* to Moses created the Jewish people; the wise and heroic lawgiver Lycurgus created the Spartans. In both cases acceptance of and reverence for the law established boundaries to moderate the passions in ways that were subsequently maintained and applied by statesmen.

Although no country can exist without some type of law-stabilized passion moderating order, Spinoza considers most claims regarding political foundings to be based on untrustworthy histories if not superstitions. Philosophical questioning reveals many reverential beliefs to be false. Laws can be given a better, more philosophical foundation. Spinoza seeks to replace superstitions with rational creation and (by means of social contract) acceptance of a new type of law by an enlightened public. This new law and public will additionally, through modern science and its engineering enabling prowess, which had previously also been restrained by false beliefs about the nature of being, be able to purse the conquest of nature for broad human material benefit. Spinoza was the first political philosopher to develop an extended defense of democracy. Yet in order to make way for an enlightened democratic public he needed to rid the public of superstitions. One of his primary concerns was thus to call out the superstitions of the Bible and its claims to divine revelation, in order to deprive political affairs of the corrupting and deforming influences of revelation.

In order to purge the political realm of the influencing power of religious superstition (while nevertheless leaving religion in place as a personal or subjective choice subordinate to civil religion), Spinoza almost single handedly created the discipline of modern biblical criticism. By means of textual and historical analysis he systematically attacked the veracity of the Bible, with his key attack resting on an assertion of the irrationality of miracles. For Strauss, however, the criticism of miracles is fundamentally incoherent.

Spinoza assumes that miracles are irrational or impossible in order to argue that miracles are impossible. To adapt one of Strauss's examples: Geological science teaches that the Earth was created by natural processes from existing materials and is more than 4.5 billion years old. The Bible teaches that God in a supernatural act created the Earth from nothing only a few thousand years ago. Therefore the Bible is false. But the "rational" conclusion assumes that supernatural creation is not possible, that God did not miraculously create the Earth a few thousands of years ago in such a manner that it would appear to scientific investigation like it is billions of years old. The apparent age is simply part of the miracle.

The only way to disprove the possibility of revelation is to prove the impossibility of miracles by having sufficient knowledge of the whole to prove that miracles are not possible. But philosophy, most especially Socratic philosophy, is forced by its own questioning to admit as its most fundamental knowledge that human knowledge is limited, that it does not have access to knowledge of the whole, that humanity is in fact constituted at its highest level by the search for rather than the possession of knowledge of the whole. Indeed, this is a principle that even the form of philosophy known as modern natural science is forced to admit when it characterizes its knowledge as fallible.

The implications for technology are all but obvious. To quote Strauss from his 1965 preface to *Spinoza's Critique of Religion*:

> The genuine refutation of orthodoxy would require the proof that the world and human life are perfectly intelligible without the assumption of a mysterious God; it would require at least the success of the philosophical system; man has to show himself theoretically and practically as the master of the world and the master of his life; the merely given world must be replaced by the world created by man theoretically and practically. (Strauss [1930] 1965, 29)

The attempt to refute miracles demands attempts to engineer the world. The stability of such an engineered and engineering state moreover depends crucially on the possibility of replacing state cohesion on the basis of divine law with strictly rational cohesion. That is, human engineers would have to design, construct, and manage the Earth as a technological artifact with the willing consent of non-engineers. As one leading student of Strauss has it, Spinoza would transfer the activity of statesmen who preserve the law "to the activity of scientists, or of politicians who have been decisively influenced by a scientific understanding of political affairs" (Rosen 1963, 414). In such a liberal technocracy, arising through its self-assertion, guidance is no longer provided by a divine or semi-divine founder and a reverence for the law that tames the passions but by mixed government of the many and the few combined to satisfy the passions. Replacing the philosopher king, technoscientist kings now instruct denizens of the cave on the realities behind its shadows and demonstrate their manipulability for public use and convenience; the denizens emerge free to use their passions to guide the use of their newly acquired if incompletely understood powers. We are today attempting to engineer a technological civilization composed of an elite few that has placed itself in participatory service to a many with little more than minimal moral or technical education.

Spinoza's separation of morals in its most emphatic form as revelation from the pursuit of knowledge in its distinctly modern scientific form sets the stage for a historically unprecedented and problematically laced engineering and technological self-assertion, that is, for the unending creation of a techno-lifeworld. The problem today may not be as immediately dramatic as were the catastrophes of two world wars and the collapse of the Weimar republic into fanatical nihilism, but they may not be far off.

To mention just three problems that have become increasingly apparent in our time: No matter how much public education, it is increasingly difficult to negotiate and/or moderate the democratic appetite for the peace, material welfare, and personal freedom promised by engineering and technoscience. Indeed, it is increasingly difficult even to educate the general public regarding the character of technoscientific knowledge and power. Finally, the democratic public is increasingly resentful of alleged technoscientific statesmen, suspecting them of functioning as just another lobby in a zero-sum game of interest groups in pursuit of their own good. Technoscientific teachings concerning the inherent instability of science and engineering, indeed the very instability of lived experience in a technoscientific lifeworld, together with the ideology of innovation in engineering and technology as virtually unlimited in their ability to solve problems, further conspire to make almost any law or moral constraint temporary if not dubious and to replace the very idea of law with self-assertions of cultural identity.

The theological-political problem is thus of particular urgency precisely because of the exclusion of religion from the structural transformation of the public sphere. Through the immoderation of contemporary engineering and technology and the manifest difficulty of taming the passions, we now live in a techno-lifeworld shot through with internal fragilities and facing unprecedented external dangers in which we too often look the other way or invest the future with superstitious hopes.

9

Strauss nevertheless recognizes that despite the inability of philosophy to disprove the possibility of revelation there is no simple return to revelation. At the conclusion of "Progress or Return?" he asks what might be possible. In response he returns to what he considers the two roots of Western civilization: Athens and Jerusalem, reason and revelation. "All alleged refutations of revelation [including Spinoza's] presuppose unbelief in revelation, and all alleged refutations of philosophy [by revelation] presuppose faith in revelation" (Strauss 1989, 269). What this means is that the philosophic way of life is not self-evidently the best way of life. The way of life manifested in the rational questioning search for knowledge, like the religious way of life of obedience to revelation, is itself based on faith.

The tension between faith and reason has been a continuous trope in the West from St. Paul and Tertullian to Anders Nygren and Ian Barbour. But unlike others, Strauss has argued that the tension is not so much something to be overcome or transcended—as projected, for instance, in the great systems of Thomas Aquinas and Hegel—as to be lived. In Strauss's words:

> It seems to me that the core, the nerve, of Western intellectual history, Western spiritual history, one could almost say, is the conflict between the Bible and the philosophic notions of the good life. This was a conflict which showed itself primarily, of course, in arguments—arguments advanced by theologians on behalf of the Biblical point of view and by philosophers on behalf of the philosophic point of view. There are many reasons why this is important, but I would like to emphasize only one: it seems to me that this unresolved conflict is the secret of the vitality of Western civilization. The recognition of two conflicting roots of Western civilization is, at first, a very disconcerting observation. Yet this realization has also something reassuring about it. The very life of Western civilization is the life between two codes, a fundamental tension. There is therefore no reason inherent in Western civilization itself, in its fundamental constitution, why it should give up life. But this comforting thought is justified only if we live that life, if we live that conflict. No one can be both a philosopher and a theologian, nor, for that matter, some possibility which transcends the conflict between philosophy and theology, or pretends to be a synthesis of both. But every one of us can be and ought to be either one or the other, the philosopher open to the challenge of theology or the theologian open to the challenge of philosophy. (Strauss 1989, 270)

The two texts on which I have relied to consider the hypothesis of political philosophy of technology both originated as addresses to a Jewish audience. "What Is Political Philosophy?" was initially a lecture at Hebrew University in Jerusalem. The first paragraph acknowledges that, "In this city, and in this land, the theme of political philosophy . . . has been taken more seriously than anywhere else on earth" (Strauss 1959, 9). Yet in the second paragraph Strauss also noted how "political philosophy and its meaningful character [first] came to light in Athens" (10).

On a third occasion, when addressing a non-Jewish audience in a public lecture at the City University of New York in 1967, Strauss revisited the Jerusalem-Athens relationship with what he refused to call anything more than "some preliminary reflections." In attempting once again to negotiate living between these two ways, he noted how Greek philosophers and poets as well as the Torah proclaim themselves to be sources of wisdom.

> We, then, must try to understand the difference between biblical wisdom and Greek wisdom . . . in the strict and highest sense. According to the Bible, the beginning of wisdom is fear of the Lord; according to the Greek philosophers, the beginning of wisdom is wonder. . . . We are confronted with the incompatible claims of Jerusalem and Athens to our allegiance. We are open to both and willing to listen to each. (Strauss 1983, 149)

Yet by expressing an openness to each, willing to consider the claims of each before deciding, "we have already decided in favor of Athens against Jerusalem." This seems to be unavoidable "for all of us who cannot be orthodox" insofar as we are heirs of Spinoza even when we reject the adequacy of Spinoza (150).

How should we interpret Strauss's enigmatic challenge to philosophers and theologians under conditions of a progressively engineered and engineering way of life—conditions in which in fact the code of revelation is marginalized, with accidental if not essential consequences that loom as more dire than any that Strauss himself experienced?

10

As indicated, in his exploration of fundamental divides between the codes of Athens and of Jerusalem, Strauss consistently takes up with reason and revelation, philosophy and faith, in what he calls their most "emphatic" versions, their strict and highest forms. For Strauss, political philosophy is not the same as political science, social philosophy, political theology, or political theory; it is political thought of an extreme character and not simply as an ideal type. Straussian political philosophy is to politics as the theoretical physics of Steven Weinberg, Lee Smolin, or Kip Thorne is to rocket engineers or the pure mathematics of Leonard Euler, Paul Erdös, and Grigori Perelman is to accounting. For Strauss, the lower is always best and only properly understood from the perspective of the higher. Strauss practices what might be called extreme philosophy by means of the extreme reading of philosophical texts in order to illuminate our political situation: a political situation that is itself so extreme we have difficulty recognizing it or figuring out how to respond.

With regard to Strauss's Athens/Jerusalem contrast we note immediately two things. One is that it is always Jerusalem that is named first, is accorded priority. Another is that it is certainly not fear that characterizes modern science, engineering, and technology, that indeed it is something closer to wonder—and wonder not just about the world but about what human beings are able to make of the world through their new forms of scientific knowledge production and engineering design creation. The idea of any "return" is equally foreign; it is "progress" to which modern philosopher-scientists are unalterably attracted—even as modern science and its engineering and technology offshoots destabilize the political order more profoundly than classical philosophy.

In such circumstances, when Strauss argues that philosophers should be open to the challenge of theology—note that *this* is the emphasis, not that theologians should be open to the challenge of reason: he does not think that Spinoza has anything to teach those who believe in miracles—it suggests that scientific, engineering, and technological reason need to be moderated if not restrained by revelation. Confronted by the wisdom of revelation, scientists are summoned to recognize the irrationality, the ultimate faith dependency, of their extreme commitment to knowledge production as an inherent good, as an end in itself, as an end without limits; and engineers who accept the challenge of living in openness to revelation would appear called on to live by codes more stringent than any formulated by professional engineers societies, indeed to practice something like Hans Jonas's heuristics of fear.

The pivotal issue here is the interpretation of revelation. As Strauss points out in the first major work of his fertile 1950s, *Persecution and the Art of Writing* (1952), and is implicit in his discussion of the problematics of return in Judaism, there is an opposition not just between revelation and reason but also between revelation as law (for Jews) and revelation as knowledge (for Christians). "For the Christian, the sacred doctrine is revealed theology; for the Jew and the Muslim, the sacred doctrine is, at least primarily, the legal interpretation of the Divine Law (*talmud* or *fiqh*)" (Strauss 1952, 19). For Strauss, revelation in its most emphatic form is law, and as revealed law is not to be questioned, or at least not openly questioned, even by philosophers, who recognize that even they need a stable political order within which to practice their questioning quest for knowledge.

A hypothetical political philosophy of technology after Strauss would thus seem to be this: Engineering and technology, if they are not to be ever more wisdom exceeding and socially destabilizing pursuits, must become subject to lawful constraint on the model of that most emphatic law that was once an essential feature of the Western tradition. Only this might constrain engineers from the unlimited pursuit of wondrous but nihilistic possibilities for management of the Earth and a re-engineering of the human. The framework of political philosophy of technology rests on recognition of this fundamental.

NOTES

1. "Progress or Return?" is the title of the last chapter in Strauss 1989. It also appears as the last chapter, in a variant version, in Hilail Gildin, ed., *An Introduction to Political Philosophy: Ten Essays by Leo Strauss*, second ed. (Detroit: Wayne State University Press). One section was given more contemporaneous publication in Hebrew, *Iyyun: Hebrew Philosophical Quarterly* (Jerusalem) 5, no. 1 (January 1954), 110–26, and subsequently as "The Mutual Influence of Theology and Philosophy," *Independent Journal of Philosophy* 3 (1979), 111–18.

2. Spaeth's edited volume includes "The Crisis of Our Time," as well as his talk "The Crisis of Political Philosophy" (91–102), along with some remarks by Strauss in discussion sections.

REFERENCES

Bergson, Henri. [1932] 1935. *The Two Sources of Morality and Religion*. Translated by Ashley Audra, Cloudesley Brereton, and W. Horsfall Carter. London: Macmillan.
Churchill, Winston S. 1932. *Thoughts and Adventures*. London: Buttersworth.
———. 1937. "Mankind Is Confronted by One Supreme Task," *News of the World* (14 November).
Ellul, Jacques. 1954. *La technique, ou L'enjeu du siècle*. Paris: Armand Colin. [Trans. *The Technological Society* (New York: Knopf, 1964).]
———. 2014. *Théologie et Technique: Pour une éthique de la non-puissance*. Edited by Yves Ellul and Frédéric Rognon. Geneva: Labor et Fides.
Kass, Leon. 1971. "The New Biology: What Price Relieving Man's Estate?" *Science* 174, no. 4011 (19 November): 779–88.
———. 1985. *Toward a More Natural Science: Biology and Human Affairs*. New York: Free Press.
Ogburn, William Fielding. 1922. *Social Change: With Respect to Culture and Original Nature*. New York: B. W. Huebsch.
Rosen, Stanley. 1963. "Benedict Spinoza." In *History of Political Philosophy*, edited by Leo Strauss and Joseph Cropsey, 413–32. Chicago: Rand McNally.
Spaeth, Harold J., ed. 1964. *The Predicament of Modern Politics*. Detroit: University of Detroit Press.

Strauss, Leo. [1930] 1965. *Spinoza's Critique of Religion*. Translated by E. M. Sinclair. New York: Schocken Books.

———. [1932] 1965. "Comment's on *Der Begriff des Politischen* by Carl Schmitt." In *Spinoza's Critique of Religion*, translated by E. M. Sinclair, 331–51. New York: Schocken.

———. 1948. *On Tyranny: An Interpretation of Xenophon's* Hiero. New York: Political Science Classics. [Revised and expanded edition, eds. Victor Gourevitch and Michael S. Roth (Chicago: University of Chicago Press, 2000).]

———. 1952. *Persecution and the Art of Writing*. Glencoe, IL: Free Press.

———. 1953. *Natural Right and History*. Chicago: University of Chicago Press.

———. 1958. *Thoughts on Machiavelli*. Glencoe, IL: Free Press.

———. 1959. *What Is Political Philosophy? And Other Studies*. Glencoe, IL: Free Press.

———. 1964a. *The City and Man*. Chicago: Rand McNally.

———. 1964b. "The Crisis of Our Time." In *The Predicament of Modern Politics*, edited by Harold J. Spaeth. Detroit: University of Detroit Press.

———. 1968. *Liberalism Ancient and Modern*. New York: Basic Books.

———. 1983. *Studies in Platonic Political Philosophy*. Chicago: University of Chicago Press.

———. 1989. *The Rebirth of Classical Political Rationalism: An Introduction to the Thought of Leo Strauss*, edited by Thomas L. Pangle. Chicago: University of Chicago Press.

———. 2006. "The Living Issues of German Postwar Philosophy (1940)." In *Leo Strauss and the Theological-Political Problem*, edited by Heinrich Meier, 115–39. Translated by Marcus Brainard. Cambridge: Cambridge University Press.

———. 2018. *On Political Philosophy: Responding to the Challenge of Positivism and Historicism*, edited by Catherine H. Zuckert. Chicago: University of Chicago Press.

Strauss, Leo, and Joseph Cropsey, eds. 1963. *History of Political Philosophy*. Chicago: Rand McNally.

Chapter 6

The Nuclear Menace and the Prophecy of Doom

Jean-Pierre Dupuy

The next time an atomic bomb will be dropped over a civilian population, breaking what has been called the "nuclear taboo," it is very likely that the event will be interpreted as the bursting forth of the possible into the realm of impossibility, as was the case with the destruction of the Twin Towers on 9/11. From now on, one heard it said, even the worst horrors have become possible. Note that if something *becomes* possible, presumably this is because it was not possible before. And yet, common sense objects, if it actually occurs, this must be because it *was* possible all along. Common sense proves here once more to be a detestable guide.

METAPHYSICS OF THE PROPHECY OF DOOM

Bergson and the Possible

In *The Two Sources of Morality and Religion*, French philosopher Henri Bergson described the sensations he felt on August 4, 1914, on learning of Germany's declaration of war on France:

> Horror-struck though I was, and though I felt a war, even a victorious war, to be a catastrophe, I experienced what William James expresses, a feeling of admiration for the smoothness of the transition from the abstract to the concrete: who would have thought that so terrible an eventuality could make its entrance into reality with so little disturbance? The impression of this facility was predominant above all else. (1935, 159–60)

Yet this disturbing familiarity stood in sharp contrast to Bergson's feelings *before* the catastrophe. The prospect of war appeared to him and his friends "as *at once probable and impossible*: a complex and contradictory idea that lasted right down to the present day" (159, emphasis in original). Some years later, Bergson managed very well to unravel this apparent contradiction in reflecting upon the nature of a work of art in an essay entitled "The Possible and the Real" ([1930] 1946). "I believe in the end we shall consider it evident," Bergson wrote, "that the artist in executing his work *is creating the possible as well as the real*" (121, emphasis added). Why is it, then, he asked, that one might "hesitate to say the same thing for nature? Is not the world a work of art incomparably richer than that of the greatest artist?" (121). The hesitation to extend this idea to acts of destruction is greater still. And yet who has contemplated the images of September 11 and not been filled with a feeling of exaltation and dread that resembles what one feels in the presence of the sublime, in the sense that Edmund Burke and Immanuel Kant gave to this word? Of the terrorists, who could hardly have failed to have sensations of the same kind, we may also say that they created the possible at the same time as they created the real. This was, as I say, the metaphysical view that most commentators spontaneously adopted.

In the same text, Bergson reports a delightful conversation with a journalist who had come to interview him, during the Great War, on the subject of the future of literature. "How do you conceive, for example, the great dramatic work of tomorrow?" he was asked. "But," Bergson objected, "the work of which you speak is not yet possible." "But it must be, since it is to take place," retorted the other. To this Bergson replied:

> No, it is not. I grant you, at most, that it *will have been possible.*" "What do you mean by that?"—"It's quite simple. Let a man of talent or genius come forth, let him create a work: it will then be real, and by that very fact it becomes retrospectively or retroactively possible. It would not be possible, it would not have been so, if this man had not come upon the scene. That is why I tell you that it will have been possible today, but that it is not yet so." "You're not serious! You are surely not going to maintain that the future has an effect upon the present, that the present brings something into the past, that action works back over the course of time and imprints its mark afterwards?"—"That depends. That one can put reality into the past and thus work backwards in time is something I have never claimed. But that one can put the possible there, or rather that the possible may put itself there at any moment, is not to be doubted. As reality is created as something unforeseeable and new, its image is reflected behind it into the indefinite past; thus it finds that it has from all time been possible, but it is at this precise moment that it *begins to have always been possible*, and that is why I said that its possibility, which does not precede its reality, will have preceded it once the reality has appeared." (118–19, emphasis added)

Before 1907, the year when Pablo Picasso painted *Les Demoiselles d'Avignon*, that painting was not possible. More than a century later, the fact that it has become real entails that it was indeed possible before 1907. The truth value of the proposition "The painting *Les Demoiselles d'Avignon* was possible in 1900" depends on the time at which it is enunciated. Hence the recourse to a grammatical tense that is not frequently used in English—the future perfect: "it *will have been possible.*"

The explanation of our inaction in the face of many looming disasters is to be found right here: anyone who wishes *to prevent* a catastrophe must believe in its possibility *before* it occurs. The paradox is that if one succeeds in actually preventing it, its non-realization keeps it firmly within the domain of the impossible, and efforts at prevention appear in retrospect to have been useless.[1]

Günther Anders and the Quandary of the Prophet of Doom

German philosopher Günther Anders (1902–1992) was the most profound and radical thinker to have reflected on the major catastrophes of the twentieth century. He is less well known than two of his companions who studied with him under Heidegger: Hans Jonas, who was his friend, and Hannah Arendt, to whom he was once married. This is probably due to his intransigence and the fragmented character of his work.

Rather than weighty systematic treatises, Anders preferred shorter pieces on current issues, sometimes written in the form of a parable. More than once, he will have told in his own way the story of the flood.

> Noah was tired of playing the prophet of doom and of always foretelling a catastrophe that would not occur and that no one would take seriously. One day, he clothed himself in sackcloth and put ashes on his head. This act was only permitted to someone lamenting the loss of his dear child or his wife. Clothed in the habit of truth, acting sorrowful, he went back to the city, intent on using to his advantage the curiosity, malignity and superstition of its people. Within a short time, he had gathered around him a small crowd, and the questions began to surface. He was asked if someone was dead and who the dead person was. Noah answered them that many were dead and, much to the amusement of those who were listening, that they themselves were dead. Asked when this catastrophe had taken place, he answered: tomorrow. Seizing this moment of attention and disarray, Noah stood up to his full height and began to speak: the day after tomorrow, the flood will be something that will have been. And when the flood will have been, *all that is will never have existed.* When the flood will have carried away all that is, all that will have been, it will be too late to remember, for there will be no one left. So there will no longer be any difference between the dead and

those who weep for them. *If I have come before you, it is to reverse time*, it is to weep today for tomorrow's dead. The day after tomorrow, it will be too late. Upon this, he went back home, took his clothes off, removed the ashes covering his face, and went to his workshop. In the evening, a carpenter knocked on his door and said to him: let me help you build an ark, *so that this may become false*. Later, a roofer joined with them and said: it is raining over the mountains, let me help you, *so that this may become false.* (Anders 1972; emphasis added)

The whole quandary of the prophet of doom, as well as the ingenious way of getting out of his predicament, is inscribed in this magnificent parable. The prophet of doom is not heard because his word, even if it brings information, does not fit with the beliefs of those to whom it is addressed. It is often said that if we fail to act in the face of catastrophe it is because our knowledge is uncertain. Yet, even when we have all the clues at our disposal, we are unable to transform this information into belief.

Noah's way out is brilliant. *It consists in the staging of mourning for deaths that have not yet occurred*, in such a way, says Anders, that time is "reversed," since the effect (mourning) comes before the cause (the deaths). This is indeed unusual since it could only make sense if the deaths to be mourned were inscribed in the future at a determinate date. What makes death bearable, for many of us, is precisely that we tend to liken an unknown end to an indeterminate end, and so to the absence of an end. "Whatever certainty there is in death," seventeenth century French moralist Jean de La Bruyère remarked, "is mitigated to some extent by that which is uncertain, by an indefiniteness in time that has something of the infinite about it" (1933, 398). By contrast Anders's aim with this parable is to stress that the catastrophe is inevitable, or rather, *will have always been* inevitable once it occurs.

The prophecy of doom purports to be the antidote of Bergson's metaphysics. Like any *pharmacon*, at the same time poison and remedy, the former retains some essential traits from the latter. In the same way that for Bergson the catastrophe is impossible before it occurs and starts having always been possible once it occurs, in the prophecy of doom the catastrophe is not necessary before it occurs and begins to have always been necessary once it occurs. In either case, it is the actualization of the event—the fact that it occurs—that retrospectively substitutes a modality for a previous one: possibility, in one case, necessity (or impossibility) in the other.

The paradox of the prophecy of doom is as follows. Making the perspective of catastrophe credible requires one to increase the ontological force of its inscription in the future. The foretold suffering and deaths will inevitably occur. But if this task is too well carried out, one will have lost sight of its purpose, which is precisely to raise people's awareness and spur them to action so that the catastrophe *may not occur*—"let me help you build an ark, *so that this may become false.*"

What prevents the metaphysics inherent in Anders's parable to boil down to an old-fashioned classical fatalism is its future-perfect structure. Sophisticated though the latter may sound, it is of common usage among us. Consider the following statement made by Yascha Mounk in the *New York Times* in 2017: "If Mr. Trump's presidency ends in humiliation, future generations may well conclude that it was bound to fail all along" ("The Past Week Proves That Trump Is Destroying Our Democracy," August 1). This statement was true if either Mr. Trump's presidency wouldn't end in humiliation or if it was bound to fail all along. The statement didn't say that Mr. Trump's presidency was necessarily doomed; only that if it turned out to be a failure, it *would have been* destined to be such all along. Necessity here is retrospective.

Being a Compatibilist

Although very suggestive, Anders's parable remains a poetic rendering of abstract ideas, that is an allegory. In my own work, I have tried to give it a formal interpretation in the framework of analytical metaphysics (and analytical theology).

My starting point has been the age-old problem of the compatibility between determinism and free will in its modern version fleshed out by such philosophers as David K. Lewis (1986) and Robert Stalnaker (1981).[2]

Lewis calls "soft determinism" "the doctrine that sometimes one freely does what one is predetermined to do; and that in such a case one is able to act otherwise though past history and the laws of nature determine that one will not act otherwise." He then defines compatibilism as "the doctrine that soft determinism may be true" (1981, 112).

Let us call C the state of the world at a time t_1. We have:

A1: C was the case at t_1

Consider a subject S whose action x at $t_2 > t_1$ is determined by the laws that govern his world according to:

A2: If C was the case at t_1, then S does x at t_2

From A1 and A2 we derive by *modus ponens*:

A3: S does x at t_2

Can x be a free although predetermined act? To defend soft determinism, it is always useful to start from the argument(s) put forward by those who deny it. The so-called "incompatibilist" thesis uses an operator □ which, applied to a proposition p, asserts that p is true in all possible worlds: it is necessary. More specific to our problem, we will call $□^S_t$ the operator of necessity such that:

$□^S_t$ (p) means: p is true and S is not free at t to perform an act such that, if he were to perform it, p would be false.

The incompatibilist argument can be written as follows:

N1: $□^S_{t2}$ (C was the case at t_1)

N2: $□^S_{t2}$ (If C was the case at t_1, then S does x at t_2)

Thus, by *modus ponens*,

N3: $□^S_{t2}$ (S does x at t_2)

N1 expresses the principle of the fixity of the past. N2 says that the laws that determine the subject's actions remain the same in all possible worlds. The conclusion N3 states that S does actually do x at t_2, but he does not act freely since it is not in his power to act otherwise.

Can this argument be refuted? Depending on the nature of the problem, there are two possibilities, neither of which has greater a priori legitimacy than the other.

a) We could accept N1, in which case we would have to reject N2. The past is fixed, and the subject, supposedly able to act otherwise, has the power to invalidate the fixity of the temporal chain which links C to x. The nature of this power must be made very clear. As Lewis puts it, we must distinguish between two versions:

Strong version: "I am able to break a law."

Weak version: "I am able to do something such that, if I did it, a law would be broken" (Lewis 1981, 113).

Obviously, there is no way that *in our world* the subject could act so that the link between C and x would be violated: this would be contrary to hypothesis A2, which indeed remains valid. The strong version is eliminated but not the weak one. To paraphrase Lewis, the way in which I was determined not to do anything other than x "was not the sort of way that counts as inability" (112). The power that this sort of ability represents is called "counterfactual."

b) Conversely, we could accept N2, in which case we would have to reject N1. This time the temporal chain A2 is held to be fixed (that is, true in all possible worlds). To maintain that the agent's action, x, is free although determined by the past and the laws that govern the world, we have to grant the agent a power to invalidate the past. This power obviously cannot be causal. Here too we must distinguish between:

a *strong version*: "I am able to change the past," which is "utterly incredible," to use Lewis' terms,

and a *weak one*: "I am able to do something such that, if I did it, the past would have been different from what it was in the actual world."

The Calvinist theologian and analytic metaphysician Alvin Plantinga, who defends the weak version, has logically dubbed "counterfactual power over the past" this kind of ability (1986).

Although, as I said, the two ways of grounding compatibilism have an a priori equal legitimacy, contemporary philosophers such as Lewis or Stalnaker, probably because of their respective stints in the domain of rational choice theory, have focused almost exclusively on the former, which preserves the fixity of the past. I have explored thoroughly the second approach and been able to show that it formalizes elegantly the properties we have discovered as characterizing the prophecy of doom.

The first thing to be noted is that there exist situations in which the counterfactual power an agent possesses over the past causally prohibits him from acting in a certain way.[3] Consider a paradigmatic illustration that has been the object of numerous cogitations from Thomas Hobbes onward: the promise case.[4] At t_1 Mary asks Peter to lend her $1,000 and she promises to pay off her debt at $t_2 > t_1$. We are in a state of nature à la Hobbes: there are no state institutions, no judicial system, no rule of law. The agents are only guided by their self-interest. If the loan could take place, it would be mutually beneficial.

In the temporality that preserves the fixity of the past, it is immediate that the loan is impossible. Reasoning by backward induction we realize along with Peter that Mary at t_2 will break her promise. Peter would be a fool to lend her anything.

In the temporality that maintains N2, that is a necessary link between past conditions and future action, at the cost of doing away with the fixity of the past, things work very differently. Let's say Peter is an omniscient predictor capable of anticipating Mary's actions in all possible worlds. If Mary held her promise at t_2 Peter would anticipate it and the mutually beneficial loan would take place. On the other hand, if she were to renege on her promise, Peter would anticipate it as well and he would not lend her the money. We see here in action the counterfactual power that Mary has on her past via her action. However, if the loan does not exist, Mary is not in a position to renege on her promise to pay off her debt. Hence a contradiction which is immediately solved by the conclusion that Mary will not renege on her promise if the loan takes place. The loan will indeed take place to the mutual benefit of Peter and Mary.

This example illustrates that in the temporality we are examining it is not true that any future goes, since "it is not the future if you stop it."[5] The future must be such that the past that it counterfactually determines does not causally prevent its occurrence. In other words, the future, far from being the outcome of the laws of nature applied to determinate initial conditions (prediction) or something that we create according to our will ("prospective"),[6] is the solution (one of the solutions) to an equation in which the unknown x—the future action—appears on both sides of the equation in the following form: $x = F[x]$, as if it were determining itself. According to the received terminology, we will say that the future appears as the *fixed point* of a certain operator F. The latter expresses the causal consequences of a past that is itself determined counterfactually by the future x. This can be represented graphically as shown in figure 6.1.

In this conception of time, the future is fixed—that is, necessary—since it is linked to the past by N2, a proposition that states that this link is true in all possible worlds. However, this is only true once the past is determined, which presupposes that the future itself is determined. In other terms, the future is necessary—it has always been necessary—but only once it has become actualized. This is the essential trait we have learned to ascribe to the metaphysics of the prophecy of doom.

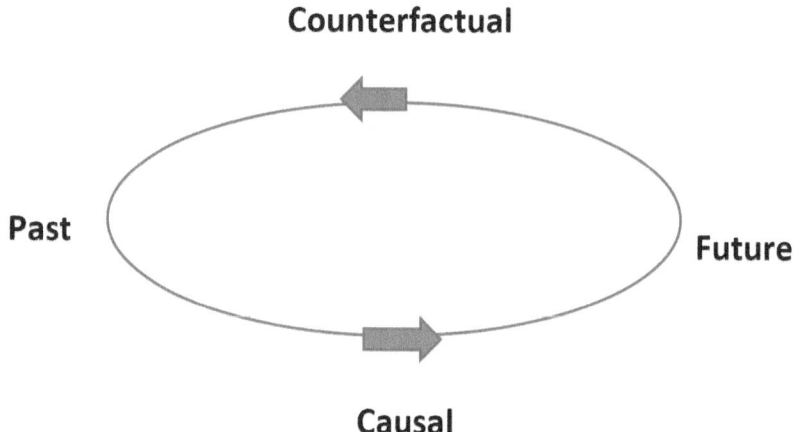

Figure 6.1. Projected Time

ON THE MULTIPLICITY OF METAPHYSICS AND THE CHOICE OF THE MOST PERTINENT

The indeterminacy of the past as long as action has not been performed and the necessity of the future once action is taken together serve to define a metaphysics of temporality that I have dubbed "Projected time." In what follows, in order to prepare the ground for my analysis of nuclear deterrence, I will introduce another metaphysics, which I name "Occurring time," the one that supports all strategic reasoning, be it carried out by an economist, a game theorist, a planner, an engineer, a designer, or a military strategist. It corresponds to a very distinct conception of free will for which the agent's actions are driven by a set of beliefs and desires rather than "pushed" by a determinism. Named the belief-desire model, its most familiar graphic representation today is the decision tree. At every node of the tree an agent has the choice between several possible future options. When he chooses among them he holds the past as fixed, that is counterfactually independent of his choosing. Fixed past, open future, the metaphysics of occurring time is obviously in sharp contrast with that of projected time.

If metaphysics is the branch of philosophy that explores the fundamental nature of reality, according to a received definition, the question arises, how can we account for the plurality of metaphysics?

In the fourth century BCE, a member of the Megarian School named Diodorus Kronos proposed an axiomatic—that is, a set of propositions held to be self-evidently true—designed to show that the actual is the only possible and that the future is already determined.

The three axioms are:

1. Every true proposition about the past is necessary.
2. The impossible does not logically follow from the possible.
3. There is a possible which neither is presently true nor will be so.

Diodorus demonstrated that they are incompatible. One of them at least must go. Axiom 3 seems self-evident to most people today. However, if they hold like Diodorus that 1 and 2 also are self-evident, then they must deny 3. That is, they must hold that an event that happens neither in the present nor in the future is an impossible event.

One of the greatest French philosophers of the twentieth century, Jules Vuillemin (1996) has written a history of Western metaphysics on the simple basis of which axiom or axioms various philosophers decided to drop. This makes a fascinating story.

The multiplicity of metaphysics finds its origin in Diodorus's theorem of incompatibility. A comparison comes to mind with the history of geometry. Once it was demonstrated fairly late in the history of mathematics that Euclid's fifth axiom, the so-called parallel axiom, couldn't be derived from the first four, it became conceivable to imagine a geometry in which this axiom wouldn't hold. The concept of a Riemannian manifold followed. And it proved extremely *useful*, as is well known, to Albert Einstein, who was in the process of elaborating his theory of general relativity. French mathematician Henri Poincaré then asked: "Is Euclidian geometry true? This question is deprived of meaning altogether.... A geometry cannot be truer than another; it is enough for it to be more convenient" ([1917] 2004, 75–76). Likewise, let's not ask whether projected time is truer than occurring time but if it is or not more useful than the latter.

It all depends on the kind of problem that is faced. Note first that projected time and occurring time are two ways of skirting Diodorus's aporia. The former denies axioms 1 and 3, the latter endorses them both and therefore denies 2.[7] In my work on catastrophes (Dupuy 2002)—including a nuclear conflict—I have shown that projected time avoids many paradoxes that occurring time—that is, strategic thinking—comes up against when it comes to conceptualizing the temporality that separates us from a looming disaster the date of which is unknown. The second part of this paper will illustrate this point. Projected time defines an attitude that is neither complacency (or voluntarism) nor fatalism. Complacency stresses that the catastrophe although possible is not inevitable: the future is open. Fatalism makes it inevitable. By granting the agent the counterfactual power to act upon the past conditions that determine him, projected time helps him navigate between the devil of catastrophism and the deep blue sea of dumb optimism.

For reasons already mentioned our *Zeitgeist* leans toward the latter. It is worth, then, remembering that the experience of projected time has accompanied humankind since time immemorial. It is intimately linked to the religious apprehension of the world. In all traditional societies, there are people called prophets (*nabis* in ancient Israel) whose function is to interpret and convey the divinity's will. The prophets of the Bible, for instance, were extraordinary men, often great eccentrics, and they did not go unnoticed by their neighbors. The influence their prophecies had on the world around them and on the course of events had purely human and social causes; but it was due also to the fact that those who heard them believed that the word of the prophet was the word of the Lord and that this word, which came to the prophet directly, from on high, had the power to bring about the very thing that it announced. We would say today that the word of the prophet had a *performative* power: in saying things, he brought them into being. However, the prophet was well aware of this. One might be tempted to conclude that the prophet had the power to which political revolutionaries aspire: he spoke so that things might change in the direction that he wished to impress upon them. But this would be to overlook the fatalistic aspect of prophecy, which reads out the names of all those things that will come to pass, just as they are written down on the great scroll of history, immutably, ineluctably.

Revolutionary prophecy, particularly in the form it came to acquire in Marxist doctrine, has preserved the highly paradoxical mixture of fatalism and voluntarism that characterizes biblical prophecy. German philosopher Hans Jonas said of dialectical materialism that it was "a most peculiar mixture of colossal responsibility for the future with deterministic release from responsibility" (1985, 113–14).

The metaphysics of projected time enables us to extend the notion of prophecy to our secular age and substitute for the obscure dialectic between voluntarism and fatalism a rigorous and non-paradoxical third way that is neither one nor the other. For the modern prophet, especially the prophet of doom, it is necessary to seek the fixed point of the loop between past and future, at which an expectation (on the part of the past with regard to the future) and a causal production (of the future by the past) coincide. The prophet, knowing that his public announcements are going to have a causal impact on the world, must take account of this fact if he wants the future to confirm what he foretold. The future is an x—that is, a solution to an equation that says that the reactions to the past anticipations of x causally bring about x.[8]

In this sense, prophets are legion in our modern democratic societies, founded on science and technology. The experience of projected time is facilitated, encouraged, organized, not to say imposed by numerous features of our institutions. All around us, more or less authoritative voices are heard that proclaim what the more or less near future will be: the next day's traffic on the freeway, the result of the upcoming elections, the rates of inflation and growth for the coming year, the changing levels of greenhouse gases, etc. The *futurists* and sundry other prognosticators know full well, as do we, that this future they announce to us as if it were written in the stars is a future of our own making, even if it is in reaction to these very announcements. We do not rebel in general against what could pass for a metaphysical scandal. We have then the experience of projected time.

METAPHYSICS OF NUCLEAR DETERRENCE

Caveat

I am writing these lines at a time when the prospect of a nuclear war between the United States and Russia is deemed by many observers stronger than it has ever been. In a book published in 2015 and titled *My Journey at the Nuclear Brink*, former Secretary of Defense William Perry wrote: "Today, the danger of some sort of a nuclear catastrophe is greater than it was during the Cold War, and most people are blissfully unaware of this danger" (2015, 17). It seems that we are back to the situation called "Mutually Assured Destruction," known by its opportune acronym MAD.

The American filmmaker Errol Morris, in his movie *The Fog of War*,[9] asks Robert McNamara, the former secretary of defense under President Kennedy, what he thinks protected humanity from extinction during the Cold War, when the United States and the Soviet Union permanently threatened each other with mutual annihilation. Deterrence? Not at all, McNamara replies: "We lucked out." Twenty-five or thirty times during this period, he notes, mankind came within an inch of apocalypse. I will show that this response is self-contradictory. All those "near-hits"[10] may have been the necessary condition for nuclear deterrence (ND) to work. To the extent that ND can be at times efficient, my objective is to show that everything occurs then *as if* the protagonists had immersed themselves in the peculiar logic of projected time.

Let me hasten to add that this is in no way meant to be a justification of nuclear deterrence in the MAD form. My conviction is that the latter is morally abhorrent. But there is a logic to it that can be discerned quite clearly. Above I referred to the religious mind as probably the cradle of the experience of projected time. In what follows it will appear that a number of features of MAD keep *the mark of the sacred* (Dupuy 2014).

My strategy will be as follows. In a first phase, I will expound the broad lines of the intellectual history of ND, following Steven P. Lee's excellent book, *Morality, Prudence, and Nuclear Weapons* (1996). There is not one argument put forward by the protagonists in that discussion that has not been questioned, disputed, challenged, refuted by some, defended by others, in an unending quest for reason and justice. I won't enter in those controversies and will be content with just reporting what the dominant arguments have been. My critical standpoint resides elsewhere, and I will expound it in a second moment. It consists in showing that confusions spoil the debate, and they stem from the fact that a good number of arguments belong to strategic reasoning and find their place within the metaphysics of occurring time while others, in general more recent, pertain to projected time and presuppose the renunciation of strategy. Two incompatible metaphysics of time clash invisibly.

A Brief History of Nuclear Deterrence Theory

For more than four decades during the Cold War, the discussion of MAD assigned a major role to the notion of *deterrent intention*, on both the strategic and the moral level. And yet the language of intention can be shown to constitute the principal obstacle to understanding the logic of deterrence.

In June 2000, meeting with Vladimir Putin in Moscow, Bill Clinton made an amazing statement that was echoed almost seven years later by Secretary of State Condoleezza Rice, speaking once again to the Russians. The antiballistic shield that we are going to build in Europe, they explained in substance, is only meant to defend us against attacks from rogue states and terrorist groups. *Therefore be assured:* even if we were to take the initiative of attacking you in a first nuclear strike, you could easily get through the shield and annihilate our country, the United States of America.

Plainly the new world order created by the collapse of Soviet power in no way made the logic of deterrence any less insane. This logic requires that each nation exposes its own population to certain destruction by the other's reprisals. Security becomes the daughter of terror. For if either nation were to take steps to protect itself, the other might believe that its adversary considers itself to be invulnerable, and so, in order to prevent a first strike, hastens to launch this strike itself. Before being a doctrine, MAD is a situation, in which nations are at once vulnerable and invulnerable: vulnerable because they can die from attack by another nation; invulnerable because they will not die before having killed their attacker—something they will always be capable of doing, thanks to a second-strike capacity, no matter how powerful the attack that will have brought them to their knees.

Throughout the Cold War, two a priori arguments were made that seemed to show that nuclear deterrence in the form of MAD could not be effective. The first argument has to do with the non-credible character of the deterrent threat under such circumstances: if the party threatening a simultaneously lethal and suicidal response to aggression that endangers its "vital interests" is assumed to be at least minimally rational, calling its bluff—say, by means of a first strike that destroys a part of its territory—ensures that it will not carry out its threat. The very purpose of this regime, after all, is to issue a guarantee of mutual destruction in the event that either party upsets the balance of terror. What chief of state having in the aftermath of a first strike only a devastated nation to defend would run the risk, by launching a retaliatory strike out of a desire for vengeance, of putting an end to the human race while committing suicide in the process? In a world of sovereign states endowed with the minimal degree of rationality that Hobbes granted to the inhabitants of the state of nature, namely the instinct of self-preservation, the nuclear threat has no credibility whatever.

The credibility question occupies the great majority of the debates about ND. Many experts conclude in particular that it is folly to make extreme threats that one is not sure one will deliver on. If your enemy calls your bluff, either you deliver and you risk what Carl von Clausewitz called the escalation to the extreme—that is, mutual annihilation—or you cave in and your credibility is down for the future. One of the best ways to keep your credibility intact is to multiply the occasions in which you show the world that your threats are not empty words: you do deliver and build a reputation of toughness.

The last remark leads to the second argument present in the literature that likewise points to the incoherence of the MAD strategic doctrine. Its premise is that, in Leon Wieseltier's words, "Nuclear deterrence is the only public arrangement that is a total failure if it is successful only 99.9 percent of the time" (1983, 75). To be effective, ND must be absolutely effective. Not even a single failure can be allowed, since the first bomb to be dropped would already be one too many. But in that case never will the adversaries be in a position to test the other's resolve to deliver on its threats. Perfect nuclear deterrence is said to be self-defeating or "self-stultifying"[11] since it undermines the very conditions that would make it efficient.

Nuclear deterrence doesn't work because the threat to retaliate is not credible. It doesn't work also because if it did, that assumption would lead to a contradiction. Those two reasons add up to the conclusion that the nuclear opponents are unable to deter one another. And yet, the Cold War, also known as Nuclear Peace, seemed to demonstrate the opposite, in spite of a significant number of "near-hits." An explanation had to be found.

Belatedly, it came to be understood that in order for deterrence to have a chance of succeeding, it was absolutely necessary to abandon the notion of deterrent *intention.* In principle, the mere *existence* of two deadly arsenals pointed at each other, without the least threat of their use being made or even implied, is enough to keep the warheads locked away in their silos. As two major philosophers put it, "The existence of a nuclear retaliatory capability suffices for deterrence, regardless of a nation's will, intentions, or pronouncements about nuclear weapons use" (Kavka 1987, 48); or: "It is our military capacities that matter, not our intentions or incentives or declarations" (Lewis 1989, 67).

Initially due to McGeorge Bundy, this doctrine has received the name of existential deterrence. The insistence on the causal power of the mere existence of nuclear weapons is a way to downplay the importance of strategy, intentions, plans, all major constituents of military thinking. If there is no need to threaten anyone it is because the weapons themselves, due to their incommensurate power, speak for us. If rationality plays a role here it is "the kind of rationality in which the agent contemplates the abyss and simply decides never to get too close to the edge" (Lee 1996, 248).

Fate and the Tiger

How exactly does existential deterrence work? Who or what deters whom? It is significant that the explanations provided by the best theoreticians rely on a non-human actor. We will consider two of them.

Let's start with David K. Lewis and the following quote: *"You don't tangle with tigers—it's that simple"* (1989, 68). The implication is that the game is no longer played between two adversaries. It takes on an altogether different form. Let's admit we are convinced that neither is in a position to deter the other in a credible way. *However, both want and need to be deterred.* The way out of this impasse is brilliant. It is a matter of creating jointly a fictitious entity that will deter both at the same time. The game is now played between one actor, humankind, whose survival is at stake, and its double, namely its own violence exteriorized in the form of a wild animal. The fictitious and fictional "tiger" we'd better not tangle with is nothing other than the violence that is in us but that we project outside of us. It is as if we were threatened *but also protected* by an exceedingly dangerous entity, external to us, whose intentions toward us are not evil, but whose power of destruction is infinitely superior to all the earthquakes or tsunamis that Nature has in store for us.

According to French anthropologist René Girard ([1972] 2005), the sacred stems from a similar mechanism of self-externalization of human violence. It used to be said of the atomic bomb, especially during the years of the Cold War, that it was our new sacrament. Very few among those who were given to saying this sort of thing saw it as anything more than a vague metaphor. But in fact there is a very precise sense in which the bomb and the sacred can both be said to *contain* violence in the twofold sense of the verb "to contain": to have within oneself and to keep in check. The sacred holds back violence through violent means, the original one being sacrifice. In the same way, throughout the Cold War, it was as though the bomb had protected us from the bomb. The very existence of nuclear weapons, it would appear, had prevented a nuclear holocaust.

One must not come too near to the sacred, for fear of causing violence to be unleashed; nor should one stand too far away from it, however, for it protects us from violence. Likewise, we cannot risk coming too close to the nuclear tiger, lest it should devour us; nor can we risk standing too far away, lest we forget the danger it represents. For deterrence to work it's all about finding the right distance from the big cat.

The second quote is from Bernard Brodie: "It is a curious paradox of our time that one of the foremost factors making deterrence really work and work well is the lurking fear that in some massive confrontation crisis it may fail. Under these circumstances one does not tempt fate" (1973, 430–31). Fate has replaced the tiger, but both images have in common that they place the deterrent in something other than human agency. We will return in the conclusion to a salient feature of this extraordinary quote, namely that it conjoins contingence (eventuality of failure) and necessity (fate), but we can pause at this stage and consider the following claim: the metaphysics of nuclear deterrence in its existential form is projected time. The renunciation of strategic thinking, the recourse to fate and the minimization of human agency, are all features that point in that direction.

Nuclear Deterrence in Projected Time

Let us admit for the sake of the discussion that the threat that underlies nuclear deterrence in its MAD form is not credible. The reasoning that supports this conclusion is strategic, and it is grounded in the metaphysics of occurring time. We reason by backward induction, and we posit that if the bluff of the menacing party is called, the latter will prefer to yield rather than being annihilated. The would-be attacker won't be deterred. The question is, doesn't projected time provide an alternative ground that would account for the efficiency of nuclear deterrence?

Given what we have learned in the first part of this paper, we can easily reach a conclusion, and it is negative. In projected time, nuclear deterrence doesn't fare better than in occurring time, but it is for entirely different reasons. The reasoning goes as follows:

1. If deterrence works, the escalation to the extreme—that is, the realization of the MAD threat—doesn't take place.
2. If the escalation to the extreme doesn't take place, then it is impossible. [Negation of Diodorus's 3rd axiom.]
3. If it is impossible, then nuclear deterrence doesn't work.
4. We have shown that if nuclear deterrence works, then it doesn't work.
5. Therefore, nuclear deterrence doesn't work.

The core of this argument is of course proposition 2, which expresses the condition that in projected time the future is necessary: an event that happens neither in the present nor in the future is an impossible event.

This reasoning gives a solid foundation to the second argument put forward by the critics of MAD. The alleged "self-defeatingness of a successful deterrence" appears to be a tortuous way of expressing a straightforward *reductio ad absurdum* (propositions 4 and 5).

The detour via the metaphysics of projected time proves unsuccessful. There is a way, however, to render it successful, and it consists in taking seriously the dialectic between contingency and necessity that is suggested in Brodie's quote. Meanwhile, we are going to realize that projected time is capable of solving the paradoxes of nuclear deterrence much more easily than strategic reasoning.

Nuclear Deterrence and the Indeterminacy of the Future

The suggestion that the manipulation of uncertainty can be a strategic tool that helps solve the credibility problem is not new. The conviction that if the agents are minimally rational they won't deliver on their threat to retaliate and launch the escalation to the extreme has led to the idea that it can be rational to pretend that one is irrational. It was first conceptualized by economist and game theorist Thomas Schelling in his landmark *The Strategy of Conflict* (1960) but made famous under the moniker "Madman Theory" by Richard Nixon during the Vietnam War. The following quote is eloquent. Nixon to his chief of staff H. R. Haldeman:

> I call it the Madman Theory, Bob. I want the North Vietnamese to believe I've reached the point where I might do anything to stop the war. We'll just slip the word to them that, "for God's sake, you know Nixon is obsessed about communism. We can't restrain him when he's angry—and he has his hand on the nuclear button" and Ho Chi Minh himself will be in Paris in two days begging for peace. (Haldeman 1978, 122)

The problem of course remains: what happens if the other side calls your bluff? In the face-off between Donald Trump and Kim Jong Un that took place during the summer of 2017, the question was, who is pretending to be mad and who is not pretending, because he is really mad?

However, in Brodie's quote, we are no longer talking strategy. The twofold reference to fate and the eventuality of failure takes us to a completely different world. The notion that it requires an accident for fate to come to pass is as old as the oldest myths of the planet. Think of Oedipus: it was proclaimed by the Oracle that he would commit parricide and incest. What precipitated the realization of this prophecy was a random encounter with a disgruntled old man who was barring his way. The merger of fate and accident is a common theme of many religious traditions. Rome had a goddess who represented at the same time luck (good or bad) and fate—or, to use the language of modalities, contingency and necessity. Her name was *Fortuna*.

Once again, the metaphysics of projected time offers a framework capable of giving a precise and formalized rendering of these intuitions. The key is a concept I have not yet introduced: the uncertainty of the future in projected time.

The uncertainty of the future in *occurring* time is approached with the usual tools. In the Madman Theory, the agent confronting some crazy behavior asks himself whether the folly is feigned, in which case the Madman will likely yield if his bluff is called, or whether he is *really* mad, in which case he may launch the escalation to mutual destruction if attacked. The agent ascribes a subjective probability epsilon, hopefully very small, to the latter possibility and the complement to 1 to the former. The way he comes to a decision is left to him—he may deem the Savage criterion of the maximization of expected utility senseless if the magnitudes are extreme: exceedingly large for the consequences, very small for epsilon—but one thing is assured: the two options make up a partition of the set of possibles that is a *disjunction* without overlap.

In projected time, uncertainty takes on a radically different form. There are no alternative possible futures, since the future is necessary. What replaces the disjunction is a *superposition* of states. Both the escalation to the extreme and its negation are part of the fixed future. It is because the former figures in the future that deterrence has a chance to work. It is because the latter figures in the future that the adversaries are not bound to destroy each other. Only the future when it comes to pass will tell.

The signature feature that distinguishes the two forms of uncertainty is the following: in occurring time, epsilon, the probability of the catastrophic scenario, can be equated to zero without that leading to a contradiction. If we continue to call epsilon the relative weight that this scenario has in the superposition of states, then it is essential that epsilon remain strictly positive. Were it to become naught, the escalation to the extreme would become impossible, for the reasons already adduced, and deterrence would fail. Superposition of states and strict positivity of epsilon are kindred concepts.

I have arrived at the notion of superposition of states via a conceptual itinerary that owes nothing to quantum theory. However, one cannot but notice the resonances. There is probably an affinity, to say the least, between the metaphysics of projected time and some of the basic concepts of quantum theory. I cannot pursue this line of inquiry here (see Dupuy 2000). However, I have proposed to name the kind of uncertainty proper to projected time *indeterminacy*. That is the correct translation of the German word *Unbestimmtheit* which Heisenberg chose to name his famous relation: *Unbestimmtheitsrelation*, infelicitously translated as "principle of uncertainty."

It is time to conclude. The nuclear deterrent that really works has been, and still is potentially, *the indeterminacy of the future in a conception of time that makes the future necessary*. It is indeed possible to provide rational foundations to the efficiency of nuclear deterrence. And that conclusion is horrendous.

NOTES

1. A semi-comical illustration: the Y2K efforts at preventing a universal computer collapse at the (false) turn of the century, a collapse that didn't take place, were deemed by many afterward to have been a waste of resources.

2. As far as modalities are concerned, let me recall that, given an adequate definition of a possible world, the possible is that which is true in at least one possible world; the necessary that which is true in all possible worlds; the impossible that which is untrue in all possible worlds; and the contingent that which is possible without being necessary.

3. This paradox is akin to the so-called "grandfather paradox" that appears to be a consequence of the assumption of time travel. If I could travel to the past and kill my grandfather "I" couldn't be. The grandfather paradox relies unnecessarily on causal connections though, which is not the case with the implications of the counterfactual power over the past.

4. Also known in game theory as the assurance game.

5. Quote from Philip K. Dick's tale, "Minority Report," a beautiful and profound illustration of some of the ideas presented here.

6. Reference to the method known in France as *Prospective*, elsewhere as the Scenario method, or, more vaguely, "futurology," invented by the French philosophers Gaston Berger and Bertrand de Jouvenel at the end of the 1950s.

7. Hence the paradoxes of backward induction. See Dupuy (2000).

8. Not any future goes. The prophet Jonah knew that if he prophesied the fall of Niniveh as God had asked him to do, the Ninivites would repent and God would forgive them. He preferred to run away from God's gaze.

9. *The Fog of War: Eleven Lessons from the Life of Robert S. McNamara*, directed by Errol Morris, Sony Classics, 2003.

10. The usual phrase is "near-misses." Interestingly enough, it literally says the opposite of the meaning it is supposed to convey.

11. Expression used by Gregory Kavka (1987) apropos of a different but kindred argument, which has for a long time been the ethical justification of the French nuclear doctrine known as deterrence "from the weak against the strong." The claim is that the deterrent intention to inflict "incommensurable" harm to the other party if it attacks you is not a genuine intention, since your true intention is to not have to carry it out. As the tortuous expression goes, "We form the deterrent intention in order to make it so that the conditions that would lead to its being acted upon are not realized." Plenty are the cases in the literature where the theory of nuclear deterrence is said to be self-invalidating.

REFERENCES

Anders, Günther. 1972. *Endzeit und Zeitenende: Gedanken über die atomare Situation.* Munich: Beck. Quoted in Thierry Simonelli, *Günther Anders: De la désuétude de l'homme.* Paris: Éditions du Jasmin, 2004, 84–85.
Bergson, Henri. 1935. *The Two Sources of Morality and Religion.* Translated by R. Ashley Audra and Cloudesley Brereton. Garden City, NY: Doubleday.
———. [1930] 1946. "The Possible and the Real." In *The Creative Mind: An Introduction to Metaphysics.* Translated by Mabelle L. Andison, 96–112. New York: Philosophical Library.
Brodie, Bernard. 1973. *War and Politics.* New York: Macmillan.
Dupuy, Jean-Pierre. 2000. "Philosophical Foundations of a New Concept of Equilibrium in the Social Sciences: Projected Equilibrium." *Philosophical Studies* 100, no. 3: 323–45.
———. 2002. *Pour un catastrophisme éclairé.* Paris: Seuil.
———. 2014. *The Mark of the Sacred.* Stanford, CA: Stanford University Press.
Girard, René. [1972] 2005. *Violence and the Sacred.* New York: Continuum.
Haldeman, H. R., 1978. *The Ends of Power.* New York: Times Books.
Jonas, Hans. 1985. *The Imperative of Responsibility: In Search of an Ethics for the Technological Age.* Chicago: University of Chicago Press.
Kavka, Gregory. 1987. *Moral Paradoxes of Nuclear Deterrence.* Cambridge: Cambridge University Press.
La Bruyère, Jean de. 1933. *Les Caractères.* Paris: H. Didier.
Lee, Steven P. 1996. *Morality, Prudence, and Nuclear Weapons.* Cambridge: Cambridge University Press.
Lewis, David K. 1981. "Are We Free to Break the Laws?" *Theoria* 47, 113–21.
———. 1986. *On the Plurality of Worlds.* Oxford: Blackwell Publishers.
———. 1989. "Finite Counterforce." In *Nuclear Deterrence and Moral Restraint*, edited by Henry Shue, 51–114. Cambridge: Cambridge University Press.
Perry, William. 2015. *My Journey at the Nuclear Brink.* Stanford, CA: Stanford University Press.
Plantinga, Alvin. 1986. "On Ockham's Way Out." *Faith and Philosophy* 3, no. 3: 235–69.
Poincaré, Henri. [1917] 2004. *La science et l'hypothèse.* Paris: Flammarion.
Schelling, Thomas. 1960. *The Strategy of Conflict.* Cambridge, MA: Harvard University Press.
Stalnaker, Robert. 1981. *Ifs: Conditionals, Belief, Decision, Chance, and Time.* Dordrecht: D. Reidel.
Vuillemin, Jules. 1996. *Necessity or Contingency: The Master Argument.* Stanford, CA: CSLI Publications.
Wieseltier, Leon. 1983. *Nuclear War, Nuclear Peace.* New York: Holt, Rinehart and Winston.

Chapter 7

The End of Technology and the Renewal of Reality

Albert Borgmann

We can think of technology as the term and the force that is characteristic of the modern era and that began with the Industrial Revolution. But as an animating power it may well have both crested in power and reached the bottom of possibility as we can conclude from Christopher Preston's *The Synthetic Age* (2018). But it will definitely not pass in time to yield a renewal of reality that will stave off the worst of climate change. The atrophy of technology and the slowly rising renewal of reality may, however, make the world more susceptible to the clear and urgent environmental tasks before us. We have to support those tasks as vigorously as we can.

The Industrial Revolution transformed reality. This transformation has brought about clearly visible and often painful changes, but by the middle of the nineteenth century it had repaid our efforts with security and comfort. Digging down one more level of analysis we can think of technology as a kind of moral commodification. *Economic* commodification is conceptually straightforward—it is the process of pulling something that's outside of the market into the market. *Moral* commodification is looking at the same process from a cultural or moral point of view—the detachment of a thing or practice from its contexts of engagement with a time, a place, and a community. Consider cloth and clothing. Once upon a time we lived in "the age of homespun," as Laurel Thatcher Ulrich (2001) has reminded us, when wool was spun at home, socks knitted, linen woven and bleached, and clothing tailored. One item after another was taken over by factories and was made available affordably and in a great variety. At the same time, the familial skills and practices disappeared.

For commodification and technology to remain vital enterprises there has to be a field of engagements outside of the market, an area that has yet to be colonized by technology. One after another of the preindustrial practices has been uprooted and replaced with a comfortable commodity. But the field of skills and practices is not unlimited. Eventually commodification runs out of commodifiable things and practices. By now, too much of the world has already been commodified and made overabundantly available. There is too much commodified food; too much information on the internet; too much stuff that is stuffing our garages and spilling over into micro-storage facilities; too much ease and comfort provided by Alexa and the Internet of Things. In fact, technology as an enterprise of research and development is stalling as Robert J. Gordon (2017) has recently reminded us.

The research and development enterprises of technology are still very much needed for the machinery that sustains our lives, for the infrastructure, for medicine, for renewable energy. But as an animating force that gets us out of bed in the morning and makes us go after more comfort and consumption, it has reached its end, except of course for the poor in this country and around the globe, for the people who are lacking the basic comforts of life. But for the upper and middle classes of the advanced industrial countries, the renewal of the world cannot come from one more iteration of the pattern of comfort and consumption. Renewal has to be the opposite of that.

The renewal of reality is not a matter of invention and prescription. It becomes visible through a process of discovery. So where do we find evidence of renewal? We find it where we see decommodification, the restoration of contexts of engagement. Such restoration must be possible *within* the technological infrastructure of reality. We cannot replace the water supply with village fountains. We cannot replace public transportation with horses and buggies.

The three best-organized and visible enterprises of renewal are the new urbanism, the artisan economy, and organic farming. The new urbanism builds neighborhoods that invite neighborly interaction and a walkable environment, where your porches make you talk to neighbors and watch the kids and where you can walk to the grocery store, to the movies, to the post office, and so on. A helpful introduction can be found in the (somewhat misleadingly titled) *Suburban Nation: The Rise of Sprawl and the Decline of the American Dream* by Andres Duany, Elizabeth Plater-Zyberg, and Jeff Speck (2010).

The artisan economy is frequently mentioned, but rarely examined and supported. It plays an important role in the developing countries, but in those countries it usually constitutes a transitional economy on the way to commodification. In the advanced industrial countries, however, it can be an enduring counterforce to mass production and consumption. It reweaves contexts of community and comprehension. An artisan has a definite location in a community. Her work is known and understood by customers and citizens. The products of artisanship are more valuable and less abundant than mass-produced commodities, and therefore less disposable. A table that is quickly and cheaply bought at Ikea will be disposed of when you move or no longer need it. A table that comes into being through an agreement between you and an artisan is likely to become an heirloom. Something broadly analogous happens throughout the artisan economy. Instances are well-described in Charles Heying's *Brew to Bikes* (2010).

Organic farming reweaves the texture of agriculture from the strands of tradition and nature. It is most clearly the countermove to the synthetic age by rejecting all synthetic fertilizers and pesticides and genetically modified plants and animals. It also renews the ties between the land and the people through farm-to-table connections. It is well organized through national and international organizations.

Yet its share of food production and consumption in this country is a mere 6 percent or so, and one may wonder if the other two movements of renewal are any more robust and influential. What inspires hope are the many people who engage or reengage the real world in their leisure and daily lives—the musicians, the runners, the bikers, the hikers, the skiers, the hobbyists, the painters, the ceramicists, the gardeners, the dog lovers, and many more. What inspires sorrow is the fact that these people are unaware of their common renewal of reality. The forces of consumption, to the contrary, dominate the media, the money, and public discourse.

What is it that the people of renewal have in common? They share

- a command of manual skills;
- and a regular exercise of those skills;
- in a local setting;
- with natural communal relations; and
- with a profound enjoyment of all of those engagements.

And how many such people are to be found in this country? Mainstream social research seems not to be interested in the set of features that the forces of renewal share. You can piece together some information from scattered sources, a task beyond my time, I say with sadness. But the mutual awareness of constructive people may yet rise, and if it does, their impact on politics and culture may bring about a wider and deeper renewal of reality.

ACKNOWLEDGMENT

Carl Mitcham has been the great founder and chronicler of the philosophy of technology. My work has been based on his support and inspiration.

REFERENCES

Duany, Andres, Elizabeth Plater Zyberg, and Jeff Speck. 2010. *Suburban Nation: The Rise of Sprawl and the Decline of the American Dream*. New York: Northpoint Press.

Gordon, Robert J. 2017. *The Rise and Fall of the U.S. Standard of Living since the Civil War*. Princeton, NJ: Princeton University Press.

Heying, Charles. 2010. *Brew to Bikes: Portland's Artisan Economy*. Portland, OR: Ooligan Press.

Preston, Christopher. 2018. *The Synthetic Age: Outdesigning Evolution, Resurrecting Species, and Reengineering Our World*. Cambridge, MA: MIT Press.

Ulrich, Laurel Thatcher. 2001. *The Age of Homespun*. New York: Knopf.

PART II

Philosophy and Engineering

Chapter 8

An Engineer Considers Technological (Non)Neutrality

"But Where Are the Values?"

Byron Newberry

In my reading, I have frequently encountered the claim that engineers have a value-neutral attitude about technology. One of the first places I read this was in Carl Mitcham's book *Thinking through Technology* (1994, 73), where he describes John Dewey's view of technology as neither opposed to value, nor "as neutral with regard to value, as scientists and engineers think." Although I am an engineer, I do not, however, consider myself a "happy technologist," as one author calls STEM experts who blithely assume that technology is morally inert (Balabanian 2006, 16). Whether that makes me an outlier as an engineer is a question that I considered in a previous article (Newberry 2007). Despite my openness to the idea that technologies embody values, the question of technological neutrality has persistently bedeviled me. I have struggled to articulate to my own satisfaction precisely what it means for a technology to embody values. Nor have I been able to completely dismiss the possibility that value-ladenness is illusory or metaphorical, or stems from some cognitive or semantic confusion.

This uneasiness on my part provides the underlying motivation for this chapter, which has two main goals. The first is to explore, and attempt to clarify, some of the conceptual issues I think bear on this question. Specifically, I will highlight ways in which I think the question of neutrality is not as clearly defined as many might assume. My second goal stems from the appearance in recent years of a raft of articles that I think have provided some new insights on this topic. I will thus try to provide a brief overview of this recent literature. A couple of these articles defend value neutrality. In fact, the subtitle of this essay is borrowed from Joe C. Pitt (2014, 95), who, failing to readily observe values when inspecting artifacts, asks, "But where are the values?" It is a fair question, and it is essentially the same question that I've struggled to answer. Other recent articles—a clear majority—argue *for* value-ladenness.

Why does this matter? This issue bears on the concerns many people have about the effect of technology on human welfare, social relations, and environmental sustainability. Such concerns lead naturally to a host of questions. Should technologies be regulated, or sometimes even banned? Who should be involved in the development and oversight of technologies? What factors should be considered in the technology development process beyond economics and functionality? Another set of questions revolve around who is to blame for negative consequences of using a technology. The designer? The maker? The user? And on what basis do we hold them accountable? The answers to these questions and more may hinge to some degree on the value-neutral/laden distinction. If value-ladenness is true, this might help justify taking collective action to regulate technologies or to assign some responsibility for the consequences of their use to the producers, and not just the users. Value-ladenness is often formulated in terms of technologies having *intrinsic value*—that is, value in and of itself. If, conversely, value-neutrality is true, some might think that any prior restraint on technology would be hard to justify. Value-neutrality is often formulated in terms of technologies having only *instrumental value*—that is, only having value as a means to achieving some other ends, and not for their own sakes. Thus, responsibility is confined to the people who use technologies, and moral judgment is applied only to the ends toward which such use is directed.

It is particularly important that engineers engage with this topic, which is why I introduce it to my students in a course I teach called Social and Ethical Issues in Engineering. As a society, we should hope that those creating technologies might also engage in contemplating the significance of them beyond just technical specification, and also to grapple with questions of human values related to technology. One of the earliest essays I read about technological neutrality was by Robert Whelchel, an engineer writing for an audience of other engineers. Advocating a value-laden view of technology, he wrote,

> No one would contend that every individual practicing engineer needs to study and contemplate the type of questions broached in this essay [about technological neutrality]. But the profession as a whole does need to address these issues. If our profession chooses to confine itself to narrow technical specialties, then we will be ignoring precisely those aspects of technology which are most significant to society. Such a stance is not only embarrassing but irresponsible. (1986, 7)

Since my native language (so to speak) is engineering and not philosophy, I am under no illusion that my efforts here will definitively capture all the philosophical nuances of what has been written about technological neutrality. Yet I hope that my engineer's-eye view of this subject will provide a framing that others, especially engineers, might find useful.

VALUE-LADEN TALK

I want to first briefly address the ubiquity of *value-laden talk* among people generally, or at least my perception that such is the case. I will provide a variety of examples to show the ways this phenomenon is manifest. When I introduce this topic in my class, I often start with a line from the 2003 film *Pirates of the Caribbean—Curse of the Black Pearl*. In explaining to his companion why he is so keen to recover his lost ship, the *Black Pearl*, Captain Jack Sparrow says,

> Wherever we want to go, we go. That's what a ship is, you know. It's not just a keel and hull and a deck and sails. That's what a ship needs. But what a ship is . . . what the Black Pearl really is . . . is freedom.

On one level, Sparrow describes the ship as a functional arrangement of parts and materials. But on another level, he defines it in terms of an embodied value—it is, for him, literally a *vessel of freedom* (if you pardon the pun). This is a concise, and somewhat poetic expression of a value-laden view of technology. Of course, you might point out that an example taken from a fictional story about ghostly pirates is not evidence of anything. In that case, consider the following passage from an article about the American obsession with automobiles.

> They are the epitome of convenience. That's the allure and the promise that's kept drivers hooked, dating all the way back to the versatile, do-everything Ford Model T. Convenience (*some might call it freedom*) is not a selling point to be easily dismissed—this trusty conveyance, always there, always ready, on no schedule but its owner's. (Humes 2016, emphasis added)

The idea here is that beyond any specific use of an automobile (e.g., driving to work, taking children to school), the very having of it reveals a more fundamental *embodied* value which, as with Sparrow's ship, the author describes as *freedom*.

The terms *embodied* and *embedded* are often used when discussing this topic, with the latter generally connoting values supposedly inserted by designers into their artifacts. But this usage is not limited to philosophical discussions of value-neutrality. For example, an online product description for a screwdriver states, "Norbar's range of Torque Screwdrivers *embodies the values* of other Norbar products: accuracy, ease of use and comfort in use" (emphasis added, see https://www.norbarusa.com/products/view/ncategory/categoryname/tts-torque-screwdrivers/category_multid/349, accessed December 15, 2020). And in a recent article criticizing COVID-19 contact-tracing apps, the authors write, "Without sound legal protections and safeguards, tracing apps will not only fail but *will embed values* that may not be those that represent the society we wish to be" (Leins et al. 2020, 8, emphasis added).

Perhaps the best demonstration of the ubiquity of value-laden talk is to look for it where it might least be expected. The saying, "Guns don't kill people, people do," is a canonical example of the value-neutral view (and a favorite trope in articles like this one). The expression is associated with gun advocates. I contend, however, that its usage by gun advocates is largely pretextual—for political and legal advantage—and that it does not correlate with any *actual belief* in technological neutrality. This may be a presumptuous claim, but I defend it on the empirical grounds that when gun advocates talk about guns, they often talk about them in the same way that Captain Jack Sparrow talks about his ship. For example, the editor-in-chief of the National Rifle Association's in-house journal, *America's 1st Freedom*, writes the following:

> While guns are, by definition, inanimate objects, they are so much more. . . . Guns aren't just steel, wood and polymer, not hardly. They have an aura, a feel, and, metaphorically speaking, they have personalities in our hands . . . this love of American freedom held in our hands, is a powerful and freeing force. (Miniter 2020)

The parallelism between this quote and the one above by Captain Jack Sparrow is remarkable. Like Sparrow's ship, on one level the gun is a functional arrangement of parts and materials. But on another level, it is the very embodiment of—once again—*freedom*. Such recurrences of the idea of freedom embodied in technologies are perhaps local manifestations of an overarching conception of technology, which Mitcham (2020, 85) describes thusly: "At its deepest level, technological design aims to enlarge human freedom."

In reference to the rural community where she grew up, another gun advocate writes, "guns were a symbol of liberty and autonomy, but they were also a profoundly practical tool. Meat doesn't just come from a grocery store" (Olmstead 2017). Here, she explicitly contrasts the *instrumental* value of a gun with some of the more enduring values it represents. She also mirrors Captain Jack's language by telling us what she thinks a gun truly *is*, in value terms. Death, she writes of a gun, is, "what it *is*, in its very essence" [emphasis in original].

Thus, it is my observation that people naturally talk and act as if technology is value laden—even in some cases when they have a vested interest not to. At the very least, many people seem to have a cognitive confusion about the value-laden/neutral distinction, even when explicitly addressing it. Consider this example taken from a report on technological literacy published by the US National Academy of Engineering (NAE). In dispelling misconceptions about technology, the authors write,

> Another common misconception is that technology is either all good or all bad rather than what people and society make it. They misunderstand that the purpose for which we use a technology may be good or bad, but not the technology itself. (National Academy of Engineering 2002, 5)

This is a textbook definition of the value-neutral viewpoint, and, given that it comes from the National Academy of Engineering, perhaps it lends credence to the claim that engineers take an instrumental view. Yet this view is repeatedly contradicted within the same report. One of the characteristics of a technologically literate citizen, the report asserts, is an understanding "that technology reflects the values and culture of society." Similarly, the authors write, "Technology mirrors our values, as well as our flaws." And the report makes favorable reference to the ideas of people like Langdon Winner and Neil Postman, who are staunch critics of technological neutrality. The lack of self-awareness of these contradictions in the report is remarkable.

In the following example, people are seen to express a value-laden perspective on technology, but this time in a more deliberate and well-considered way. Amish communities, rather than being anti-technology, as some might naively assume, may be described more aptly as being highly discerning about technology (Wetmore 2007; Ems 2014). While they do favor technologies we might label as *old fashioned*, they also utilize some modern technologies (e.g., telephones, automobiles), but only in a highly circumscribed fashion. They are intentional about which technologies they choose to use, and under what circumstances they will use them. Such decisions are communally made based on the alignment between the community values and the values they believe are intrinsic to the technology.

> First they choose technologies that they believe will best promote the values they hold most dear—values like humility, equality, and simplicity. Thus they have rejected the speed, glamour and personal expression of automobiles

in favor of modest, slow, and community-building horse-drawn buggies. (Wetmore 2007, 21)

For the Amish, a belief in the value-ladenness of technology seemingly underpins one of the key organizing principles of their communities. They are, perhaps, archetypal *critical theorists*, resolute in making "a choice at a higher level determining which values are to be embodied in the technical framework of [their] lives" (Feenberg 2006, 14).

Let me be clear in saying that none of the foregoing examples that illustrate how people think, speak, and act in ways that suggest technology is value laden prove, on their own or collectively, that technology is value laden. But even if value neutrality is true, this phenomenon is fascinating and warrants explanation. Does th*is* occur because technology really *is* value laden, and the way people talk reflects that reality? Or is it a semantic confusion akin to *function talk* in biology, whereby biological features like organs, behaviors, or processes are routinely described as having *functions* or *purposes*—which are teleological words—despite the fact that those teleological descriptions are arguably misleading with respect to the non-goal-directed nature of the processes by which such features arise?

THE FUZZINESS OF NEUTRALITY/VALUE-LADENNESS

One thing that becomes apparent when reading the literature on this topic is that there is a wide diversity of thought about it—not just about whether technology is neutral/value-laden, but also about how that question gets asked and about what approaches are taken to answering it. Do technologies embody values? Are technologies morally good or bad? Do technologies impinge upon human agency and action? Do artifacts have politics? These and other framings of the question are all surely related in some way. But are they really the same question? Are they distinct facets of a larger question? Are they commensurable at all? This is not a clear-cut issue to me. And this has played no small part in why I have struggled to fully digest the question of technological neutrality/value-ladenness. So, in this section I will briefly review some efforts, including my own, to organize the varieties of argument related to neutrality/value-ladenness.

Andrew Feenberg (1999) posits four broad schools of thought with respect to how to think about humanity's relationship to technology, and he represents them as existing in four quadrants defined by the two axes of *neutrality* and *autonomy*. The neutrality axis locates one's view of the relationship between means and ends. On the neutral end, ends are independent of the means used to achieve them, and means stand ready for use toward any end.

On the value-laden end, means and ends are intimately intertwined. The autonomy axis locates one's view of the human ability to exercise control over technology. On the autonomous end, technology advances by its own internal logic. On the controllable end, humans can steer technologies according to their unfettered will.

With Feenberg's model, someone expressing a belief in neutrality would fall into either the *determinist* quadrant (technology is neutral and autonomous), or the *instrumentalist* quadrant (technology is neutral and controllable). The former leads to a technocratic mindset whereby technology is seen as an ineluctable and naturally evolving force that provides a neutral foundation for human activity. The latter leads to a view that sees humans in control of their own destiny via their command of technology, which may include what Feenberg calls a "liberal faith in progress," the view often ascribed to engineers.

By contrast, those who express a belief in value-ladenness would fall into either the *substantivist* quadrant (technology is value laden and autonomous), or the *critical theorist* quadrant (technology is value laden and controllable). In the former case, we get a view in which traditional human values are ultimately overridden by the juggernaut of technological imperatives, possibly leading to a pessimistic outlook—or *dystopianism*, to use Feenberg's term. In the latter case, while technologies (as means) profoundly shape human values (and thus vie with human values to become ends-in-themselves), ultimately, with sufficient thought and effort, humans retain the ability to exert control over both the means and the ends. Feenberg places himself in the critical theorist camp.

Feenberg's two-axis model implies that neutrality and autonomy are two independent variables. In treating them as such, Feenberg draws distinctions between the four quadrants that are both instructive and insightful. Yet, I cannot help but ask, are neutrality and autonomy really independent? If neutrality is what exists when ends are independent of means, then would not technological autonomy be the antithesis of neutrality, rather than being orthogonal to it? After all, if the means are beyond our control, then surely the ends are as well. Conversely, wouldn't having total control over our means allow us to determine our ends freely? I'm not the first person to ask these questions. For example, Hans Oberdiek (1990) sets the two concepts in opposition to one another in just this way in an article titled "Technology: Autonomous or Neutral?" He criticizes both views for absolving engineers from having any "responsibilities for what they invent and develop" (67). He construes *neutrality* to be synonymous with technology being *fully controllable*, and thus the antithesis of being *autonomous*.

Christian Illies and Anthonie Meijers (2009) make a similar point. They highlight two distinct debates in the philosophy of technology that parallel

Feenberg's two axes, the *autonomy debate* and the *moral relevance debate*. In the former, at the limits, technologies act as either *instruments* or *agents*. In the latter, they are either *morally neutral* or *morally charged*. Illies and Meijers note that being an instrument is closely linked to being morally neutral, while being an agent is closely linked to being morally charged. While Illies and Meijers ultimately treat autonomy and neutrality as separate things, they suggest they are more correlated than Feenberg's axes imply.

This makes me wonder whether the question of neutrality is a question about a single idea or about a conjunction of related ideas. Many authors whose writings have been widely influential in the philosophy of technology in the past three-quarters century or more, perhaps representing the majority of classic accounts, comprise a genre I call *critique of technological culture*. Many of the ideas about value-ladenness familiar to students of philosophy of technology stem from this literature. But ideas about value-ladenness are not necessarily the explicit focus of this literature so much as they are a byproduct. Rather, this literature generally aims to offer wide-ranging critiques of technological culture, often involving multidimensional analyses that adduce evidence from sociology, psychology, politics, and economics, in addition to philosophy, and it is within this context that a stance on neutrality/value-ladenness emerges, perhaps implicitly, as an element of the much broader story; it may even require some interpretation to tease it out the specifics.

An example is Jacques Ellul, whom Feenberg classifies as a substantivist and thus in the value-laden camp. Ellul (1964, 141) writes, "It [technique or technology] is not a neutral matter. . . . It refracts in its own specific sense the wills which make use of it and the ends proposed for it." By defining neutrality in terms of both *wills* and *ends*, Ellul explicitly melds both of Illies and Meijer's notions—autonomy (or agency) and neutrality (or moral relevance)—into one. But he does not offer a concise defense of this. Rather, his justification must be understood within the totality of his lengthy critique of technological society, which sees the technological world as distorting human values and practices into an all-consuming quest for efficiency in every aspect of life.

Don Ihde (1990) also seems to fold together the ideas of *moral relevance* and *autonomy* under the umbrella of *neutrality*, but with a different twist. He defines neutrality in terms of technologies being "mere things . . . like inert matter" (4). Then he goes on to say, "At an even greater extreme of the neutrality/non-neutrality debate, there are those who hold . . . that once created and put in place, technology takes on a life of its own and becomes autonomous" (6). Here he seems to be suggesting that *autonomy* is an extreme form of non-neutrality, and thus the former is a subset of the latter.

This combination of *control* (autonomy) and *values* (moral relevance) in the context of defining *neutrality* also appears in Mary Tiles and Hans Oberdiek (1995, 13), who write, "On this optimistic view we are firmly in control of the technologies we produce. . . . As such, any technology is value neutral: we impose our values in deciding which technology to use and how."

Another taxonomy of thought about neutrality/value-ladenness, and one that also blurs the lines between autonomy and neutrality, is found in Martin Peterson and Andreas Spahn (2011). Like Feenberg, they posit four possibilities for stances on neutrality. But unlike Feenberg, rather than occupying quadrants in two-dimensional space, they are ordered along a line as follows: the *strong view* of value-ladenness, the *moderate view* of value-ladenness, the *weak neutrality* thesis, and the *strong neutrality* thesis. This format reminds me of a Likert scale ranging from strongly agree to strongly disagree.

They describe the *strong view* of value-ladenness as the claim that "both humans and technological artefacts can be moral agents." Here, they are responding primarily to phenomenological arguments such as those of Ihde and Peter-Paul Verbeek, which might be more accurately characterized as viewing moral agency as being distributed in some sense over the human/technology composite. Surely if one could justifiably ascribe moral agency to a technological artifact or to a hybrid human-artifact relation, that artifact would qualify as morally value-laden. Next, they write that the *moderate view* of value-ladenness, "does not entail that technological artefacts are, or can be a part of, a moral agent," but it does entail that, "the role of (some) technological artefacts is quite different from that of other artefacts." This latter entailment is quite vague, which Peterson and Spahn attribute to an inherent vagueness that is characteristic of *moderate view* accounts. Thus, it is not obvious what comprises the conceptual mortar that binds together ideas in the *moderate view* category, other than that they comprise those accounts that defend value-ladenness without claiming that artifacts have moral agency.

Starting from the other end of the scale, the *strong neutrality* thesis, as defined by Peterson and Spahn, rests on the following three conditions: "technological artifacts (i) never figure as moral agents, and are never (ii) morally responsible for their effects, and (iii) never affect the moral evaluation of an action" (2011, 423). In other words, technological artifacts are morally inert in a thoroughgoing way. This is in contrast to the *weak neutrality* thesis, which accepts requirements (i) and (ii) but rejects requirement (iii). In other words, this thesis agrees that technologies are not agents, nor are they responsible, but it does grant that "technological artefacts sometimes affect the moral evaluation of actions" (423). Presumably, this *affecting* is done in a way that does not entail the embodiment of moral values.

Of the options they define, Peterson and Spahn endorse the *weak neutrality* thesis. Their taxonomy provides a way to identify and think about some issues upon which the question of neutrality may turn. However, it is a rather coarse-grained measure that doesn't reveal distinctions between various nuanced arguments within categories. And, as I will argue in a subsequent section, their demarcation between the *moderate view* of value-ladenness and the *weak neutrality* thesis is not completely clear. Another feature to note about this taxonomy is that the neutrality of technology is couched almost completely in terms of *autonomy* as opposed to *moral relevance*. The distinctions Peterson and Spahn draw between the different categories have to do with agency, responsibility, and action, and not about value embodiment, means, and ends.

Finally, there is a smaller and relatively more recent corpus of literature that more narrowly, and more explicitly, addresses the neutrality/value-ladenness question in more analytical ways using the philosophical tools of *value theory*. An example is Ibo van de Poel and Peter Kroes (2014). Their interest is in the practice of value sensitive design (VSD), which is a strategy that attempts to intentionally embed desired values in technologies by design. They seek to formally show doing so is possible by way of value-theoretic arguments. In their conclusion they write, "the central outcome of our analysis is that the neutrality thesis does not hold and that it is possible for technical artifacts to embody values . . . of a specific kind, namely extrinsic final values. Values may be designed into technical artifacts and therefore VSD is possible" (121).

The segment of the literature of which van de Poel and Kroes's work is a part, along with some other relatively recent work, will dominate my review in the next section. I will end this section by pointing out that while I suspect many readers will be familiar with the concept of neutrality, as well as with the literature I've cited, I think that such familiarity belies a certain underlying fuzziness of the concept, including some potentially incommensurable ways that people think about and attempt to answer the question. At least this seems to be the case from my engineer's vantage point. And to the extent that it is important for engineers to grapple with this issue, this take on the fuzziness of neutrality proves instructive.

SURVEYING RECENT THOUGHTS ABOUT NEUTRALITY/VALUE-LADENNESS

Pitt, a philosopher of technology who perhaps is in the minority in his discipline as a defender of technological neutrality, argues against the notion of value embodiment within artifacts. In asking, "But where are the values?" (2014), he demands that any values embodied by an artifact be objective and empirically identifiable, seemingly akin to measuring the artifact's mass or its spatial dimensions. "Point to . . . the value," he exhorts. "Do they have colors? How much do they weigh?" (95) I think Pitt is right to ask for some reasonable way to recognize any values supposed to be embodied in an artifact. I have found myself at times resonating on an affective level with this or that author's narrative account of the value-ladenness of technology, only to be ultimately dissatisfied with a vagueness about the mechanism of value embodiment or frustrated by one-off examples of embodiment that seem belabored, contrived, and not obviously generalizable. Yet the way Pitt frames *embodiment* is puzzling to me. He proffers two possible accounts of what embodiment might mean, one in which technologies *have values*, and another in which all the myriad value-driven decisions made by anyone associated with the development of a technology are somehow cumulatively reflected in it. The first he rejects by arguing that "values are the sorts of things that only humans have" (90). The second he is willing to grant, but he argues the result is trivial because the plethora of values thereby embedded would be impossible to parse, and therefore this account "says nothing significant" (101).

I agree with Pitt on both counts. But I also think the alternatives he considers for what embodiment might mean do not actually capture the full spectrum of possibilities. As for the first account, Pitt essentially argues that the idea that technologies *have values* or *contain values* is a category error—*having values* (in the sense of being motivated to care about things) is by definition unique to humans. Yet accounts of value embodiment do not typically suggest that technologies literally *have* or *contain* values in the sense that a person has values (notwithstanding any lax wording that might superficially suggest as much). Rather, they are either premised on the assumption that technologies *have value*, in the sense that *people value them*, or else on the assumption that technologies can be *good* in some intrinsic sense. That artifacts *have value* to people, or can *be valued* by people, is clearly plausible, if not downright commonsensical. What may be controversial is whether artifacts can be valued in particular ways that make them subject to moral assessments. Whether technologies can have intrinsic goodness is even more controversial.

Boaz Miller has recently provided a detailed rebuttal of Pitt's argument about empirical identification of values, arguing in the first place that values need not be empirically discernable to be embodied, and in the second place that they sometimes are empirically discernable. Miller then offers the following criterion for embodiment:

> If a certain function is value laden, and certain physical features of an artifact are required to effectively perform it, and the existence of these features in the artifact has no other reasonable justification, then the artifact may be said to embody the respective values. (2021, 10)

Miller's emphasis here on function and on justifying an artifact's features as being for that function, seems to ground embodiment in designer intent, which makes this formulation untenable, I think, for reasons discussed in more detail below.

For van de Poel and Kroes (2014, 109), an artifact "bears or embodies a value" just whenever it is valuable to someone. But whether that makes the artifact morally value-laden depends on the reasons someone has for valuing it, which for van de Poel and Kroes means whether it is valued in a *final* rather than *instrumental* sense. My point here is not to defend the account of van de Poel and Kroes. Rather, it is to suggest that Pitt's demand that embodied values be readily discernable by way of empirical measurements taken from the physical artifact would not seem sufficient for assessing the validity of van de Poel and Kroes's account, since there is no reason to believe on their account that embodied values would be testable in that way. Though some physical features of an artifact will likely contribute to understanding the reasons someone values it, a full understanding of that valuing would surely entail also understanding something about the particular context of the artifact's creation and use, and perhaps also something about human nature in general. As Feenberg notes,

> From a common-sense perspective a technical device is simply a concatenation of causal mechanisms. No amount of scientific study will find anything like a purpose in it. But perhaps common sense misses the point. After all, no scientific study will find money in a $100 bill. Not everything is a physical or chemical property of matter. (2006, 10)

If any values were discernable solely from the physical properties of an artifact, they would, I think, correspond to what van de Poel and Kroes call *intrinsic final values*. Van de Poel and Kroes readily concede that if value-ladenness depended on the existence of such values, then neutrality

> appears hard if not impossible to deny. The idea that a technical artifact has a form of value that remains the same independent of its relation to anything else, in particular of its design context or its context of use is very implausible. ... [This] construal of the idea that technology is value-neutral is more or less a truism. (2014, 110)

Van de Poel and Kroes go on to argue that another category of values—*extrinsic final values*—is the more appropriate candidate for value embodiment, and these depend (at least in part) on factors external to the artifact. Specifically, they formulate the value-neutrality thesis as follows: "The designed properties of technical artifacts cannot form the resultant base of moral extrinsic final values" (112). By defining value neutrality explicitly in terms of extrinsic—relational—values, Pitt's defense of neutrality seems to have been short-circuited. In a more recent article, Kroes (2020) attributes this to Pitt mistakenly thinking of technical artifacts as purely physical, rather than having a dual nature comprising both physical and intentional components.

As an aside, as an engineer who is used to analytical problem solving, there is something satisfying about the analytical nature of the value-theoretic arguments made by van de Poel and Kroes, as well as by others I will discuss below. But, like most things in philosophy, it also opens up its own can of worms. For example, in the case of value embodiment, this question has traditionally turned on the distinction between intrinsic and instrumental values, with value-ladenness ostensibly depending on the embodiment of intrinsic values. But the very existence of intrinsic values has been called into question by some (e.g., Olson 2015), while others make a further distinction between intrinsic values and final values, the latter of which might be neither instrumental nor intrinsic (e.g., Rabinowicz and Rønnow-Rasmussen 2000) and which therefore might be the key to value-ladenness (e.g., van de Poel and Kroes 2014). Yet others dispute the meaningfulness of final value (e.g., Tucker 2019). And even if we concede the existence of intrinsic or final values, along with instrumental values, there is no clear consensus about what specific values belong to one or the other category. For instance, van de Poel and Kroes (2014) take *safety* as a final value, whereas others (Flanagan, Howe, and Nissenbaum 2008) seem to consider it instrumental. To top it off, some might avow that any such distinctions between the instrumental and the final, or means and ends, while useful, are ultimately contrived (Dewey 1939). So, value-theoretic arguments are not as straightforward as an engineer might like.

Van de Poel and Kroes develop their account in the context of demonstrating the possibility of engineers using the VSD approach to embody values in new technologies in an intentional way. Roughly, they argue that if the designed features of a technology, qua being designed features, have the potential to contribute to some extrinsic final value, then the technology embodies that value. Michael Klenk (2021) details objections to this approach, both epistemological and metaphysical: in short, we are not guaranteed to know a designer's intent, and we cannot account for changing circumstances of use, respectively. I agree with Klenk's objections in principle. Like Miller's account discussed above, by focusing on designer intentions I consider the van de Poel and Kroes approach to be too limited to provide a general account of value embodiment. But a general account of value embodiment may be more than van de Poel and Kroes intended to achieve, as their goal seems more narrowly to have been to demonstrate the possibility of VSD as one path to value embodiment.

There is one aspect of van de Poel and Kroes's account that I believe is important: value embodiment, they argue, is related to the *potential* for a technology to contribute to some value, rather than on an actual contribution. That is, they make a distinction between an *embodied value* (potential) and a *realized value* (actualization of the potential).

As noted above, Klenk challenges the approach of van de Poel and Kroes, and then offers his own alternative account of embodiment, as follows:

> Artefact x embodies value V if x affords to a set of subjects S in conditions C an ability A and there is reason to positively respond to A (positive value), or there is reason to negatively respond to A (negative value). (2021, 534)

Roughly speaking, if an artifact affords an ability that someone has reason to value (or disvalue), then the artifact embodies that value (or disvalue). Unlike the accounts of Miller and van de Poel and Kroes, rather than being designer-centric, value embodiment in this case hinges on what the artifact has the potential to do that a user may value, and hence is user-centric. This does not, however, negate an important role for designers in value embodiment, since designers will strive to anticipate what users find valuable and thus will design an artifact's features accordingly. Like the account of van de Poel and Kroes, however, Klenk's focus on affordances puts the emphasis on an artifact's potential to do something, rather than on it actually doing something. Klenk's account, like van de Poel and Kroes's, ultimately turns on accepting that extrinsic final values exist and are sufficient to render an object morally value-laden. I find Klenk's account particularly compelling and think it could be strengthened by expanding his use of affordances to also include constraints—that is, not only might an object's affordances provide reasons to value (or disvalue it), but its constraints (what is doesn't allow you to do) might too.

Klenk's account has been further elaborated by Fabio Tollon (2022), who seeks to enhance the concept of affordances in two ways, both psychological in nature. In the first, Tollon distinguishes between what he calls *mere* affordances and *meaningful* affordances. The distinction has to do with the extent to which an affordance aligns with an agent's interests. In the second, Tollon introduces the concept of the *force* of an affordance, which has to do with the psychological pressure an affordance might exert on an individual to take it up, perhaps by design. While these contributions by Tollon might help us understand more about the psychology involved in whether a given person might or might not respond to a given affordance, I am not convinced that they provide any additional strength to Klenk's basic argument for value-ladenness.

Another defense of value-neutrality has been offered by Peterson and Spahn (2011), whose taxonomy of neutrality positions was described in the previous section. As the title of their article—"Can Technological Artefacts Be Moral Agents?"—suggests, the primary target of their criticism is the idea that technologies can somehow have moral agency which, as mentioned earlier, they call the *strong view*. They pick the work of Peter-Paul Verbeek (2005, 2006, 2008) as an example of this line of thought, and then offer a rebuttal to it. Verbeek pursues a phenomenological argument—drawing inspiration from Ihde—that the mediating character of technology in our lives blurs the line of moral agency between the human and the artifact, distributing it over both. Peterson and Spahn freely admit that the presence of technologies influences human decision-making, but they reject that they do so in any other than a passive, non-agentic way.

As described in the previous section, Peterson and Spahn identify a *moderate view* of value-ladenness that I find rather vaguely defined, and hard to distinguish from their own preferred position of *weak neutrality*. Recall that weak neutrality met two criteria: (i) technological artifacts are never moral agents, and (ii) technological artifacts are never morally responsible for their effects. With respect to these conditions, it is not clear to me that (i) and (ii) are distinct criteria. On a common understanding of moral responsibility, it is moral agents that can have it. So, accepting (i) would seem to entail accepting (ii). In any case, however, these conditions do not appear to be sufficient for rejecting the *moderate view*, as I will explain below.

Peterson and Spahn highlight the work of Illies and Meijers (2009) as an example of the moderate view and offer their rebuttal. Illies and Meijers draw from the ideas of mediation found in Ihde and Verbeek, but they modify their account to reject any claim of agency on the part of an artifact, while simultaneously trying to retain claims to its moral relevance. Thus, they explicitly accept both conditions (i) and (ii), writing on the one hand that, "there are no compelling arguments to attribute agency to artefacts," and asserting on the other hand that their approach "attributes moral relevance to artefacts without making them morally responsible or morally accountable for their effects" (176). They develop an argument based on what they call *action schemes* to show that artifacts have a type of *second-order responsibility* by virtue of altering the sets of actions a person has available to choose from at a given time. This second-order effect does not rise to the level of moral agency, but it does entail moral relevance according to Illies and Meijers.

Peterson and Spahn seem to acknowledge that the acceptance by Illies and Meijers of conditions (i) and (ii) poses a problem for differentiating the *moderate view* from *weak neutrality*. They surmount this difficulty by adding a new condition for the *weak neutrality thesis*. To wit, "Actions, unlike sets of actions, are morally right or wrong, or good or bad" (2011, 423).

Their reference here to *sets of actions* is specifically in response to Illies and Meijers's concept of *action schemes*. This new condition seems designed to strip the action schemes concept of the moral relevance Illies and Meijers assign to it. Without further unpacking the action schemes concept here, or Peterson and Spahn's criticisms of it, I will just note that even granting for the sake of argument that Peterson and Spahn's criticism of action schemes has merit, this extra condition would seem to be particular to Illies and Meijers's account. Thus, I do not think their arguments succeed in defeating moderate views of value-ladenness in general, or even differentiating them from weak neutrality. Peterson later adapted this line of argument about neutrality as the chapter "Are Technological Artifacts Mere Tools?" in a book (2017). The title suggests a broader scope than just rebutting the notion of technological agency, but I find the arguments to be largely the same.

Like Pitt, I think Peterson and Spahn have left part of the solution space unexplored, and in at least a couple of ways. First, as already mentioned, their arguments against the moderate view seem more to be arguments against Illies and Meijers's views specifically, and thus are not necessarily generalizable against the wide variety of accounts that might fall into the moderate-view category. Second, Peterson and Spahn focus on the rightness and wrongness of actions, specifically, which is not surprising given their ostensible motivation was to rebut claims about the moral agency of technology. Yet many accounts of value-ladenness or value-embodiment do not make specific claims about agency, about the rightness and wrongness of actions, or about moral responsibility. Rather they make claims about the ways in which technological artifacts are valued by people, and the ramifications of such valuing. In their article, Peterson and Spahn (2011) use an example of a terrorist possessing a suitcase that he believes contains a small nuclear bomb, and which also has a button to detonate it. They use this example to illustrate how the presence of technologies can affect a moral evaluation of actions. If the terrorist pushes the button and the bomb explodes, there is one set of consequences. If the terrorist's belief that the suitcase contains a bomb is mistaken, and pushing the button does nothing, then there is a much different set of consequences. Thus, a moral evaluation of the terrorist's actions may vary depending on the technology contained in the suitcase.

This is a reasonable example for what it demonstrates, but I think it also elides consideration of a crucial aspect of the scenario—the very aspect upon which many moderate view accounts of value-ladenness might seek to focus. The example does not consider the question of how we should feel about the fact that there exists an unsecured nuclear bomb in a suitcase with a detonation button. It seems at least plausible to consider this a bad state of affairs that derives (at least in part) from the nature of the technology, and this is so regardless of anything we might know or think about the past, present, or future actions of any specific person. I do not claim that simply highlighting this aspect of the scenario demonstrates the value-ladenness of technology. Rather, I claim that it suggests the debate space for questions about whether technologies embody values, or whether we can sensibly label technologies good or bad in certain contexts, extends well beyond the issue of whether technologies mediate how we morally evaluate the actions of specific individuals. The very existence of a nuclear bomb in a suitcase is morally fraught.

My foregoing criticism of Peterson's and Spahn's work also highlights the conundrum I raised in the previous section. In that section I suggested that debates about autonomy and debates about neutrality (moral relevance) were closely linked. Peterson and Spahn would seem to agree, since they approach neutrality from an autonomy/agency perspective. This supports the notion that the autonomy question is not orthogonal to neutrality, but rather is some facet of it. By counterarguing that they have overlooked the embodied value aspect of neutrality, I have implicitly suggested that while perhaps linked, the two issues are clearly distinct.

A couple of additional recent works are also worth mentioning. Although it does not seem to have gotten much attention, David Morrow (2014) has contributed a very interesting article provocatively titled, "When Technologies Makes Good People Do Bad Things." Its point of departure is the aforementioned work of Illies and Meijers (2009), specifically with respect to the mediating role of technology in human activity. Rather than positing action schemes, however, Morrow offers an argument against technological neutrality that is rooted in behavioral economics. It makes a compelling argument for why some technologies might aptly be considered bad, even though they are quite useful, and even though everyone uses them with good, or at least neutral, intentions. The central observation is that the availability of a technology lowers the cost to people of some options for action relative to others (where cost is interpreted more broadly than just money). Thus, technologies mediate peoples' behavior, not through some mysterious technological agency, but rather by simply provoking a reordering of their preferences for what actions have the most value to them at a given moment in time. In many cases, Morrow argues, this can lead to a tragedy-of-the-commons situation in which many people who are all rationally using a technology for what, when considered in isolation, seem like good or benign reasons, can create a bad global outcome. Morrow suggests that, in principle, this has implications for the responsibilities of designers, and for the social control of technology, but he also cautions against assuming that the fact of non-neutrality alone provides justification for any particular policies.

Related to Morrow's ideas with respect to the relationship between behavioral economics and non-neutrality of technology, but coming at it from a much different angle, Trine Antonsen and Erik Lundestad (2019) reinterpret the work of Albert Borgmann in behavioral economics terms. They argue that while the examples Borgmann uses to reject the idea of the "normative neutrality of our material contexts" (87) are limited in scope, there is nonetheless ample empirical evidence available from behavioral economics to support Borgmann's non-neutrality thesis, evidence that "not only explains and justifies the non-neutrality thesis but also brings forth why the non-neutrality thesis does not entail that choices can never be free or autonomous" (87). To make their argument, Antonsen and Lundestad draw from the famous work on cognitive biases by Daniel Kahneman and Amos Tversky. Specifically, they point to the ubiquitous tendency of people to default to System 1 thinking; that is, to take mental shortcuts that are mediated by cues taken from their surrounding environment, including the technological environment. In this sense, the technological milieu provides a particular framing for the options available to us for action. It is not a deterministic framing. We still have the ability to use System 2 thinking to make considered choices that can counteract the technological framing, and sometimes we do. Yet, the reality is that we don't often do so, mainly out of convenience. As a result, much of our action is biased by our technological environment, sometimes in ways that are not to our ultimate benefit. Consistent with Borgmann's call to reform technology to "make the technological universe hospitable to focal things" (Borgmann 1984, 211), Antonsen and Lundestad argue that, "We should seek to arrange our material context so that we will make [System 1] decisions which are to our own good" (90). Like Morrow's account above, this account of Antonsen and Lundestad provides a non-mysterious, causal mechanism for morally relevant technological mediation of human action, and one in which human agency is fully retained.

In the foregoing discussion I attempted to do two main things. The first was to explore some of the reasons I think that with respect to the question of technological neutrality, the question itself is less well-defined than one may think, notwithstanding the fact that most people familiar with the philosophy of technology literature may have a fairly robust understanding of it. As an engineer who teaches a course on the social and ethical implications of engineering, the question of whether technology is value neutral versus value laden is one that I have long included in my classroom discussions. Like most educators, I have a strong desire to explain things clearly and to answer my students' questions well. This topic has at times challenged my ability to do both. My perceptions of the fuzziness of the meanings of neutrality have emerged out of years of presenting this topic to—and being challenged on it by—my engineering students, who not only are quite bright as a rule, but who also tend to be inquisitorial sticklers for precision!

Thus, I try to read whatever I can find on this topic in hopes of developing better ways to present it, or new examples to use. This led me to my second goal for this chapter, which was to survey some of the more recent contributions to the literature on this topic—primarily from the past decade or so. Unlike some of the more classic works that address neutrality/value-ladenness as but one element within a larger conversation about the pathologies of technological society, much of this more recent literature is more narrowly focused and employs arguments that are more analytical or empirical, for example, from value theory and behavioral economics. I hope that this brief review might also prove a useful resource for others teaching, or otherwise thinking about, issues of technological values.

When, after having discussed it in class, I ask my students to write reflections on the topic of technological values, I get a wide range of responses, from the stereotype-reinforcing (for engineers), "At the end of the day, technology is nothing without the user," to the stereotype-busting, "Every technology is designed with a certain purpose in mind, so a technology cannot be neutral," as well as everything in between. Many of my students offer quite nuanced perspectives that provide me with new insights—if not into thinking about the question of neutrality itself, at least into how nascent engineers think about it. But regardless of where a student lands on the neutrality axis, the most common cross-cutting theme across all my students' responses is illustrated by this response: "Before our discussion on technology, I can honestly say I have never considered the morality of technology." From a pedagogical perspective, then, the primary value in asking the question is not, I think, in getting to a definitive answer, but rather in raising awareness that this question exists at all, and that it matters. Fortunately, I have found that even the most instrumentally minded of my students usually come to agree that it does matter.

REFERENCES

Antonsen, Trine, and Erik Lundestad. 2019. "Borgmann and the Non-Neutrality of Technology." *Techné: Research in Philosophy and Technology* 23, no. 1: 83–103. https://doi.org/10.5840/techne201951497.

Balabanian, Norman. 2006. "On the Presumed Neutrality of Technology." *IEEE Technology and Society Magazine* 25, no. 4: 15–25. https://doi: 10.1109/MTAS.2006.261460.

Borgmann, Albert. 1984. *Technology and the Character of Contemporary Life*. Chicago: University of Chicago Press.

Dewey, John. 1939. *Theory of Valuation*. Chicago: University of Chicago Press.

Ellul, Jacques. 1964. *The Technological Society*. New York: Vintage.

Ems, Lindsay. 2014. "'Amish Workarounds': Toward a Dynamic, Contextualized View of Technology Use." *Journal of Amish and Plain Anabaptist Studies* 2, no. 1: 42–58.

Feenberg, Andrew. 1999. *Questioning Technology*. New York: Routledge.

———. 2006. "What Is Philosophy of Technology?" In *Defining Technological Literacy: Towards an Epistemological Framework*, edited by John R. Dakers, 5–16. New York: Palgrave MacMillan.

Flanagan, Mary, Daniel C. Howe, and Helen Nissenbaum. 2008. "Embodying Values in Technology: Theory and Practice." In *Information Technology and Moral Philosophy*, edited by Jeroen van den Hoven and John Weckert, 322–53. Cambridge: Cambridge University Press. https://doi.org/10.1017/CBO9780511498725.017.

Humes, Edward. 2016. "The Absurd Primacy of the Automobile in American Life." *The Atlantic*, April 12. https://www.theatlantic.com/business/archive/2016/04/absurd-primacy-of-the-car-in-american-life/476346/. Accessed December 24, 2020.

Ihde, Don. 1990. *Technology and the Lifeworld*. Bloomington: Indiana University Press.

Illies, Christian, and Anthonie Meijers. 2009. "Artefacts without Agency." *The Monist* 92, no. 3: 420–40.

Klenk, Michael. 2021. "How Do Technological Artefacts Embody Moral Values?" *Philosophy & Technology* 34: 525–44. https://doi.org/10.1007/s13347-020-00401-y.

Kroes, Peter. 2020. "Moral Values in Technical Artifacts." In *Feedback Loops: Pragmatism about Science and Technology*, edited by Andrew Wells Garnar and Ashley Shew, 127–40. Lanham, MD: Lexington Books.

Leins, Kobi, Christopher Culnane, and Benjamin Rubinstein. 2020. "Tracking, Tracing, Trust: Contemplating Mitigating the Impact of COVID-19 with Technological Interventions." *The Medical Journal of Australia* 213, no 1, 6–8.e1. https://doi.org/10.5694/mja2.50669.

Miller, Boaz. 2021. "Is Technology Value-Neutral?" *Science, Technology, & Human Values* 46, no. 1: 53–80. https://doi.org/10.1177/0162243919900965.

Miniter, Frank. 2020. "Marlin Turns 150 and Reminds Us Why a Love of American History Is a Good Thing." *America's 1st Freedom*, April 17, 2020. https://www.americas1stfreedom.org/articles/2020/4/17/marlin-turns-150-and-reminds-us-why-a-love-of-american-history-is-a-good-thing/. Accessed December 16, 2020.

Mitcham, Carl. 1994. *Thinking through Technology: The Path between Engineering and Philosophy*. Chicago: University of Chicago Press.

———. 2020. *Steps toward a Philosophy of Engineering: Historico-Philosophical and Critical Essays*. Lanham, MD: Rowman & Littlefield.

Morrow, David R. 2014. "When Technologies Makes Good People Do Bad Things: Another Argument against the Value-Neutrality of Technologies." *Science and Engineering Ethics* 20: 329–43. https://doi.org/10.1007/s11948-013-9464-1.

National Academy of Engineering and National Research Council. 2002. *Technically Speaking: Why All Americans Need to Know More about Technology*. Washington, DC: National Academies Press. https://doi.org/10.17226/10250.

Newberry, Byron. 2007. "Are Engineers Instrumentalists?" *Technology in Society* 29, no 1: 107–19. https://doi.org/10.1016/j.techsoc.2006.10.004.

Oberdiek, Hans. 1990. "Technology: Autonomous or Neutral." *International Studies in the Philosophy of Science* 4, no. 1: 67–77. doi: 10.1080/02698599008573346.

Olmstead, Gracy. 2017. "The Gun as Symbol: Who Owns It?" *The American Conservative*, November 13, 2017. https://www.theamericanconservative.com/articles/the-gun-as-symbol-who-owns-it/. Accessed January 1, 2021.

Olson, Jonas. 2015. "Doubts about Intrinsic Value." In *The Oxford Handbook of Value Theory*, edited by Iwao Hirose and Jonas Olson, 44–59. New York: Oxford University Press. doi: 10.1093/oxfordhb/9780199959303.013.0004.

Peterson, Martin. 2017. "Are Technological Artifacts Mere Tools?" In *The Ethics of Technology: A Geometric Analysis of Five Moral Principles*. Oxford: Oxford University Press. https://oxford.universitypressscholarship.com/view/10.1093/acprof:oso/9780190652265.001.0001/acprof-9780190652265-chapter-9.

Peterson, Martin, and Andreas Spahn. 2011. "Can Technological Artefacts Be Moral Agents?" *Science and Engineering Ethics* 17: 411–24.

Pitt, Joe C. 2014. "'Guns Don't Kill, People Kill': Values in and/or around Technologies." In *The Moral Status of Technical Artefacts*, edited by Peter Kroes, and Peter-Paul Verbeek, 89–101. Springer: Dordrecht. https://doi.org/10.1007/978-94-007-7914-3_6.

Rabinowicz, Wlodek, and Toni Rønnow-Rasmussen. 2000. "II—A Distinction in Value: Intrinsic and for Its Own Sake." *Proceedings of the Aristotelian Society* 100, no. 1: 33–51. https://doi.org/10.1111/j.0066-7372.2003.00002.x.

Tiles, Mary, and Hans Oberdiek. 1995. *Living in a Technological Culture: Human Tools and Human Values*. New York: Routledge.

Tollon, Fabio. 2022. "Artifacts and Affordances: From Designed Properties to Possibilities for Action." *AI & Society* 37: 239–48. https://doi.org/10.1007/s00146-021-01155-7.

Tucker, Miles. 2019. "From an Axiological Standpoint." *Ratio* 32: 131–38. https://doi.org/10.1111/rati.12219.

Van de Poel, Ibo, and Peter Kroes. 2014. "Can Technology Embody Values?" In *The Moral Status of Technical Artefacts*, edited by Peter Kroes and Peter-Paul Verbeek, 103–24. Dordrecht: Springer. https://doi.org/10.1007/978-94-007-7914-3_7.

Verbeek, Peter-Paul. 2005. *What Things Do: Philosophical Reflections on Technology, Agency and Design*. University Park: Pennsylvania State University Press.

———. 2006. "Materializing Morality: Design Ethics and Technological Mediation." *Science Technology & Human Values* 31: 361–80.

———. 2008. "Obstetric Ultrasound and the Technological Mediation of Morality: A Postphenomenological Analysis." *Human Studies* 31: 11–26.

Wetmore, Jameson M. 2007. "Amish Technology: Reinforcing Values and Building Community." *IEEE Technology and Society Magazine* 26, no. 2: 10–21. doi: 10.1109/MTAS.2007.371278.

Whelchel, Robert. 1986. "Is Technology Neutral?" *IEEE Technology and Society Magazine* 5, no. 4: 3–8. doi: 10.1109/MTAS.1986.5010049.

Chapter 9

How Engineers Can Care from a Distance

Promoting Moral Sensitivity in Engineering Ethics Education

Janna van Grunsven, Lavinia Marin, Taylor Stone, Sabine Roeser, and Neelke Doorn

Moral (or ethical) sensitivity is widely viewed as a foundational learning goal in engineering ethics education.[1] A recent literature review of US-based engineering ethics interventions, for instance, found that twenty-five out of the twenty-six reviewed articles listed "ethical sensitivity or awareness" as a learning goal (Hess and Fore 2018). One particularly prominent account of the nature of moral sensitivity and what it means for professionals to acquire and exhibit it arises in a series of articles by Kathryn Weaver and Carl Mitcham (Weaver et al. 2008; Weaver and Mitcham 2016), who focus on nurses but maintain that their account equally applies to engineers. They argue that engineers—much like nurses—are engaged in an activity in which moral sensitivity plays a central role, namely an activity of *caring* for others. They readily acknowledge, however, that the target of care in the engineering context is crucially different from the target of care in the nursing context. The nurse attempts to be sensitive to the particular needs of individual patients. We call this *particularized care*. The engineer, by contrast is tasked with designing and maintaining structures and systems that, ideally, help take care of society at large. Though Weaver and Mitcham acknowledge this contrast, it is drawn too quickly. We wager that fostering *care* within engineers-in-training requires that close attention be given to the phenomenological difference between "the to-be-cared-for Other" in the world of the engineer [the *engineer's Other*] and the "to-be-cared-for Other" in the world of the nurse [the *nurse's Other*]. The phenomenological approach interrogates the manner in which the Other is experientially manifest in one's activities. As the phenomenological tradition has emphasized, our experience of other people is shaped by shared practical contexts, or worlds, which are textured by background practices, norms, and ideologies (see Gallagher 2007). By explicating the phenomenological specificity of the to-be-cared-for-Other in the world of the engineer, we can identify more precisely what kind of care we can expect from and help cultivate within engineers (in training).

We focus on two dimensions of the engineer's world that can have a formative effect on how the engineer's Other is constituted. First, at a practical level, the relationship between the engineer and her Other is characterized by *distance*, which is unavoidably built into the engineer's practices and activities. This is contrasted with the nurse's Other, who is characterized by *proximity*. This contrast brings out a problem or challenge: what exactly does it mean to care for an Other who is marked by distance? Second, at an ideological level, the world of the engineer is prone to a particular discourse—a particular way of understanding what engineering is, what it produces, and what it means to be a good engineer. This discourse is marked by an *ideology of neutrality*, an umbrella term we use to refer to several interconnected background commitments. When internalized, these background commitments exacerbate the distance between the engineer and her Other, complicating the engineering student's understanding of herself qua care-taker.

But engineering and design students can also be exposed to an alternative image of the engineer: as someone tasked with caring for society at large through thoroughly normative, value-laden activities. Yet while notions of care and care ethics have slowly permeated through scholarship (e.g., Adam and Groves 2011; Vallor 2016; van Wynsberghe 2016), the explicit delineation of care-centered approaches in engineering ethics education is still relatively novel (e.g., Russell and Vinsel 2019).

In this chapter, we aim to further develop such approaches by proposing that the pedagogical endeavor of cultivating moral sensitivity—via the notion of care-taking—should include two aims. It should dispel the ideology of neutrality while offering a *positive* image of what it means for an engineer to care for her Other. This requires that we take up the question of "What does it means to care for an Other who is marked by distance?" Relatedly, we must ask what it means to promote such care through our educational endeavors. While the type of care characteristic of the nurse is particularized care, the two types of care relevant for the engineering context are what we call *generalized care* and *universalized care*. Generalized care refers to the practice of attending to the needs, concerns, emotions, and desires of those who are affected by the products and activities of engineering by taking one or some individuals as representatives of a larger cohort of stakeholders. Universalized care is exhibited when engineering activities of design and maintenance reflect a responsiveness to a universally shared feature of the engineer's Other, namely her vulnerability as a technology-dependent being. While generalized care has received a fair amount of attention in engineering ethics education, both at our own institution and beyond, universalized care seems to have remained largely under the radar. Though our main contribution aims to be conceptual, we illustrate these different notions of care, and how they might be operationalized in engineering ethics education, via a recently developed pedagogical exercise, which we term a tinkering workshop.

MORAL SENSITIVITY: ITERATIVE MOVEMENT BETWEEN PERCEPTION, AFFECTIVITY, AND DIVIDING LOYALTIES

The notions of moral sensitivity and care are intimately related. Psychologist James R. Rest has developed an influential account of the idea that moral sensitivity involves an awareness of how others are affected by one's own actions which he argues makes moral sensitivity a foundational ethical competency for professionals like nurses and engineers. He presents moral sensitivity as the ability to "interpret" or "perceive" a given situation in terms of how one's own actions may or may not affect "the welfare of someone else either directly or indirectly" (1982, 29). Rest builds on research from the field of psychology to argue that, although basic, moral sensitivity is hardly automatic. Identifying salient ethical features of a situation and seeing it in terms of how one's own actions may affect both proximal and distant others are complicated by the fact that "many people have difficulty in interpreting even relatively simple situations" and that "individuals exhibit striking differences in their sensitivity to the needs and welfare of others" (29). We will expand on this idea, proposing in the next two sections that how one interprets "even relatively simple situations" and exhibits "sensitivity to the needs and welfare of others" is, in professional contexts, partially determined by the practical and ideological ways in which one's professional world and one's role as a professional in that world are shaped.

While engaging with Rest's influential proposal, Mitcham et al. propose their own model of moral sensitivity. This model, which they develop via an engagement with the professional nursing context, consists of three interconnected moments: *moral perception, affectivity,* and *dividing loyalties* (Weaver et al. 2008, 8; see also Weaver and Mitcham 2016).

1. *Moral perception* is defined as an "intuitive discrimination of cues and patterns," that awakens a professional to "client and situational needs." Phenomenologically speaking, they wager that moral perception is experienced as "a gut level 'jolt'" in response to "some cue" that something in one's routine activities—often performed unreflectively and habitually—warrants immediate attention (Weaver et al. 2008, 609).
2. *Affectivity* bears a strong conceptual similarity to empathy. It offers a "vivid rendering of what it means to be human . . . based on the professional putting of oneself in the place of clients," which "increases responsiveness and preserves client dignity and caring" (609).
3. The act of *dividing loyalties* captures the importance of adopting "strategies of interpretation, justification, and reflection" (609) in order to arrive at different perspectives on a relevant issue—which includes perspectives held by relevant stakeholders as well as those articulated in textual "sources of knowledge (e.g., expert opinions, policies and professional conduct codes)." The act of dividing loyalties has the potential to "expose assumptions and privileges" through "critical scrutiny of the larger social system," and to detach "from privileged relationships and the immediacy of the situation long enough to distinguish personal biases and assumptions" (609–10). As we will later suggest, a key "bias and assumption" that must be targeted pedagogically in the fostering of moral sensitivity in engineers-in-training is the ideology of neutrality.

Weaver et al. (2008) believe each of these three components is a necessary condition for genuine moral sensitivity. A nurse must possess an openness to letting unreflective routine actions be interrupted by "a gut-level jolt," the sense that something or someone in her perceptual environment requires an immediate response. But, by itself, moral perception falls short of moral sensitivity. After all, Weaver et al. note, "spontaneous recognition of the moral issue can be inadequate or misleading" (610). Similarly, affectivity, though crucial for a nurse's empathic grasp of her patient's humanity, can be "subject to personal motives and misunderstanding." Think, for instance, of the evidence that implicit biases in health care providers can suppress moral perception of and affectivity toward Black women during pregnancy, labor, and the postpartum period (Roeder 2019). A wider reflective perspective, through which a nurse can "solicit breadth and depth" about these issues, for instance by talking to advocacy groups or learning about the nature of implicit biases and their pernicious consequences, can circle back into moral perception and affectivity, thus widening the scope of her moral sensitivity. At the same time, Weaver and Mitcham warn that the sources professionals turn to in the process of dividing loyalties can also be capable of "uphold[ing] the hierarchy of more powerful stakeholders" or of "address[ing] only issues prior to code or policy development" (2016, 610). Hence, engaging in dividing loyalties without perceptual and affective attunement to situational demands will fall short of establishing robust moral sensitivity. As Weaver and Mitcham conclude: "When combined, the individual limitations of the attributes are overcome. In moving back and forth between moral perception, affectivity and dividing loyalties, the professional modulates a situation through interpretive understanding and evaluation" (610).

THE WORLD OF NURSING AND THE WORLD OF ENGINEERING

The iterative nature of this view of moral sensitivity serves as the foundation for the argument we make, namely that the specific world in which a professional is embedded—in our case the world of engineering—shapes how that professional can morally perceive and affectively respond to her to-be-cared-for Other. In homing in on the world of engineering and how moral sensitivity can be enacted there, we are critically examining Weaver and Mitcham's claim that their account of moral sensitivity, though developed in the nursing context, can be extended to the engineering context. As mentioned, Weaver and Mitcham (2016) and Weaver et al. (2008) invoke a care analogy between the professional life of the nurse and that of the engineer. In both instances, they suggest, proper care is dependent on the three moments of moral sensitivity they have identified:

> Through moral perception, the professional distinguishes and appreciates the client's unique situation amid its complex context. Affected by the encounter, the professional is motivated to anticipate and alleviate the suffering (in nursing) or protect safety and welfare (engineering). To inform a reasoned and appropriate course of action, the professional explores and interprets the often divided perspectives and competing demands of involved stakeholders which can include clients, social institutions, and the public. (Weaver and Mitcham 2016, 6–7)

Note how Mitcham and Weaver focus on the engineer's relation to the *client* in establishing an analogy between moral perception in nursing and engineering contexts. But this move is problematic. For as they readily acknowledge, "Engineers need to practice ethical sensitivity . . . not just with regard to their immediate clients or employers but also with respect to all those who may be affected by their work" (14). This, then, raises the question: *What does it mean to care not only for the welfare of a particular individual client with whom one stands in an in-person dyadic relationship, but also, perhaps first and foremost, to care for the welfare of "all those who may be affected" by one's activities?* When we talk about the engineer's Other, it is precisely the other in this sense ("all those who may be affected by their work") that interests us. From here on it is this Other that we refer to when we use the term *the engineer's Other*. We now argue that there are important phenomenological differences between "the to-be-cared-for Other" in the world of the engineer (i.e., the *engineer's Other*) and the "to-be-cared-for Other" in the world of the nurse (i.e., the *nurse's Other*). We believe one must pay close attention to these differences in order to identify fruitful ways of promoting moral sensitivity in engineers-in-training.

The Other in the Practical World of Nursing

As we saw, Rest maintained that moral sensitivity is hard to achieve, with "many people" finding it difficult to "interpre[t] even relatively simple situations" (1982, 29). Of course, one could ask what counts as a relatively simple situation, particularly in complex professional settings? As Weaver and Mitcham note, "The comprehensive recognition of ethical issues is difficult because such issues are almost always embedded in webs of social custom, personal and relational histories, and competing needs and interests of professional practice" (2016, 12). Perhaps it is better to say, then, that the difficulty of moral sensitivity lies not merely in the difficulty of "interpreting even relatively simple situations," but also in the complexity of situations themselves—where situations that have the appearance of simplicity are in fact multilayered and framed by a variety of relationships, norms, beliefs, institutional practices, demands, and ideologies. Ideological assumptions are the topic of the next section. In this section we look at the worlds of nursing and engineering from a practical point of view, as enacted through the practical day-to-day goals and activities of its professionals.

Mari Skancke Bjerknes and Ida Torunn Bjørk's ethnographic sketch, derived from a study of newly qualified nurses, provides an entry into the practical world of nursing and how the nurse's Other is manifested as a target of care in this world. The study contains the following vignette of Tina, a nurse who just started working at a Norwegian university hospital:

> In one room, a young woman lay with her eyes closed. . . . Tina quietly approached the bedside and bent down, whispering: "Are you awake? Can I get you something to drink?" The patient only grunted in reply and did not open her eyes. Her face was pale and she had a kidney bowl on her chest holding some absorbent tissue. While looking at the intravenous catheter and checking the intravenous injection, Tina kept an eye on the patient's face. She stood there

for some minutes. After a period of silence, the patient replied, "My mouth is dry, so perhaps just a swallow." A glass of fresh water was on the bedside table, and Tina carefully supported the patient and helped her take a small mouthful. The patient swallowed some water and spat out the rest. "Just a small swallow, yes, that's good," Tina said. After helping the patient into a comfortable position, Tina said that she would be back again soon. On the way out, I noticed a soft light in the single room, a glow from the lamp beside the bed, positioned in such a way that it would not bother the patient. I asked Tina about this patient afterwards, and she told me that the patient had a serious disease, hyperemesis gravidarum, which meant that she was constantly sick with nausea and vomiting. The patient would have to stay in bed for months or possibly for her entire pregnancy, and she was not supposed to have anything to drink or eat. "I really feel sorry for this patient, so young and being so sick day after day," she said. Later in the afternoon, and in between phone calls and her responsibilities for other patients, Tina slipped in to see the young pregnant woman. "I must not disturb her," she whispered to me. "I have to follow up and see if she is all right, and whether she has vomited." She helped the patient and offered her a special moisture stick in a glass of iced water to cleanse her mouth, which would give her a sense of tasting water. Tina's caring and sensitive attitude when taking care of the patient was in marked contrast with her much more determined and brisk manner of walking as soon as she was back in the corridor. (Bjerknes and Bjørk 2012, 3)

Tina's actions reflect both the capacity of putting her acquired professional medical and technical knowledge into practice and of providing fine-grained, context-sensitive care for her patient's particular needs, reflected in embodied and linguistic registers. In adjusting her movements and the volume of her voice to accommodate her patient's condition, and in offering words of encouragement ("just a small swallow, yes, that's good"), Tina exhibits a "vivid rendering" of her patient's particular needs and experiences. Some would argue that providing such care, which we have termed *particularized care*, is interwoven with the telos of nursing, with nurses-in-training typically referring to the care-providing component of their profession as the main motivation for pursuing this line of work (Bjerknes and Bjørk 2012; Halperin and Mashiach-Eizenberg 2014). Of course, as already noted, providing such patient-centered care is not the only responsibility of a nurse. Nurses operate in complex socio-technical systems shaped by different medical, technical, organizational, bureaucratic, social, and temporal demands. As such, we can understand nursing as involving a constant balancing-act of internalizing, implementing, and abiding by third-person professional norms and knowledge and exhibiting a continual responsiveness to the first-person experiential point of view of the patient, of attending to her expressed needs and emotions, weighing what these might mean and the best response given the larger context of the patient's medical predicament. Despite the asymmetrical epistemic relationship between the nurse and her Other, with the nurse possessing a body of professional knowledge about the patient's medical situation that the patient does not have, the nurse must simultaneously possess a readiness to take the patient's verbal and bodily expressions as capable of being authoritative and of determining the appropriate course of action in providing particularized care. This balancing act typically involves being able to "provide the technical aspects of practice *in an* [often] *unreflective manner*" while also allowing for this to be "interrupted by some cue." The "awakening" to this cue, which Weaver, Morse, and Mitcham characterized as moral perception, is said to be experienced by nurses as "a gut-level jolt . . . or worry." We propose that if *anyone* can bring about such a gut-level jolt, disrupting absorption in habitually executed professional activities, it is a concrete particular human person, visible, present, and capable of changing the course of any routine interaction.

The Other in the Practical World of Engineering

In the practical world of engineers, a to-be-cared-for Other rarely plays this "jolt-producing" role because in-person interactions between engineers and their Other are far more restricted, and the kind of care engineers provide is almost never particularized. The activities of engineers in general are not aimed at the needs of particular others and a "vivid rendering" of their humanity, but at society at large or large cohorts within it.[2] One cannot build or maintain a bridge by "awakening [to] and particularizing" the "situational needs" of everyone who will be crossing it or who will be otherwise affected by its presence. It seems, then, that Weaver, Morse, and Mitcham's characterizations of moral perception and affectivity miss the mark here. *We posit that seeing the Other as a particular person and meeting that person's particular needs is antithetical to the activity of engineering.* Maintaining and designing products made for mass-scale use or mass-scale reproduction requires, by definition, that one treats the needs of the Other in a more homogenous uniform way. As such, we propose that where the nurse's relation to her Other is characterized by *proximity*, the engineer's relation to her Other is marked by *distance* (see figure 9.1).

The distance between the engineer and her Other can be brought out more concretely by taking a look at an ethnographic sketch of the engineer, involved in her day-to-day activities. In the following scene, described by Louis L. Bucciarelli, we ask you to trace to what extent the engineer's Other becomes manifest, is attended to and cared for, throughout the engineer's day-to-day practical activities:

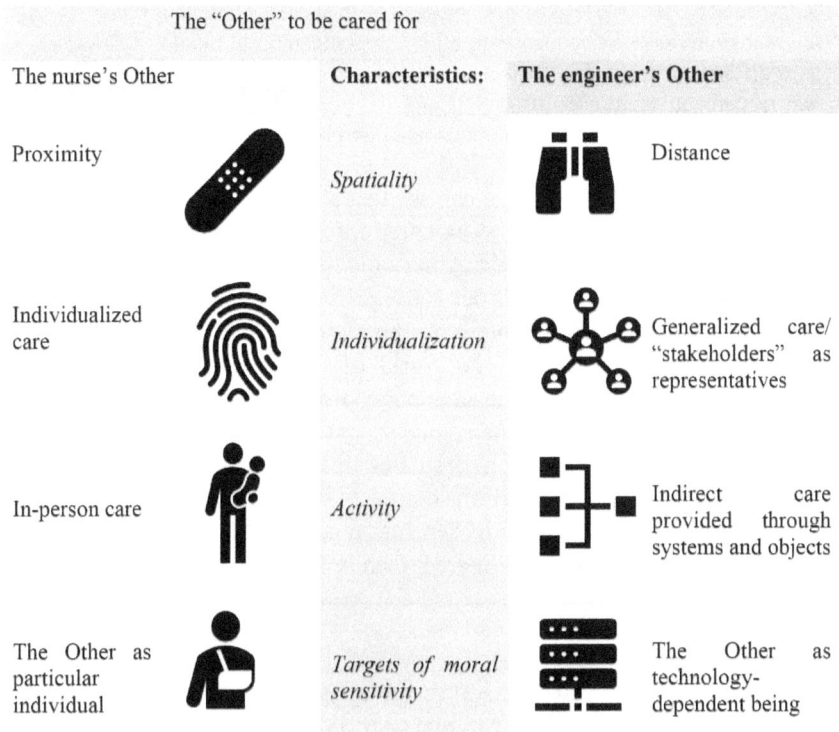

Figure 9.1. The Nurse's Other and the Engineer's Other

I observe members of the firm engaged in a variety of activities. Some of these are solitary: sketching at the board, running a computer analysis at the terminal, putting a prototype sub-system through its paces down in the lab, phoning a subcontractor back in the office, checking out codes in the library, conceptualizing, dreaming, detailing, cost-estimating, cursing, etc. Others are collaborative: sitting in on a design review in the board room, leading a meeting in the project manager's office, consulting with purchasing or production, advising the new hire, celebrating the shipment of the new product, negotiating time on the next design task, laughing, bantering, bickering, cursing, etc. (1988, 162)

The engineer's Other is nowhere to be found here. Much, though not all, of what the engineer attends to is what Bucciarelli terms her "object-world":

The mechanical engineer, designing a structure to hold the plates used to collimate an X-ray beam, moves within an object world of beams, of steel, of geometric constraints, of stress levels, of close tolerances, of bearing surfaces, of positioning errors, of fasteners, and of metal machining practice. The electrical engineer designing a photo voltaic module works in terms of voltage potentials, and of current flows. He sketches networks with special symbols for diodes and current sources, resistive elements, all within a meaningful topology. These are two different worlds. The project manager casts out schedules and milestones, worries about manpower allocation and development cost trends, speaks of interface constraints and critical paths: another object world. (162)

From these brief sketches, one can see that an engineer's proximal other is not in any immediate sense the to-be-cared for Other, but her colleagues with whom she collaborates, her direct clients, and the technological artifacts to be maintained or designed through her activities.

But the emphatic distance between the engineer and her to-be-cared-for Other is not just a result of the world of engineering understood in a day-to-day practical sense. The challenge of promoting a self-conception of the engineer as care-taker in engineering students is exacerbated by an ideology of neutrality that often circulates in the world of engineering and engineering education. Although the degree to which the ideology of neutrality is internalized by students will depend on cultural and institutional variations, we maintain that traces of it are typically found at technical universities around the globe.

IDEOLOGICAL MECHANISMS SHAPING THE WORLD OF ENGINEERING

As we sketched in the previous section, being a good nurse involves a balancing act between internalizing and applying third-person expert knowledge (knowledge of clinical technology, diseases and bodily processes, institutional rules and protocols) and taking seriously a patient's first-person reports and expressions of their needs, feelings, and sensations. Of course, individual nurses may fail in performing this difficult balancing act. For instance, moral perception and affectivity may be hindered by an overreliance on a third-person medical-technical apprehension of a situation and an undervaluing of a patient's first-person reports.[3] We already mentioned, for instance, that the first-person reports of pregnant, birthing, or postpartum Black women are disregarded at a disproportionate and deadly rate (Roeder 2019). Crucially, though, we categorize these occurrences *as* failures. We think something has gone wrong in how the nurse executes precisely their role *as* a nurse. A good nurse, a nurse who performs their role well, is typically understood as someone who exhibits not only the practical medical-technical know-how characteristic of their field, but also as someone who is responsive to the humanity in their patients, someone who puts themself "in the place of clients" with the aim of preserving "client dignity and caring" (Weaver et al.

2008, 609). Attunement to a patient's expressed needs, feelings and emotions is as much a part of being a good nurse as exhibiting the required medical and technical knowledge.

Engineers and engineers-in-training are judged in strikingly different terms. Despite the fact that nearly every professional code of conduct for engineers characterizes engineering as an activity that "must be dedicated to the protection of the public health, safety, and welfare" and that "has a direct and vital impact on the quality of life for all people," the idea that engineers are *at their very core* in the business of care-taking is largely absent from the engineer-in-training's self-conception.[4] Anecdotally, when we ask TU Delft students what it means to be a good engineer, the answers we typically receive contain references to technical, scientific, and managerial skills. Being a good engineer in the care-taking ethical sense of the word tends to be seen as a separate category from being a good engineer in the functional sense of the word: it is of course viewed as commendable if an engineer falls into both categories at once, but concerning oneself with the ethical care-providing dimensions and consequences of one's line of work is generally not seen as *constitutive* of what it means to be a good (well-performing) engineer.

These anecdotal observations about how engineering students understand "the good engineer" are confirmed by engineering ethics educators at institutions across the globe. In a focus group in 2019, we discussed the challenges of teaching ethics to engineering with thirteen engineering ethics lecturers and professors from a variety of institutions in Europe, Australia, and the United States.[5] One of the main findings was that "simply" getting students to see ethical issues as deeply intertwined with their practice (i.e., getting them to see how their technical activities might impact on the well-being of distant others) was widely seen as a primary marker of success in teaching ethics to engineering students. Even at institutions that explicitly foreground ethics in their engineering curricula, the idea that engineers are *at heart* involved in an irreducibly ethical endeavor of caring for direct and indirect stakeholders does not seem to feature prominently among students.

We believe this is true in part because engineering students either explicitly or implicitly accept an ideology of neutrality, or one or more of the background assumptions of which this ideology consists (see also Cech 2013).[6] We take this ideology to consist of the following views:

1. The *neutrality thesis* of technology
2. The *applied-science-view* of engineering
3. The *rationalistic view* of engineering-relevant knowledge
4. The *technocratic view* of the engineer as social actor

According to the neutrality thesis of technological artifacts, a technology's ability to promote or undermine human welfare is determined by the intentions of its users. Indeed, facial recognition technology, nuclear energy, and social media platforms are all products of engineering that can be used in ways that are good or bad for human beings and the planet at large. But the neutrality thesis states that there is nothing about these technologies *themselves*—nothing in their "nuts and bolts" to use Langdon Winner's phrase (1980, 28)—that undermines or promotes well-being.[7] As such, the designing of technological artifacts too is deemed to be neutral with respect to ethics.

The idea that technological objects are value-neutral is related to a way of understanding the domain or activity of engineering itself, namely as the practical application of rigorous scientific theorizing. As Vermaas et al. explain, this "applied-science-view cannot be separated from the way in which engineers perceived themselves. From the dawn of the Industrial Revolution until far into the last century, traditional engineering disciplines such as civil and mechanical engineering shifted increasingly from continuations of practical and traditional craftsmanship to the scientific end of the spectrum" (2011, 55–56). As they further point out, this shift directly impacted on the education of engineers: "engineering curricula were organized in such a way that students were taught, above all else, just how to apply the theories gained from applied scientific research" (56).

If students are taught that engineering is an activity that takes "pure" descriptive scientific knowledge and churns that knowledge into value-neutral artifacts, this will inform students' conception of what it is to perform that activity well, what it means to become a good engineer. Good engineers are seen as

> no-nonsense problem solvers, guided by scientific rationality and an eye for invention. Efficiency and practicality are the buzzwords. Emotional bias and ungrounded action are anathemas. Give them a problem to solve, specify the boundary conditions, and let them go at it free of external influence (and responsibility). If problems should arise beyond the work bench or factory floor, these are better left to management or (heaven forbid) to politicians. (Herkert 2001, 410)

Exhibiting care for and moral sensitivity toward the well-being of others is quite clearly not an integral part of the way in which the world of the engineer and the engineer's role in the world at large are framed here. Although the applied-science-view of engineering has come under pressure in recent decades (see Vincenti 1990), and although countless engineering curricula have adopted a broader perspective on what it means to be a good engineer, the effects of this view still continue to reverberate through the world into which engineers-in-training are enculturated.

Sabine Roeser (2012) has challenged the view of engineers "as the archetype of people who make decisions in a rational and quantitative way." Instead, she argues that what we need is "a new understanding of the competencies of engineers: they should not be unemotional calculators; quite the opposite, they should work to cultivate their moral emotions and sensitivity" (103). Attempts to cultivate this in engineers-in-training may be met with some reluctance. As an engineering ethics professor from TU Eindhoven pointed out in our focus group: "the students have this rationalistic view of 'these are the facts' and the things besides the facts are emotions and are irrelevant. We bring something else and there is resistance" (Marin, van Grunsven, and Stone 2022).

One way to tackle this is by confronting students with the questionable theoretical assumption that rationality and emotions are antithetical notions. This view rests on an oversimplified view of emotions that fails to take into account how emotions often help bring out morally salient features of a situation. To *feel* indignation when discovering that your children and you have been knowingly, avoidably, or even purposely exposed to dangerous toxic waste or psychologically damaging technologies is to exhibit a *rational* response to the morally salient features of those situations; it is a way of getting things right, in a moral sense, about what should or should not have occurred here. By contrast, someone who doesn't feel indignation is having a different cognitive understanding of the situation, an understanding that misses part of what is morally relevant: "By caring about certain things we are able to perceive evaluative aspects of the world that we would otherwise not be able to be aware of" (Roeser 2012, 106).

Our Ethics and Philosophy of Technology Section at TU Delft invites its engineering students to explicitly reflect on the value-neutrality thesis and the applied-science-view of technology (see van Grunsven et al. 2021). To give a simple example, by explicating the difference between descriptive versus normative statements and by asking students to categorize common engineering-statements, such as "this high-frequency switch is *safe*," or "this is a *good* high-frequency switch," students begin to attend to the ineluctably normative-evaluative judgments that they make all the time, in a manner that problematizes the "applied science view" of engineering.[8] And by asking students to use the value-neutrality thesis in order to make sense of commonplace innovations and artifacts such as the speed bump or hostile design (e.g., park benches designed so as to prevent homeless people from sleeping on them) (see Rosenberger 2020; Verbeek 2011), students become aware of the limits of the value-neutrality thesis and the ways in which technological innovations profoundly shape the possibilities for action, the values, and the social infrastructure of society.

One reason why it is particularly important to challenge engineering students' often unexamined dichotomous view of rationality versus emotions is that this view, when left in place, can exacerbate the already distant relation between the engineer and her Other. In the previous section, we argued that the moments when individual nurses prioritize third-personal knowledge wholly at the expense of a perceptual affective attunement to the needs and the emotions of her patients are moments of failure and believe that the nurse has failed in her role qua nurse. In the context of engineering, the ideological tenets we have identified can enact a world in which the engineer's Other's feelings and emotions tend to not just be improperly cared for but purposefully sidelined, as the emotionally charged responses of laypersons to technological innovations and interventions are discredited for reflecting a "non-rational" ill-informed stance. As mentioned, we can invite students to question the legitimacy of this *technocratic* outlook by challenging the theoretical separation between rationality and emotion. Role-play activities, for example, in the form of stakeholder meetings, offer an embodied experiential way of trying to cross the distance that is created when engineers-in-training assume that her Other's emotional responses to a technology can be discredited in virtue of their alleged irrationality. Weaver and Mitcham concur, proposing that "Role taking opportunities . . . help students identify blind spots that limit them from seeing others' perspectives" (2016, 7–8).

Indeed, without recognizing those other perspectives, engineers often incorporate their own assumptions and biases about what their users need into designs. To name just one of many examples: engineers and designers have been known to build augmentative and alternative communication (AAC) technologies for non-speaking children without actually consulting children in order to establish what symbols would optimally promote expressivity, thus unintentionally but effectively silencing a whole cohort of people whom they were precisely aiming to equip with effective expressive resources (Van Grunsven and Roeser 2022). This example illustrates that the link between being a good (well-functioning) engineer and being an engineer who cares for her stakeholder's needs, desires, and experiences is an intimate one, such that a failure to care in this sense is just a failure to do one's work well.

In a recently developed engineering ethics exercise, we encouraged students to experience this link by asking them to redesign an existing technological health care artifact (Van Grunsven et al. 2022). These redesigns were meant to promote ethical values, particularly inclusivity, by considering first-personal user-testimonials. In groups of four, students used scrap material to alter their selected artifact, explicating throughout the process how those material alterations bore on the needs and desires of their stakeholders. This "tinkering workshop" proved to be an effective hands-on way to challenge the ideology of neutrality and to bring out that a dimension of care for human persons was embedded within the activity of material tinkering. It facilitated reflection on the relationship between the specific needs and emotions of stakeholders and the designing of good technology in the functional sense. As one student, who altered a tricycle to improve mobility, put it:

> So we look at the tricycle. . . . We actually could move around with it and discuss with each other, like, what certain kind of situation would we be in or would this person be in. You can have more empathy. . . . We had to imagine that there was this person that had to use the tricycle to get up on a hill or something like that. That was the whole point. And we all were imagining how that would turn out. . . . Eventually just, we were all thinking . . . what if it was my grandma, how would she react? (Franssen 2022, interviewee 1, 6–10)

This passage highlights that the distance between the engineer and her Other can be partially overcome via exercises that connect concrete design choices with the expressed needs and desires of relevant cohorts of stakeholders. However, the passage also exposes the challenges of care in this context. The students eventually tended toward particularized care, imagining the specific needs, desires, and challenges of their grandmothers. In reality, though, the engineer offers what can be labeled as *generalized care*—care that aims to close the distance between the engineer and her Other by taking seriously the needs, emotions, concerns, and desires of a group of stakeholders (e.g., aging adults with certain mobile disabilities, or non-speaking children who depend on AAC for their daily communication needs). Of course, by taking one or some individuals as representatives of a larger cohort, the engineer's efforts of providing generalized care, though undeniably essential, will at the same time fail to live up to the standards of care as identified by Weaver, Mitcham, and Morse. Generalized care by definition cannot exhibit the kind of particularized, situation-specific moral perception and affectivity toward persons that they have built into their account. The other in her particularity remains, to a degree, at a distance in engineering contexts.

If our phenomenological analysis is right in suggesting that distance is to some degree inevitably built into the engineer's relation to her Other, it questions the move to transpose a particularized notion of care (characteristic of the nursing context) into the engineering classroom. When contrasted with particularized care, the generalized care that we want to encourage in our engineering students may feel like a fundamentally flawed type of care (which in turn could de-motivate students from prioritizing such care). But perhaps there is an additional approach to introducing care into the engineering education context. In addition to generalized care, we propose the usefulness of what we call *universalized care*. This notion of care offers an alternative positive view of what it means to develop a sensitivity to an Other marked by an irreducible distance.

ENGINEERS AS MAINTAINERS AND THE NOTION OF UNIVERSALIZED CARE

In the previous section we cited Vermaas et al. (2011), who describe the historical shift away from engineering understood as a "practical and traditional craftsmanship to the scientific end of the spectrum" (56). This shift in the understanding of engineering points to the availability of an alternative discourse, a different way for thinking about what engineering is and what engineers (should) do. As Bucciarelli brings out, such an alternative is necessary if we want to make sense of a wide range of activities and judgments that do not fit in comfortably with the ideologized picture of the engineer as the applied scientist moving about in a pristine object-world:

> What engineers do, and are expected to do, includes much more than rational problem solving and constructing efficient means to reach desired, externally specified ends. In engineering practice, value judgements are made all the time, often not explicitly—about the user, about robustness, about quality, about responsibilities, safety, societal benefit, risks and cost. However, it is object-world work that is [often] seen as primary by engineering faculty—and consequently seen as such by our students. (Bucciarelli 2008, 143)

Here, then, is a sketch of the state of affairs as it pertains to engineering students being socialized into the world of engineering. There are at least two distinctly different discourses, two ways of framing what the world of engineering is like, what becoming a good engineer involves, what engineers-in-training should practice and focus their attention on, what they should become sensitive to. One is the picture of the engineer as moving about in an ethically neutral world applying emotion-free scientific knowledge to produce value-neutral technical artifacts. The other is the picture of the engineer as essentially in the business of care, of attending (even if implicitly) to issues concerning a technology's degrees of safety, risk, and ability to promote well-being in its users.[9]

As Andrew Russell and Lee Vinsel have recently suggested, we can get a firm handle on the idea of engineering as a form of care via a realistic look at some of the "basic facts of ordinary life with technology" and of the role that engineers play in maintaining this life (2019, 256). *Maintaining* is the crucial notion here. For as Russell and Vinsel emphasize: "Much of modern life depends on well-functioning technological systems, and the vast majority of human work will always be aimed at maintaining them—that is, the labor is oriented towards taking care of the world and its inhabitants" (261). Indeed, over 70 percent of trained engineers dedicate their professional lives not to innovating and designing but to maintaining, to taking care of the technical systems that quietly support our daily functioning.[10] However, Russell and Vinsel note that in engineering education and "in most technology studies, maintenance, repair, and upkeep are largely ignored, rendered invisible" (256; see also Young 2021).

Russell and Vinsel draw a comparison between the quiet supporting role that women have traditionally played in maintaining the daily functioning of the family and engineers as quiet care-takers of the technical systems upon which we all depend. With this analogy, they present care ethics as the central normative ethical theory to be incorporated into engineering education. For those familiar with care ethics, this suggestion may come as a surprise. After all, the emphasis in care ethics is typically on particularized care and "the compelling moral salience of attending to and meeting the needs of the particular others for whom we take responsibility" (Held 2006, 10). Not surprisingly, then, care ethics has played an important role in the education of nurses. It is perhaps fairly straightforward to imagine a positive role for care ethics in the evaluation of a particular subset of technological artifacts, such as care and service robots, which fulfill a central role in a dyadic relationship (see Van Wynsberghe 2016). But what Russell and Vinsel are proposing is that care ethics should be considered central for the education of engineers *tout court*. But, how can care ethics—as a theory that is in the first instance focused on dyadic, close personal relationships with and responsibilities to particular others—play such a central role?

We suggest that Russell and Vinsel's proposal, with its emphasis on maintenance, repair, upkeep, rehabilitation, and care for technical structures, opens up a positive approach to what it means to care *precisely* for a distant Other. Specifically, we propose that one important form of care in the world of engineering is what we call *universalized care*. Universalized care is care for the engineer's Other in her universally shared standing as a vulnerable technology-dependent being, as someone whose daily functioning and well-being depends on well-functioning technological systems. Ensuring the safety of bridges and trains that support our daily travel, ensuring the availability and reliability of the digital infrastructures that increasingly enable many of our daily human activities and interactions, ensuring the integrity of pacemakers, wheel chairs, contraceptives and other health care technologies supporting our bodily health and autonomy—these activities can all be legitimately framed as ways of caring for people qua technology-dependent beings. Such care is perhaps most emphatically performed by maintaining, preserving, and rehabilitating the technological structures that work well in quietly supporting our daily functioning. But it can also be exhibited, or precisely undermined, through the processes and products of innovation. Framing engineering as an activity of universalized care can underscore the profound moral failure of innovators who purposely exploit the ways in which human beings can become dependent on technologies. For instance, think of how engineers working for Facebook, Instagram, and Apple intentionally operationalized such dependency in order to advance corporate financial interests, by building addictive properties into their innovations (Bhargava and Velasquez 2021; Wu 2017). By linking an in-class evaluation of these software engineers to a wider conception of engineering as care-taking, we can invite students to consider that, despite their technical skill in meeting design objectives, these engineers nevertheless fall short of being good engineers. Through the lens of the value-neutrality thesis, these engineers will likely be seen as good—that is, well-functioning—engineers who just happen to have morally bad intentions. We propose a different take on the situation. Whereas the nurse fails in her role as a nurse when she lacks responsiveness to her Other's particular needs and well-being, we suggest that the engineers in this example fail in their role as engineers when they lack responsiveness to their Other by purposely exploiting the universal human feature of technology-dependence to the detriment of their well-being.

We wager that the aforementioned tinkering workshop, though initially targeting generalized care, can also be utilized as a bridge toward reflecting on universalized care. Consider the following observation from one of the students who participated in the workshop:

> We don't want to just see how a tool can help people, but we want to also see how tools can be *embedded inside the life* of people. . . . I didn't actually think about that before the project. I was just thinking that tools like this just help us, but its more than that. *It's there to be a part of our lives*. If we have a certain disability, *you cannot do anything without that* [artifact]. (Franssen 2022, interviewee 2, 2.62, italics added)

It is only a small step from here to encourage students to reflect on their own forms of technology dependency. After all, while technology dependency is perhaps more obvious in contexts of illness and disability, we are all technology-dependent beings. The workshop, then, could be followed by a discussion in which students identify technological artifacts and systems in their own lives that highlight that dependency, exploring the different ethical implications thereof: do pervasively implemented lock-in technologies generate specific care duties to its technology-dependent users? What are the ways in which technology dependency has been purposely exploited and/

or unjustly distributed? And can we think of cases where technology dependency has been tackled with care, be it through activities of maintaining or innovating the technological environment in which we all are embedded? Questions such as these could sensitize students to the dimension of universalized care at stake in engineering endeavors.

In sum, engineers are not in the business of caring for her Other as a unique individual with whom she stands in a close personal relationship. We suggested that generalized gestures toward overcoming the distance between the engineer and her Other, for instance through exercises that explicate the link between stakeholder testimonies and material design choices, are undeniably important. This pertains both to making sure one's designs optimally fulfill the needs of cohorts of stakeholders (recall the AAC technology example), and for exposing engineers-in-training to the idea that they are in the care-taking business. However, we also added to this that care for the Other—in the context of engineering—involves recognition of a more universal feature of the engineer's Other, namely their deep dependency on the technological systems that engineers design, build, and maintain. Recognizing this is fully compatible with, and puts forth, a positive view of what it means to care for an Other in her distance.

As we have seen, Mitcham and his collaborators understand moral sensitivity as an iterative movement between moral perception, affectivity and dividing loyalties. What we have argued in this paper is that this view of moral sensitivity cannot be readily transported from the nursing context to the engineering context on the basis of a care analogy. The particularized care characteristic of the nursing context is decisively different from the generalized and universalized forms of care characteristic of the engineering context. That said, we agreed with Mitcham and his collaborators that care should be foregrounded as a central notion for engineering (ethics) education. As Russell and Vinsel furthermore suggest, this points to a key role for care ethics in engineering (ethics) curricula. While we do not disagree with Russell and Vinsel, we want to conclude by making the tentative suggestion that a bottom-up approach might be desirable, particularly in the male-dominated engineering world, where insights from a feminist ethics of care might be met with significant resistance. Though we should of course never insulate students from ideas they might initially feel uncomfortable with, there seems to be a pedagogical advantage to gradually getting students to see care-related concepts as being central to rather than external to engineering itself. This can be achieved through a pedagogical exercise that we have termed a *tinkering workshop*. Through this workshop students can discover that improving, altering, fixing, and augmenting technology is hardly a descriptive move in a value-neutral world, but that it can instead be seen as an activity of care-taking, both in its generalized and universalized forms. We think that this active learning exercise in which students discover, on their own, the prevalence of care-dimensions in engineering has the potential to introduce notions of care in the ethics classroom, while "meeting students where they are," to speak in the terms of Mary Sunderland (2014).

To harken back once more to the picture of moral sensitivity with which we started our analysis: by linking activities of care with choices regarding material and design, the engineering student's moral sensitivity can be refined, opening up a perceptual awakening and affectivity toward the complex nature of the engineer's Other. This awakening is in part promoted through an understanding of the ideology of neutrality as a moment in the history of engineering. Becoming aware of this ideology *as* an ideology can then be seen as an activity of *dividing loyalties* that allows for a reflexive and critical view of the biases and presuppositions inherited within the world of engineering. This process of deepening the engineering student's moral sensitivity is perhaps as much a process of the student becoming aware of her professional world, how it shapes her understanding of herself, and what it means to be a good engineer.

ACKNOWLEDGMENTS

This research was supported by the 4TU Centre for Engineering Education, grant number TBM_ERE_2021_02_4TU.CEE.TUD.

NOTES

1. We will use the terms moral sensitivity and ethical sensitivity interchangeably (see also Weaver et al. 2008).

2. Though one could argue that engineers do sometimes cater to the specific needs of others, for example, when catering to the needs of (often corporate) clients. Thanks to Glen Miller for pointing this out.

3. Of course, other factors come into play as well, such as bureaucracy—with nurses getting bogged down in institutional rules and protocols–or time-pressure and an overload of patients to care for. And the balancing act goes both ways. A nurse could overvalue the expressed experience of a patient at the expense of situation-relevant third-person knowledge and protocol.

4. https://www.nspe.org/resources/ethics/code-ethics.

5. Admittedly, the data gathered from our focus group may reflect a Western bias.

6. Cech speaks of an ideology of depoliticization, which refers to the assumption that engineering is a strictly "'technical' space where 'social' or 'political' issues such as inequality are tangential to engineer's work" (2013, 67).

7. Winner is emphatically critical of the value-neutrality thesis. For a thorough overview of this literature, see Byron Newberry's "An Engineer Considers Technological (Non)Neutrality: 'But Where Are the Values?,'" chapter 8 of this book.

8. These examples are taken from the course *Philosophy of Engineering Science and Design*, taught by our colleague Maarten Franssen (for a more comprehensive picture of the Delft approach to philosophy and ethics of engineering teaching see https://www.tudelft.nl/ethics/).

9. One might even want to argue that there are three pictures: (1) the picture of engineering as it is framed from the perspective of the ideology of neutrality; (2) a picture that recognizes ethics as important for engineers and which focuses on those ethical theories most in line with some of the rationalistic, calculative postures built into the first picture (in other words, the emphasis in the second view is on rule-based ethical theories and how they can help engineers make better ethical decisions); and then (3) the picture that foregrounds the role of emotions and attitudes of care as central to engineering qua activity.

10. As Russell and Vinsel note, "Most civil engineers work on keeping up existing physical infrastructures, like roads and bridges. Even in 'cutting-edge' fields, like software, about 70% of budgets go into maintenance and upkeep, whereas only about 8% of budgets go into new design" (2019, 257).

REFERENCES

Adam, Barbara, and Chris Groves. 2011. "Futures Tended: Care and Future-Oriented Responsibility." *Bulletin of Science, Technology & Society* 31, no. 1: 17–27.

Bhargava, Vikram R., and Manuel Velasquez. 2021. "Ethics of the Attention Economy: The Problem of Social Media Addiction." *Business Ethics Quarterly* 31, no. 3: 321–59.

Bjerknes, Mari Skancke, and Ida Torunn Bjørk. 2012. "Entry into Nursing: An Ethnographic Study of Newly Qualified Nurses Taking on the Nursing Role in a Hospital Setting." *Nursing Research and Practice*, article id 690348. https://doi.org/10.1155/2012/690348.

Bucciarelli, Louis L. 1988. "An Ethnographic Perspective on Engineering Design." *Design Studies* 9, no. 3: 159–68. https://doi.org/10.1016/0142-694X(88)90045-2.

———. 2008. "Ethics and Engineering Education." *European Journal of Engineering Education* 33, no. 2: 141–49. https://doi.org/10.1080/03043790801979856.

Cech, Erin A. 2013. "The (Mis)Framing of Social Justice: Why Ideologies of Depoliticization and Meritocracy Hinder Engineers' Ability to Think about Social Injustices." In *Engineering Education for Social Justice*, edited by Juan Lucena, 67–84. Dordrecht: Springer.

Franssen, Trijsje. 2022. Interviews about the educational exercise tinkering with technology. 4TU. ResearchData. Dataset. https://doi.org/10.4121/20020154.v1.

Gallagher, Shaun. 2007. "Simulation Trouble." *Social Neuroscience* 2, nos. 3–4: 353–65.

Halperin, Ofra, and Michal Mashiach-Eizenberg. 2014. "Becoming a Nurse: A Study of Career Choice and Professional Adaptation among Israeli Jewish and Arab Nursing Students: A Quantitative Research Study." *Nurse Education Today* 34, no. 10: 1330–34.

Herkert, Joseph R. 2001. "Future Directions in Engineering Ethics Research: Microethics, Macroethics and the Role of Professional Societies." *Science and Engineering Ethics* 7: 403–14. https://doi.org/10.1007/s11948-001-0062-2.

Held, Virginia. 2006. *The Ethics of Care: Personal, Political, and Global*. Oxford: Oxford University Press on Demand.

Hess, Justin L., and Grant Fore. 2018. "A Systematic Literature Review of US Engineering Ethics Interventions." *Science and Engineering Ethics* 24, no. 2: 551–83.

Marin, Lavinia, Janna van Grunsven, and Taylor E. Stone. 2022. Transcripts of focus group with educators on the topic of teaching ethics to engineers. 4TU.ResearchData. Dataset. https://doi.org/10.4121/19657161.v1.

Rest, James R. 1982. "A Psychologist Looks at the Teaching of Ethics." *Hastings Center Report* 12, no. 1: 29–36.

Roeder, Amy. 2019. "America Is Failing Its Black Mothers." *Harvard Public Health*, Winter. Accessed August 23, 2022. https://www.hsph.harvard.edu/magazine/magazine_article/america-is-failing-its-black-mothers/.

Roeser, Sabine. 2012. "Emotional Engineers: Toward Morally Responsible Design." *Science and Engineering Ethics* 18, no. 1: 103–15. https://doi.org/10.1007/s11948-010-9236-0.

Rosenberger, Robert. 2020. "On Hostile Design: Theoretical and Empirical Prospects." *Urban Studies* 57, no. 4: 883–93.

Russell, Andrew L., and Lee Vinsel. 2019. "Make Maintainers: Engineering Education and an Ethics of Care." In *Does America Need More Innovators?*, edited by Matthew Wisnioski, Eric S. Hintz, and Marie Stettler Kleine, 249–69. Cambridge, MA: MIT Press.

Sunderland, Mary E. 2014. "Taking Emotion Seriously: Meeting Students Where They Are." *Science and Engineering Ethics* 20, no. 1: 183–95. https://doi.org/10.1007/s11948-012-9427-y.

Vallor, Shannon. 2016. *Technology and the Virtues: A Philosophical Guide to a Future Worth Wanting*. Oxford: Oxford University Press.

Van Grunsven, Janna, Lavinia Marin, Taylor Stone, Sabine Roeser, and Neelke Doorn. 2021. "How to Teach Engineering Ethics? A Retrospective and Prospective Sketch of the TU Delft Approach." *Advances in Engineering Education* 9, no. 3. https://advances.asee.org/wp-content/uploads/vol09/Issue3/Papers/AEE-Innovative-Doorn-2.pdf.

Van Grunsven, Janna, and Sabine Roeser. 2022. "AAC Technology, Autism, and the Empathic Turn." *Social Epistemology: A Journal of Knowledge, Culture and Policy* 36, no. 1: 95–110. doi: 10.1080/02691728.2021.1897189.

Van Grunsven, Janna, Samantha Marie Copeland, Trijsje Franssen, Lavinia Marin, Cristina Richie. 2022. "Tinkering with Healthcare Technologies. Interactive

Redesign of Bio-Medical Artefacts from a Crip Techno-Science Perspective." edusources.nl. doi: 10.48544/22469997-a9a9-4157-a4b4-33c5a54992f8.

Van Wynsberghe, Aimee. 2016. "Service Robots, Care Ethics, and Design." *Ethics and Information Technology* 18, no. 4: 311–21.

Verbeek, Peter-Paul. 2011. *Moralizing Technology: Understanding and Designing the Morality of Things.* Chicago: University of Chicago Press.

Vermaas, Pieter, Peter Kroes, Ibo van de Poel, Martin Franssen, and Wybo Houkes. 2011. *A Philosophy of Technology: From Technical Artefacts to Sociotechnical Systems.* San Rafael, CA: Morgan and Claypool.

Vincenti, Walter G. 1990. *What Engineers Know and How They Know It: Analytical Studies from Aeronautical History.* Baltimore: Johns Hopkins University Press.

Weaver, Kathryn, Janice Morse, and Carl Mitcham. 2008. "Ethical Sensitivity in Professional Practice: Concept Analysis." *Journal of Advanced Nursing* 62: 607–18. https://doi.org/10.1111/j.1365–2648.2008.04625.x.

Weaver, Kathryn, and Carl Mitcham. 2016. "Prospects for Developing Ethical Sensitivity in Nursing, Engineering, and Other Technical Professions Education." *British Journal of Education, Society & Behavioural Science* 18, no. 2: 1–18. https://doi.org/10.9734/BJESBS/2016/27485.

Winner, Langdon. 1980. Do Artifacts Have Politics? *Daedalus* 109, no. 1: 121–36.

Wu, Tim. 2017. *The Attention Merchants: The Epic Scramble to Get Inside Our Heads.* New York: Vintage.

Young, Mark Thomas. 2021. "Maintenance." In *Routledge Handbook of the Philosophy of Engineering,* edited by Diane P. Michelfelder and Neelke Doorn, 356–68. New York: Routledge.

Chapter 10

Parallel Steps toward Philosophy of Engineering in China and the West

Nan WANG and LI Bocong

Before the 1990s, few scholars believed that it was necessary to establish philosophy of engineering as a subdiscipline of philosophy. However, the situation changed rapidly in the following ten years. While philosophy of science and philosophy of technology were initiated in the West and then spread to China, philosophy of engineering emerged in China and the West simultaneously. A few Chinese and Western scholars, working independently but in parallel, began to develop the field of the philosophy of engineering, which has developed across the globe during the first two decades of the twenty-first century.

The development of a new field is an academic paradigm shift. According to Thomas Kuhn, two main characteristics of an innovative paradigm are that it is "sufficiently unprecedented to attract an enduring group of adherents away from competing modes of scientific activity" and "sufficiently open-ended to leave all sorts of problems for the redefined group of practitioners to resolve" (Kuhn 1960, 10). The main characteristics of the adoption and institutionalization of a paradigm are "the formation of specialized journals, the foundation of specialists' societies, and the claim for a special place in the curriculum" (16).

This chapter examines two aspects of the emergence and development of the philosophy of engineering, as a new discipline, in China and the West. One is the degree to which the paradigm of the philosophy of engineering has been developed, and the other is the degree to which the new discipline has been institutionalized, which is assessed by examining the academic community, specialized journals, professional societies, and so on.

Our inquiry builds on several scholarly works in this area. LI Bocong wrote four articles on the formation of the philosophy of engineering in China and the West (LI Bocong 2008, 2010, 2021a; LI Bocong and Sumei CHENG 2011). Carl Mitcham briefly described the Western formation of the discipline in the introduction to a collection of his essays in 2020 (Mitcham 2020). Nan WANG organized a special issue of the journal *Technology in Society* titled "Engineering, Philosophy and Ethics: Emerging Chinese Discussions on Technology in Society" and wrote its introduction "The Emergence of Chinese Discussions of Engineering, Philosophy, and Ethics" (2015). *Philosophy of Engineering, East and West* (Mitcham et al. 2018), a proceeding of the "Forum on Philosophy, Engineering and Technology" in 2012 (fPET-2012), provided a brief overview of the differences between Chinese and Western scholars on the philosophy of engineering.

These earlier studies only briefly compared general topics; there have been no systematic reviews of the philosophy of engineering in particular cultural contexts including Chinese and Western contexts, and no systematic comparison between contexts. Much of the prior work occurred before the emergence of many new trends in China and the West in the last decade. Our chapter provides a detailed review on the parallel steps toward the philosophy of engineering in China and West in three stages, assesses their parallel development, and reflects on future prospects.

THE EMBRYONIC STAGE, 1990–1999

In the last decade of the twentieth century, although philosophy of science and philosophy of technology were in the mature stage of development, philosophy of engineering was still in its embryonic stage. Only few Chinese and Western pioneers advocated for the formation of philosophy of engineering as a new subdiscipline. Although there was no academic exchange on philosophy of engineering between Chinese and Western scholars, they had some shared understanding of the space as they worked on other subfields of philosophy such as philosophy of science, philosophy of technology, philosophy of economics, and so on.

The Embryonic Stage in the West

Paul Durbin's *Critical Perspectives on Nonacademic Science and Engineering* (1991a) was the first valuable collection of essays on philosophy of engineering. The fourth volume in the Lehigh University Press series *Research in*

Technology Studies, it clearly represented Western scholars' different attitudes toward this new subdiscipline. In the foreword, the two coeditors-in-chief of the book series Steven L. Goldman and Stephen H. Cutcliffe wrote, "Taken together, the essays begin to define the parameters of an as yet virtually nonexistent discipline, namely, philosophy of engineering. Their publication will, we hope, spur a continuing conversation that will make philosophy of engineering part of the ongoing study of technology" (Goldman and Cutcliffe 1991, 7).

In his introduction, Durbin presents a different view. Its first section is titled "The Need for a Philosophy of R&D," and at its very beginning he said: "When I first conceived this project, I had in mind a narrower focus—the so-called R&D community" (Durbin 1991b, 11). He argued for the necessity of this new subdiscipline, concluding that "What I would argue is that this set of issues—epistemological, ethical, social, even economic—ought to constitute the core of a new sub-discipline, philosophy of R&D" (17). For him, philosophy of R&D was the main theme, as it "often gets neglected in studies of science, technology, and engineering" (11). So while Durbin realized that a new situation was emerging in philosophical research on engineering, he thought it should be a component of a philosophy of R&D rather than one that stands alone. Many years later, when reflecting on what he wrote in this introduction, he still thought of philosophy of engineering as a subfield, "a mostly missing but much needed part of philosophy of technology," even as he praised the value of philosophical research on engineering: "If I were to do justice to the topic here, I should include materials from all the authors included in that volume since each one has at least one book-length study of engineering from his or her philosophical perspective" (Durbin 2006, 141).

In addition to having declared the importance of the philosophy of engineering in the foreword, Goldman also had a couple of contributions to this theme. One is the chapter titled "The Social Captivity of Engineering" (Goldman 1991) in the same volume. Another, titled "Philosophy, Engineering, and Western Culture" (Goldman 1990), was included in *Broad and Narrow Interpretations of Philosophy of Technology* (also edited by Durbin), the seventh volume in the series *Philosophy and Technology* published by The Society for Philosophy and Technology (SPT).

In "Philosophy, Engineering, and Western Culture," Goldman explored the worldview and rationality of engineering, which, in Western culture, was widely assumed to be the same as that of science. He believed that an obstacle for misunderstanding engineering and neglecting the philosophy of engineering in Western intellectual tradition lay in "a persistent Western cultural prejudice that subordinates engineering to science" (Goldman 1990, 125), which took the form of "a clear preference for understanding over doing, for contemplation over operation, for theory over experiment" (127). Under this wrong assumption, Goldman lamented that "today, then, philosophy of science is a fully accepted and highly respected branch of philosophy, while philosophy of engineering carries as much professional distinction as philosophy of parapsychology" (140). As for the relationship between philosophy of science and philosophy of engineering, he stated that "philosophy of engineering should be the paradigm for philosophy of science, rather than the reverse" (140), because the value-laden character of engineering was much more evident than that of science, making science a special form of engineering in nature. In the 1990s, he was a trendsetter for his bold proposition for the importance of the philosophy of engineering, which was followed in China and the West in later decades.

In "The Social Captivity of Engineering," Goldman argued that engineering was captive to society, both intellectually and socially, and the two complementary captivities were responsible for the fundamental usurpation of engineering as an autonomous discipline. Based on this, he argued that a philosophy of engineering was necessary because "engineering poses a formidable challenge for philosophy. . . . It is quite clear that philosophers cannot 'solve,' in any traditional sense of the term, the comprehensive epistemological cum metaphysical cum political, moral and aesthetic problems that engineering poses" (Goldman 1991, 142). According to him, "a philosophy of engineering inevitably opens out to philosophical anthropology; moreover, it cannot take the primarily intellectual and disengaged form characteristic of mainstream Western philosophy and science" (141–42).

Goldman's arguments for a philosophy of engineering in the 1990s were undoubtedly pioneering. He was joined by a couple of philosophers and engineers, such as Taft H. Broome Jr., Billy Vaughn Koen, and Carl Mitcham, all contributors to Durbin's *Critical Perspectives on Nonacademic Science and Engineering*, who took similar views.

Broome (1991) proposed three modes in which philosophy engages engineering, including "philosophy *and* engineering," "philosophy *in* engineering," and "philosophy *of* engineering" (italics in original). However, all three modes met with a resistance rooted in the cultures of philosophy and engineering that frustrated attempts to engage. He proposed that "bridges over the gaps on either side will be the essentials of an emerging culture that I label *science, technology, and human values* (STHV)" (266). Obviously, he had a strong sense of various *philosophical engagements with engineering*, but he thought that they should be part of STHV rather than a separate subdiscipline.

Koen wrote "The Engineering Method" for the volume. While he did not mention philosophy of engineering in the text, he focused on engineering method, now one of the most important topics in the philosophy of engineering. He pointed out that "unlike the extensive analysis of the scientific method, little significant research to date has sought the philosophical foundations of engineering" (1991, 33). He argued that engineering method was quite distinct from the scientific method, and it had two unique features: "engineering is the use of engineering heuristics" and "engineering method is related in fundamental ways to human problem solving at its best" (59).

Mitcham's contribution to the volume was "Engineering as Productive Activity: Philosophical Remarks" (1991), in which he argued that engineering was the productive activity of making. He extended this argument in *Thinking through Technology: The Path between Engineering and Philosophy* (1994). It is generally recognized as one of the most important books in philosophy of technology, but, as he later noted, it still gave engineering a certain prominence (Mitcham 2013, vi). Mitcham investigated the etymology of "engineering," and discussed "the action of making" and "the process of using" under the name of "technology activity" in the tasks of crafting, inventing, designing, manufacturing, working, operating, and maintaining (Mitcham 1994, 144–149, 210–246). At this point, his ideas had not received much attention in the West.

In 1998, Mitcham published an article titled "The Importance of Philosophy to Engineering," in which he underscored an urgent need for interaction between philosophy and engineering. The core thesis of this article, which challenged common presumptions, was that philosophy is centrally important to engineering. Due to the inherently philosophical nature of engineering, he argued that philosophy, especially engineering ethics, would help engineers deal with professional problems and understand and defend themselves against philosophical criticism. At the end, he made an inspirational announcement: "Engineers of the world philosophize! You have nothing to lose but your silence!" (1998, 43), an imitation of a famous saying at the end of the *Manifesto of the Communist Party*: "Workers of the world unite! You have nothing to lose but your chains" (Marx and Engels [1848] 1969).

The Embryonic Stage in China

The early version of philosophy of engineering in China was sketched in a short book *Rengong lun tigang* (人工论提纲, *An outline for the theory of human making*) by LI Bocong in 1988. As the title shows, he emphasized the importance of viewing "human making" as a central theme of philosophical research. In the preface, he claimed that "the philosophical system should be transformed from a system with epistemology as the focus into one with human making as the focus" (LI Bocong 1988, 2). This book established a foundation for his subsequent monograph *Gongcheng zhexue yinlun* (工程哲学引论, *An introduction to the philosophy of engineering*) in 2002.

In 1992, in "Wo zaowu gu wozai: Jian lun gongcheng shizai lun (我造物故我在: 简论工程实在论, I create, therefore I am: On engineering realism)," LI Bocong first used the term *gongcheng zhexue* (工程哲学, philosophy of engineering), and pointed out the need to pursue philosophy of engineering as a new area. And he put forward "I create, therefore I am" as a maxim for philosophy of engineering, which stands in sharp contrast with Descartes's maxim "I think, therefore I am." He divided engineering activities into planning, implementation, and use stages, which shows some similarities to Mitcham's later division of technological activity, and he conducted a brief philosophical analysis of each stage. LI Bocong was the only scholar who published essays on philosophy of engineering in China in the 1990s; in fact, his work establishing the discipline was ignored or opposed by most Chinese scholars.

In 1995, LI Bocong published an article "Nuli xiang jingji zhexue he gongcheng zhexue lingyu kaituo: Jian lun 21shiji zhexue de zhuanxiang (努力向经济哲学和工程哲学开拓: 简论21世纪哲学的转向, [Working hard to initiate philosophy of engineering and of economy: On the turn of philosophy of the 21st century])." Reflecting on the linguistic turn in the twentieth century, in which philosophy of language became central, he argued that the philosophical turn of the twenty-first century would be driven by philosophy of economics and philosophy of engineering, as the economic factor is the most important element of engineering activity, and engineering activity has become the most important human activity. As the title shows, he emphasized the great prospects for the future of the philosophy of engineering. He argued that "the first theme of philosophical research, the science of wisdom, should be research on the wisdom of creating and using things. However, both Chinese and Western philosophies have neglected this theme in history. Philosophical research in the new century will be of great significance if it explores research on 'creating and using artifacts.' This work can also be called the study on philosophy of engineering" (1995, 87).

It is helpful to briefly compare Goldman and LI Bocong, the pioneers of philosophy of engineering in the West and China, respectively. As mentioned above, each wrote two important essays on this theme in the 1990s. They advocated that engineering should not be regarded as the application of science, but as a unique practice that was closely connected with society and economy. They both examined the necessity of philosophy of engineering, and the significance of philosophy of engineering to the development of philosophy in future. Goldman explored the obstacles to the misunderstanding of engineering and the neglect of philosophy of engineering in Western philosophical tradition since Plato, and clarified the nature characteristic of engineering and the revolutionary meaning of philosophy of engineering, whereas LI Bocong stated that "human making" was a missing theme in the philosophical tradition and philosophy of engineering would be the turn of philosophy in the twenty-first century. Furthermore, LI Bocong developed a brief conceptual framework of philosophy of engineering and proposed a maxim for it.

THE EMERGING STAGE, 2000–2009

At the beginning of the twenty-first century, philosophy of engineering entered the emerging stage in terms of both theoretical development and institutionalization. The greater emphasis on philosophy of engineering in China and the West reflected a "significant increase" in the number of researchers and the "remarkable progress" of research development (LI Bocong 2008). More publications appeared—four books titled "philosophy of engineering" or "engineering philosophy" and a few journals—and professional societies and academic conferences were set up in China and the West. Academic exchanges on philosophy of engineering between China and the West gradually increased, and the development in this stage can be seen as collaborative rather than separate.

Theoretical Research in China and the West

Goldman's third article on philosophy of engineering, "Why We Need a Philosophy of Engineering: A Work in Progress" (2004), repeated that philosophy of engineering should be an emerging discipline, as it had been "an as yet virtually nonexistent discipline" in his assessment in 1991. He argued that a main feature of engineering reasoning lies in "a contingency-based form of reasoning that stands in sharp contrast to the necessity-based model of rationality that has dominated Western philosophy since Plato and that underlies modern science" (2004, 163). "A contingency-based philosophy of engineering ... might enable more effective technological action" because "a contingency-based conception of rationality" embodied in engineering "is an advance over any conception of rationality that sets values and action aside as not within the scope of rationality, knowledge and truth" (163).

At the very beginning of the twenty-first century, two monographs on philosophy of engineering came out. One was LI Bocong's *Gongcheng zhexue yinlun* (工程哲学引论, *An introduction to the philosophy of engineering*) (2002), written in Chinese and published in China; a revised edition was published by Springer in English in (LI Bocong 2021b). The other was Louis L. Bucciarelli's *Engineering Philosophy* (2003), written in English and published by Delft University Press. Both initially used the term "philosophy of engineering" or "engineering philosophy" as a title of the book.

The two authors came from different professional backgrounds. As an engineer, Bucciarelli regarded engineering design as a core subject. His stated intention was "to show that philosophy can matter, does matter, to engineers. I want to explore in what ways it might contribute to doing a better job of designing" (2003, 1). As a philosopher, LI Bocong focused on the ontological foundation for the philosophy of engineering and argued that "whether engineering is an independent object has become a prerequisite question to this book" (2002, 3). He found this foundation in the trichotomy of science, technology, and engineering: discovery is the core of scientific activity, invention the core of technological activity, and making the core of engineering activity.

Their aims differed as well. The aim of Bucciarelli's book "is not to try to construct a formal philosophical treatise, but to draw upon my experiences—as designer, as consultant, as researcher, as teacher—in setting out and exploring some hopefully fruitful connections" (2003, 3). He had unique insights on related issues of design from perspectives of philosophy, for example, the definition of design, the relation between scientific knowledge and engineering knowledge, and engineering education, but his "engineering philosophy" was a narrow field. Rather than an experiential focus, LI had an ambitious agenda to develop the theoretical system of philosophy of engineering, "a sub-discipline of philosophy with substantial content by making a detailed discussion on 50 categories related to philosophy of engineering" (LI Bocong 2002, 24). The categories distinctive to engineering itself come not only from fields in philosophy, but also from economics, management, sociology, and psychology. According to his three stages of the engineering process, these categories were divided into three groups: goal-setting, planning, and decision-making; operation, institution, and micro-production mode; and value, alienation, and life.

Meanwhile, both China and the West began to realize that separate studies conducted by engineers and philosophers were insufficient, and further development of philosophy of engineering must rely on the close communication and collaboration between the two groups. In 2007, two important fruits of cooperation between the two circles on philosophy of engineering appeared. They were *Gongcheng zhexue* (工程哲学, *Philosophy of engineering*) (YIN Ruiyu, WANG Yingluo, and LI Bocong 2007) in Chinese in China and *Philosophy in Engineering* (Christensen, Meganck, and Delahousse 2007) in English published by Academica in Denmark: most of the contributors are still active in the field philosophy of engineering today.

Gongcheng zhexue "is not a collection of 'scattered' and independent research essays of each author, but the result of organized and close cooperative discussion and research" (LI Bocong 2008, 15) among more than twenty authors including engineers, entrepreneurs, managers, and philosophers. As a symbol of the alliance between philosophical and engineering circles, its first part was on theory and its second part on practice, an unusual combination for a philosophical book. In the "Theory Part," it put forward ten basic viewpoints of philosophy of engineering, but also discussed the essence and characteristics of engineering, engineering thinking, engineering methodology, and the engineering viewpoint. In the "Practice Part," it carried out philosophical analysis on seven large Chinese engineering projects, such as construction of the Daqing oilfield, the Three Gorges Dam, a high-speed railway construction project, and the Baoshan Iron and Steel project. As the postscript said, on the one hand, it will enrich the content of philosophy and promote the development of philosophy itself; on the other hand, it will deepen the understanding of engineering and promote the development of engineering practice (YIN Ruiyu, WANG Yingluo, and LI Bocong 2007).

In contrast, *Philosophy in Engineering* was originally oriented to be a course book. The first sentence of the preface is "This book is intended for courses in philosophy of science for engineers at the bachelor's level in engineering studies," and, especially, "engineering students in their third year of studies" (Christensen et al. 2007, 6). But it offered a broad perspective on the role of engineering in the twenty-first century and contributed to a critical reappraisal of engineering as a profession. With a "metadisciplinary approach" encompassing historical, epistemological, ethical, and global aspects of engineering, it covered four important areas—history, epistemology, ethics, and engineering education—of philosophy of engineering.

According to Mitcham (1994, 21–29), it took fifty years, from 1877 to 1927, for four books to have used the phrase "philosophy of technology" as the title. However, based on the above historical review of the development of philosophy of engineering, it took only five years, from 2002 to 2007, to have four books titled "philosophy of engineering" or "engineering philosophy." Furthermore, the authors of the four books of philosophy of technology were all from Europe (German or Russia), whereas those of philosophy of engineering were from Europe, the United States, and China.

Institutionalization

In 2003, LI Bocong initiated the Center for Research of Engineering and Society in the Graduate University of Chinese Academy of Sciences (GUCAS). Its core task is to study philosophy of engineering, sociology of engineering, and history of engineering. At the end of that year, the academic working committee of the Chinese Society for the Dialectics of Nature (CSDN) held the first academic conference on philosophy of engineering at Xi'an Jiaotong University and started to promote the institutionalization of philosophy of engineering in China.

The publication of *Gongcheng zhexue yinlun* in 2002 attracted great attention from XU Kuangdi, then president of the Chinese Academy of Engineering (CAE), and YIN Ruiyu, then director of the Division of Engineering Management (DEM) of CAE. They both believed philosophy of engineering would be a great opportunity to promote theoretical research on engineering and engineering management. As YIN Ruiyu (2010) recalled, when DEM was founded in 2000, most of its academicians felt that it was necessary to strengthen the basic theoretical research on engineering and engineering management. As there are many profound and important philosophical problems inherent in engineering activities, the understanding of the essence and law of engineering activities is therefore inseparable from philosophical thinking. Thus a consensus on the research of philosophy of engineering was reached. In June 2004, at the behest of XU Kuangdi and YIN Ruiyu, DEM held a symposium on philosophy of engineering, with seven academicians of CAE and more than ten philosophers in attendance. XU Kuangdi delivered a long speech on his views on philosophy of engineering. He said: "Philosophy of engineering is very important, because engineering is full of dialectics which are worth exploring. The understanding of engineering should be raised to the philosophical level, and engineers' philosophical thinking should be promoted" (quoted in ZHAO Jianjun 2004, 213). This symposium served as a high-level starting point for the alliance between engineering and philosophical circles in China.

Shortly after the symposium, The First Chinese Conference on Philosophy of Engineering was held. During the conference, the Chinese Society for Philosophy of Engineering (CSPE) of CSDN was officially established. The establishment of CSPE was an important step toward the institutionalization of philosophy of engineering in China. After holding two annual national conferences on philosophy of engineering, CSPE has held biennial conferences since 2007.

The concept of "engineering studies," which means interdisciplinary research on engineering, was first proposed by Gary L. Downey and Juan C. Lucena in 1995 (Downey and Lucena 1995), and, in 2004, they founded the International Network for Engineering Studies (INES) in Paris. That same year, LI Bocong and Cheng DU set up an annual Chinese journal titled *Gongcheng yanjiu: Kua xueke shiye zhong de gongcheng* (工程研究: 跨学科视野下的工程, *Engineering studies: Engineering in interdisciplinary perspectives*), which became quarterly in 2009 and bimonthly in 2016. (LI Bocong and Cheng DU knew nothing about the work of Downey and Lucena.) In 2009, INES published its official journal, also titled *Engineering Studies*, and its founding editors-in-chief were Downey and Lucena. It was published three times a year originally and became a quarterly in 2012.

In 2004, Broome's program on philosophy of engineering was funded by the US National Academy of Engineering (NAE). Two years later, he launched the first American Symposium on Philosophy of Engineering at MIT which was funded by the NAE. During this meeting, American scholars decided to hold the first international Workshop on Philosophy of Engineering (WPE) in the Netherlands in the following year.

Also during this time, scholars in Britain and Denmark were making progress. The Royal Academy of Engineering conducted a series of seminars on philosophy of engineering in 2006 and 2007. The presentations from the series were released in two volumes *Philosophy of Engineering: The Proceedings of a Series of Seminars Held at the Royal Academy of Engineering*. Later, an International Research Conference on "Occupational *Bildung* and Philosophy in Engineering" was held in 2007 in Aarhus, Denmark. The reporters came from Denmark, the United States, Ireland, Belgium, and the Netherlands.

In October 2007, the first WPE was held at Technical University Delft. The theme of WPE-2007, "Engineering Meets Philosophy and Philosophy Meets Engineering," was eye-catching and memorable. It "was the largest organized activity bringing together engineers and philosophers in the last two decades" to discuss philosophical questions on engineering (van de Poel 2010, 2). Three years later, a book stimulated by this meeting titled *Philosophy and Engineering: An Emerging Agenda* was published by Springer as the second volume in their "Philosophy of Engineering and Technology" series.

WPE-2008 was held at the Royal Academy of Engineering in 2008. Its essays were included in a book titled *Philosophy and Engineering: Reflections on Practice, Principles and Process* (volume 15 of Springer "Philosophy of Engineering and Technology" series) in 2013. As "a companion to *Agenda*" (Mitcham 2013, v), the two volumes contributed to establishing three aspects of philosophy of engineering, namely philosophy, ethics, and reflection. But there is an obvious contrast between the two books: "31 of the *Reflections* authors did not contribute to *Agenda* and thus bring new perspectives," and "*Reflections* appears slightly more interested in drilling into the particulars of engineering" (v).

There was a track on reflective engineering at the 2009 meeting of the Society for Philosophy and Technology (SPT-2009) held at the University of Twente in the Netherlands. The 2010 Forum on Philosophy, Engineering, and Technology (fPET-2010) was held in Golden, Colorado, at the Colorado School of Mines. After that, fPET gradually became an extraordinary international stage of collaboration between philosophers and engineers from China and the West.

THE PARALLEL DEVELOPMENT STAGE, 2010–PRESENT

The second decade of the twenty-first century can be considered the development stage of philosophy of engineering. With sustained and rapid development, the new subdiscipline began to be accepted around the world and to continue to grow. Not only were some essays published in a "scattering" of places, but also a number of conference proceedings, handbooks, collection of essays, monographs, and book series. With a broad scholarly scope all over the world, a very wide range of topics were discussed in the West and China, and academic exchange and cooperation became frequent and deepened.

The Development Stage in the West

Springer started the book series "Philosophy of Engineering and Technology" (POET), edited by Pieter E. Vermaas, in 2010. In its first decade, the series published thirty-nine volumes. Although the main purpose of the series, to provide "the multifaceted and rapidly growing discipline of philosophy of technology with a central overarching and integrative platform," does not mention engineering, fourteen volumes have had "engineering" in their titles or subtitles. These include two monographs, four conference proceedings, and eight collections of essays. The proceedings and collections of essays draw from contributors from many countries writing on a wide range of subjects, which draw a clear outline of the central issues and debates, established themes, and new developments in philosophy of engineering. *Philosophy and Engineering: An Emerging Agenda*, the second volume, is a good example. In Mitcham's assessment, it is, "to its date the most concise, broad-spectrum introduction to philosophy and engineering interaction" (Mitcham 2020, 5).

Some editors or coeditors of these volumes are the pioneers of philosophy of engineering in the West, such as Carl Mitcham, Diane Michelfelder, Steen Hyldgaard Christensen, Anthonie Meijers, Ibo van de Poel, and David Goldberg. Notably, the first three scholars have each edited three to five books. These pioneers made considerable contributions to the proceedings and collection of essays in the POET series and also published other works.

In 2020, Mitcham's book *Steps toward a Philosophy of Engineering: Historico-Philosophical and Critical Essays*, a compilation of his essays originally published from the late 1990s to 2020 on philosophy of engineering, was published. The first sentence of this book is an affirmation of his attitude toward philosophy of engineering. He said: "This *in media res* collection records a series of halting steps in a philosophical encounter with engineering" (Mitcham 2020, xiii, italics in original). As mentioned above, he has engaged in philosophical research on engineering since the

1990s, and a couple of articles greatly developed these ideas. However, he acknowledged that he could not make a firm and quick philosophical judgment of engineering until 2020. In the preface, he said: "Although I have been thinking about engineering and technology for many years, I remain unsure of a final judgment and swing between trying to think with and think against engineering" (xiii). Mitcham's hesitating attitude to philosophy of engineering never stopped him from developing it, in terms of theory as well as disciplinary institutionalization, over the past three decades. He organized the 2010 Forum on Philosophy, Engineering and Technology (fPET-2010) in Golden, Colorado, and was one of the main organizers of fPET-2012 in Beijing. In addition to writing articles and monographs, he also has coedited four books on this theme, notably *Philosophy of Engineering: East and West* (2018), volume 330 of the Boston Studies in the Philosophy and History of Science series.

Mitcham also devoted himself to cultivating the new generation of scholars, especially Chinese scholars, of philosophy of technology and philosophy of engineering for a long time. A number of Chinese doctoral students, postdocs, and visiting scholars have studied at Colorado School of Mines, in other universities in the United States, and in universities in other countries through his introduction. Nan WANG, one of the authors of this article, and Qin ZHU, one of the editors of this volume, were among the beneficiaries. He obtained such high recognition in China that a number of Chinese universities invited him to be a guest professor. Among these invitations was a very rare honor from Renmin University of China, the top Chinese university in humanities and social sciences, for Mitcham to be an "International Distinguished Professor of Philosophy of Technology" from 2015 to 2021, which was a chance for him to further deepen the ties and cooperation between China and the United States. Mitcham is one of the most influential American scholars in China in the field of philosophy of technology and philosophy of engineering due to his great contributions to both academic research and Sino-American exchanges.

Besides Springer, a couple of other international publishers, such as Elsevier and Routledge, have paid attention to the publication of works on philosophy of engineering. Elsevier published the *Handbook of the Philosophy of Science: Philosophy of Technology and Engineering Sciences* (Meijers 2009). The book does not focus specifically on philosophy of engineering, but five of six parts have had "engineering" in their titles: it includes philosophical investigations on some fascinating issues in engineering, such as the relationship between technology, engineering, and the sciences; philosophy of engineering design; modeling in engineering sciences; norms and values in engineering; and philosophical issues in engineering disciplines.

In 2021, *The Routledge Handbook of the Philosophy of Engineering,* edited by Diane P. Michelfelder and Neelke Doorn, was published. It is considered a "huge" representative work of the latest progress in philosophy of engineering. With a broad scholarly scope and fifty-five chapters from a mix of established experts and fresh voices in the field from fifteen countries, this book focused on seven kinds of topics: foundational perspectives, engineering reasoning, ontology, engineering design processes, engineering activities and methods, values in engineering, responsibilities in engineering practice, and reimagining engineering. As the editors remark in the first sentence of the book, "As recently as twenty years ago, this *Handbook* would have been unimaginable" (Michelfelder and Doorn 2021, 1). The editors' words imply that the extremely rapid development of philosophy of engineering over the past two decades is beyond all expectations. The editors made a striking judgment on the significance of philosophy of engineering at the end of their introduction, stating that "philosophy of engineering will always be in a dynamic and beta-state, marked by imagining and reimagining, and making important contributions not only to the understanding of engineering, but also to philosophy itself" (7).

The Development Stage in China

In the theoretical research process of philosophy of engineering in China, the Division of Engineering Management (DEM) of the Chinese Academy of Engineering (CAE) and the College of Humanities and Social Sciences (CHSS) of University of Chinese Academy of Sciences (UCAS) have played key roles.

The basic impetus for DEM to initiate philosophy of engineering was the need to strengthen basic theoretical research on what is engineering and engineering management. With sustainable support from CAE, DEM continued research on philosophy of engineering, collaborating with the Chinese Society for Dialectics of Nature, especially, and with UCAS. Regarding to the importance of the joint research, then president of CAE XU Kuangdi said: "It aims to concentrate the wisdom of Chinese scholars and experts in engineering and philosophical circles, and to jointly write a special theoretical work that is helpful and enlightening for engineering, engineering management, and philosophy" (YIN Ruyu, WANG Yingluo and LI Bocong 2007, I).

In 2004, DEM began the first research program on the theory of philosophy of engineering, which led to the publication of *Gongcheng zhexue* in 2007. (The main points of this book have been described in "The Emerging Stage, 2000–2009" section.)

DEM started the second research program on the theory of engineering evolution in 2009, and published *Gongcheng yanhua lun* (工程演化论, The theory of engineering evolution) in 2011. *Evolution* is "the stretching and deepening" of philosophy (YIN Ruiyu, WANG Yingluo, and LI Bocong 2011, 3), because the evolution of engineering is not only an important part of philosophy of engineering, but also a unification of history and theory based on the history of engineering and philosophy of engineering. Lakatos famously proposed that "philosophy of science without history of science is empty; history of science without philosophy of science is blind" (Lakatos 1970, 91). One could similarly say that "philosophy of engineering without history of engineering is empty; history of engineering without philosophy of engineering is blind." The core idea of *Evolution* is "the history of engineering development is the history of direct productivity development, and the theory of engineering evolution is the theory of productivity evolution" (YIN Ruiyu, WANG Yingluo and LI Bocong 2011, 25) as "engineering is a direct and realistic productivity" (YIN Ruiyu, WANG Yingluo, and LI Bocong 2007, ii).

After the publication of *Evolution*, a greater understanding of some issues in philosophy of engineering was gained, and in 2011, DEM established the third program which resulted in the publication of the second edition of *Gongcheng zhexue* (*Philosophy of Engineering*) in 2013. Compared with the first edition, the major changes consisted of "the new viewpoint of 'engineering ontology'" and "the new understanding of problems related to engineering thinking, engineering design, engineering epistemology, and so on" (YIN Ruiyu, WANG Yingluo, and LI Bocong 2013, i–ii). The book is important because it put forward the basic view of engineering ontology, which means engineering is a direct and realistic productivity, and because it affirmed that engineering ontology is the core of the theoretical system of philosophy of engineering.

After clarifying the basic viewpoints of engineering ontology, based on DEM's two research programs in 2014 and 2018, Chinese philosophers and engineers started research on engineering methodology and engineering epistemology, because they believed that just as methodology and epistemology play an important role in the field of philosophy, engineering methodology and engineering epistemology also play a key role in the field of philosophy of engineering. Their findings led to the publication of *Gongcheng fangfa lun* (工程方法论, *The theory of engineering methodology*) (YIN Ruiyu, WANG Yingluo, and LI Bocong 2015) and *Gongcheng zhishilun* (工程知识论, *The theory of engineering knowledge*) (YIN Ruiyu, WANG Yingluo, and Enjie LUAN 2020), which comprehensively elaborated on the two important branches of philosophy of engineering. Research on the theory of engineering methodology became the major new contents of the third edition of *Gongcheng zhexue* (YIN Ruiyu, WANG Yingluo, and LI Bocong 2018).

Since epistemology and methodology have always been important branches in the field of philosophy, it seems that engineering epistemology and engineering methodology should also have received great attention. But this has not been the case. It was once thought that technological or engineering knowledge would be a new theme for the historians of technology in 1980s, but "after the publication of Walter Vincenti's *What Engineers Know and How They Know It* (1990), research concerning the nature of technological knowledge seems to have come to a standstill" (Houkes 2009, 309). What is worse, while "historians of technology have lost interest in the topic," "philosophers have not rushed in to fill the gap left by historians" (309). The same thing happened to the research on engineering methodology.

Methodology argued that engineering methodology should be regarded as an independent branch of methodology, not as a variant of scientific methodology. "Theoretically, the principles of synthesis, integration, coordination, and trade-off are not the principles to solve scientific problems. . . . Among many solutions to engineering problems, the one that 'wins' is often the result of relative optimization through synthesis, integration, coordination, and trade-off. . . . The importance of these principles and methods will be highlighted only in the vision of engineering methodology and in the commonness of the overall structure of engineering methodology" (YIN Ruiyu, WANG Yingluo, and LI Bocong 2015, 11–12). Based on this, the overall structure of engineering method consists of "generalized hardware (tools, machines, etc.), software (machine operation procedures, etc.), and orgware (engineering organization and management system, etc.)" (33). The book also specifically studied the methodology of the whole life cycle of engineering projects. The whole life cycle of engineering projects usually includes the planning and design stage, the production and construction stage, the operation stage, and the retirement stage. Each stage has its own engineering methodology.

Knowledge stated engineering epistemology or theory of engineering knowledge should not be regarded as a derived component of scientific epistemology, but as an independent branch of epistemology. The basic starting point for this statement is the difference and connection between "two kinds of material world" and "two kinds of knowledge." The natural and artificial are the two kinds of the material world, and in turn, one kind of knowledge is scientific knowledge of natural objects, and the other kind is engineering knowledge of artificial objects (YIN Ruiyu, WANG Yingluo, and LI Bocong 2020, 17–20). This book elaborated on the basic viewpoints of the theory of engineering knowledge, the types of engineering knowledge, and the systematic integration of engineering knowledge. It also investigated the knowledge of engineering decision-making and evaluation, engineering planning and design, engineering management, and the knowledge of interaction among engineering, nature, and society, as well as the inheritance, communication, and evolution of engineering knowledge.

Through almost twenty years continuous and systematic research, the DEM has organized cooperation among Chinese engineers and philosophers in the study of philosophy of engineering by means of substantive research programs, and successively published six books on philosophy of engineering that are the results of organized and close cooperative discussion and research (not "scattered" and independent essays of diverse authors). An integration of these books, the latest achievement of Chinese philosophy of engineering, the fourth edition of *Gongcheng zhexue* will be published in 2022. With the six publications, Chinese scholars have put forward a theoretical system of philosophy of engineering that includes "five theories"—the trichotomy of science, technology, and engineering; a theory of engineering ontology; a theory of engineering methodology, a theory of engineering epistemology; and a theory of engineering evolution—that all take the theory of engineering ontology as the core. At the eighth National Conference on Philosophy of Engineering held in Suzhou, Jiangsu Province, in 2017, Chinese scholars believed that a Chinese School of philosophy of engineering had been formed, in view of the comparison of the research progress of philosophy of engineering at home and abroad, and the understanding of the distinctive characteristics of Chinese philosophy of engineering (YIN Ruiyu, WANG Yingluo, and LI Bocong 2020, iii).

The revised version of LI Bocong's *Gongcheng zhexue yinlun* was published in English by Springer in 2021. As the only translated work in the Philosophy of Engineering and Technology series, it indicates that the series editor wished to introduce one of the latest and most important Chinese achievements in the philosophy of engineering and engineering studies to the West. The revised version includes LI Bocong's analysis and discussion in sociology of engineering and interdisciplinary studies on engineering since 2004, such as the concept of "engineering community" and the two modes of philosophy of engineering. In Mitcham's assessment, "LI's work constitutes the best effort to date, China or West, at drafting a comprehensive philosophy of engineering and technology sensitive to the practical lifeworld of the engineering community itself" (Mitcham 2021, viii).

A BRIEF COMMENT ON PARALLEL STEPS TOWARD PHILOSOPHY OF ENGINEERING

Looking back at the establishment and development of philosophy of engineering, it is clear that a parallel development emerged in China and the West over the past three decades. This is not a coincidence. The parallel development in the two cultures reflects that the emergence of the philosophy of engineering was simply a matter of time. (See table 10.1.)

However, there are still some differences between the development of philosophy of engineering in China and the West. First, different emphases on the philosophy of engineering. Since the 1970s, when China implemented the policy of Reform and Opening-up (program of economic reforms), a number of engineering achievements and mega-projects, such as the Three Gorges Dam, high-speed railways, manned spaceflight, and exploration of the moon, have brought rapid development of the Chinese economy and society. The national context in China has contributed to a much broader approach to the philosophy of engineering and issues such as trichotomy of science, technology, and engineering, and case studies on mega-projects have been discussed. In the West, beginning in the second half of the twentieth century, the philosophical engagement of engineering began to raise new themes, such as engineering ethics. "After World War II, the first efforts of general philosophical reflection on technology among engineers in Germany gave rise to early engagements between engineers and philosophers on this theme" (Mitcham and Nan WANG 2015, 307). This trend challenged the romantic view of engineering common since the Industrial Revolution.

Second, interaction between engineering circles and philosophy circles differs. The Chinese Academy of Engineering (CAE) has played a key role in guiding the development of the philosophy of engineering in China, and "no other national academy of engineering has sponsored philosophical reflection to the same extent as the Chinese CAE" (Mitcham 2021, viii). As one of the institutions directly under the State Council, the services that CAE provides are related to state macroscopic planning and decision-making, which are inseparable from philosophical thinking. With sustainable support and organization from CAE, Chinese engineering circles and philosophical circles established an "alliance" relationship that consists of a joint research program and a joint academic conference. It strongly promoted "organized" systematic research on the philosophy of engineering and engineering studies, and their disciplinary institutionalization in China.

In the West, some engineering societies, such as the British Royal Academy of Engineering and the U.S. National Academy of Engineering, played a pioneering role for a short time, but they did not form an organized system. Western scholars adopted a "scattered," "individual," and "spontaneous" approach. They have also paid much attention to the international exchange and cooperation at the very beginning, which resulted in many significant cross-cultural studies on the philosophy of engineering.

Third, research topics of the philosophy of engineering differ. Based on the above two differences, Chinese and Western scholars work on different issues. Chinese scholars focus on the differences and connections between

Table 10.1. Timeline of the Parallel Development of Philosophy of Engineering in China and the West

Year	China	The West
1988	Rengong lun tigang 人工论提纲 [An Outline for the Theory of Human Making] (by LI Bocong)	

Embryonic Stage, 1990–1999

1990		"Philosophy, Engineering, and Western Culture" (by Steven L. Goldman)
1991		*Critical Perspectives on Nonacademic Science and Engineering* (ed. Paul Durbin)
1992	"Wo zaowu gu wozai: Jian lun gongcheng shizai lun" 我造物故我在: 简论工程实在论 [I create, therefore I am: On engineering realism]" (by LI Bocong)	
1994		*Thinking through Technology: The Path between Engineering and Philosophy* (by Carl Mitcham)
1995	"Nuli xiang jingji zhexue he gongcheng zhexue lingyu kaituo: jian lun ershiyi shiji zhexue de zhuanxiang" 努力向经济哲学与工程哲学开拓: 简论二十一世纪哲学的转向 [Exploiting the philosophy of economics and the philosophy of engineering: On the turn in philosophy in the 21st century] (by LI Bocong)	
1998		"The Importance of Philosophy to Engineering" (by Carl Mitcham)

Emerging State, 2000–2009

2002	*Gongcheng zhexue yinlun* 工程哲学引论 [An Introduction to Philosophy of Engineering] (by LI Bocong)	

2003	Graduate University of Chinese Academy of Sciences established the Center for Research of Engineering and Society.	*Engineering Philosophy* (by Louis L. Bucciarelli)
2004	(1) A symposium on philosophy of engineering organized by the Division of Engineering Management (DEM) of Chinese Academy of Engineering (CAE); (2) the Chinese Society for Philosophy of Engineering of Chinese Society for Dialectics of Nature was established; (3) The First Chinese Conference on Philosophy of Engineering; (4) The annual Chinese journal *Gongcheng yanjiu: Kua xueke shiye zhong de gongcheng* 工程研究: 跨学科视野下的工程 [Engineering Studies: Engineering in Interdisciplinary Perspectives] was set up; (5) The DEM began the first research program on the theory of philosophy of engineering.	(1) "Why we need a philosophy of engineering: A work in progress" (by Steven L. Goldman); (2) The International Network for Engineering Studies was founded; (3) A program on philosophy of engineering funded by the National Academy of Engineering.
2006		A series of seminars on philosophy of engineering by the Royal Academy of Engineering.
2007	*Gongcheng zhexue* 工程哲学 [Philosophy of Engineering] (by YIN Ruiyu, WANG Yingluo and LI Bocong)	(1) *Philosophy in Engineering* (by Steen H. Christensen); (2) An International Research Conference on "Occupational Bildung and Philosophy in Engineering" in Denmark; (3) The Workshop on Philosophy and Engineering (WPE) in the Netherlands.
2008		The Workshop on Philosophy and Engineering (WPE) in the United Kingdom

2009	(1) *Engineering Studies* journal was established; (2) *Handbook of the Philosophy of Science: Philosophy of Technology and Engineering Sciences* (ed. Anthonie Meijers)

Development Stage, 2010–present

2010	(1) The Forum on Philosophy, Engineering and Technology in the US; (2) The "Philosophy of Engineering and Technology" book series by Springer (ed. Pieter E. Vermaas)
2011	*Gongcheng yanhua lun* 工程演化论 [The Theory of Engineering Evolution] (ed. YINin Ruiyu, WANG Yingluo, and LI Bocong)
2013	*Gongcheng zhexue* 工程哲学 [Philosophy of Engineering] (2nd ed.) (ed. YIN Ruiyu, WANG Yingluo, and LI Bocong)
2015	*Gongcheng fangfalun* 工程方法论 [The Theory of Engineering Methodology] (ed. YIN Ruiyu, WANG Yingluo, and LI Bocong)
2018	*Gongcheng zhexue* 工程哲学 [Philosophy of Engineering] (Third Edition) (ed. YIN Ruiyu, WANG Yingluo, and LI Bocong)
2019	*Steps toward a Philosophy of Engineering: Historico-Philosophical and Critical Essays* (by Carl Mitcham)
2020	*Gongcheng zhishilun* 工程知识论 [The Theory of Engineering Knowledge] (ed. YIN Ruiyu, LI Bocong, and Enjie Luan)
2021	English translation of *An Introduction to Philosophy of Engineering: I Create, therefore I Am* (by LI Bocong) *The Routledge Handbook of the Philosophy of Engineering* (ed. Diane P. Michelfelder and Neelke Doorn)

engineering, science, and technology; the essence and characteristics of engineering; engineering case studies of mega-projects; and the categories of the philosophy of engineering. Western scholars primarily explore the philosophy of engineering from the perspective of engineering ethics, engineering design, engineering methods, and engineering education.

At present, the philosophy of engineering has shown a trend of rapid development, which likely will continue, and this area will receive unprecedented focus and support from academia and society in China and the West. As it has a good foundation in theory and institutionalization, the phenomenon of "strong momentum" and "accelerated development" is likely to continue. During the past three decades, China and the West have developed their own characteristics and advantages in parallel steps toward the philosophy of engineering. In the future, they should strengthen academic exchanges and cooperation and jointly promote the philosophy of engineering from the margins to the central area of philosophy.

ACKNOWLEDGMENTS

The authors wish to acknowledge Dr. Glen Miller and Dr. Qin ZHU for their assistance in the final English editing.

REFERENCES

Broome Jr., Taft H. 1991. "Bridging Gaps in Philosophy and Engineering." In Durbin 1991a, 265–77.
Bucciarelli, Louis L. 2003. *Engineering Philosophy*. Delft: Delft University Press.
Christensen, Steen Hyldgaard, Martin Meganck, and Bernard Delahousse, eds. 2007. *Philosophy in Engineering*. Aarhus: Academica.
Cutcliffe, Stephen H. 2000. *Ideas, Machines, and Values: An Introduction to Science, Technology and Society Studies*. New York: Rowman & Littlefield.
Downey, Gary L., and Juan C. Lucena. 1995. "Engineering Studies." In *Handbook of Science and Technology Studies*, edited by Sheila Jasanoff, Gerald E. Markle, James C. Petersen and Trevor Pinch, 167–88. Thousand Oaks, CA: SAGE.
Durbin, Paul T., ed. 1990. *Broad and Narrow Interpretations of Philosophy of Technology*. Dordrecht: Kluwer Academic.
———, ed. 1991a. *Critical Perspectives on Nonacademic Science and Engineering*. Bethlehem, PA: Lehigh University Press.
———. 1991b. Introduction to Durbin 1991a, 11–23.
———. 2006. *Philosophy of Technology: In Search of Discourse Synthesis*. Special issue of *Techné: Research in Philosophy of Technology* 2, no. 2: 1–283.

Goldman, Steven L. 1990. "Philosophy, Engineering, and Western Culture." In Durbin 1990, 125–52.

———. 1991. "The Social Captivity of Engineering." In Durbin 1991a, 121–45.

———. 2004. "Why We Need a Philosophy of Engineering: A Work in Progress." *Interdisciplinary Science Reviews* 29, no. 2: 163–76.

Goldman, Steven L., and Stephen H. Cutcliffe. 1991. Foreword to Durbin 1991a.

Houkes, Wybo. 2009. "The Nature of Technological Knowledge." In Meijers 2009, 309–50.

Koen, Billy Vaughn. 1991. "The Engineering Method." In Durbin 1991, 33–59.

Kuhn, Thomas S. 1996. *The Structure of Scientific Revolutions*. Third edition. Chicago: University of Chicago Press.

Lakatos, Imre. 1970. "History of Science and Its Rational Reconstructions." In *PSA: Proceedings of the Biennial Meeting of the Philosophy of Science Association*, 91–136. Chicago: University of Chicago Press.

LI, Bocong. 1988. *Rengonglun tigang* 人工论提纲 [An outline for the theory of human making]. Xi'an: Shaanxi Science and Technology Press.

———. 1992. "Wo zao wu, gu wo zai: Jianlun gongcheng shizailun" 我造物, 故我在: 简论工程实在论 [I create, therefore I am: On engineering realism]." *Ziranbianzhengfa yanjiu* 自然辩证法研究, no. 12: 9–19.

———. 1995. "Nuli xiang jingji zhexue he gongcheng zhexue lingyu kaituo: Jianlun 21shiji zhexue de zhuanxiang" 努力向经济哲学和工程哲学开拓: 简论21世纪哲学的转向 [Working hard to initiate philosophy of engineering and of economy: On the turn of philosophy of the 21st century]. *Ziranbianzhengfa yanjiu* 自然辩证法研究, no. 2: 1–6; 22.

———. 2002. *Gongcheng zhexue yinlun: wo zao wu, gu wo zai* 工程哲学引论: 我造物, 故我在 [An Introduction to Philosophy of Engineering: I Create, Therefore I Am]. Zhengzhou: Daxiang Press.

———. 2008. "Ershiyi shiji zhi chu gongcheng zhexue zai dong xi fang de tongshi xingqi" 二十一世纪初工程哲学在东西方的同时兴起 [The rise of the philosophy of engineering in East and West at the beginning of the 21st century]. *Zhongguo gongcheng kexue* 中国工程科学 10, no. 3: 13–16.

———. 2010. "The Rise of Philosophy of Engineering in the East and the West." In *Philosophy and Engineering: An Emerging Agenda*, edited by Ibo van de Poel and David E. Goldberg, 31–40. Dordrecht: Springer.

———. 2021a. "Gongcheng zhexue: Huigu yu zhanwang" 工程哲学回顾与展望 [Philosophy of engineering: Retrospect and prospect]. *Zhexue dongtai* 哲学动态, no. 1: 37–39.

———. 2021b. *An Introduction to Philosophy of Engineering: I Create, Therefore I Am*. Translated by Nan WANG and Shunfu Zhang. Dordrecht: Springer.

LI, Bocong, and Sumei CHENG. 2011. "Gongcheng zhexue de xingqi ji dangqian fazhan" 工程哲学的兴起及当代发展 [The rise and development of philosophy of engineering: An academic interview with Professor LI Bocong]. *Zhexue fenxi* 哲学分析, no. 4: 146–62.

Marx, Karl, and Frederick Engels. (1848) 1969. "Manifesto of the Communist Party." In *Marx/Engels Selected Works*, vol. 1, by Karl Marx and Frederick Engels.

Moscow: Progress Publishers, 98–137. https://www.marxists.org/archive/marx/works/1848/communist-manifesto/ch04.htm.

Meijers, Anthonie, ed. 2009. *Handbook of the Philosophy of Science: Philosophy of Technology and Engineering Sciences*. Amsterdam: Elsevier.

Michelfelder, Diane P., and Neelke Doorn. 2021. Introduction to *The Routledge Handbook of Philosophy of Engineering*, edited by Diane P. Michelfelder, and Neelke Doorn, 1–7. Oxfordshire: Routledge.

Mitcham, Carl. 1991. "Engineering as Productive Activity: Philosophical Remarks." In Durbin 1991a, 80–117.

———. 1994. *Thinking Through Technology: The Path between Engineering and Philosophy*. Chicago: University of Chicago Press.

———. 1998. "The Importance of Philosophy to Engineering." *Teorema* 17, no. 3: 27–47.

———. 2013. "Foreword: Prospects in the Philosophy of Engineering: An Exchange between the Editors and Carl Mitcham." In *Philosophy and Engineering: Reflections on Practice, Principles and Process*, edited by Diane P. Michelfelder, Natasha McCarthy, and David E. Goldberg, v–xi. Dordrecht: Springer.

———. 2020. *Steps toward a Philosophy of Engineering: Historico–Philosophical and Critical Essays*. New York: Rowman & Littlefield.

———. 2021. "Making Engineering into Philosophy: Foreword to the English Edition." In LI Bocong 2021b, vii–ix.

Mitcham, Carl, LI Bocong, Byron Newberry, and ZHANG Baichun, eds. 2018. *Philosophy of Engineering, East and West*. Dordrecht: Springer.

Mitcham, Carl, and Nan WANG. 2015. "From Engineering Ethics to Engineering Politics." In *Engineering Identities, Epistemologies and Values*, edited by Steen Hyldgaard Christensen, Christelle Didier, Andrew Jamison, Martin Meganck, Carl Mitcham and Byron Newberry, 307–24. Dordrecht: Springer.

Van de Poel, Ibo, and David E. Goldberg. 2010. "Philosophy and Engineering: Setting the Stage." In *Philosophy and Engineering: An Emerging Agenda*, edited by Ibo van de Poel and David E. Goldberg, 1–14. Dordrecht: Springer.

WANG, Nan. 2015. "The Emergence of Chinese Discussions of Engineering, Philosophy, and Ethics." *Technology in Society* 43: 51–56.

YIN, Ruiyu. 2010 (December 7). "Gongchengyuan guanli xuebu jiji tuidong gongcheng zhexue yanjiu" 工程院管理学部积极推动工程哲学研究 [The Division of Engineering Management of the Chinese Academy of Engineering actively promotes research in the philosophy of engineering]. *Keji ribao* 科技日报.

YIN, Ruiyu, LI Bocong, and Enjie LUAN, eds. 2020. *Gongcheng zhishilun* 工程知识论 [The theory of engineering knowledge]. Beijing: *Gaodeng jiaoyu chubanshe* 高等教育出版社.

YIN, Ruiyu, WANG Yingluo, and LI Bocong, eds. 2007. *Gongcheng zhexue* 工程哲学 [Philosophy of engineering]. Beijing: *Gaodeng jiaoyu chubanshe* 高等教育出版社.

———, eds. 2013. *Gongcheng zhexue* 工程哲学 [Philosophy of engineering]. 2nd ed. Beijing: *Gaodeng jiaoyu chubanshe* 高等教育出版社.

———, eds. 2018. *Gongcheng zhexue* 工程哲学 [Philosophy of engineering]. 3rd ed. Beijing: *Gaodeng jiaoyu chubanshe* 高等教育出版社.

YIN, Ruiyu, LI Bocong, and WANG Yingluo, eds. 2011. *Gongcheng yanhua lun* 工程演化论 [The theory of engineering evolution]. Beijing: *Gaodeng jiaoyu chubanshe* 高等教育出版社.

———, eds. 2015. *Gongcheng fangfa lun* 工程方法论 [*The Theory of Engineering Methodology*]. Beijing: *Gaodeng jiaoyu chubanshe* 高等教育出版社.

ZHAO, Jianjun. 2004. "Gongchengjie yu zhexuejie xieshou gongtong tuidong gongcheng zhexue fazhan" 工程界与哲学界携手共同推动工程哲学发展 [The engineering and philosophical field work together to promote philosophy of engineering]. In *Gongcheng yanjiu* 工程研究 [Engineering Studies] (vol. 1), edited by Cheng DU and LI Bocong, 209–12. Beijing: Beijing Institute of Technology Press.

Chapter 11

The Development of the Philosophy of Engineering in China

Engaging the Scholarship of Carl Mitcham

Tong LI and Yongmou LIU

This chapter takes a cross-cultural perspective to examine the development of the philosophy of engineering in China including the unique themes that have arisen in the Chinese context. The scholarship of American philosopher Carl Mitcham serves as a frame of reference to consider how the major concerns in the philosophy of engineering in China are related to and distinct from those in the West. Mitcham's scholarship is crucial in this project as: (1) he was one of the first Western philosophers of technology read by Chinese scholars, and his work played a significant role in creating and institutionalizing Chinese research in philosophy of technology and philosophy of engineering; and (2) his work has had widespread influence in the West in philosophy of technology and has been central in Western discourse on philosophy of engineering.

Mitcham's ideas have been part of the Chinese dialectic since they were introduced to Chinese scholars in the 1980s. His definition of engineering and his views on the philosophy of engineering have often been treated as an important Western paradigm. As the problems of engineering in China have their own backgrounds, Chinese scholars often took a comparative and pragmatic approach to Mitcham's work. They examined the differences in the cultural and historical contexts that shaped Mitcham's work from what existed in China and interrogated to what extent and in what ways his work could be used for creating a philosophy of engineering responsive to Chinese realities.

They identified three crucial differences. First, the history of engineering in China is quite different from that in Western countries. Instead of furthering economic "use and convenience," Chinese engineering was initiated during the Self-Strengthening Movement (1860s–1890s) that aimed to prevent China's colonization. So the Chinese word *gongcheng* (工程) always has social and political meanings. Second, in understanding engineering and philosophy of engineering, Mitcham's view "engineering is philosophy" has been widely accepted in China. While acknowledging the reflective and critical aspects of engineering, Chinese scholars have further highlighted the transformative and ideological dimensions of engineering that include Marxist philosophical ideas such as technology or engineering as a way of "changing the world." Third, when considering the responsibility of engineers, Chinese scholars look to traditional Chinese ethics and thus create ethical dilemmas and pathways distinct from Western ones. In addition, there are some other themes in the philosophy of engineering in China that were influenced by Mitcham's scholarship but are responsive to the Chinese social and cultural contexts, including historical studies, science and technology studies (STS), comparative philosophy of engineering, and so on.

INTERPRETING MITCHAM'S WORK IN CHINESE CONTEXTS

As in Western countries, the development of the philosophy of engineering in China started with the question: what is the relationship between science, technology, and engineering? Similarly, the first scholars to explore this question in the West and China were engineers. In 1957, QIAN Xuesen (1911–2009), the engineer who was well known for leading the Chinese national initiative *liangdan yixing* (Two Bombs and One Satellite), distinguished natural science, technological science, and engineering technology in his paper "Jishu kexue zhong de fangfalun wenti (技术科学中的方法论问题, Methodological issues in technological science)" (1957). He noticed that engineers relied more on experiences and analogies to solve practical problems, which could not be solved by rigid scientific method. Meanwhile Marxist philosophers also realized the need to distinguish those concepts, for engineering had become one of the most essential forces for national development. In the same year, CHEN Changshu (1932–2011) pointed out that Chinese scholars should pay attention to the philosophical issues not only in science, but also in technology. Given the lack of resources for conceptualizing and philosophizing science, technology, and engineering, CHEN Changshu argued that Chinese philosophers had the moral imperative to introduce and translate Western, especially English, materials to jumpstart their discussions (CHEN Changshu 1957).

In 1978, the Reform and Opening Up policy emphasized the critical role of science and technology in improving China's economy, social development, and national competitiveness. The government encouraged the development of science and technology and called for efforts to learn science and technology from the West. As part of the initiative, Western scholarship concerning technology was translated into Chinese, and scholars theorized, institutionalized, and constructed new areas of philosophical research, including the philosophy of engineering, attuned to the socioeconomic situations of China, while conducting critical and comparative analysis of the Western scholarship.

Philosophy of engineering was thus developed in China to critically examine social and philosophical implications of engineering—a most important contributing factor for China's ascent in the global economy. In his 1993 article "Wo zaowu, gu wozai: Jianlun gongcheng shizailun (我造物, 故我在: 简论工程实在论, I create, therefore I am: On engineering realism)," LI Bocong conceptualized engineering as a philosophy of *zaowu* (造物, making or creating). Unlike Descartes, who highlighted the connection between existence and thinking, LI Bocong argues that the nature of the human is not thinking but creating. As he pointed out, we are not existing in a natural state but in a world we created, and through this creation, we become who we are. So philosophy of engineering is about doing and living, which emphasizes changing the world over knowing it (LI Bocong 1993). His book *Gongcheng zhexue yinlun: Wo zaowu gu wo zai* (工程哲学引论: 我造物故我在, *The introduction of philosophy of engineering: I create, therefore I am*) (2002) represented a shift in the philosophical studies of engineering from the philosophy of technology concerning engineering, to the philosophy of engineering (Fan CHEN and Qianhe CAI 2009).

Also in the 1990s, Mitcham became one of the most well-known philosophers of technology and engineering in China, mainly on account of the distinctions he drew between science, technology, and engineering; his STS approach; and his emphasis on ethics. Major Chinese philosophers of technology and engineering such as CHEN Changshu, LI Bocong, and Fan CHEN cited him as a major figure in the Western philosophy of technology. In 1992, Mitcham came to China and attended the "Sino-American Seminar on STS" hosted by the Chinese Society for the Dialectics of Nature and Tsinghua University. Since then, his interactions with Chinese scholars have become more frequent. His books and papers were translated into Chinese and used as textbooks for students in philosophy of science and technology. *Thinking through Technology: The Path between Engineering and Philosophy* (1994a, translated into Chinese in 2008) and the anthology *Philosophy and*

Engineering: Historical-Philosophical and Critical Perspectives (edited and translated in 2013) are considered required reading for Chinese scholars and students in the philosophy of technology and engineering. Universities in China have invited him to give lectures, and he has also taught classes as an International Distinguished Professor at Renmin University (see Ping YAN 2017).

To a large extent, discussions on Mitcham's philosophy of engineering started as a way for the Chinese scholars to understand the major arguments and debates in the Western philosophy of engineering and to reflect on how to construct the Chinese approach to engineering and philosophy. These discussions mainly focus on three aspects. First, Chinese scholars paid a lot of attention to Mitcham's work on the theoretical justification for the philosophy of engineering. Based on his comprehensive survey of the history of engineering, Mitcham has emphasized the differences between science, technology, and engineering, and pointed out the necessity of studying the philosophy of engineering. In "The Importance of Philosophy to Engineering" (1998), Mitcham maintained that for the purposes of self-defense, self-interest, and self-knowledge, engineers need a philosophy of engineering. Chinese scholars realized that the situation in China was the same, and since the modernization period, engineering has become increasingly important in China. More and more scholars have realized the value of philosophical inquiry into engineering for justifying the role of engineering in societal and economic development. From 2004, the Chinese Academy of Engineering has sponsored several research projects, headed by philosophers working with engineers, to study the philosophy of engineering in Chinese contexts (LI Bocong 2021).

Second, Chinese scholars discussed Mitcham's work on the distinction between the two traditions in the philosophy of technology: the engineering philosophy of technology (EPT) and the humanities philosophy of technology (HPT) (Mitcham 1994a). Most Chinese scholars have been influenced by this distinction and have used it to critique their own work and situate it in the community of the philosophy of engineering. As CHEN Changshu (1957) said, his book *Jishu zhexue yinglun* (技术哲学引论, *An introduction to the philosophy of technology*) was closer to the EPT approach but lacked HPT. Indeed, the "Northeastern School" was inspired by CHEN Changshu's preferred EPT studies, while the approaches of scholars in Beijing were mostly closer to HPT (LIU Zeyuan and Fei WANG 2002). In response to CHEN Changshu's self-critique, LIU Dachun (2007) pointed out that philosophical studies of science, technology, and engineering were quite young in China, which had not made a clear distinction between the two traditions. Despite CHEN Changshu's humble self-evaluation of the lack of humanistic assessment of technology in his work, his book still touched on at least some social or humanistic aspects of engineering, such as the influence of state intervention and issues of sustainable development. Historically and culturally, most Chinese scholars are familiar with the problems caused by the dualism between science and humanities that started in the early twentieth century, so in the area of philosophy of engineering, almost all of them are sensitive about the distinction between the two traditions and view bridging the gap as one of their essential missions.

Third, Chinese scholars were attracted by Mitcham's work on the importance of ethics and ethics education in engineering. In *Thinking through Technology*, Mitcham emphasized that it is critical for engineers to learn about the social influences and outcomes technology may generate. Nowadays, engineering has generated even more power to change the world and the future, but the specialization of engineering makes it harder than ever to morally assess a project from nonexpert perspectives. Furthermore, the stratification of modern engineering design makes it so complicated that even engineers themselves can hardly forecast and evaluate the project as a whole (see Qin ZHU 2009). From the perspective of the philosophy of technology, Mitcham is clear that engineering should not be seen as value free, and values often conflict with one another. Since the Reform and Opening Up initiative, the market economy with Chinese characteristics has brought not only wealth but also value conflicts to Chinese people. Against such a philosophical background, Chinese scholars have realized the value of teaching ethics to engineers and engineering students in helping them develop critical attitudes toward the role of engineering in changing the world.

CHINESE APPROACHES TO PHILOSOPHY AND ENGINEERING

This section provides a more comprehensive survey of the major unique themes developed by Chinese scholars in the philosophy of engineering based on: (1) the inspirations they received from their critical examinations of Western scholarship, again with Mitcham's work serving as paradigmatic; and (2) their reflections on the cultural and historical foundations of the philosophy of engineering in China.

Varied Etymologies of Engineering Concepts

Historical differences have led to different understandings of engineering in different cultures. Mitcham has emphasized the importance of historical studies of engineering, especially of key engineering concepts, which usually were presented ahistorically. He analyzes the etymology of engineering concepts such as "engineering," "technology," and "making" to illuminate how these concepts have evolved in and been transformed by different cultural contexts. The concepts of science, technology, and engineering are quite modern (or foreign) to the Chinese, and when they distinguish between technology and engineering, they take different considerations into account than most Western scholars do.

In 2010, LI Bocong started leading a research project called "The History of Engineering in Modern China" and called on scholars from engineering, philosophy, history, and other areas to collaborate (LI Bocong, LI Sanhu and Bin LI 2017). By teasing out the timeline of the development of engineering in China, researchers suggest that what they are studying is not a history of engineering per se, but a multidimensional history project that includes scientific, economic, policy, military, and even transnational dimensions.

Many scholars have discussed the etymology of Chinese terms *gongcheng* (工程, engineering), *gongchengshi* (工程师, engineer), and other related words, which were developed from more ancient words *gong* (工, work or worker, artistic designing, multifunctional tool) and *cheng* (程, measurement, transportation). *Gongcheng* literally means a task that demands careful gauging and calculation (Miller, WANG, Sethy, and Atsushi 2021). These terms were first used by British missionary John Fryer (1839–1928), who wanted to translate scientific books into Chinese (YIN Wenjuan 2017). Chinese engineering studies were ignited as part of the Self-Strengthening Movement (again, 1860s–1890s), an embrace of science and technology for military purposes supported by the Qing government. "Engineering" became one of the five subjects taught at Beiyang West Learning College in Tianjin in 1895, and scholars believed that engineering could save China from being colonized, even after the collapse of the Qing government. As a result, the social and political aspects in engineering have always been emphasized in Chinese contexts.

A group of scholars led by former vice minister of education WU Qidi have written a book in three volumes called *Zhongguo gongchengshi shi* (中国工程师史, *The history of Chinese engineers*) (2017), which divides this history into three parts: *gongcheng* in ancient times; from the Self-Strengthening Movement to the founding and early decades of the People's Republic of China; and after the "Reform and Opening Up" in 1978. The historiography here is quite political. In this book, they claim that we should pay attention to craftsmen as well as to the administration of related activities, for example, the role of the ministry of *gong* (工部) in the government and the society (WU Qidi 2017). The ministry of *gong* was one of the six ministries in feudal China, which managed all kinds of ancient *gongcheng* including metallurgy, construction, water conservancy, ceramics, shipbuilding, textiles, and so on. In 1734, the ministry of *gong* in the Qing government published a seventy-four-volume book called the *Gongcheng zuofa zeli* (工程做法则例, *Patterns of gongcheng*), in which *gongcheng* mainly referred to construction or civil engineering. Interestingly, this book is not merely about how to build houses, but also how to make the architecture beautiful, and, most importantly, to build them so that they conform to standards of Confucian principles such as propriety and morality.

After the founding of the People's Republic of China, *gongcheng* gained an even broader meaning to include social and political projects. Many social and economic projects led by the central government have been called *gongcheng*, such as the *xiwang gongcheng* (希望工程, Hope Project) (a social project to support the children in poverty) and the *lianzheng gongcheng* (廉政工程, Government Integrity Project, a political task to stamp out corruption). As YIN Wenjuan concludes, *gongcheng* has at least three special characteristics in the Chinese context: (1) instead of designing, operating is viewed as the central component of *gongcheng*; (2) *gongcheng* refers to the whole life cycle of the project, which calls for a more holistic ethical foundation; (3) the concept of *gongcheng* is not subordinate to technology, and Chinese scholars tend to define them as two separate categories that are relatively independent of each other (YIN Wenjuan 2017).

As a result, the philosophy of engineering also gains its independent status from the philosophy of technology. LI Bocong has argued for a tripartite relationship between science, technology, and engineering, which has been welcomed by both engineers and philosophers in China and which allows differences to be explained. For example, LI suggests that innovation in science and technology only occurs the first time a theory is articulated or a new artifact has been made. But engineering innovation is present in each project because each is unique. For example, each railway is built in different geographic and social situations, so the second Tibet railway could not be accomplished simply by repeating the first one. As such, engineering innovation becomes a distinct topic that should be studied with different methods (LI Bocong 2010a).

Confucianism and Marxism as Intellectual Foundations

In China, there is widespread agreement that Mitcham is right when he says that engineers need a philosophy of engineering. But when discussing Mitcham's famous phrase that "engineering is philosophy," the Chinese contexts also bring in some additional dimensions. We may say, as Mitcham suggests, engineers in the United States are "the unacknowledged philosophers of the postmodern world" (Mitcham 1998), whereas in China, engineers are widely seen as philosophers by nature. When hearing "engineering is philosophy," most Chinese would quickly be reminded of the famous slogan by Karl Marx: "philosophers have hitherto only interpreted the world in various ways; the point, however, is to change it." From the very beginning in modern China, engineering was seen as a pragmatic means to change the world. This has fundamentally influenced Chinese people's understanding of philosophy as well as engineering. Think again of QIAN Xuesen: most Chinese view him as a great engineer as well as an excellent philosopher.

As mentioned above, LI Bocong has proposed that engineering is a philosophy of *zaowu*. Although LI's *zaowu* is sometimes translated into English as making, it means something different than Mitcham's idea that engineering is a philosophy of making. *Zao* (造, making) is a practical process of creating something consciously, and *wu* (物, objects or artifacts) refers to artifacts that are the products of purposeful design and labor. Through *zaowu*, humans apply what they know about the world into practice, and, in turn, their knowledge about the world can be tested. Therefore, the philosophy of *zaowu* creates a Marxist, dialectic relationship between practice and theory. In other words, practice is the sole criterion for testing truth. As LI Bocong (2001) emphasizes, a philosophy based on knowing the world is epistemology; philosophy also must consider changing the world, which refers to the philosophy of engineering. The power of engineering to change the world is widely acknowledged, and engineering students are encouraged to adopt it as a life quest. In 2017, the Chinese Ministry of Education released several policy guidelines to create the Emerging Engineering Education (3E) initiative. ZHONG Denghua, the president of Tianjin University, as well as a member of the Chinese Academy of Engineering, explained the 3E's aims as cultivating talented engineers with high moral standards who can cope with changes and shape the future (2017). In other words, instead of adapting to the society, engineers should consciously and proactively shape the society.

Besides Marxism, traditional Chinese thoughts such as Confucianism are another intellectual source for the Chinese philosophy of engineering. LI Bocong and many other scholars prefer to use Confucian concepts such as *Dao* (道, way) and *qi* (器, objects or artifacts) to explain the philosophical characteristics of engineering. In traditional Chinese philosophy, *Dao* is seen as the highest form of way or thought, while *qi* is the artifact that contains and represents Dao. As interpreted in *I Ching* (易经, *The book of change*): "*Dao* is antecedent to the material form that exists, as an ideal method, while *qi* is subsequent to the material form that exists, as a definite thing (形而上者谓之道□形而下者谓之器)." He argues that the word *wu* in *zaowu* is exactly the same as *qi*, which refers to artifacts in contrast to natural objects. *Qi* is not merely a material object; the design of *qi* represents its maker's beliefs and thoughts. The perfect *qi* only can be made by someone who has a profound understanding of *Dao*. In other words, *qi* is value-laden. Or as indicated in a Chinese slogan: *di yi zai dao* (器以载道, *qi* carries *Dao*).

Yuk Hui has traced this pair of concepts in Chinese history in his book *The Question Concerning Technology in China* (2016). As he concludes, *Dao* and *qi* were frequently used when discussing issues concerning technics in ancient China, but, starting with the Self-Strengthening Movement, were gradually replaced by Western concepts of theory and technology, and the holistic cosmological view in China was thus dismantled by modernization. If the problem in ancient times was valuing *Dao* more than *qi*, the major concern in modern times has become valuing *qi* more than *Dao*, or even detaching *qi* from *Dao*. The collapse of the *Dao-qi* cosmology has caused most moral crises nowadays. That's one of the reasons why Chinese scholars want to reinterpret the concept of *Dao-qi* as an introspection into modernity, especially in areas concerning technology and engineering. As LI Bocong (2018) argues, the ideal of *dao qi heyi* (道器合一, unification of *Dao* and *qi*) is a good model to show what the unification of philosophy and engineering should be like. *Dao* needs *qi* to carry it in order to be manifested in sensible forms; *qi* needs *Dao* in order to become perfect (in Daoism) or sacred (in Confucianism). And so the core issue of engineering education is the *Dao-qi* relationship (LI Bocong 2017).

Engineering Identity, Expertise, and Ideology

The emphasis on social and political aspects of engineering also explains why it is natural for the Chinese scholars to accept Mitcham's idea that engineering is value laden as well as his STS approach. Mitcham has divided the development of engineering ethics into three stages, first loyalty, next technocracy, and then social responsibility (Mitcham 1994b). Through a historic inquiry into the revisions of "*Zhongguo gongchengshi xintiao* (中国工程师信条, The code of ethics of the Chinese Institute of Engineer)" from 1933 to 1996, Junbin SU and CAO Nanyan (2008) point out that the ethical ideologies of engineers in China developed along a different way. Under the pressures of colonization, social responsibility was the core idea of engineering at the very beginning. The first engineers were not soldiers, but intellectual elites who wanted to enlighten their people and defend their country. Chinese engineers began with the mission to reform society. And political factors have always been important sources of influences on Chinese engineers. This has later catalyzed studies in the areas of Science, Technology, and Public Policy (STPP) and Engineering and Public Policy in China. These approaches concern individuals involved in engineering activities, and the relationships between them.

One of the most remarkable facts here is the rise of the *red engineers* as Joel Andreas (2009) has noticed. Red engineers are those who have been cultivated by two highly selective credential systems, one academic and the other political or ideological. After the Reform and Opening Up in 1978, red engineers have become the stable core in government, enterprises, and many public institutions in China. Therefore, we argue that the term *red engineers* basically emphasizes the intrinsic political obligations of engineers and it is these political obligations that define who Chinese engineers are.

In fact, there is a subject called social engineering derived from QIAN Xuesen's system theory. According to QIAN, society is a huge system, and social engineering is to use the method of system engineering to build a better society. As BAI Shuying (2009) points out, the name social engineering may not be a proper translation, for its theories and practices are more like social policy. Nevertheless, it is a term that shows how officials are eager to solve social issues with engineering thinking and methods. Taking QIAN Xuesen again as an example: as a top engineer at the time, he was elevated as the vice chairman of the Chinese People's Political Consultative Conference from 1986 to 1998. A political-technocratic alliance seems to have appeared in history.

But how to define red engineers is also contested. Despite the increasing numbers of officials with engineering backgrounds, it is hard to say that China has totally become a technocracy. As Yongmou LIU and Zhou QIU (2016) point out, an engineering background does not mean governmental officials will all apply engineering methods to make policy decisions, and often the political capital is what really matters after all. That is to say, the political sense of "red" is much more operative than other values such as expertise.

Within this context can we understand why so many Chinese scholars pay attention to engineering communities. This is an idea related to the science community theories of Karl Polanyi and Thomas Kuhn, although the engineering community has unique characteristics. Members of a science community have many similarities, while an engineering community is more heterogeneous, complex, and localized. Chinese scholars have distinguished two kinds of engineering communities. One is professional community, which refers to labor unions and academic societies related to engineering. The other is practical community, which contains all kinds of people participating in the engineering project. Although they share the same goal, members in a practical community may encounter many conflicts of values, beliefs, interests, and so on (LI Bocong 2010b). So engineers, especially the ones leading the project, have to deal with different relationships. As YIN Ruiyu (2018) points out, engineering is a combination of technical and nontechnical factors: the former form the basis of engineering, the latter delimit its boundaries (YIN

Ruiyu, WANG Yingluo, and LI Bocong 2018). On some occasions, if not always, political factors become the most important determinant. And this ideological concern has also been embedded in engineering education.

From Theory to Practice in Engineering Ethics Education

Engineering ethics education in China has been extensively promoted by scholars from the philosophy of engineering. Engineering ethics education can thus be seen as one of the practical results of the philosophy of engineering. Since the establishment of the philosophy of engineering in China, engineering ethics education has changed in many different ways: from the curricular perspective, the standalone course of engineering ethics has gained its independent position; from the pedagogical perspective, there is more emphasis on practice instead of theory, and a student-centered approach is encouraged. In this process, Mitcham's thoughts about engineering ethics have influenced Chinese approaches to engineering ethics education in both theoretical and practical ways.

At the theoretical level, Mitcham's work has helped to justify the necessity of engineering ethics education. As engineering is becoming more and more powerful in changing the world, engineering ethics education helps engineers and engineering students develop a self-awareness of the challenges and risks they are facing. But for a long time, Chinese engineering ethics education was subordinate to ideological and political education. The spread of technocracy and scientism in modern China also enhanced the thought that engineering ethics can be reduced to engineers' self-discipline. Textbooks mostly told stories of moral exemplars instead of explaining civic virtues and professional morality (CAO Nanyan 2004). Many scholars called for an independent, standalone course of engineering ethics. By learning from the theories from the Western traditions, Chinese scholars have constructed some frameworks for engineering ethics education. In 2007, LI Bocong and two members of the Chinese Academy of Engineering, YIN Ruiyu and WANG Yingluo, coedited a textbook *Gongcheng zhexue* (工程哲学, *Philosophy of Engineering*), which provided the basis for a required course for engineering students. And the idea of teaching engineering students and engineers ethics is highlighted as a major goal of engineering education in the textbook. Recently, Tsinghua University, the cradle of red engineers, organized a group of scholars to edit a textbook *Gongcheng lunli* (工程伦理, *Engineering ethics*) (Zhengfeng LI, Hangqing CONG, and Qian WANG 2016). They also held training sessions for teachers in engineering on how to use their textbook to teach engineering ethics. Both textbooks have clearly distinguished technology and engineering: the former concerns skills and artificial objects, while the latter is an activity as well as a dynamic system. Chinese scholars claim that engineers should have their own standards of ethics.

At the practical level, Mitcham has influenced the Chinese agenda to keep in mind the mutual influence between engineering and ethics: engineers have constructed bridges between engineering and philosophy, especially to that branch of philosophy constituted by ethics (Mitcham 1998). After the Reform and Opening Up, many new problems in engineering appeared. Most Chinese scholars used to classify them as either economic (e.g., wage) or technological (e.g., poor project quality) problems, and ignore the ethical issues behind them. Many scholars advocate for a more practical approach in engineering ethics and professional education. As LI Bocong (2006) points out, scholars of ethics should care more about practice. As engineering activities are the most vital, universal, and basic kind of practice, engineering should be put at the center of practical ethics research. And engineering ethics education should be closely connected to practice. As one of the core ideas of the new initiative Emerging Engineering Education, engineering students are asked to pay more attention to capacity building and to learn how to discover and address ethical issues in real-world engineering practices.

CONCLUSION

In the Chinese community of the philosophy of engineering, Carl Mitcham is among the most influential philosophers. His ideas and methods have inspired many Chinese scholars. But instead of accepting his theory as it is, many scholars adapt it to the Chinese context. Nevertheless, it is true that both Chinese scholars and Mitcham are advocates for a philosophy of engineering. In 2018, Mitcham edited *Philosophy of Engineering, East and West*. Alongside papers by Western scholars, Chinese philosophers LI Bocong and LIU Dachun, engineer YIN Ruiyu, and other scholars contributed chapters discussing their ideas, including theoretical and practical issues related to the philosophy of engineering, history, case studies in engineering, and the ethical and social studies of engineering. As Mitcham said, the unifying theme of this collection is an effort to intensify a twofold encounter: between engineering and philosophy and between China and the West (Mitcham, LI, Newberry, and ZHANG 2018). The philosophy of engineering is a transdisciplinary as well as transnational area of study: scholars from different academic and cultural backgrounds can inspire each other and generate a comprehensive and global perspective of engineering. Perhaps, rather than a subject which only has one route and one answer, the philosophy of engineering itself is a broad area that should be studied from diverse perspectives.

REFERENCES

Andreas, Joel. 2009. *Rise of the Red Engineers: The Cultural Revolution and the Origins of China's New Class*. Stanford, CA: Stanford University Press.

BAI, Shuying. 2009. "Qiubian, rentong yu rongru: Shehui gongcheng de kunjing yu chaoyue" 求变、认同与融入：社会工程的困境与超越 [Changes, identity, and integration: Overcoming dilemmas in social engineering]. In *Gongcheng yanjiu: Kuaxueke shiye zhong de gongcheng* 工程研究：跨学科视野中的工程 [Engineering studies: Engineering in the interdisciplinary context], edited by Cheng DU and LI Bocong, 181–89. Beijing: Beijing Institute of Technology Press.

CAO, Nanyan. 2004. "Dui zhongguo gaoxiao gongcheng lunli jiaoyu de sikao" 对中国高校工程伦理教育的思考 [Reflections on engineering ethics education at Chinese universities]. *Gaodeng gongcheng jiaoyu yanjiu* 高等工程教育研究, no. 5: 37–39.

CHEN, Changshu. 1957. "Yao zhuyi jishu zhong de fangfalun wenti" 要注意技术中的方法论问题 [Paying attention to the methodological issues in technology]. *Ziran bianzhengfa yanjiu tongxun* 自然辩证法研究通讯, no. 2: 24.

———. 1999. *Jishu zhexue yinlun* 技术哲学引论 [An introduction to the philosophy of technology]. Beijing: Science Press.

CHEN, Fan, and Qianhe CAI. 2009. "Zhongwai gongcheng zhexue yanjiu zhi bijiao" 中外工程哲学研究之比较 [A comparative study of the philosophy of engineering research in China and the West]. *Ziran bianzhengfa tongxun* 自然辩证法通讯 31, no. 4: 82–87.

Hui, Yuk. 2016. *The Question Concerning Technology in China: An Essay in Cosmotechnics*. Falmouth, UK: Urbanomic.

LI, Bocong. 1993. "Wo zaowu, gu wozai: Jianlun gongcheng shizailun" 我造物，故我在：简论工程实在论 [I create, therefore I am: A brief commentary on engineering realism]. *Ziran bianzhengfa yanjiu* 自然辩证法研究 9, no. 12: 9–19.

———. 2001. "Wo si gu wo zai yu wo zaowu gu wo zai: Renshilun yu gongcheng zhexue chuyi" 我思故我在与我造物故我在：认识论与工程哲学刍议 [I think therefore I am and I create therefore I am: An essay on epistemology and the philosophy of engineering]. *Zhexue yanjiu* 哲学研究, no. 1: 21–24.

———. 2006. "Gongcheng yu lunli de hushen yu duihua" 工程与伦理的互渗与对话：再谈关于工程伦理学的若干问题 [The mutual penetration and dialogue between engineering and ethics: A second essay on issues concerning engineering ethics]. *Huazhong kejidaxue xuebao (shehui kexue ban)* 华中科技大学学报 (社会科学版), no. 4: 71–75.

———. 2010a. *Gongcheng chuangxin: Tupo bilei he duobi yanjing* 工程创新：突破壁垒和躲避陷阱 [Engineering innovation: Breaking through barriers and avoiding traps]. Hangzhou: Zhejiang University Press.

———. 2010b. *Gongcheng shehuixue daolun* 工程社会学导论：工程共同体研究 [Introduction to the sociology of engineering: Studies on the engineering community]. Hangzhou: Zhejiang University Press.

———. 2017. "Yi *Dao qi* heyi *qi* zai *Dao* zhong de linian chongshu gongcheng jiaoyu: Gongcheng jiaoyu zhexue biji zhiyi" 以道器合一、道在器中的理念重

塑工程教育: 工程教育哲学笔记之一 [Reshaping engineering education based on the ideas of the unification of *Dao* and *qi* and the embeddedness of *Dao* in *qi*]. *Gaodeng gongcheng jiaoyu yanjiu* 高等工程教育研究, no. 4: 22–29.

———. 2018. "Wo zaowu gu wozai: LI Bocong jiaoshou zai zhexuejia yu gongchengshi de duihua zhong de yanjiang baogao" 我造物故我在: 李伯聪教授在哲学家与工程师的对话中的演讲报告 [I create therefore I am: Professor LI Bocong's speech at the Dialogue between Philosophers and Engineers Forum]. *Zhongguo sanxia* 中国三峡, no. 7: 81–85.

———. 2021. "Gongcheng zhexue: Huigu yu zhanwang" 工程哲学: 回顾与展望 [Philosophy of engineering: Retrospect and prospect]. *Zhexue dongtai* 哲学动态, no. 1: 37–39.

LI, Bocong, Sanhu LI, and Bin LI. 2017. *Zhongguo jinxiandai gongchengshi gang* 中国近现代工程史纲 [An introduction to the history of engineering in Modern China]. Hangzhou: Zhejiang University Press.

LI, Zhengfeng, Hangqing Cong, and Qian Wang. 2016. *Gongcheng lunli* 工程伦理 [Engineering ethics]. Beijing: Tsinghua University Press.

LIU, Dachun. 2007. "Guanyu jishuzhexue de liangge chuantong" 关于技术哲学的两个传统 [On the two traditions in the philosophy of technology]. *Jiaoxue yu yanjiu* 教学与研究, no. 1: 33–37.

LIU, Yongmou, and Zhou QIU. 2016. "Jizhizhuyi yu dangdai zhongguo chuyi" 技治主义与当代中国关系刍议 [On the relationship between technocracy and contemporary China]. *Changsha ligong daxue xuebao (shehui kexueban)* 长沙理工大学学报 (社会科学版) 31, no. 5: 16–20.

LIU, Zeyuan, and Fei WANG. 2002. "Zhongguo jishulun yanjiu ershinian (1982–2002)" 中国技术论研究二十年 (1982–2002) [Twenty years of research in technology studies in China (1982–2002)]. In *Gongcheng jishu yu zhexue: 2002 nian juan zhongguo jishu zhexue yanjiu nianjian* 工程、技术与哲学: 2002年卷中国技术哲学研究年鉴 [Engineering, technology and philosophy: The 2002 year book of research in Chinese philosophy of technology], edited by LIU Zeyuan, WANG Xukun, and Qian WANG, 299–314. Dalian: Dalian University of Technology Press.

Miller, Glen, Xiaowei WANG, Satya Sundar Sethy, and Fujiki Atsushi. 2021. "Eastern Philosophical Approaches and Engineering." In *The Routledge Handbook of the Philosophy of Engineering*, edited by Diane P. Michelfelder and Neelke Doorn, 50–65. London: Routledge.

Mitcham, Carl. 1994a. *Thinking through Technology: The Path between Engineering and Philosophy*. Chicago: University of Chicago Press.

———. 1994b. "Engineering Design Research and Social Responsibility." In *Ethics of Scientific Research*, edited by Kristin Shrader-Frechette, 153–68. Lanham, MD: Rowman & LIttlefield.

———. 1998. "The Importance of Philosophy to Engineering." *Teorema* 17, no. 3: 27–47.

———. 2013. *Gongcheng yu zhexue: Lishide, zhexuede, he pipande shijiao* 工程与哲学: 历史的、哲学的和批判的视角 [Philosophy and engineering:

Historical-philosophical and critical perspectives]. Translated by Qian WANG et al. Beijing: People's Press.

Mitcham, Carl, LI Bocong, Byron Newberry, and ZHANG Baichun, eds. 2018. *Philosophy of Engineering, East and West*. Cham: Springer.

SU, Junbin, and CAO Nanyan. 2008. "Zhongguo gongchengshi lunli yishi de bianjian: Guanyu 'zhongguo gongchengshi xintiao' 1933–1996 nian xiuding de jishu yu shehui kaocha" 中国工程师伦理意识的变迁：关于《中国工程师信条》1933–1996年修订的技术与社会考察 [The evolution of the ethical awareness of Chinese engineers: A social and technological investigation into the revisions of the code of ethics of Chinese engineers during 1933 and 1966]. *Ziran bianzhengfa yanjiu* 自然辩证法研究 30, no. 6: 14–19.

QIAN, Xuesen. 1957. "Jishu kexue zhong de fangfalun wenti" 技术科学中的方法论问题 [Methodological issues in technological science]. *Ziran bianzhengfa yanjiu tongxun* 自然辩证法研究通讯, no 1: 33–34.

WU, Qidi, ed. 2017. *Zhongguo gongchengshi shi* 中国工程师史 [The history of Chinese engineers]. Shanghai: Tongji University Press.

YAN, Ping. 2017. "Miqiemu gongcheng lunli sixiang yanjiu" 米切姆工程伦理思想研究 [Mitcham's scholarship on engineering ethics]. PhD diss. Dalian University of Technology.

YIN, Ruiyu. 2018. "From Engineering to the Philosophy of Engineering: Philosophical Reflections of an Engineer." In Mitcham, LI Bocong, Newberry, and ZHANG Baichun, 2018, 75–81.

YIN, Ruiyu, WANG Yingluo, and LI Bocong. 2018. *Gongcheng zhexue* 工程哲学 [Philosophy of Engineering]. Third edition. Beijing: Higher Education Press.

YIN, Wenjuan. 2017. "Gongcheng zhexue shiyu xia gongcheng yu engineering de zhuanyebuduicheng fenxi" 工程哲学视域下工程与 engineering 的转译不对称分析 [An analysis of the asymmetry issue in translating from *gongcheng* to engineering in the context of the philosophy of engineering]. *Ziran bianzhengfa yanjiu* 自然辩证法研究 33, no. 8: 33–38.

ZHONG, Denghua. 2017. "Xin gongke jianshe de neihan yu xingdong" 新工科建设的内涵与行动 [The essence and actions for constructing emerging engineering education]. *Gaodeng gongcheng jiaoyu yanjiu* 高等工程教育研究, no. 3: 1–6.

ZHU, Qin. 2009. "Miqiemu gongcheng sheji lunli sixiang Pingxi" 米切姆工程设计伦理思想评析 [A commentary essay on Mitcham's philosophical thoughts on engineering design ethics]. *Daode yu wenming* 道德与文明 (1): 88–92.

PART III
Religion, Science, and Technology

Chapter 12

Christianity, Power, and Technological Domination
A Typological Approach to the Church

José Antonio Ullate

There is a large body of literature on the subject of the relationship of Christianity or the Church to power in general and to technology, both from a secular and a theological point of view. The historical connection between Christianity and technological development has favored the attribution of a univocal intellectual responsibility of the Church regarding a technological culture, and therefore, on an anti-ecological culture of domination. The main divergence among secular social theorists in this regard is whether there is an intellectual causality of Judeo-Christian thought and worldview with respect to technological culture and violence on nature,[1] or whether it is simply a matter of latent possibilities in all humans that are triggered when societies reach a certain level of complexity, regardless of the predominant religious-cultural matrix.

The main controversy within the theological world revolves around whether the biblical *ethos* of creation is one of domination or one of custody and care. Since the second half of the twentieth century, there has been a growing biblical investigation into the true biblical *ethos* of stewardship and care over reality. The causal link investigated by secular theorists has been developed from many Christian thinkers who have misinterpreted the *missio divina* of Genesis (Gn 1:27)[2] as the assignment of absolute dominion over the earth, with the only limit being the moral rectitude of the agent's intention. Yet others, including myself, believe that the sacred texts express a commission to administer, a stewardship, that is not absolute power over things.

Regardless, debates that focus solely on interpretation ignore a critical element of the reality of what we call Christianity and especially of its incarnation in a visible society that we call "Church" or "Churches." This element is the relationship between Christian doctrine and practice (or *morality*) and its historical-temporal expression through institutions. History should not be thought of as "past times," but rather a certain way of bringing the past back to present by reducing it to stories, always partial, often irreducible to each other, and based upon a few episodes—events—that the narrator uses as explanatory keys of a myriad of far more complex events (see Chiaromonte 1985), so it can be said that in one sense, history serves as a technology. Institutions also constitute a technology, in another sense, a technology of doctrine and of the *acted* word. The Church is an institution that "embodies" doctrine and practice. But the reciprocal relation also holds: shared doctrines and practices always generate some form of institutionality. To understand Christianity, one needs to interpret not only its texts, but also its institutional histories, with particular attention focused on their origins. For Christianity, this includes the ecclesiastical (*Church*) and also political (*Christendom*) forms. The institutional realities are expressed through what Continental legal doctrine names as *personnalité juridique*, the ability for an entity to have rights and duties.

To understand how the institutional character of the Church contributes to its view on a proper relation to nature and technology, I propose that one should use the typological method to identify the *type* that anticipates the *antitype* that the Church embodies. This method proposes that the people and events of Christianity can be better understood by recognizing their Old Testament antecedents. The Israelite monarchy, which can be said to be the constitutional moment for the Jewish people, is the key biblical figure that helps to develop a nuanced understanding of the Church's perspective, first with respect to power and then to technology and nature. What emerges is a worldview of the Church that is complex and ambivalent.

A TYPOLOGICAL UNDERSTANDING OF THE CHURCH

Typology is "one specific mode of the larger category of interpretation, one in which a present event, person, situation, or thing suggests a likeness to an event, person, situation, or thing of the past" (Cahill 1982, 267). Typology establishes relationships between a preceding historical figure (*typos*) that imaginatively prefigures, anticipates, or announces another later historical figure (*antitypos*) that brings to a full degree what is prefigured by the *typos*. Adam, the "*typos* of the one who was to come" (Rom 5:14), prefigures Jesus. Typological interpretation is often used in oral transmission, which has a figurative and primarily formulary character. Typological interpretation does not claim to decipher an essence or provide an account of a nature. Or what is the same: between the *antitypos* and the *typos* there is no relation of univocity, but of approximation through analogy, of proportion.

Northrop Frye has persuasively argued that the entire Bible is typologically structured from a hermeneutical point of view, an idea expressed by Tibor Fabini as "typology is not only a Christian view of history but also a principle that orders the Christian Bible" (2009, 145). It can be said that typology is a way in which intelligence, making use of imagination, contemplates history and historical processes. This way of looking rests on the conviction that God speaks and acts in history through a creative power that reaches down to the last details of events. Precisely, the conviction that everything is ordered by that power supports the possibility of relating one historical fact to another with the purpose of penetrating more deeply into those facts, illuminating each other. As Cahill puts it, typology is "basically an imaginative vision of history and historical process ultimately grounded on the conviction of a creative power of God who speaks and acts," "not an exegetical method but rather the result of a conviction that salvation had taken place in the end-time through Jesus Christ" (1982, 265, 275).

The importance of typology matters for a second reason, namely the way the Church views itself in relation to its Jewish origins. The dogmatic constitution *Lumen Gentium* declares that the Church of Christ "subsists in the Catholic Church" (Second Vatican Council 1965, I, § 8; cf. Agamben 2000; CDF 2007). Christianity, before being called so,[3] from its very beginning in the preaching of Jesus, has recognized—as the author of the Epistle to the Hebrews does—that God had spoken in ancient times through the prophets "on many occasions and in various ways" before doing it by His own Son in the end times (Heb 1:1–2). Augustine of Hippo, pointing out the mystical continuity between the two Testaments, says that "what was hidden in the Old Testament is revealed in the New" ([419] n.d., Book II, § 73).

Therefore, the Church materializes a double mystery of subsistence: the old Israel subsists in the Church—the *ekklesia*, gathering—of the disciples of the Messiah Jesus, and the messianic *ekklesia* subsists in the legal institution that we call the Catholic Church.

To *subsist in* is not merely *to be*. To subsist *in something* else implies a duality: something that subsists and some other thing in which operates that subsistence, because if one thing *subsists in* another it is necessary that they be two different things. The thing in which another *subsists* gives continuity to that other thing. However, in doing so, that in which something else *subsists* not only performs that function, but conditions that which continues: the thing that subsists in another remains what it once was, but manifests itself in a new different way, a conditioned way by its "host." The phenomenon of subsistence, then, shows, but at the same time veils, that to which it gives subsistence.[4]

What is of the greatest importance here is that the Church, and before it Jesus himself, recognizes a peculiar authority in the texts of the Old Testament. The Church preserves the entire canon of the Old Testament as the Word of God, and at the same time claims for itself the prerogative of making explicit the content of these biblical teachings, of updating what is virtual, obscure, or latent in them: *occultum et involutum aperire, latens explicare*. The magisterium of the Church thus functions similarly as the ancient rabbinical oral traditions (*misnáh*) and learned commentaries (*gemara*), but in a diachronic and centralized way.

The *Catechism of the Catholic Church* teaches that "Christians venerate the Old Testament as the true Word of God. The Church has always vigorously rejected the idea of dispensing with the Old Testament on the pretext that the New has rendered it void" (1993, § 123). A critical interpretive link between both testaments is, therefore, its typology. I argue that this is true not just for individuals and events, but also for the institution of the Church itself, which can be traced back to the Israelite monarchy.

UNDERSTANDING THE ISRAELITE MONARCHY

Biblical stories are a blend of myth and interpretive (nonscientific) history. That is to say, through a memorial form, they have a predominantly performative and parenetic character. The post-exilic "edition" of these texts is the overlay of different oral traditions and, therefore, in general the texts do not convey a linear, perfectly traced and unidirectional purpose. They collect many voices that had previously transmitted stories (*re-latus*, past participle of *re-fero*, bring back) that emerged from the people, and the stories of innumerable bards are interspersed with later *peripeteia*.

Regarding the political and bio-architectural question (of organizing the relationship with nature), the oldest layers of the Deuteronomic stories, those that have to do with the foundation of the community, have a clearly anarchic slant. This distrust of a strong government keeps a striking continuity with the *ethos* of custody toward nature—toward creation—characteristic of the Pentateuch.

The Mosaic cycle contains a warning against tyrannical monarchy as much as against the destructive chaos which comes from an anarchy which does without morality. Aaron Wildavsky (2005) refers to the prohibition of the excess of power that enslaves the people, and to the rejection of the disorder that leads to the debauchery of polytheism. But Moses's task is more similar to a divine mission than to political leadership, especially as the latter is understood in modernity. The leadership of Moses prefigures and prepares, in an exceptional and exalted way, the degree of compaction and social stability necessary for the fixation of a proper idea of the "political." The conquest of the land of Canaan—without forgetting the parenetic-epic and non-historical-scientific character of the text—inaugurates the founding stage of the social idea in Israel.

Moshe Halbertal and Stephen Holmes (2017) identified "the beginning of politics" in the transition between the world of the judges of Israel and the birth of the Israelite monarchy. The literary hinge between the two worlds is the prophet Samuel, last of the judges and in charge of finding and anointing the first of the kings in the name of Yahweh.[5] He is the witness of the end of a world and the birth of another, very different. The book of Judges describes the period "when the tribes of Israel gathered as a united people in Canaan" (Gottwald 1985, 143), and describes a social life based on a *political theology* that differs completely from other political arrangements in that region of the world at that time.[6]

In broad lines, "monarchy was understood as part of the permanent furniture of the cosmos itself" (Halbertal and Holmes 2017, 5). The monarchies of the area were thought to embody divine order: "The king was either a god, an incarnation of a god, or a semimythic human king who was elected by the gods to serve as a necessary mediator between the divine order and the human world" (4).

By contrast, the political theology of the pre-monarchical Israelite tribes constituted a radical novelty: "Rather than declaring that the *king is a god*, the new theology postulated instead that *god is the king*" (5), and their anomalous way of social organization that we know as the time of the Judges, "a state of divinely supervised anarchy," "divinely inspired anarchy," or "divine anarchy" (7, 82). During the period of about two centuries that elapsed between the conquest of Canaan and the election of Saul as the first king of Israel, "there was no king in Israel and each one did what seemed right (in his own eyes)" (Jgs 17:6, 21, 25, but also 18:1 and 19:1). The absence of law does not imply moral arbitrariness, but rather personal responsibility for the whole, which sets the conditions for prudence but carries a great risk. Many exegetes consider this a reprobation of the state of social and moral anarchy of an anomic order: Pinchas Kahn, for instance, interprets this statement as "a pro-monarchy position as a corrective to moral deterioration, evident of societal anarchy" (2016, 21).

The sacred text does not evaluate what type of regime achieves better results in improving the ethical conduct of individuals, understood as pure compliance with a set of rules. The determining criterion that underlies these apparently historical stories about the succession of favorable and adverse events of the People of Israel is to show that it is always possible to remain faithful to the alliance between Yahweh and his People. This criterion can also be interpreted as the progressive discernment of the touchstone of the Covenant, which is love of neighbor: how, overcoming their own desire for security and control, the Israelites learn gradually that the manifestation of love for God is love for one's fellowman.

Time and again the Israelites themselves express their desire for a predictable and secure order, which in their eyes is embodied in a monarchy. The episode of Gideon's victory over the Midianites is particularly eloquent of the *divine-political* logic that characterizes the period of the judges: "Then the men of Israel said to Gideon, 'Rule over us, you and your son and your grandson also; for you have delivered us out of the hand of Midian.' Gideon said to them, 'I will not rule over you, and my son will not rule over you; the Lord will rule over you'" (Jgs 8:22–23).[7]

The prophet Samuel, the last of Israel's judges, reluctantly receives the mission to choose and anoint the first king of the Israelites. Samuel closes the time of the judges and inaugurates the time of the monarchy. This transition, far more than a simple replacement of regimes, contains a mystery of continuity and discontinuity.[8]

When Samuel is declining, he has the idea of assigning the judiciary to his children, a confused and probably unconscious manifestation of the tendency toward monarchy. The elders of Israel gathered to ask Samuel instead to choose a true king to lead them and give them security, "as other peoples have." The idea of a king for Israel displeases Samuel. But the prophet-judge receives an oracle from the Lord:

> Listen to all that the people are saying to you; it is not you they have rejected, but they have rejected me as their king. . . . Now listen to them; but warn them solemnly and let them know what the king who will reign over them will claim as his rights. (1 Sm 8:7–9)

This passage contains the keys to understand the meaning of the transition from divinely supervised anarchy or divine monarchy to human monarchy and the political regime of law. It is Yahweh himself who declares the dissonance between the human monarchy and the divine one. It is not a question of effectiveness, for the balance of the period of judges was not so good. The point is, precisely, that the very adoption of a consequential criterion, without properly weighing other deeper factors, to dismiss the relevance of the time of the Judges, means thinking already from within the immanent logic of the monarchy. In other words: the practical failure of the "regime" of the Judges does not legitimately override the theological reasons that were at its origin. By asking for a human king, Yahweh says, "It is me whom they reject as their king." Samuel, then, before giving them the human king they ask for, solemnly and prophetically warns them: "This is what the king who will reign over you will claim as his rights: . . . you yourselves will become his slaves" (1 Sm 8:4–18).

In the oracle of Yahweh, "the king" is a metonym for all the kings who are to rule Israel. Granting the primacy of social life to security and predictability reorganizes and redefines power in such a way that it intends to have a radical dominance over its subjects. A perspective that prioritizes security and predictability transforms the gaze on social reality. Things and people, then, only acquire the status of "reality" to the extent that they conform to those prejudices (paraphrasing the phrase attributed to Hegel: "So much worse for things themselves!"). The social identity and communal understanding of the Hebrews will change from members of Yahweh's people—a kingdom of priests and a holy people (cf. Ex 19:6)—to subjects of a human power.

Samuel's editors have left an ambivalent and paradoxical portrait of the Israelite monarchy. Samuel's harangue—following the same pattern as Jotham's—perpetuates the need to return, again and again, to the origin of the institution in order to understand it. The dark prophecy will be fulfilled in an incomprehensible and unexpected way for men: the curse will be, at the same time, a blessing, an election, a predilection. All this without ceasing to be a curse. Although he advises against it, and although he warns of the consequences that it will entail, Yahweh accepts the People's choice and transforms it into a new way of intervening in the community life of the Israelites.

The social artifact—a social technology—that is the monarchy acts thus as a *pharmakon*, as a poison and as a remedy, as a curse that causes a blessing, all at the same time and without hiding its origin. The singular Israelite is, at the same time, inside and outside that *pharmakon*: he has enough exteriority about the monarchy to criticize it, but he is so penetrated by it that he cannot do without it to "bypass" his relationship with Yahweh or with reality. The relationship of the particular subject with the monarchy is a permanent *in fieri* in which everything already is, but nothing is yet fully.[9] This is the unique and paradoxical condition of the monarchy.

The emblematic figure of the monarchy is the second king, David, who is closely associated with his son Solomon, although the layers of oral traditions make them both appear paradoxical. For the sake of brevity, I will focus on Solomon, specifically on 1 Kings 4. The usual interpretation is that Solomon overcomes his moral dissipation and in his abominable religious promiscuity at the end of his days, which accepts "Judah and Israel were as numerous as the sand that is on the seashore; they all ate, drank and lived happily" (1 Kgs 4:20) in a literal sense. That interpretation overlooks the fact that such popular satisfaction, undoubtedly what the Israelites aspired to when they asked Samuel to give them a king, came at an unprecedented price. This verse is the last of 1 Kings 4. All the rest of the chapter details the rationalistic logic of government, which leads to a praxis erosive of the ancient Israelite tribalism:

> King Solomon was king over all Israel, and these were his high officials: Azariah the son of Zadok. . . . Solomon had twelve officers over all Israel, who provided food for the king and his household; each man had to make provision for one month in the year. (1 Kgs 4:1–7)

It is usually taken for granted that the distinction between the tribes is first lost with the invasion of the kingdom of Israel by the Assyrians (2 Kgs 17, 18) and later with the invasion of the kingdom of Judah by the king of Babylon (2 Kgs 25). However, a careful reading of the first book of Kings shows that Solomon is the one who distorts and hollows out the genuine idea of tribe in Israel (Gn 49) introducing, de facto, the notion of subject, which anticipates the modern sense of citizen and citizenship in Israel. A subject is an individual thought of as a social atom subordinated to an end in whose definition he does not participate. From Solomon on, it is the idea of "the Israelite character" or "Judaity" that builds the identity of the members of Yahweh's people, instead of the particular immediate mission of each tribe, each of them different, but all of them mutually required for completeness.[10]

The editor-compilers of the Torah emphasize the lack of institutionality of the government of the foreseen king by putting in the mouth of Moses a description that fits better with the judges than with the historical kings. To assess whether the kings achieve institutional austerity, Moses mentions two measures emblematic of excessive, despotic power. A king will not have horses in abundance, nor will he have many wives (Dt 17:16–17). Solomon had four thousand stables for the horses of his chariots and twelve thousand horses (1 Kgs 5:6). As for women, Solomon "had seven hundred women with the rank of princesses and three hundred concubines" (1 Kgs 11:3).

The transition from the judiciary to the monarchy highlights significant aspects of Yahweh's way of intervening in the lives of men. The writers-editors of Judges, Samuel, Kings, and Chronicles clearly and repetitively exalt the paradigmatic character of the monarchy of Israel and thus prepare the expectation of its mystical sequel in a Messiah that will purify that monarchy of all imperfections. However, underneath this exaltation is a network of oral traditions whose preservation shows the permanence of an even more fundamental ideal, that of the exaltation of personal responsibility in a community of equals. This subterranean and fragmentary flow, initially found more in what is omitted rather than said, emerges in a tremendous, explicit, and surprising way in the warning Samuel gives before granting Israel the king they had requested. But even his explicit warning is so seemingly incongruous with the dominant message that it becomes invisible to the "literal" reader, especially if unfamiliar with the ironic character of the entire oral tradition.

> The not saying of something (or the pretending not to say it) is an ancient rhetorical device. . . . Often enough the reticence is intended to increase the impact of what it purports to conceal while making it inevitable that a properly informed reader will at once, and with the added emotion attendant on discovery, recognize what is really meant. (Alter 1988, 155–56)

The preceding description of how the Israelite monarchy was constituted and its characteristics, especially in regard to power and domination, yields a nuanced understanding of the *typo* that anticipates the Christian Church. In summary, the Israelite monarchy, as a peculiar social artifact or social technology, is the crystallization of some ethical concerns and ways of perceiving the meaning of collective and individual life. It shows a tendency in the Israelite social group to want to accept the revelation and, at the same time, with equal or more strength, to contradict the epistemic bases (understood as complete ways of perceiving reality, to *see* it) implicit in the same revelation. That is to say: the Israelites say they want the radical and liberating novelty of Revelation (salvation), but yet struggle to accept, or even outrightly reject, it and the institutions that arise to bring it to fruition.

The tendency to interpret revelation from the perspective or against the background of security and predictability is reflected, then, in an institution that has a hybrid nature: on the one hand it presents itself as a champion of faith and on the other it makes it difficult for people to appropriate that same faith. Here we find the link between institutionalization and the culture of despotic domination over created reality or nature. The concrete form of institutionalization, of creating an institution at the service of the assurance and predictability of the future (abolition of the non-programmable uncertainty of the real, abolition of the present by the management of the future) is the same epistemic basis that leads to the logic of accumulation without an intrinsic limit.

Biblical texts declare over and over again things seemingly irreconcilable regarding the manifestation of the Israelite monarchy: it always degenerates into abuses and obstacles to the liberating novelty of Revelation, and yet, this *skandalon* never deters God's salvific plan for His People. This institution only shows even more the character of gratuitousness and surprise of that very novelty of the revelation: of its unpredictability.

The liberating novelty revealed by Yahweh demands from his people a responsibility that they shake off (thus the practical failure of the "regime of the judges"). The people want to remain the recipients of Yahweh's blessings, but, not realizing the radical transformation those blessings demand, they reject the regime of divinely supervised anarchy and defy Yahweh by claiming a vigorous institution from which they expect security and predictability for the blessings. In part, the monarchy will provide them with both, but at the price of

1. aggravating the blindness or deafness that in a first moment made the people unaware of the contradiction in their aspirations and that in a second moment made them unable to reconsider their project after the oracle received from Samuel, and
2. reducing their prominence, a responsibility that to the same extent that was deprived from them was transferred to the institution itself. The erection of the monarchical institution required the renunciation of many of the responsibilities and rights that they had received with the fulfillment of the promise of the land of Canaan.

Finally, the promise does not cease to be effective despite this weakness or collective betrayal. Yahweh performs an unpredictable act, which constitutes a further revelation of the depth of his love for his people: he adopts as his own way of taking care of his people precisely the human monarchy, that social technology that men have built against the divine monarchy.

Yahweh/Elohim had envisioned the Israelites to organize themselves without subordination to a centralized power, but that meant a freedom and responsibility that allowed their misuse and misguidance.[11] This *corruptio* of prudence opens the doors of the monarchy, but not before making clear what the price will be for the security and predictability that the king's power will bring. What is unexpected is that Yahweh/Elohim makes that choice of the people his own and thus endorses the monarchy, integrating it within his providence. But, at the same time, He preserves the validity of the principles of freedom and personal responsibility. These principles, which apparently contradict the way of understanding life and religion imposed by the monarchy, remain underneath the messianic-monarchical order and establish a hope of greater fulfillment in the present. This unpredictable way of acting encourages an "ironic" way of understanding life, in which seriousness and play go hand in hand.

THE CHURCH AND TECHNOLOGICAL POWER

The analogy between the Israelite monarchy and the Christian Church helps us to glimpse the great ambivalence that pervades both institutions in the light of Revelation and in relation to power. The scriptures show a way of intervening in the lives of men (not in history, which are retrospective interpretative views) that we can qualify as paradoxical but that have results fully consonant with the contingent condition of the created (see Cayley 2005, esp. chapters 3 and 4). The sacred author's criticism of the monarchy does not push him to demand a return to the previous state of personal responsibility. However, this previous state remains, underlying the present reality as an ideal factor for understanding the meaning of life itself. It follows the same dynamism that later will be manifested in the Church: *subsistence in*. The novelty that these institutions (monarchy and later the Church) embody necessarily adopts a contingent mode of expression. A mode in which that novelty subsists. A way that could always be different and could always be better. But the only way to experience this novelty of life is in the present, thus liberating the future. For this reason, this experience is inseparable from the historical form that this institution adopts. Thus, a prudential way of dealing with the Church or the monarchy is established. A radical fidelity that, at the same time, is radically critical. That is to say: it will be necessary to live within and under the monarchy, but maintaining always an ironic distance from the monarchy's claim to exhaustively explain reality, that is, always bearing in mind the memory of that divinely supervised anarchy:

> Pointing to this inherent contradiction in the human political project did not lead the author to recommend abolishing sovereign authority or reverting back to the divine anarchy or the weak and decentralized social order that preceded the monarchy, when laws went unenforced and every man did what was right in his own eyes. . . . The author turned a penetrating gaze onto the punishing costs of sovereign power as such. (Halbertal and Holmes 2017, 166–67; cf. Jgs 21:26)

Regarding the relationship between the kingdom of God and the Church, I will point out that Jesus in the Gospel speaks of the Kingdom of God (βασιλεία τοῦ θεοῦ) or, using a metonymy, of the kingdom of heaven (βασιλεία τῶν οὐρανῶν). Matthew is the only evangelist who puts the word *assembly* in the mouth of Jesus (Mt 16:18; 18:17) and there can be no doubt that this "assembly" has to be understood in relation to that "kingdom of God," which is the continuous object of parables and teachings of Jesus.

In that sense, it is highly significant that the term *Kingdom of God* evoked in its listeners the latent tension throughout scripture between the kingdom of God as proclaimed in Judges and a kingdom under human rule, as it is described by Samuel and later evolves to his messianic sublimation. To speak of the Kingdom of God for a first-century Israelite was to mention the restoration of a project that had been considered failed but that, at the same time, had preserved a prophetic aura. Jesus speaks of the Kingdom of God and associates it with all the marks that accompany the liberation that he announces (cf. Lk 4:16–30).

That offers a key to clarify the relationship of Christianity to technology as an ever-growing form of power. Witnessing this intimate ambiguity does not mean concluding that these institutions are dispensable, but it alerts us to the inevitable excesses that fatally accompany them.

On the one hand, the very idea of the institution of the Church contains the tendency to construct the future, to organize the existence of its faithful. But, on the other hand, and at the same time, the Church preserves and transmits an ideal of care for creation and hospitality toward one's neighbor—even when its individual agents seek other, and often opposed, ends (a "heterogeny of ends," see Vico 2012, 679). That ideal of care is only effective in the present, and, contrary to an institution, disregards future concerns. The vision of care and the ideology of domination correspond, then, to the struggle between the present and the future.

The Church's ambiguity regarding power and technological power is not part of its identity, but is something like a *proprium* (*idion*), a predicate that accompanies its existence as a continuous risk. This ambiguity is not the essence of the Church, but it is always at its side, like a shadow. The temptation to ensure the goods that God has promised it as graces manifests in it (as happened in the monarchy) as an unresolvable relationship with power, of which technological power is a condensed expression.

The Church, like the monarchy, has in itself the tendency to secure itself and therefore to take hold of power. For that to happen, it is necessary, to the same extent, that it deprive its members of a large part of their responsibility.

This affinity with power as a means to assure its mission determines that the Church (as a *congregatio*) is inclined to interpret the task of creation as a domain only limited by the ethics of the agent (*finis operantis*) instead of prioritizing concern for the action itself (*finis operis*). At the same time, the Church as *convocatio*[12] contains the continuous renewal of surprise, of the originality of the Revelation above our ideological constructions and therefore always contains the underlying sense of gratuitous care. In the Church—as the *locus* of Christianity—there is thus a tension between a domineering and virtually despotic ethos and an ethos of care and stewardship. This ethos of care reaches everything created but is not identified with an ecological concern in contemporary sense, since it does not have any dependency on a discourse related to the logic of scarcity and fear, but rather is the expression of a sense of creatureliness, to live joyfully an existence full of mystery.

Thus, the ideal of care and stewardship of creation is more radical, deeper, and prophetic in character than the managing of security and the project of transmutation of the anomalies of reality into a predictable order.

The heart of the Church is the prophecy of the *Kingdom of God*. Although Christians cannot ignore their relationship to the ideology of domination and *hubris*, and to the excesses it produces, including the destruction of nature and, above all, of the very meaning of human life, at the same time, but in a more radical sense, this *ekklesia* (*qahal*) of Jesus is a privileged sphere of the gratuitousness and care for others and for the whole of creation. It is both a curse and a blessing.

CONCLUSION

Ultimately, the typological approach does not yield a simple answer regarding the Church and the culture of (technological) domination, but rather an understanding of the ambivalent logic that is carried to the extreme by incarnation.

The incarnation of the Logos, the entry of the Logos into human time, "transecting, bisecting the world of time,"[13] means that the heavenly truth of God agrees to submit to the limitation of the contingent. In this way, the incarnation entails the permanent possibility of misrepresentation and frustration. That is to say, God runs the risk of expressing Himself as a man, with a human word, always at the mercy of the good or bad will of the listeners, but also of the constitutive limitation of the human word, which lends itself to misinterpretation (sometimes even from the part of the most sympathetic listeners). The accepted possibility of failure, in the sense of nonacceptance of the life in fullness that Jesus offers (of which the stories collected in the Gospel are the main testimony), is not an "accident" or an "evolutionary" mistake or something that went wrong, but an essential part of the reality of the Gospel and of the Church.

To pose the problem of the moral and cultural responsibility of the Christian vision with respect to the culture of despotic domination over nature and with respect to technological *hubris* is an opportunity for a greater theological deepening of what the incarnation means and implies. In other words: it puts us in a position to undertake an investigation that is still much more radical, more fundamental to understanding our world today.

The typological image of the monarchy, as the archetype of a completely new and surprising way of intervening in the lives of men, frees us from taking sides in the debate for and against the responsibility of the Church as *magistra technologiae*. The *typo* illustrates the way of belonging to the *antitypo*. The "monarchology" illustrates the "ecclesiology." Just as the Israelite is, at the same time, inside and outside the monarchy, so the Christian is with the Church. The idea that an institution can be, at the same time, on the one hand, an instrument of choice and liberation, of blessing, and on the other, also be an instrument of punishment, invites us to go beyond an excessively univocal epistemic framework in which our choices are binary. It is an investigation that is still to be done.

What I have exposed corresponds to a fact of common experience: it is the same culture of the Church that has fueled in some the awakening of an awareness of care for nature while in others it has legitimated the accumulation of material wealth and therefore of a relationship with nature as a resource for exploitation. This state of affairs did not result by pure chance, from which nothing can be deduced, but, on the contrary, it is proof of a constitutive ambivalence also regarding the role of the Church in the relationship with creation.

NOTES

1. "The Jewish-Christian tradition and through it the modern world, says Hillman, conceive instead the world as a purely objective space, absolutely inanimate, empty of any venerable meaning, and therefore totally available to the conquest and exploitation by man. But since ultimately a world without a soul is already a dead world, ours is a world that we cannot but feel destined for consumption, for destruction" (Quinzio 1984, 43). The fear and terror of animals in front of man (Gn 9:2) is only understood some chapters ahead (Gn 9:9–17): the covenant is made *with all life*, not only with humans. Therefore, that fear is the fear of someone that depends on another, but that other, who has the power to arbitrarily dominate, has not received it to use it in this way.

2. All Bible quotations are from the New Revised Standard version, Catholic EPub edition (New York: Harper, 2013).

3. Acts 11:26. Up to this time the believers were known as disciples of Jesus Messiah and formed the assembly (*ekklesía*) of the disciples.

4. A distant analogy with Aristotle's original idea according to which the accident does not have being (*is not*) in itself, but in another, in a substance: *subsistit in* ("That which is primarily, and to which all the other categories of being are referred, has been discussed—namely substance, for the others are called beings in accordance with the account of substance, i.e. quantity, quality and the others which are so called: for they will all involve the account of substance" (Aristotle 2006, 1045b).

5. The cycle that begins with the entry into the promised land and constitutes the political foundation of Israel is described in Joshua, Judges, 1 Samuel, 2 Samuel, 1 Kings, and 2 Kings.

6. "Political theology" is a phrase that Carl Schmitt put into circulation in 1922 to describe the study of the correspondence between the political concepts and religious ideas (or metaphysical ones) that prevailed at different historical moments: he thought religious ideas are transposed to the social order. For Schmitt, religious ideas give foundation to political life, and they allow rationalist modernity to escape from the quagmire in which he thought it had plunged social life. Halbertal and Holmes use the term in a looser but similar sense, to refer to the social implications of a certain religious worldview.

7. Gideon adopted some of the ways of a monarch, and one of his sons, Abimelech, proclaimed himself king of the cities of Shechem and Beth Miló after killing all except one of his seventy male siblings. The admonitory parable harangue that the surviving son, Jotham, addresses to those cities contains a foretaste of Samuel's admonition against the monarchy (Jgs 9:19–20). Buber called Jotham's parable "the strongest anti-monarchical poem in world literature" (1973, 75).

8. For Joseph Ratzinger it is precisely this articulation of continuity and discontinuity that constitutes the idea of "reform" (Benedict XVI 2005).

9. The tension between eschatology already fulfilled and hope in a full manifestation that we cannot yet anticipate is a classic subject of the hermeneutics of the Pauline and Johannine Theology of the New Testament (see Manini n.d.).

10. This idea is most clearly expressed in Ivan Illich's idea of di-symmetry, "of opposing domains that fit each other and are answerable to each other without being an identical image of each other . . . that only by opposition do things hold together, only by a plurality of powers can freedom be nourished, only by limitation can language live," as explained by David Cayley in "Part Moon, Part Travelling Salesman: Complementarity in the Thought of Ivan Illich," *DavidCayley.com* (blog), November 7, 2014, https://www.davidcayley.com/blog/2014/11/7/part-moon-part-travelling-salesman-complementarity-in-the-thought-of-ivan-illich.

11. Deviance synthesized in the gruesome episode of the concubine of the Levite of Ephraim in Gibeah (Jgs 19).

12. The mass of those who have been called, in which the determining factor is always the initiative of God who calls. For the two aspects of the Church, *convocatio*-sanctifier and *congregatio*-sanctified, see Henri de Lubac (1952, 87).

13. "Then came, at a predetermined moment, a moment in time and of time, / A moment not out of time, but in time, in what we call history: transecting, bisecting the world of time, a moment in time but not like a moment of time, / A moment in time but time was made through that moment: for without the meaning there is no time, and that moment of time gave the meaning. / Then it seemed as if men must proceed from light to light, in the light of the Word, / Through the Passion and Sacrifice saved in spite of their negative being; / Bestial as always before, carnal, self-seeking as always before, selfish and purblind as ever before, /Yet always struggling, always reaffirming, always resuming their march on the way that was lit by the light; / Often halting, loitering, straying, delaying, returning, yet following no other way" (Eliot 1963, VII, 163).

REFERENCES

Agamben, Giorgio. 2000. *Il tempo che resta*. Torino: Bollati Boringhieri.

Alter, Robert. 1988. "Anteriority, Authority, and Secrecy: A General Comment," *Semeia* 43. Quoted in Carolyn J. Sharp, *Irony and Meaning in the Hebrew Bible*. Bloomington: Indiana University Press, 2019.

Aristotle. 2006. *Metaphysics, Book Θ*. Oxford: Clarendon Press.

Augustine. (419) n.d. *Questionum in Heptateuchum*. [Questions on the Heptateuch.] Accessed June 2022. http://www.augustinus.it/latino/questioni_ettateuco/index2.htm.

Benedict XVI. 2005. Address to the Roman Curia offering them his Christmas Greetings, 22 December. Accessed June 2022. http://www.vatican.va/content/benedict-xvi/it/speeches/2005/december/documents/hf_ben_xvi_spe_20051222_roman-curia.html.

Buber, Martin. 1973. *Kingdom of God*. New York: Harper & Row.

Cahill, Joseph P. 1982. "Hermeneutical Implications of Typology." *Catholic Biblical Quarterly* 44, no. 2: 266–81.

Catechism of the Catholic Church. 1993. Accessed August 9, 2022. https://www.vatican.va/archive/ENG0015/_INDEX.HTM.

Cayley, David. 2005. *The Rivers North of the Future. The Testament of Ivan Illich as Told to David Cayley*. Toronto: House of Anansi Press.

Chiaromonte, Nicola. 1985. *The Paradox of History*. Philadelphia: University of Philadelphia Press.

Congregation for the Doctrine of the Faith (CDF). 2007. *Responses to Some Questions Regarding Certain Aspects of the Doctrine of the Church*. Accessed June 2022. https://www.vatican.va/roman_curia/congregations/cfaith/documents/rc_con_cfaith_doc_20070629_responsa-quaestiones_en.html.

de Lubac, Henri. 1952. *Méditation sur l'Église*. Paris: Aubier.

Eliot, T. S. 1963. "Choruses of the Rock." In *Collected Poems, 1909–1962*. New York: Harcourt, Brace & World.

Fabini, Tibor. 2009. "Typology: Pros and Cons in Biblical Hermeneutics and Literary Criticism (from Leonhard Goppelt to Northrop Frye)." *RILCE* 25, no. 1: 138–52.

Gottwald, Norman K. 1985. *The Hebrew Bible. A Socio-Literary Introduction*. Philadelphia: Fortress.

Halbertal, Moshe, and Stephen Holmes. 2017. *The Beginning of Politics. Power in the Biblical Book of Samuel*. Princeton, NJ: Princeton University Press.

Kahn, Pinchas. 2016. "Shofetim—The Book of Judges: Anarchy vs. Monarchy." *Jewish Bible Quarterly 44*, no.1: 21–28.

Manini, Filippo. n.d. *Tra "già" e "non ancora."* Accessed June 2022. http://dimensionesperanza.it/aree/formazione-religiosa/teologia/item/5951-tra-gi%c3%a0-e-non-ancora-filippo-manini.html.

Quinzio, Sergio. 1984. *La croce e il nulla*. Milan: Adelphi.

Second Vatican Council. 1965. Dogmatic Constitution on the Church, *Lumen Gentium*. Accessed June 2022. https://www.vatican.va/archive/hist_councils/ii_vatican_council/documents/vat-ii_const_19641121_lumen-gentium_en.html.

Vico, Giambattista. 2012. *Scienza Nuova*. Milan: Biblioteca Universale Rizzoli.

Wildavsky, Aaron. 2005. *Moses as a Political Leader*. Jerusalem: Shalem Press.

Chapter 13

Technology in Cosmic Terms
The World Council of Churches in Amsterdam, 1948

Jennifer Karns Alexander

"Modern industrial society was unable to establish a tolerable justice or to give the vast masses, involved in modern industry, a basic security" (Niebuhr 1948, 17).[1] Thus argued the American theologian Reinhold Niebuhr at the founding meeting of the World Council of Churches in Amsterdam in 1948. He added another point: that the failings of modern technological society were also religious failures of the Christian churches. Niebuhr, professor at Union Theological Seminary and former pastor in the Reformed Church, was one of the most influential theologians of the twentieth century—he has enjoyed recent fame for his influence on former American president Barack Obama—and he was not alone in taking up technology as a religious issue. Technology was in fact a primary focus of the founding meeting of the World Council of Churches.

The founding World Council meeting offers a chance to consider technology in cosmic and existential terms, and it illustrates in response to technological crisis a radical posture of human humility and obedience before the divine. The World Council, a Christian body Protestant in actuality but not by design, was not the only religious institution to mount a critique of industrial society; Pope Francis did so in *Laudato Si* (2015), calling for climate action as a moral imperative, and the Catholic International Humanum Foundation has long been interested in the impact of technology on social justice; religious critiques of technology, including in Judaism and Islam, were the focus of a 2020 special issue of the journal *History and Technology*.[2] A serious examination of the World Council's early critique can show us limitations in how scholars have treated technology by revealing entangled within it a very non-technological orientation away from human-willed agency.

The founding of the World Council was entangled with technology. I refer not to the transportation and communication technologies that made the Amsterdam gathering possible, nor to the rubble and disrupted infrastructure that reminded delegates en route of the increasingly destructive potential of technologies of war. At Amsterdam, to speak of technology was to speak of things that furthered or impeded the will of God. The metaphor of entanglement comes from textiles, of fibers, cords, and threads, knotted and felted together, and impossible to tease apart. Technology and God-talk resisted being pulled apart at the conference because, there, technology and religion were inseparable (see Alexander 2020). It was impossible to tease apart the call of the churches for unity in the postwar world, still in crisis, and the backdrop to that call: the charge that people had come to have faith in themselves and in their own works—in their own technologies—rather than in God. This was the charge: that self-will and self-reliance substituted for God's power, as people had come to rely on their own techniques to organize, administer, and build. As one of the four official reports from the meeting put it, "Even in the present-day confusion, there are still many who believe that man, by wise planning, can master his own situation" (World Council of Churches 1948c, 213). The report of Section III stated that "the deepest root of [our current] disorder is the refusal of men to see and admit that their responsibility to God stands over and above their loyalty to any earthly community and their obedience to any worldly power" (World Council of Churches 1948d, 189). Section III, one of four working sections of the Council, was devoted to social concerns and explicitly took technology as a subject (see Mitcham 2020, 595; Alexander 2013). Beyond giving evidence of human hubris, technology also enabled exploitation of others and their dehumanization, which, argued members of Amsterdam's Section III, denied the God-given value of persons and contradicted the biblical command that people love one another.

The World Council's critique of technology emerged from extended consideration of technology in terms both broad and deep, through a process of deliberation and discernment by scores of church people in far corners of the globe, writing and reviewing study papers, holding local meetings, and corresponding with each other through the Geneva office of the World Council in Formation.[3] It was from the Amsterdam meeting and the preparatory work of Section III that the French lay theologian and jurist Jacques Ellul became internationally known as a critic of technology (Alexander 2013).

The critique that emerged from the World Council built upon a hard-won unity, an agreement on a very basic statement of faith: that the World Council was a fellowship of churches that accepted "our Lord Jesus Christ as God and Savior." This statement contained two distinct relations in frequent play in the meeting: how people should relate to God, and how they should relate to each other. Entanglement is found in the simultaneity of meaning of these two relations. One, Christ as God and Savior was an attestation of religious belief; the other implied enacting that belief, by treating people as fellow creatures, created and valued by God. This meant that however else one may analyze social dislocation, poverty, imperialism, racism, and authoritarianism—and these all appeared in the World Council's critique of technology—they were, at the World Council, simultaneously and inseparably issues of obedience before God.

The World Council's critique of technology illustrates the idea of the entanglement of technology and religion, and suggests a posture of humility and listening before technological catastrophe. It also makes an empirical point: that religious critique of technology and technological society was not marginal. The World Council critique did not come from the fringes of Christian belief and practice, but from the center and the mainline.

CONTEXT OF THE ASSEMBLY

The World Council of Churches was officially constituted at an assembly in Amsterdam in August 1948. The World Council was a culminating achievement of the international ecumenical movement of the twentieth century, which sought to bring together the Christian denominations that had formed, reformed, and splintered following the Protestant Reformation.[4] Many of the differences to be overcome were great, historic, and painful; some denominations considered others heretical. Nonetheless, nearly fifteen hundred church people gathered for the assembly, representing 44 countries and more than 140 churches and denominations. Technology was a central interest at the inaugural gathering; although people in all four main sections of the meeting discussed technology, especially atomic weapons and human hubris, it was Section III, the division devoted to social issues, that took technology as a focus.

Deep divides had prevented Christian denominations from worshipping and working together. Christian churches and denominations had different doctrines and different liturgies; standards for church membership, understandings of personal piety and salvation, and interpretations of the authority of the Bible and of church traditions differed significantly between some denominations. Underlying the Council's founding were the missionary movement of the late eighteenth and nineteenth centuries and the interchurch organizations that had begun bridging such divides, including the London Missionary Society (founded 1795), the American Home Mission Society (1826), the Evangelical Alliance (1846), and the YMCA (1844), culminating in the World Missionary Conference in Edinburgh in 1910. The World Council also expressed the international concern for unity across boundaries that underlay, in politics, the short-lived League of Nations and the more recently founded and more successful United Nations.

The World Council in Formation had begun planning the Amsterdam assembly during the war, at its Geneva office and at the Chateau de Bossy, north of the city along the lake; significant financial support came from John D. Rockefeller Jr. (Zeilstra 2020, 275). In Amsterdam, delegates of the first Council assembly knew they stood at an important point in history. Citizens of Amsterdam thronged streets decorated to celebrate Queen Wilhelmina's Golden Jubilee and the reopening of the city after the war. The great paintings of the Rijksmuseum—Vermeer, Rubens, Rembrandt—had been rehung, retrieved from their sheltering places in sealed barges along the canal bottoms. Reminders of the war lingered in ration-books issued to World Council delegates and in the gaunt figures of many there who bore testament to the effects of war: in their threadbare clothes, their hesitation in speaking, their sometimes air of weariness even in the excitement of new work. One publication hailed it as the most representative Protestant gathering since the Reformation,[5] and while most delegations came from Europe and North America, others arrived from Australia, Burma, Brazil, Ceylon, China, Czechoslovakia, East Africa, Egypt, Ethiopia, Iceland, India, Indonesia, Japan, Korea, Mexico, New Zealand, Philippines, Rhodesia, Russia (from exile), Siam, South Africa, West Africa, and the West Indies (Gaines 1966, 225–30). Denominations represented included the Society of Friends (Quakers), Orthodox (Coptic, Greek, and Russian), Anglican, Congregational, Episcopal, African Methodist Episcopal, Methodist, Baptist, Presbyterian, Moravian, Mennonite, Old Catholic, Reformed, Church of the Brethren, and Church of Christ; several non-church bodies, including the Salvation Army and the Ecumenical Patriarchiate of Constantinople, also attended. Not all Christian denominations participated. The Roman Catholic Church rejected the ecumenical movement as antithetical to its conception of itself as the one true and universal church. Some conservative evangelical churches in the United States, most notably the Southern Baptist Convention and Lutherans of the Missouri Synod, rejected the World Council, which they saw as a liberal, and perhaps even non-Christian, body, because some of its members embraced liberal theology and biblical criticism and because conservatives expected the World Council to put greater emphasis on social programs rather than focusing on the conversion of non-believers (Gaines 1966, 225–30).

World Council founders aimed only to provide an association of churches and sought no authority. The institution was, by design, to be cooperative and largely without power: the Council was to have no authority over the churches represented within it, and any actions would be carried out by its constituent churches. The World Council agreed that it would not promulgate doctrine, and that it would make only the most foundational statements of faith, reflecting agreement of the churches attending and serving as a bedrock basis for unity.

Unity at the World Council's founding meeting was hard-won. Language differences were a significant barrier to ecumenical work, and denominational differences remained great enough to fracture unity even among those who agreed on a foundational statement of faith (Tatlow 1967, 423). Although the World Council translated preparatory materials and main addresses into French, German, and English, French and German speakers still found themselves at a disadvantage in the many face-to-face meetings held in English. The Orthodox churches had played an essential early role in translating ecumenical goals into actual practice, yet World Council preparations typically did not include translations into Greek or Russian (Held 2003, 295–96; Vasilevich 2020). Churches from Africa were served through colonial languages, and churches in Central and South America often lacked translations into even the Spanish and Portuguese of their colonial pasts. Language difficulties were not merely practical; they underscored how deep other differences were. Delegates acknowledged officially that even in the forum of the World Council, they had not been able "to present to one another the *wholeness* of our belief in ways that are mutually acceptable" (World Council of Churches 1948b, 205). In some cases denominational orientations were unintelligible to members of different traditions, and some denominations saw evidence of heresy in others. Some delegations were wary of worshipping together with others, and some would not celebrate the sacrament of communion with Christians outside their own traditions.

Despite these difficulties, agreement emerged at a foundational level. The primary statement of unity adopted at Amsterdam was so basic that it became known as "the basis"; it described the World Council as a fellowship of churches that accepted "our Lord Jesus Christ as God and Savior." This was simultaneously a statement of God as something other than and beyond the human, and of the value of persons, because God wished to save them. The agreement expressed a concern with relationships in two basic forms: between persons and God, including between churches and societies and God; and between person and person. Considerable disagreement existed about precisely what constituted right relationships to God and between persons, but the fundamental significance of the issues was not in dispute.[6] The World Council critique charged that technology had deranged both these fundamental relationships.

The critique developed at the World Council was not just earnest rhetoric; it had a purchase on real-world technology. W. A. Visser 't Hooft, World

Council General Secretary, was uncompromising in tasking the study department: "Study can have meaning only as a preparation for action," he wrote (1948, 184–85). There was indeed a worry that talk was all that the Council would do; English rector Frank Bennett worried about "a piling up of more dry bones" (Bennett 1948, 69). Yet a wide variety of endeavors that went beyond rhetoric and touched the hard technological world came under the purview of the World Council's critique, and embodied responses to the issues it raised. Communications technologies (film, radio) had a role both in evangelism and, worrisomely, in propagandizing, people noted. Some celebrated a world more accessible to the gospel; others saw technological developments as carrying the problems of Europe to the non-European world (Pauck 1948, 37).[7] Members of Britain's Industrial Mission Movement joined delegates in Amsterdam; other delegates reported on agricultural worship services in Uganda with offerings and blessings of the tools of daily work: hoes, seeds. Women in the French Reformed Church established worshipping communities in industrial neighborhoods, and Presbyterian missionaries in Manchuria described revivals and full Bible classes made possible by the work of translation and printing and developing the exchange of goods (Wyon 1948, 113, 121).

An atmosphere of crisis nonetheless flavored the Amsterdam assembly. Delegates traveling to the meeting came from cities still in rubble, along roads nearly impassable and often alongside rail lines skewed and twisted; the last refugees were not resettled until well into the 1950s, and food remained rationed, livestock herds decimated, and farms reduced to barren mud (Judt 2005). The war had not solved the great issues the world faced, including the dislocation and oppression of workers of the land and manual and industrial laborers; potent legacies of imperialism and racism continued; and the war had introduced yet another, perhaps even greater peril, the new and apocalyptic human powers of destruction found in the bomb. Joseph Hromadka of the John Hus Faculty, Prague, described the postwar years as still volcanic, "pregnant with destructive explosions and earthquakes" (1948, 114). Niebuhr identified the problems in people's and nations' inability to create community and their inability to build and achieve justice. Both failures were rooted in conditions created by developing technical civilization. Henry van Dusen, president of Union Theological Seminary in New York and chair of the World Council's Study Department, in welcoming attendees to the Amsterdam conference, identified "the greatest dread of all": human mastery of the atom, "foreshadowing annihilation of man and all his works" (1948, n.p.). Technology was thus central to how the World Council understood the crisis it faced.

STUDY MATERIALS IN PREPARATION FOR AMSTERDAM

The World Council's critique of technology emerged from study materials developed and circulated before the 1948 founding meeting by the Geneva office of the World Council in Formation. Reinhold Niebuhr and J. H. Oldham led Section III; Oldham, formerly a missionary in India and active in international mission organizations for decades, was known as a critic of Christian racism and a theorist of Christianity and labor (Bliss 1984; Oldham 1935, 1955, 1961). Niebuhr and Oldham commissioned clergy, academic theologians, missionaries, and committed church members to write articles explaining what they saw as the pressing social issues facing the Christian churches. The resulting papers were reviewed over several years prior to publication by an extensive international network. The study department translated papers from their original languages into English, French, and German, and sent them around the world for review by individuals and by local study groups; some papers were reviewed by more than ten commenters and study groups, and many went through multiple rounds of review and revision. Final papers were bound into pamphlets for each of the four working sections of the World Council and distributed before the Amsterdam meeting. Papers came from well-known theologians and pastors, including Paul Devanandan of Bangalore, M. Searl Bates of Nanking, Karl Barth of Switzerland, and the brothers H. Richard Niebuhr and Reinhold Niebuhr of the United States; editors of Christian publications including Kathleen Bliss of the *Christian Newsletter*; and interested lay people, including John Foster Dulles of the United States. Pre-circulated study materials were also published shortly after the meeting, bound together with the official reports that had emerged in conference.

Study materials reveal criticisms that did not make their way into what became the official report of Section III, most notably objections from authors in Asia to its European focus. They also reveal, on the part of several authors, an attempt to balance the goods and evils of technology by recognizing positive roles for technology, often in a dialectic with serious criticism. They reveal criticism that technology had led to social disintegration and increased poverty, that it was implicated in imperialism and racism, and that it caused dehumanization and made authoritarianism more possible.

Examples of balancing good and evil come from M. M. Thomas, Bliss, and Oldham. Thomas, formerly of Mar Thoma Syrian Church of Kerala, India, and in 1948 on the staff of the World Student Christian Federation (Thomas 1989; Wilson 2016), drew a contrast between communications and transportation technologies that supported human community and had enabled five hundred attendees of the Third International Missionary Council to meet in 1938 at Tambaram, South India, and atomic weapons technology that militated against humanity and had leveled Hiroshima and Nagasaki in 1945. Thomas saw a dialectic at work: the machine as instrumental to God's design, and also instrumental in human disorder. Bliss, editor of the *Christian Newsletter*, saw a similar dialectic in the extension, far beyond industry, of principles of industrial organization that widened opportunities for women to engage in professional welfare work and social organizing, which were new forms of professional labor that included many of the traditional caring and support roles previously performed by women in family homes; developing forms of organization also increased impersonalization, created divisions between living and working, and undercut neighborliness (Thomas 1989, 71–72;1948, 83–90; Conway 1990, 68–70).

The most extensive defense of technology came from Oldham, who also analyzed its evils in a pair of pre-circulated study articles. Oldham was heavily influenced by the work of American architectural critic and analyst of technology Lewis Mumford, whose influential *Technics and Civilization* (1934) had appeared just as Oldham began serious theological study of technology. Oldham called the machine a gift of God because it had lessened the burdens of the most brutal forms of work and made possible the forms of planning and control that had brought victory in war. Industrial developments had also made possible the trade union movement and development of what Oldham recognized as a "genuine industrial culture." The machine did indeed have evil aspects, as Oldham termed them. It was intimately associated with war, and bound up with imperialism, capitalism, and the powerful modern state. Oldham identified the momentum of technology, a dynamic later investigated by the eminent historian of technology Thomas Hughes (1993). Technics had come to have a power and momentum of their own, Oldham wrote: "Science and technics determine the whole ethos of modern life. They create the atmosphere in which we live. They dictate the prevailing ideas and attitudes of modern society." The modern age had no need of God, Oldham charged, as people in power chose to run things according to their own ideas and to rely on their own abilities. Oldham's chief indictment of technological society lay in its derangement of a right relationship with God, in its association with self-will and self-sufficiency (Oldham 1948a, 34, 43; Oldham 1948b).

Bates and Thomas wrote from the perspective of China and India. Bates was a professor of history at the University of Nanking; he was American with a Yale PhD in Chinese history, and he taught in China as a member of the United Christian Missionary Society. He had been in Nanking during the Japanese assault initiated in 1937 and was prominent in the International Committee for Nanking Safety Zone (Bates 1938; MacInnis 1998; McLean

1984). Thomas, in addition to his dialectical critique of technology above, also offered a critique of the foci of Section III.

Both Bates and Thomas were concerned with poverty, especially rural poverty, and imperialism and racism, and both took exception to much of the pre-circulated material on technology and the machine because it focused largely on European experiences and did not address poverty. Poverty was a more immediate and pressing issue for the people of their regions than the challenges of industry and urban life. Bates charged that study materials focused on mechanized societies and neglected people whose lives were neither dominated by technology nor served by it. Thomas argued that much of Asia remained under the sway of European imperialism and had been reduced to an "agricultural farm" for European industries. The missionary movement could not be separated from the imperialism of dominant European nations, Thomas claimed, and he rejected the notion that the Christian churches were in a position to mount an adequate response to imperialism and racism. As he pointed out, Dutch Christians in South Africa used doctrines of creation to justify apartheid, and Dutch political parties defended imperialism in Indonesia on similar grounds (Bates 1948, 61–70; Thomas 1948, 72–73).

Where machines and machine society did have influence, Thomas saw dehumanization: in rationalistic interpretations that treated people as calculable and manipulable. The machine age had atomized people into members of a mass society who faced an impersonal fate, and the machine, or technics, had become an idol, seemingly part of the natural order and thus irresistible. Thomas hoped for a stronger doctrine of creation and human vocation that would counter technology's force: a true doctrine of creation would recognize "man as man"; this, for Thomas, meant persons created and valued of God, and able and called to make decisions in a realm of freedom "poised between divine grace and natural necessity" (Thomas 1948, 75–76).

The authoritarian potential of technical organization was most concerning for Bates. Asian workers encountered the industrial nations when they were swept into their organizational reach: Japanese farmers adopted chemical fertilizers, Indonesian plantation systems supplied overseas markets, mining exploited people's bodies as labor, for materials used elsewhere. People in authority might appropriate technical efficiencies to develop and maintain centralized and authoritarian states, Bates argued, citing authoritarian Japan and centralized Germany, which had both accomplished much more than their resources would seem to allow. Technical efficiencies that had increased industrial yields during the war might also be turned to the management of populations, especially in regions without traditions opposing arbitrary authority and supporting individual and group rights (Bates 1948, 66).

The circulation of papers for multiple rounds of review and revision led to published study materials that were more than individual opinions: they were the grounds from which a consensus was emerging. They treated issues of concern to all people, not just Christians: racism, social disintegration, poverty, imperialism, authoritarianism. These issues reflected the basic faith statement agreed by World Council members: these were issues of right relationships, between person and person, and between person and God.

REPORT OF SECTION III

The developing critique of technology at the World Council moved toward consensus in the official Report of Section III. Delegates debated and approved the report, and it was then endorsed by the plenary. The official report reflected discussions in person at the assembly about pre-circulated materials. It demonstrates the centrality of the critique of technology to the early World Council.

Four elements of the report, three statements and a directive, were especially important to the critique of technology. The first was a clear statement of the root of the postwar crisis: people's refusal to admit that their responsibility to God lay above their responsibility to any power or community on earth. This was a foundational point; it was not a charge against technology. The second important element was a statement that technology was crucially implicated in particular factors at work in the crisis, in the vast concentrations of power technology made possible, and in the momentum technology had itself developed. The report identified the momentum of technical society explicitly. The third element was a statement of harm and possibility: that people had God-given value, and they did not exist merely to produce things at work but "to participate in the shaping of society." The fourth important element was a directive: that "the Christian churches should reject the ideologies of both communism and laissez-faire capitalism." This last point was the most contentious of the whole assembly (World Council of Churches 1948d, 190, 193, 195; Alexander 2013, 195–97). In what follows I examine these elements at length.

The first statement created the foundation for the critique of technological society that emerged at Amsterdam. It affirmed that when people refused to recognize the primacy of their responsibility to God as greater than any other responsibility, they were prone to elevate human technical mastery over the will of God. This reflected the contribution of Reinhold Niebuhr, who argued that there existed "basic conditions set by God to which human life must conform," conditions that could not be mapped onto any particular type of social or political organization; his vision was "a tolerably just social order," a practical, pragmatic order, provisional, decided politically as people related to each other uncoerced and in the uniqueness of their personalities, balancing order and freedom under the Christian command that people love one another. Modern techniques overran this command, in their use of rules and procedures intended to ensure particular outcomes, and in the rewards they offered to people, institutions, and states willing to exercise technical power for their own ends. The failure of Christianity and the churches was historic; late medieval churches and political organizations had failed to shape and constrain developing commercial cultures, and early modern churches and societies had similarly failed in understanding and directing industrial energies. Technologies with the potential to build global community, such as techniques of communications and transportation, were used instead for imperialism. The economic theory of *laissez-faire* failed to recognize that human freedoms could be expressed through destruction as well as through creation, Niebuhr argued, as people underestimated both the evil and the freedom of which they and others were capable. Despite the role of technology in making possible new forms of human community, including genuine working-class culture, modern industry had failed in the larger task of creating a livable society (Niebuhr 1948, 17, 20). The solution was no particular social or political order, nor could it be achieved by willful and deliberate human effort, but only in continuing obedience to the divine law of love and continuing assessment of what such obedience meant.

The second important element was the report's statement identifying two technological causes of crisis: techniques of concentrating power, and the momentum of technological developments. This critique reflected pre-circulated papers of Ellul, a lay theologian with the Reformed Church of France who had served in the French resistance and at the time of the Amsterdam meeting was teaching law at Bordeaux; Bliss, a British theologian trained at Cambridge who had served as a missionary in southern India before war broke out, and at the time of the Amsterdam meeting was editor of the *Christian Newsletter*; and Oldham, who had joined with Niebuhr in leading Section III. Ellul did not attend the Amsterdam meeting, having recently lost his son in an accident. Ellul's pre-circulated study paper had been widely read, and Bliss presented his address in Amsterdam. Ellul and Bliss analyzed the spread of techniques of concentrating power into new areas, including the social and home spheres, and of the power of standardizing functions and modern techniques of organization and administration. Ellul analyzed the situation in Europe, drew attention to the heightened prestige of production and management, and identified technics as a primary unifying feature of society as other forms of community collapsed. Ellul, alone among the contributors to Section III, also identified propaganda as an organizational technique of great power. Oldham had described technology's momentum in the study materials discussed above, which he saw in starkest form in technology's connection to war and in links between regimentation in the military and in industry. Science and technology had brought into human hands forces and organizations people could not control, Oldham argued, extending the reach of some people and not of others; the concentration of power afforded by organizational techniques was thus contrary to democratic and equitable society (Ellul 1948, 50–60; Bliss 1948, 83–90; Oldham 1948a, 44; Oldham 1948b, 122–23, 126–27).

A third important element of the report was its statement that people were more than their labor and did not exist to fill production quotas. This reflected contributions from J. C. Bennett, professor of theology and ethics at Union Theological Seminary, New York, and from Oldham. Bennett charged the churches with driving away industrial workers, by supporting unjust economic arrangements in Great Britain and, especially in the United States, catering to the complacency of middle-class members. Existing arrangements denied people the opportunity to shape the societies within which they lived, in violation of what the report identified as "a duty implied in man's responsibility toward his neighbor" (World Council of Churches 1948d, 193). Oldham noted that technical development marched on, "there is no slackening of the pace" and people increasingly became subject to the impersonal workings of mechanical processes (J. C. Bennett 1948; Oldham 1948a, 44; Oldham 1948b, 122–23, 126–27).

The directive that the churches reject both laissez-faire capitalism and communism, the fourth element identified above, occasioned one of the few acrimonious exchanges of the meeting. It was greeted with shouted dissent as the report was submitted to the full gathering; the term *laissez-faire*, added during debate in an effort to blunt criticism, failed to appease several American delegates, who objected most strongly. The directive directly reflected the contributions of Reinhold Niebuhr. Niebuhr had described capitalism as an optimistic understatement of the destructive potential of human freedom; equally problematic was the "totalitarianism of communism," into which socialism, though offering useful interim approaches to justice issues, shaded when consistently and rigidly applied. The report used a passage on communism and capitalism to take up issues of poverty and racism, charging Christians in capitalist systems to recognize the appeal of communist visions of racial equality and deliverance from poverty, and to recognize the complicity of their own churches in "economic injustice and racial discrimination" (Niebuhr 1948, 19, 22; World Council of Churches 1948d, 193).

The Report, in criticizing both communism and capitalism, thus reflected a critique Niebuhr had mounted in study materials: that technical society had helped political religions to emerge. Reinhold Niebuhr and H. Richard Niebuhr both addressed the topic. Reinhold Niebuhr identified fascism and socialism as political religions, in that they promised a more integrated community and engendered fierce loyalty; fascism he identified as cynical and cruel, and socialism as utopian and sentimental. He had not criticized capitalism as a political religion because he associated it largely with the United States, whose technical achievements and natural wealth had so far spared it the social convulsions that heightened the appeal of immersive political ideologies. H. Richard Niebuhr implicated churches in the political religion of nationalism and found sin in the doubt and disobedience to God of people's turning to their own powers and in "their worship of civilization and nation." Erich Voegelin had identified political religions in his 1939 book of that name; the concept referred to rituals and commitments taken up by followers of secular ideologies that followed patterns typically associated with religious enthusiasms and adherences. Reinhold Niebuhr argued that political religions of the sort he identified had emerged in response to the "new conditions of a technical age," which exacerbated circumstances in which people turned to political ideologies (R. Niebuhr 1948, 21–22; H. R. Niebuhr 1948, 78, 87).[8] People had come to think that specific laws and rules governed how justice was achieved, and thus sought rule-based remedies in which they saw no room for God's order. Specific rules and concepts were always historically contingent, Niebuhr argued, and in that contingency God's work and order appeared. Wilhelm Pauck, Chicago professor of historical theology, named scientific humanism as the rival faith that most foundationally incorporated technology, in its belief that human goals could be reached through the natural abilities of humankind and "in the energies and resources of nature which can be made accessible to those abilities." This faith saw "fabulous possibilities for good in the man-controlled technological world" through planned economies and political organization. The compiling of human knowledge and human techniques entailed "estrangement from the Christian faith" and enthrallment to worldly human powers (1948, 39, 37).

A discussion of rival religions and faiths thus cycled back to the issue of human will. A tension existed in Section III's report between the indictment of technological society as having replaced reliance on God with reliance on human technological powers and the freedom of humans created in the image of God to act. The tension was written into the report, which concluded with a call to responsible action, governed by obedience to God. The commitment to the will of God as primary was more than a value of submission, however; it was also a value of discernment and of agency. H. Richard Niebuhr noted the difficulties of such discernment when he identified a subtle and deceptive assertion of agency, disguised as self-criticism: people often saw their own failure of will as the cause of harm and thus in fact celebrated their own power. He identified this as faith in the human self rather than faith in God (H. R. Niebuhr 1948, 78).

The World Council of Churches' critique of technology calls us to consider technology in cosmic and existential terms, and to ask seriously what was the meaning of a call, in crisis, toward humility and obedience before God. The Amsterdam response was a radical rejection of human will and self-directed agency, an exception to prevailing norms, developed deliberately by a large body of people of varied backgrounds, following reflection, study, and discussion. Entangled with technology in the Amsterdam critique—and thus as part of and inseparable from technology—was an orientation that looked decidedly non-technological, that turned away from deliberate and purposeful human action and toward the divine.

In more concrete terms, the World Council has come to be known for social justice work, which was heavily influenced by Section III, and the legacy of the ecumenical critique of technology can be seen in organizations such as the Industrial Mission Movement, whose early members had ties to the World Council, and the United Farm Workers of the United States, organized by Cesar Chavez, whose salary was paid by the California Council of Churches. The World Council did not address environmental effects of technology until more recently. Pressing issues identified by the World Council remain unresolved, including racism, economic exploitation, and legacies of imperial subjugation, and the Council has made efforts to address them all, most notably in its controversial Program to Combat Racism begun in the late 1960s. The World Council's critique thus offers resources and models for confronting the existential threats created by technology today, most apparently in practical actions, but perhaps more deeply in an underlying orientation away from human-willed activity and toward a posture of humility and discerning listening in the face of catastrophe.

NOTES

1. On Reinhold Niebuhr, see Lemert (2011), Holder and Josephson (2012), and Lovin and Mauldin (2021).

2. *Laudato Si,* accessed May 16, 2022. https://www.vatican.va/content/dam/francesco/pdf/encyclicals/documents/papa-francesco_20150524_enciclica-laudato-si_en.pdf; Preston (1971); Special Issue on Religion and Technology, *History and Technology* 36 (2020).

3. The literature on the World Council of Churches is voluminous. See Gaines (1966) and Fredericks and Nagy (2021).

4. On ecumenism, see Rouse and Neill (1967), Fey (1970); and Bouwman (2018).

5. Gaines (1966, 230–31, 250, 225–26) and "Religion: The First World Council," *Time Magazine*, August 30, 1948. Accessed June 27, 2022. https://content.time.com/time/subscriber/article/0,33009,799114,00.html.

6. The Report of Section II identified God's purpose as "to reconcile all men to Himself and to one another in Jesus Christ his Son," and the Report of Section III identified each person as "responsible to God and his neighbour" (World Council of Churches 1948c, 212; World Council of Churches 1948d, 192).

7. Pauck noted that his paper incorporated contributions taken directly from members of the Chicago Ecumenical Study Group.

8. On H. Richard Niebuhr, see Kliever (1977) and Beach-Verhey (2011).

REFERENCES

Alexander, Jennifer Karns. 2013. "Radically Religious: Ecumenical Roots of the Critique of Technological Society." In *Jacques Ellul and the Technological Society in the 21st Century,* edited by Helena M. Jerónimo, José Luís Garcia and Carl Mitcham, 191–203. Dordrecht: Springer.

———. 2020. "Introduction: The Entanglement of Technology and Religion." *History and Technology* 36, no. 2: 165–86.

Bates, Miner Searle. 1938. "Miner Searle Bates Letters: The View from Ginling" digitization project, Barnard College, 2021. https://mct.barnard.edu/memories-of-the-nanjing-massacre/miner-searle-bates.

———. 1948. "The Situation in Asia–I." In World Council of Churches, *The Church and the Disorder of Society*, Vol. III, 61–70.

Beach-Verhey, Timothy A. 2011. *Robust Liberalism: H. Richard Niebuhr and the Ethics of American Public Life.* Waco, TX: Baylor University Press.

Bennett, Frank. 1948. "The Church's Failure to Be Christian." In World Council of Churches, *The Universal Church in God's Design*, Vol. I, 63–72.

Bennett, J. C. 1948. "The Involvement of the Church." In World Council of Churches, *Church and the Disorder of Society*, Vol. III, 91–102.

Bliss, Kathleen. 1984. "The Legacy of J. H. Oldham." *International Bulletin of Mission Research* 8: 18–24.

———. 1948. "Personal Relations in a Technical Society." In World Council of Churches, *Church and the Disorder of Society*, Vol. III, 83–90.

Bouwman, Bastiaan. 2018. "From Religious Freedom to Social Justice: The Human Rights Engagement of the Ecumenical Movement from the 1940s to the 1970s." *Journal of Global History* 13, no. 2: 252–73.

Conway, Martin. 1990. "Kathleen Bliss, 1908–1989." *Ecumenical Review* 42, no. 1: 68–70.

Ellul, Jacques.1948. "The Situation in Europe." In World Council of Churches, *The Church and the Disorder of Society*, Vol. III, 50–60.

Fey, Harold E., ed. 1970. *A History of the Ecumenical Movement, vol. 2, 1948–1968.* Philadelphia, PA: Westminster Press.

Fredericks, Martha, and Dorottya Nagy, eds. 2021. *World Christianity: Methodological Considerations.* Boston: Brill.

Gaines, David P. 1966. *The World Council of Churches: A Study of Its Background and History.* Peterborough, NH: R. R. Smith.

Held, Heinz Joachim. 2003. "Orthodox Participation in the World Council of Churches: A Brief History." *Ecumenical Review* 55, no. 4: 295–312.

Holder, R. Ward, and Peter Josephson. 2012. *The Irony of Barack Obama: Barack Obama, Reinhold Niebuhr, and the Problem of Christian Statecraft*. Farnham, UK: Ashgate.

Hromadka, Joseph L. 1948. "Our Responsibility in the Post-War World." In World Council of Churches, *The Church and the International Disorder*, Vol. IV, 114–42.

Hughes, Thomas P. 1993. *Networks of Power: Electrification in Western Society, 1880–1930*. Baltimore: Johns Hopkins University Press.

Judt, Tony. 2005. *Postwar: A History of Europe since 1945*. New York: Penguin Books.

Kliever, Lonnie D. 1977. *H. Richard Niebuhr*. Waco, TX: Word Books.

Lemert, Charles C. 2011. *Why Niebuhr Matters*. New Haven, CT: Yale University Press.

Lovin, Robin W., and Joshua Mauldin, eds. 2021. *The Oxford Handbook of Reinhold Niebuhr*. Oxford: Oxford University Press.

MacInnis, Donald E. 1998. "Miner Searle Bates." *Biographical Dictionary of Christian Missions*. Macmillan. http://bdcconline.net/en/stories/bates-miner-searle.

McLean, Cynthia. 1984. "The Protestant Endeavor in Chinese Society, 1890–1950: Gleanings from the Manuscripts of M. Searle Bates." *International Bulletin of Missionary Research* 8, no. 3: 109–12.

Mitcham, Carl. 2020. "The Ethics of Technology: From Thinking Big to Thinking Small—and Back Again." *Axiomathes* 30: 589–96.

Niebuhr, H. Richard. 1948. "The Disorder of Man in the Church of God." In World Council of Churches, *The Universal Church in God's Design*, Vol. I, 78–88.

Niebuhr, Reinhold. 1948. "Our Present Disorder." In World Council of Churches, *The Church and the Disorder of Society*, Vol. III, 13–28.

Oldham, J. H. 1935. *The Modern Missionary: A Study of the Human Factor in the Missionary Enterprise in Light of Present-Day Conditions*. London: SCMP.

———. 1955. *New Hope in Africa*. London: Longmans Green.

———. 1961. *Work in Modern Society*. Richmond, VA: John Knox Press.

———. 1948a. "Technics and Civilisation." In World Council of Churches, *The Church and the Disorder of Society*, Vol. III, 29–49.

———. 1948b. "A Responsible Society." In World Council of Churches, *The Church and the Disorder of Society*, Vol. III, 120–54.

Pauck, Wilhelm. 1948. "Rival Secular Faith." In World Council of Churches, *The Church's Witness to God's Design*, Vol. II, 37–52.

Preston, R. H., ed. 1971. *Technology and Social Justice: A Symposium Sponsored by the International Humanum Foundation*. Valley Forge, PA: Judson Press.

Rouse, Ruth, and Stephen Charles Neill, eds. 1967. *A History of the Ecumenical Movement, 1517–1948*. Second edition. London: SPCK.

Tatlow, Tissington. 1967. "The World Conference on Faith and Order." In Rouse and Neill 1967, 405–44.

Thomas, M. M. 1989. "My Pilgrimage in Mission." *International Bulletin of Missionary Research* 13, no. 1: 28–31.

———. 1948. "The Situation in Asia—II." In World Council of Churches, *The Church and the Disorder of Society*, Vol. III, 71–79.
van Dusen, Henry P. 1948. "General Introduction." In World Council of Churches, *Man's Disorder and God's Design*, n.p.
Vasilevich, Natalia. 2020. "The 1920 Encyclical of the Ecumenical Patriarchate and the Proposal for a 'League of Churches.'" *Ecumenical Review* 72, no. 4: 673–82.
Visser 't Hooft, W. A. 1948 "The Significance of the World Council of Churches." In World Council of Churches, *The Universal Church in God's Design*, Vol. I, 177–99.
Voegelin, Erich. 1939. *Die politischen Religionen*. Stockholm: Bermann-Fischer Verlag.
Wilson, Lois M. 2016. "Reflections on an Intentional Life." In *The Life, Legacy and Theology of M. M. Thomas*, edited by Jesudas M. Athyal, George Zachariah, and Monica Melancthon, 43–54. New York: Routledge.
World Council of Churches. 1948a. *Man's Disorder and God's Design: An Omnibus Volume of the Amsterdam Assembly Series*. New York: Harper Brothers.
———. 1948b. *The Universal Church in God's Design*, Vol. I of World Council of Churches, 1948a.
———. 1948c. *The Church's Witness to God's Design*, Vol. II of World Council of Churches, 1948a.
———. 1948d. *The Church and the Disorder of Society*, Vol. III of World Council of Churches, 1948a.
———. 1948e. *The Church and the International Disorder*, Vol. IV of World Council of Churches, 1948a.
Wyon, Olive. 1948. "Evidences of New Life in the Church Universal." In World Council of Churches, *The Church's Witness to God's Design*, Vol. II, 110–33.
Zeilstra, Jürjen. 2020. *Visser 't Hooft, 1900–1985: Living for the Unity of the Church*. Amsterdam: Amsterdam University Press.

Chapter 14

Beyond Tools, Means, and Ends

Explorations into the Post-Instrumental Erewhon

Jean Robert

The technological or instrumental epoch was the period of history in which artifacts separated from the body, called tools or instruments, could embody personal intentions. This epoch is now coming to an end. Hence, most contemporary discussions of technology and its predicaments are little more than interpretations of a new epoch in the terms of one that has passed.

During the instrumental age, the most important judgments concerned the rationality of means-to-ends relationships. Such judgments were standard in economic, political, and, above all, in techno-logical discourses. Today, these judgments generally miss the point. To be meaningful, ends must carry the intentions of *concrete* historical subjects—whether persons or communities. To function as such because embedded in a concrete reality, means must be *limited* in number, power, and size. This inherent relationship between meaningfulness and means also entails that means can be limited without being "scarce." The notion of "scarce means" includes outsized technologies aimed at "ends" that reflect bureaucratic imputations of human needs. Under the regime of economic scarcity that dominates contemporary understanding of the relationship between alleged "means" to supposed "ends," goods are scarce without being bounded in number, size, power, and location. The availability of piped water is the paradigmatic example of the distinction between "scarce means" and limited means that can convey the ends of concrete subjects.

When "means" are scarce without being limited and "ends" are imputed to constructed subjects, the very possibility of evaluating the relationship of means-to-ends collapses. Moreover, since the category "tools" or "instruments" is identified as the "means" when conceptualizing the means-end relationship, the collapse of the latter also marks the waning of the instrumental era. It is to hasten recognition of the demise of the tool as such that Ivan Illich, the author of *Tools for Conviviality* (1973) pronounced a surprising "goodbye to tools" in the last years of his life. Illich gave the name "convivial" to that feature of artifacts which express the just proportion of a tool with the natural powers of its user's body, and he defined proportionality as the fit of the tool to the body's autonomous powers. We still use the word "tool" or "instrument" to describe such things as jets, the internet, and kidney dialysis machines, though these have broken all proportional links with the powers of the body and the senses. Even as the instrumental age mutates into a no-man's-land between known technology-related predicaments and the unnamed horrors of worse to come—one can think of the causes and consequences of climate change—Illich suggested it is also an invitation to a liberating *metanoia*.

Actually, two authors have moved on parallel tracks in their efforts to name the post-instrumental Erewhon and to devise concepts to understand its elusive new threats. Both Jacques Ellul (1980) and Illich (2005) departed from their previous analyses of "the technological society" and of "convivial tools," respectively, and proposed the word "*System*" to name what lies beyond the age of instruments. Both understood that a unique historical mutation had rendered obsolete the very concepts that had previously allowed them to be unusually acute analysts of the late Technological Age. Both saw in the mutation of the technological society into the system a betrayal of the vocation of the West, by the West. This vocation is a call to freedom. Tools are compatible with freedom if they are available to both be taken up and put down. This double possibility can only be preserved when tools are strictly limited in power, size, and number. Such limits not only prevent my tools from encroaching upon your freedom (and ultimately mine too) but also so not require their operation to function as pseudo-ethical imperatives.

Today, while the threats that once lurked beneath the veneer of industrial optimism and the ideology of progress are becoming daily realities, yesteryear's wishes for political self-limitation have given way to a techno-sophic logorrhea. In contrast, Ellul believed in the possibility of a new start for freedom, and Illich had the courage to maintain hope in the face of bleak expectations. Ellul understood that the absence of such traditional exteriorities as God's judgment against man's overweening pride necessitated an immanent examination of technology that made clear that the only way of living with "systems" was to practice an ethics of renunciation to power (*une éthique du non-pouvoir*). Accordingly, he had admonished his contemporaries "not to do all what they were [technically] capable of doing" (Ellul 1980). Similarly,

Illich (2005) called for *ascesis* to avoid the disembodying consequences of "systems." For Illich, it is the threefold predicament of disincarnation, the loss of the senses, and the eclipse of personal ends that characterizes the future beyond tools. Illich invites his readers to "throw their nets beyond all planning objectives," toward the not-yet, the *nondum* that the poet Paul Celan (1998) located "North of the Future," the unexplored territories beyond the future.

NEW CONCEPTS FOR THE END OF AN EPOCH

It is only within the limits of proportionality that tools can serve personal ends. In the era of the System, by which Ellul meant the fact that technology has become the ineradicable milieu of man, what one might still, by error, call "tool" has broken all proportional bounds with the body and senses. From intellectual laziness one may still believe that humans control the System, but it, in fact, holds them. To avoid the confusion that comes from using "tool" to describe artifacts that can serve no one's ends and make a stronger claim on individuals than the individuals do on them, I propose, in a first approximation, to call such artifacts "systemic pseudo-tools." They simulate what, in the System, cannot be: a proportional relation, a mutual fit between bounded means and personal ends. Systemic pseudo-tools break the double availability that characterized classical tools: they can be, and soon, *must* be taken, but they can hardly be left. Once you're "wired," it is difficult to "unplug": the System cannot be de-computerized.

Systemic pseudo-tools are but the nodes through which the System turns people qua users into subsystems. But they are not really tools and even less technical means: they are *interfaces*. In the System, sensory perception is understood as resulting from an interface between two systems, one being an artifact and the other a person. Such insertion within interactive systems can be glimpsed in a life enmeshed by five hundred high-definition television channels. Thirty years ago, Illich analyzed the counterproductivity—the unintended movement away from an original goal—of service institutions and carefully defined three kinds of counterproductivity. Today, we must speak of the paradoxical efficiency of counterproductivity. Medicine, for instance, is increasingly iatrogenic as suggested by the number of medically caused deaths in US hospitals. However, medicine has been so successful at inserting the iatrogenic body into popular perception that patients now effectively integrate themselves into the biomedical system. As Ellul (1980) has suggested, every aspect of existence within the System is subject to control, manipulation, and experimentation. All institutions that produce services also reproduce the conditions for control, manipulation, and experimentation and the main service agencies (medicine, education, transportation) have become the principal intermediaries by which their clients are transformed into functional subsystems of the System.

As interactive systems are increasingly ubiquitous, adaptation to them becomes not just recommended but a necessity. Illich (1995) proposed to call the mediation proper to the age of the System the *show*. The show is the transductor or program that allows the interface between systems. Contrary to the image, the show does not situate the spectator by providing a standpoint—that is, a place for his feet—but makes reality recede to a remote shore inaccessible to the hands, the feet, the senses and, finally, to common sense and imagination. Whether one optimizes "health" prompted by the injunctions of Fitbit or "drives" an autonomous car, it is the show on the screen that facilitates the interface between the human subsystem and technical system. What systems consider real are databases, measurements, and programs for building models. However, since the System disguises its interfaces as tools and its shows as images, it is still (mistakenly) interpreted according to the known concepts of instrumentality and perspective. Systemic threats are thereby misinterpreted rendering humans more vulnerable.

This essay is an attempt to unveil the specificity of the System as a historically unprecedented reality to foster a liberation from its enthralling shows. To understand the *hexis* (state) and the *praxis* (action) of the present age, a new set of concepts is necessary. According to Illich (2005), these concepts must answer two profoundly troubling historical realities:

1. Comparisons between today's System and previous or other material cultures are mostly meaningless.
2. The System is the ultimate result of a perversion of the gospel and of its transformation into the fundamental ideology called Christianity.

Ellul responded to these historical realities by offering a phenomenology and characterology of "technological Systems." Even though he continues to use the term technological, his description of the System emphasizes how technique has become the milieu of man and as such the real environment of his thought and actions. In this encapsulating way, the System marks a break with all past techniques, which did not mesh into an integrated and integrating whole.

Illich's response consisted in also unearthing the historicity of instrumentality, but by observing the demise of technology in the mirror of its origins. According to Illich (2005), the concept of instrumentality that sustained the technological age had its origin in the concept of *causa instrumentalis*, which was elaborated during the eleventh and twelfth centuries as an outgrowth of the Aristotelian *causa efficiens*. To understand the changes in conceptual and perceptual modes that took place then, Illich proposed to focus on a *"mouvance de longue durée"*—a long-term historical movement he called the Great Tradition of Proportionality that had developed prior to technological instrumentality.

In other words: Illich thought the end of the instrumental or techno-logical era, increasingly manifest in our days, could not be fully appreciated without being studied simultaneously with its beginning. Accordingly, with his friends and students of Bremen and State College, he reflected on the dusk and dawn of the Technological Age (Illich 2005). Illich died before completing this project. However, the conversations with him are still fresh in his friends' memories. Besides, he generously distributed among them tens of notes that constitute paths to the study of what he called the *instrumentum separatum* and which was considered a novelty of the late twelfth century: an artifact that embodies intentionalities but is separated from the body. The idea of "the tool" refers to this separated artifact which flourished during the technological epoch, and to which the Age of Systems bids goodbye.

INSTRUMENTALITY FRAMED BY PROPORTIONALITY

That the hammer is separate from the hand that wields it even if it replicates, more powerfully, one of the uses of a fist is a modern conceit. The separation of the *organon* from the body as *instrumentum separatum* and the concomitant disembedding of *causa instrumentalis* from the four Aristotelian causes are the first steps toward the historical break with the tradition of proportionality, a break that characterized modernity in the broad sense. But what is proportionality? The word *proportio* is the Latin translation of the Greek term *logos* that means relation, but also speech, balance, or proportion. In its mathematical sense, the *logos*, since Euclid (about 300 BC), refers to the relation of two elements (a:b), while the similitude between two or more relations (a:b::c:d::e:f) is called *analogia* in Greek. Nine centuries after Euclid, Boethius in *De institutione musica* translated *logos* and *analogia* from Greek to Latin as *proportio* and *proportionalitas* (2009). For the ancients, *techné* (art or artifice rather than technique) was proportional to *phusis* (nature) and in this sense were both manifestations of the Logos, both are part of a superior harmonic order. The proportionate relation between *techné* and *phusis* did not mean that mechanical tools imitated bodily *organa*, as they have come to do in modern times. Instead, both hammer and hand were only of the four causes of Antiquity, which were, themselves, a manifestation of order, harmony or *Logos*.

Classical quadri-causality slowly gave way to the uni-causality of modern techno-science. Over the Technological Era, *causa materialis*, *causa formalis*, *causa finalis*, and *causa efficiens* were progressively reduced to an outgrowth of *causa efficiens* that the Schoolmen called *causa instrumentalis*. To say that "B is a means to C" no longer implies a proportional mean term—that is, the mutual fit or adequacy (*adequatio*) between one cause and another. For the Ancients, gold is adequate to the form *ciborium*, but not to the form table. In contrast, wood is adequate to a table. In Antiquity and the Middle Ages, the mean term was proportion itself, as noted by Plato in the *Timaeus*: "That two elements be united by themselves, without a third, is impossible, for a certain third in the middle must provide the link between both. But the most beautiful of all links is the one that becomes one with what it unites to itself. That, proportion, by its very nature, does best" (31c). Proportion was essential to knowing itself, which was essentially an analogical relation. Concerning that, Lee Hoinacki—inspired by Aquinas—writes (1998):

> Knowing is an action, and every action can be examined in terms of proportion. Indeed, every action must be proportionate, both to its object and its initiating agent (source or cause). From its object, it will receive its species or character, and from the strength of the agent, its precise intensity ([Aquinas, 1998]II-II, q. 26, a. 7). External things, which can be the object of an action, are, in themselves, good, but they do not always have the proper proportion to this or that action (I-II, q. 18, a. 2). For an action to be good, then, it needs a good object and a proper or proportionate relationship to that object. One can act badly—use evil means—to reach a good end. Such an action cannot be good.
>
> More fundamentally, action is a movement tending toward some end. And whatever moves in this way has an aptitude or proportion to that end, 'for nothing tends toward a non-proportionate end.' This aptitude or proportion can be named: it's an appetite for the good; as such, it is love. When the action reaches its object, there is a resting in the good (I-II, q. 25, a. 2). Here the reader's mind must go in two directions: one, to recall that any such argument is based upon the order of the universe; two, to imagine those particular actions in one's life where one acted to achieve or obtain something. The argument can be tested.

In contrast, the Technological Age was increasingly pervaded by the instrumental mentality for which the relation between A and B was no longer the proportion (A:B), but the instrumental relation (AB) in which A is an instrumental means to B and in which the mean(s) was no longer the matrix that unites two beings in the here and now, rendering itself one with them. When the mean as *proportio* was ignored and replaced by *causa instrumentalis*, then every B of which an A is the means became itself the means of a C, in an unending instrumental chain: ABCD. . . .

RELIGIOUS AND COSMOGONICAL ORIGINS OF INSTRUMENTALITY

Heidegger (1977) argued that technology is founded upon a disposition which is not itself technological. He called it *Gestell* and he conceived it as a breach with quadri-causal or "tetra-etiological" thinking. Since the thirteenth century, the administration by the Church of a "power of divine redemption" seems to have provided the context in which Westerners imputed to the most diverse objects an "instrumentality" increasingly separated from the body's proportional powers. Shall we see here "the disposition that founded technology without being technological itself"? If not, where and when did instrumentality start to prevail over proportionality? When did the proportional mean term give way to the instrumental means? From 1990 on, in conversation with Hoinacki, Illich asked the following question: Is sacramental theology the flowerbed of instrumentality? At the beginning of his research, he surmised that, at the origin, medieval theologians forged the concept of *causa instrumentalis* in the hope of distinguishing the seven sacraments from a diffused sacramental order. However, it must be stressed that this intuition remained for him a working hypothesis and that he never ceased to consider another possibility, namely that the concept of instrumental cause first nested in cosmology. In his thought, these two possible origins often intertwined: Already before 1100, theologians had started to give importance to *causa instrumentalis*. According to them, God, the *primum movens* of the universe, delegated to the angels the task of moving the spheres of the world, and these did it by means of instruments called *corpora coelestia*. Sometime after that—how much later is an object of ongoing debates—the sacraments themselves came to be seen as God's instruments for human salvation. Despite all the uncertainties concerning "priorities" (cosmology or theology first?) and "periodization" (when did the instrumental cause prevail over the other classical causes?), it seems that Illich has put here his finger on a major metamorphosis. Yet, to understand the magnitude of this metamorphosis, one must attempt to tune one's intuition to perceptions anterior to that change, until it is in consonance with a resolutely non-instrumental understanding of the sacraments. What follows is a personal exercise. First, I have looked for expressions of proportionality in the twelfth century. Then, after a leap of about a century, I comment on technological and theological expressions of instrumentality.

THE CATHEDRAL BUILDER, THE PLOWMAN, AND THE PRIEST: FROM PROPORTIONAL MEAN TERM TO TECHNICAL MEANS

The Eucharistic celebration is the event of Christ's presence. It renders the believers as participants in his flesh and blood (Jn 15:7,16) and, with it, makes of their community the (visible) body in which the Verb becomes Flesh. The other visible body of the Assembly is its external shell of defenses, traditions, possessions, rules, laws, and dogmas. The early medieval mind associated the spiritual and the physical body in a proportional relation: the former was to the latter as the Verb to the Flesh, the heart to the feet, spirituality to the chores of the earth, heaven to the soil, the flesh that suffers and feels to the body visible to others in the world. The heart and the feet, prayer and the plights of the earth are different, but one cannot be without the other. Yet, their union, consonance, harmony, or "symphony" cannot and must not be instrumentalized, synthesized, "designed": it is free as grace itself. The Promethean urge to pull harmony into the realm of feasibility, to design the concordance of the discordant by the imposition of a mechanical "temperament," has been a great temptation for and constant provocation to the Church. In the event of the Eucharist, the present moment or liturgical time is the proportional mean term between the temporality of two distinct "coming-abouts": Christ's promise to his disciples confirmed in the Last Supper and the Kingdom to come. The liturgical event as *Copula* (relation) between both leads the community of the believers from their memory to their hope.

The fact that the real presence of the incarnated verb has been elucidated in the language of proportionality is evident in the *incipit* of the report on the construction and inauguration of the first Gothic Basilica by its builder, Abbot Suger of Saint-Denis, written shortly after 1140. I quote him in Latin, his language, because all translations that I consulted were misleading: they violated the thought style of the twelfth century and obscured the epistemic landslide that occurred between then and now (Abt Suger 2000).

Divinorum humanorumque disparitatem unius et singularis summaeque rationis vis admirabilis contemperando coaequat: et quae originis inferioritate et naturae contrarietate invicem repugnare videntur, ipsa sola unius superioris moderatae armoniae convenentia grata concopulant.

The disparity between the divine and the human is made consonant [literally: "contemperata," "tempered"] thanks to the admirable virtue of a unique and superior proportion [ratio]: Beings that due to the inferiority of their origin and the contrary characters of their respective nature seem to deem each other mutually repulsive [repugnare] themselves join [concopulant] in the gracious proportional relation [convenentia] of a superior harmony.

Like a symphony, the Abt's report ends with a counterpoint of its initial phrase:

> *Benedicta gloria Domini de loco suo; benedictum et laudabile et superexaltatum nomen tuum, Domine Jesu Christo, quem summum Pontificem unxit Deus Pater oleo exsultationis prae participibus tuis. Quae sacramentali sanctissimi Chrismatis delibatione et sanctissimae Eucharistiae susceptione materialia immaterialibus, corporalia spiritualibus, humana divinis uniformiter concopulans, sacramentaliter reformas ad suum puriores principium; his et hujusmodi benedictionibus visibilibus invisibiliter restauras, etiam praesentem in regnum coeleste mirabiliter transformas, ut cum tradideris regnum Deo et Patri, nos et angelicam creaturam, coelum et terram, unam rempublicam potenter et misericorditer efficias; qui vivis et regnas Deus per omnia saecula saeculorum. Amen.*

Blessed glory of God in His place, exalted be your name, lord Jesus Christ, whom God the Father, with the oil of exultation anointed as the supreme link [pontifex: bridge maker] between those who participate in you. In the degustation of the holy sacramental oil and the reception of the holy Eucharist, while you join [concopula(n)s] the material and the immaterial, the body and the spirit, the human and the divine, you sacramentally regenerate their original purity; in the grace of visible blessings, you transform them invisibly into the first fruits of heaven, so that, when you will establish the Kingdom by God, you will potently and mercifully join us to the angels and Earth to Heaven into one republic; you who live and reign for the centuries of centuries, Amen.

The most beautiful of all links is the one that becomes one with what it unites: proportion. Christ's presence in the Eucharist is this link. Such a "musical" interpretation is part of the medieval expression of the faith in the Incarnation of the Word. The most beautiful of all links harmoniously joins the soil and heaven, the body and the sentient flesh, the human and the divine. It is wisdom, the second person of the Trinity, the incarnated word. It is the *copula* that unites (concopulat) "the beings that by the inferiority of their origin and the contradictions of their different natures seem deemed to repulse one another." In the Latin text, the words *contemperans, concopulans, concertatio similium et dissimilium* (called *convenentia disparium* by later authors), *charitate ministrante, sacramentali sanctissimi Chrismatis delibatione* occur again and again. They describe an act (linking, uniting, joining, harmonizing) or designate the mean term, the link or copula (for instance caritas, love) that harmoniously unites dissimilars.

At about the time that Sugerius inaugurated the basilica, a new instrument was transforming the art of tilling the land. It was the asymmetric plow, equipped with a moldboard for turning over the earth (*Wendepflug* in German). The previous plow (*Hakenpflug* in German, *araire* in French) had a symmetrical plowshare that did not turn over the sod. To understand the reach of this innovation, one must think of the work of the plowman: he cuts and opens the topsoil, lifts the sod detaching it from its base, and turns it over, "so that the dead roots become dung for the new ones." With the primitive plow without an asymmetric moldboard, the plowman had to take care of each one of these operations and to handle his tools accordingly. He and not his tool was the *causa efficiens* and this required great skills, for his capacity of turning the sod depended more on his wrists than on the strength of his draft team. Plowing was an art. With the progressive introduction of ever better moldboards since the eleventh century, the husbandman's intentionality migrated, so to speak, to a "machine" that took care of the three operations almost independently of its user's skill: even an unskilled peasant could now till the fields. A tool that so incorporates or "materializes" an intentionality becomes a kind of *causa efficiens*, or better, the concept of *causa instrumentalis* was perhaps invented to define this novelty: *an object endowed with a mechanical intentionality.*

Let's now turn to theological expressions of instrumentality. I will first comment on the following passage from Thomas's *Summa Theologica* (1964–1981):

> *sicut dictum est, ministri Ecclesiae instrumentaliter operantur in sacramentis, eo quod quodammodo eadem est ratio ministri et instrumenti. Sicut autem supra dictum est, instrumentum non agit secundum propriam formam, sed secundum virtutem eius a quo movetur. Et ideo accidit instrumento, inquantum est instrumentum, qualemcumque formam vel virtutem habeat, praeter id quod exigitur ad rationem instrumenti, sicut quod corpus medici, quod est instrumentum animae habentis artem, sit sanum vel infirmum; et sicut quod fistula per quam transit aqua, sit argentea vel plumbea. Unde ministri Ecclesiae possunt sacramenta conferre etiam si sint mali.* (III, q. 64, a. 5, c)

> As stated above (Article 1), the ministers of the Church work instrumentally in the sacraments, because, in a way, a minister is of the nature of an instrument. But, as stated above (62, 1,4), an instrument acts not by reason of its own form, but by the power of the one who moves it. Consequently, whatever form or power an instrument has in addition to that which it has as an instrument, is accidental to it: for instance, that a physician's body, which is the instrument of his soul, wherein is his medical art, be healthy or sickly; or that a pipe, through which water passes, be of silver or lead. Therefore the ministers of the Church can confer the sacraments, though they be wicked. (III, q. 64, a. 5, c)

For Thomas Aquinas, around 1265, the sacraments are still embedded in classical quadri-causality. As causes, they are formal and final. Inasmuch as they are efficient, it is only as signs. Yet he calls the minister's action instrumental. According to him, an instrument—unlike the new plow—does not act following its proper form but following the virtue of he by whom it is moved: except for what is inherent in its constitution, any form or quality can be conferred to a tool. Just as a physician's body, which is the tool of a mind endowed with the medical art, can be healthy or sick, and just as a water pipe can be a good duct regardless if it is made of silver or of lead, so also the Church's ministers can administer the sacraments even if they are bad.

Sometime after Thomas, Christ's presence was increasingly perceived as being expressed by an object, the host. Still nominally called *Corpus Christi*, the assembly came to expect the instrumental acts of the priest to manifest Christ's presence. Not only artifacts like the new plow but also assemblies of people could now be the bearers of "mechanized intentions" and thereby became institutions. Professor Ludolf Kuchenbuch, to whom I owe the preceding reflections on the plow, also suggested that the progressive transformation of the Eucharistic assembly into a "church" can be correlated with the construction of an apparatus that incorporated and mechanized salvific intentions. Yet, for both Kuchenbuch and this author, the presence of Christ in His assembly remains a mystery. What might be less mysterious for the historian to investigate more fully is how this presence was "instrumentalized." Once instrumentalized, the Eucharist was endowed with a mechanized or institutionalized intentionality which mediated Christ's presence and its distribution by the priest. The priest stood now between what had been the table of *agapē* and the assembly. He drank alone while his parishioners observed him.

What did the church goers see? They saw the priest ingest an essence in the form of the host. How did they see him? Their gaze was no longer Aristotle's haptic gaze that, not unlike an erectile member, threw itself toward the *visibilia* to bring their color back to the eye. They were now invited to capture the *species* of what they saw. And what is the species of the almost immaterial host? It is Christ's body and blood, in an object that can be distributed. The *officium* made its salvific substance present to all who attended. Unlike the participants in the old *agapē*, believers were now spectators who could not touch or smell and barely tasted what they were partaking of.

The instrumentalization of the Incarnated Verb's presence as distribution of a salvific substance was paradoxical for still another reason. In the Middle Ages, Christ's body was the tip of an iceberg whose immersed mass consisted in a long familiarity with relics. The relics were, very materially, bones of holy dead. They were the link of everyday life to the beyond and were, consequently, behind or beneath all ceremonies and rites in which it was invoked, that means in about all celebrations and rituals. The more ubiquitous holy bones became, the less incarnate the Verb's presence was perceived. Did the drift toward uni-causal instrumentality that made the technological age possible have roots in this paradox? I think it did. It is as far as I can go in exploring the theological roots of the age of tools. I invite others to flesh out the historical matrix of the instrument so that we can better grasp its demise.

REFERENCES

Abt Suger von Saint-Denis. 2000. *Ausgewählte Schriften: Ordinatio, de consecratione, de administratione,* edited by Andreas Speer und Günther Blinding. Darmstadt: Wissenschaftliche Buchgesellschaft.

Boethius. 2009. *De institutione musica libri quinque—liber primus, VIII and IX,* edited by Hans Gebhardt. Munich: Grin Verlag.

Celan, Paul. 1998. "North of the Future." In *A Voice . . . Translations of Paul Celan,* edited and translated by Muska Nagel. Orono, ME: Puckerbrush Press.

Ellul, Jacques. 1980. *The Technological System.* Translated by Joachim Neugroschel. New York: Continuum.

Heidegger, Martin. 1977. *The Question Concerning Technology and Other Essays.* Translated by William Levitt. New York: Harper & Row.

Hoinacki, Lee. 1998. *A Statement on Tools.* Unpublished manuscript.

Illich, Ivan. 1973. *Tools for Conviviality.* Berkeley, CA: Heyday Press.

———. 1995. "Guarding the Eye in the Age of Show." *RES* 28 (Autumn): 47–61.

———. 2005. *The Rivers North of the Future: The Testament of Ivan Illich as Told to David Cayley,* edited by David Cayley. Toronto: Anansi Press.

Plato. 2008. *Timaeus.* In *Timaeus and Critias.* Translated by Desmond Lee. New York: Penguin Classics.

Thomas Aquinas. 1964–1981. *Summa Theologica.* 61 vols. London: Blackfriars.

Chapter 15

Understanding Bureaucratic Order

The Theological Paradigms of Modern Hierarchy

Sajay Samuel

What Max Weber (1978) described as the form and dynamic of bureaucracy has become a universal template for organizing human affairs. As an ordered arrangement of interrelated offices, bureaucracy has become the preferred instrument to accomplish seemingly every human purpose. Whereas conventional wisdom held that bureaucratic administration is the exclusive preserve of the state apparatus, Weber argued that both public agencies and private enterprises, both governments and corporations, exhibit the bureaucratic form. A particular set of offices may be organized to make profits and be named Amazon. Another set can be organized to deliver health care and be called National Health Service, or to entertain the masses and be named Netflix, or to deliver advertising and be called Google.

Today the bureaucratic form is perhaps the one constant across the many variations of economic modes and political regimes. In mature market economies such as Italy or emerging mixed economies such as India, in neoliberal nations such as the United States or communist countries such as China, the bureaucratic form is a ubiquitous template for organizing all kinds of human activity. In the twenty-first century, the overwhelming proportion of what human beings touch, smell, hear, taste, or see is bureaucratically made: little is self-made or communally done. The techno-scientific research and development that produces both satellites and coronavirus vaccines are no less the products of bureaucratic management than the food people eat and the pleasures they enjoy. Denizens of technologically advanced countries, long exposed to bureaucracy, can be excused for forgetting that life has never been so ordered and managed; the recent transmogrification of millions of self-sufficient peasants in India and China into employees and consumers shows the novelty and power of bureaucratic order.

Weber expected the bureaucratic form to persist because it reduced humans to relatively powerless cogs in a machine. My argument explores an alternative hypothesis, that the growth and spread of the bureaucratic order continues because of its religious roots. It is informed by prior work by Ivan Illich, Michel Foucault, and Giorgio Agamben, who locate the paradigm for the contemporary bureaucratic order in the millennia-long elaborations of Christian theology and the Catholic Church. It is also inspired by Carl Mitcham, who suggested that the instrument—understood as a neutral object subsumed by the intention of its user—was rooted in theological elaborations of the notion of *causa instrumentalis*, especially in the form of angels, sacraments, and the clergy (Mitcham 2018).

After briefly explaining Weber's conception of bureaucracy, I trace three of its most important characteristics—hierarchical arrangement, instrumental logic, and universal scope—back to their theological paradigms. The explication of *hierarchy* by Pseudo-Dionysius not only sacralizes power but also articulates it in a series of interrelated and ranked offices. That the occupant of each office is regarded as a mere *instrument* in the service of another, usually larger, purpose anticipates the instrumental rationality of bureaucracy, wherein each is only supposed to do his job. The *universal* scope of modern bureaucracy can be found in the Church's attention to all aspects of its members' lives, and its universal mandate to offer salvation to all people. Insofar as contemporary bureaucracies—whether religious, corporate, political, or anything else—are hierarchically organized, instrumentally designed, and universally applicable, they carry the signature of these constitutive theological paradigms.

I conclude by almost closing the circle that began with the bureaucratic order. The theological template is paradigmatic of the bureaucratic order, with one difference. Whereas the church attempted to conduct every aspect of this worldly existence for the sake of the next, the bureaucracy can only promise salvation in this world. But because there is no escaping the bureaucracy there is no salvation from it. Perhaps Ivan Illich was right: the bureaucratic order is the ongoing inversion of an order not for this world.

THE BUREAUCRATIC ORDER

While many economists, sociologists, and psychologists may have usefully corrected some details of Weber's explanation of bureaucracy, none have fundamentally challenged his general characterization of the modern bureaucracy or his prediction of its metastasizing growth. While it was widely thought that only a state apparatus is truly bureaucratic, Weber argued that both public agencies and private enterprises, both governments and corporations, express the bureaucratic form. Neither the purpose nor the legal ownership of an organization—whether profit-seeking corporations or NGOs, governmental agencies or charitable trusts—determine the form of the bureaucracy.

Weber noted that the modern bureaucracy differs from its precursors in its "continuity, stability, and permanence" (Weber 1978, 964). Modern bureaucracies are legally incorporated entities, and their right to exist is codified in law. Their continuity does not depend on the preferences or largesse of political patrons as did premodern bureaucracies, such as in Egypt during the New Kingdom period and in China during the Qin Dynasty. According to Weber, premodern bureaucracies are decisively characterized by such feudal or patrimonial features as nepotism and the purchase of offices by the highest bidder. When mandarins, functionaries, or clergy are compensated by taxes, tithes, or a share of agricultural produce, their positions are only as stable as the fluctuating fortunes of the emperor, the patron, or food production. Much like political appointees in the United States government today, premodern bureaucrats who were paid in kind served at the pleasure of the political authority in power. In contrast, the permanence of the modern bureaucracy is cemented by the regular payment of a salary which, in turn, is supported by an entrenched money economy and legitimate taxing authority (968).

As Weber emphasized, the modern bureaucracy is also distinguished from its historical precursors by its means-end or instrumental rationality. A bureaucracy is an instrument to divide labor and assign authority through the arrangement of its offices. The total activity of an organization is divided into tasks, a subset of which is assigned to a particular office. Each office is given authority over a set of tasks to be completed by those employed by it. For example, in the United States, only the Treasury Department is authorized to print currency, whereas only the Federal Reserve Bank is permitted to determine the amount to be printed. The conduct of authorized activities typically follows written impersonal rules and methodical processes of execution. Penalties are assessed on those who fail to properly fulfill duties or follow procedures. The organization chart depicts this distribution of disciplinary power, which is enforced by organizational policies ultimately supported by the law. Moreover, superordinate and subordinate offices are arranged so that commands are efficiently transmitted to those who must carry them out and the results of such actions are, in return, made known to the command centers. Files, whether electronic or paper, not only describe the jobs attached to an office but also constitute the material traces to verify the specifics—who, when, where, how—of the jobs done. The terabytes of data now amassing in servers on sea and on land constitute the virtual traces of the bureaucratic order in the digital era.

The instrumental rationality of a bureaucracy is not limited to the rational organization of means alone. Rather, both means and ends are instruments of the bureaucratic form. The instrumentalization of "ends" is evident when considering the mission of any bureaucracy. Organizations explicitly state their "mission statement," understood as the purpose for which they are constituted. For instance, "to deter war and ensure our nation's security" is the stated mission of the United States Department of Defense. Such ends or missions are not permanently written in stone. Bureaucracies periodically revise their missions as circumstances warrant. For example, the green clothing company Patagonia is now "in the business to save the planet" instead of once narrowly finding "solutions to the environmental crisis." It is in this sense that bureaucracies are rightly understood as rational means to achieve instrumental ends. That both means and ends are instruments reveals the fuller sense in which bureaucracies are instrumentally rational. All bureaucracies function in the name of whatever else—stockholders, citizens, refugees, consumers, planet—and it is this vicarious action that constitutes a bureaucracy as a total instrument.

The instrumental rationality of a bureaucracy is not limited to the arrangement of offices and the imputation of the interest to be served through a mission statement. Crucially, the occupant of every office is also an instrument, and it is this thoroughgoing separation of the office from the person of the office holder that is the *sine qua non* of bureaucratic management. Bureaucrats are expected to do their jobs *sine ira et studio*. Neither anger nor affection should affect the prosecution of a bureaucrat's task. Managers are trained to be impersonal and dispassionate in executing the functions of the office they temporarily occupy. The identity of an individual holding an office, be it the chief executive officer or the janitor, does not matter: an organization works best when its members do their jobs, whatever these are. All bureaucrats are instruments, valued for "precision, speed, unambiguity, knowledge of files, continuity, discretion, unity, strict subordination, reduction of friction and of material and personal costs" (Weber 1978, 973).

Some years ago, the well-known management consultant Tom Peters enviously invoked the church bureaucracy as a structure worth copying by secular organizations trying to get flatter and more networked.[1] What he did not know is that theological paradigms have, over millennia, shaped the key features of the present-day bureaucratic order: its hierarchical structure, its instrumental form, and its universal scope.

THE HIERARCHICAL ORDER: POWER SANCTIFIED

Weber has already pointed out that modern bureaucracies are hierarchical in the sense that they are a ranked arrangement of offices where usually all offices are subordinated to one other office, except for the highest office (often president or CEO).[2] The president is positioned above the cabinet minister who is over the secretary; the assistant professor reports to the department chair who reports to the dean who reports to the provost. Offices and their occupants are subordinated for reasons of effectiveness. "Unity of command" avoids conflicting orders from different supervisors. More generally, the "principles of management" insist that unified leadership articulated through ranked offices is a technical requirement of effective organization—for speed of decision, clarity of communication, defined locus of responsibility, and other reasons.[3] These characteristics are essential in understanding the predominantly practical sense of the term hierarchy commonly held today.

Yet the technically justified structure of a bureaucracy is erected on a theological ladder of angels. The literal translation of "hierarchy" (ἱεραρχία) is "sacred power" (*hieros* = sacred). It was coined by Pseudo-Dionysius the Areopagite to describe the angelic structure of rank-ordered activities thought necessary to allow God's word to work in the world.[4] Unlike democracy or plutocracy, hierarchy does not describe a particular social group that holds the reins of power. Instead, because God is the source of the power which acts on the world, it is power itself that is sacred. Attending to the Areopagite's explication of the organization by which God's word informs the world will reveal how a theological precursor serves as a template for the modern bureaucratic hierarchy.

According to Pseudo-Dionysius, the celestial (angelic) and the ecclesiastical hierarchies have parallel and interlocking structures, and "hierarchy" has three senses: "a sacred order, a state of understanding, and an activity approximating as closely as possible to the divine" (*CH*, 164D). The order of a hierarchy refers to its structure or arrangement. The celestial hierarchy is structured in three ranks of heavenly beings—angels—each of which comprise, in turn, three ordered ranks. For example, the first hierarchy of angels contains *seraphim*, *cherubim*, and *thrones*, in that order. These are the angels closest to God, whereas the *principalities*, *archangels*, and *angels* occupy the hierarchical triplet of immaterial beings who occupy the third and farthest rank from God.[5] This heavenly hierarchy has a parallel in the earthly world. According to Pseudo-Dionysius, God made the ecclesiastical hierarchy to be "a ministerial colleague" of the celestial hierarchy (*CH*, 124A). Consequently, the ecclesiastical hierarchy is also organized in a three-tiered hierarchical order composed of the *sacraments*, *the clerical orders*, and the *initiates*. As with the celestial hierarchy, each of these is a triplet.[6] For example, whereas *baptism*, *communion*, and *consecration* make up the sacraments, the clergy is composed of the *hierarchs*, the *priests*, and the *deacons*. *Monks* and *the sacred people* comprise two of the three ranks of those being initiated into the sacred structure. The lowest rung of these "are made up of all those members who are dismissed from the sacred acts (of liturgy) and rites of consecration" (*EH*, 532A). In the Pseudo-Dionysian scheme not everyone is within a hierarchy. There are those "who are *stone-deaf* to what the sacred sacraments teach . . . who have rejected the saving initiation . . . and [who say] . . . I do not want to know your ways" (*EH*, 432C; *CH*, 260C).

The second referent of hierarchy as a state of understanding or knowledge becomes apparent when considering the principle by which beings are ordered in the ranks of the hierarchy as structure. While those ignorant of the sacred teachings are cast outside the hierarchy, it is their state of knowledge that defines the rank of those in the hierarchy. The celestial and ecclesiastical hierarchies are connected by the angels and the hierarchs. As the lowest rank of the immaterial beings, it is the angels who "are closer to the world" and therefore "take care of our own [i.e., the human] hierarchy" (*CH*, 260A–206B). In a similar way, as the highest rank of the clerical order, it is the hierarchs who have the greatest human capacity to receive and pass on the divine word. The ranking of angelic and clerical orders prompts the question about the criteria by which a particular being—material or immaterial—was assigned to a specific rank in the two-fold hierarchy. Why did the seraphim enjoy the highest rank among heavenly beings while the deacons occupied the lowest rung of the clerical hierarchy?

In brief, it was the attainment of divine knowledge based on the God-given capacity to receive it that determines the rank enjoyed by a specific being. Even if "everything in some way partakes of the providence flowing out of this transcendent Deity," not everything participates in the Divine Godhead to the same degree (*CH*, 177C). Just as light passes more easily through translucent beings and is completely blocked by opaque matter, it is the capacity for divine knowledge possessed by different beings that determines their position in the hierarchy (which corresponds with their level of godlikeness).[7] Unlike the human deacons who occupy the lower ranks of the ecclesiastic hierarchy, the seraphim are ranked closest to God because they possess the "highest native intelligence," which allows them "to be raised up directly to him, a capacity which compared to others is the mark of their superior power and their superior order" (*CH*, 208D). Moreover, among celestial beings, knowledge is passed on immaterially from the superior to the inferior ranks "from mind to mind" (*EH*, 376C). In contrast, and as befits the human nature of those of the ecclesiastical hierarchy, enlightenment between the human ranks requires help from "the sacred veils" of scriptures and liturgy (*CH*, 121C). Accordingly, beings are rank-ordered in the hierarchy by their relative God-given powers of understanding.

The third meaning of hierarchy as activity becomes clear when considering the purpose of the celestial and ecclesiastical hierarchies in Pseudo-Dionysius. The transcendent deity, he writes, has created everything "out of goodness" and wants to "summon everything to communion with him to the extent possible" (*CH*, 177D). However, "for reasons unclear to us but obvious to itself," the blessed Deity also wants to specifically save "every being endowed with reason and intelligence" (*EH*, 376B). The salvation of such rational beings depends on them "being as much as possible like and in union with God," which is to say, to be divinized (*EH*, 376A). To be divinized, all must receive and accept the divine teaching. There are two mysteries that the hierarchies are structured to reveal. The lesser mystery is why God wants to save us. The deeper mystery is what God is and how he acts in the world. At best, what can be said of God is that he is beyond the sayable, beyond the comprehensible, beyond the visible.[8] As for his relationship to the world, it is Jesus through "whom we have obtained access to the Father" (*CH*, 165C). This "divine ray of Jesus" generously proceeds out of itself to enlighten and to lift up to itself "those beings for which it has a providential responsibility" (*CH*, 121C). To make his aspiring creatures more like himself, the divine beam of light purifies their ignorance, illuminates the purified with divine knowledge, and perfects the illuminated with understanding. "For every member of the hierarchy, perfection consists in . . . becom[ing] a 'fellow-worker for God'" (*CH*, 165B). It is the activity to purify, illuminate, and perfect the task or mission of working for God that constitutes the third sense of hierarchy. In this way the hierarchical order—which is simultaneously structure, knowledge, and activity—exhibits "the mystery of its own enlightenment" (*CH*, 165B).

We can now properly appreciate the Pseudo-Dionysian hierarchy. Hierarchy does not refer to a social group that has power over others but rather to the sacralization of power itself. Moreover, hierarchy refers not only to the structure or arrangement of beings, both spiritual and material, but also to their state of understanding and the activity they engage in. The Aeropagite's hierarchical ranks are composed of interrelated beings who are graded according to their knowledge of the divine Word of which they are the messengers or angels (ἄγγελοι). Though they belong to heterogenous spaces, the detailed description of Pseudo-Dionysian hierarchies allows us to also recognize their similarity to contemporary organizations fifteen hundred years later. Both comprise structures invested with power. Both are populated by beings who are rank-ordered by their capacities and knowledge and interrelated by the unified mission or purpose they serve.

THE LITURGICAL ORDER: INSTRUMENTALITY

The hierarchy ensures that each rank will "actually imitate God in the way suitable to whatever role it has" (*CH*, 165C). The role played by each member of the hierarchy, whether celestial or human, is to receive and pass on the divine teaching. As such, the hierarchical order positions its members to effectively execute specific functions for which they are selected, based on capability and desire. The hierarchies structure the actions of its members to collectively act on the initiates, the lowest level in the hierarchy, in the name of God, the head of the hierarchy. Consequently, the hierarchical order can be understood as a template to structure actions of a specific sort: acts by X (clergy) on Y (laity) for Z (God). This structure is the core of liturgical practice, and more generally of service. Understanding the liturgical order—acting on someone in the name of someone else—will shed light not only on the instrumental form of the bureaucracy but also on the impersonal face of the bureaucrat.

The liturgy did not always mean what it does today. The liturgy as a religious service performed for the laity began as *leitourgia*, a Greek term referring to a public service financed by citizens of wealth and status. The *leitourgia* of antiquity was a civic duty rather than an economic obligation. These civic duties regularly undertaken included financing naval vessels during wartime, supporting performers at festivals, and training teams for athletic competitions: all were paid for by some citizens, who received honor and esteem, for the sake of all. Andrew Carnegie's many public gifts—universities, opera halls, libraries—freely given are a latter-day example of the ancient *leitourgia*.

Over time, the liturgical practice for financing public works became so widespread in the Roman Empire that it was "transformed into a vast bureaucratic apparatus that obliged almost everybody, regardless of personal status or fortune" (Heron 2018, 47). It was also expanded to include all kinds of service, not just those that benefited the community at large. In this broad context, liturgy included "cultic service to the divinity"—building temples, financing sacrifices—which is the only meaning of the term today. When Christianity became a state religion under the reign of Constantine, however, the clerical hierarchy of bishops, priests, and deacons were exempted from these public liturgies that so burdened the population. The religious liturgy conducted by the priestly class satisfied the public service requirement because it was considered higher in rank. The narrowing of the liturgy from "public service" to "cultic services to the divinity" not only raised the rank of the clergy but also redefined its ground. No longer would liturgy be services performed by the people for the people. Instead, it became a service performed by the clergy for the laity.

The terms *clergy* and *laity* descend from the Greek κλῆρος and λαός, respectively. As Heron remarks, "In the New Testament, the whole Christian people had been defined as *kleros* to the extent that they were all equally considered 'heirs' (*kleronomoi*)" (2018, 69).[9] There was no separation or distinction within the Christian flock (*kleros*) since all were equal beneficiaries of divine love. In contrast, the laity (*laikoi*) refers to only some part (*laos*) of the whole people. When translators had to find a suitable Greek word to translate the Hebrew *'am* (people), they chose *laos* instead of *demos*. That is, they chose a word that referred to a portion or part of the whole people, instead of all the inhabitants of a political territory (*demos*). The choice of *laos* was apposite: *'am* referred to a specific people—the "people of Sodom," the "people of the Pharaoh" and even the "people of Yahweh"—just as *laos* did. Though *'am* carried a stronger sense of ownership or possession than did *laos*, the reference to a portion of some whole was central to both. By the time of Augustine, the people of God (*laos tou theou*) referred exclusively to all Christians, to the Church. It is this undivided people of God that will, in time, divide into clergy and laity. All church members are still Christians, and yet the clergy and the laity are incomparable parts of the whole because the latter serves as an object for the former. While the clergy ministers to the laity through specific functions, the laity has only one task, which is not really a task: to submit to the ministrations of the clergy and pay for the privilege.

It is in clarifying this role of the cleric that Thomas Aquinas formulates the idea of *instrument* that constitutes the kernel of the liturgical order. Just as angels serve as the ministers of God to bridge the chasm between divinity and humanity, so also the ministers of the Church serve to bridge the gap between divinity and the laity. Thomas suggests several reasons why an omnipotent God might want to delegate the execution of his order to inferior beings. Having a retinue of ministers and helpers frees God from sullying his hands with mundane tasks; allows God a medium to extend and amplify his power; showcases his dominance and grandeur; and finally ensures the perfection of divine providence (Heron 2018, 83–84). In explaining this last reason for the need for ministers as mediating causes (*causae mediae*) of divine government of the world, Thomas argues there are two ways in which men can imitate God: they can be good, and they can be the cause of good in others. As administrators of God's work, the clergy are neither the governor nor the governed but intermediaries between the two. Such clerical work is fundamentally vicarious since it is done on behalf of someone else (God) and directed toward yet another (the laity). Through such vicarious action, church officials imitate God in being the cause of good in others. Notably, it is in elaborating this ministerial task of vicarious activity that Thomas formulates what will shape the contemporary Western understanding of technology. Today, technology is often understood to neutrally mediate human purposes into the world. That notion is perhaps most clearly expressed in popular culture by the US gun lobby's favored line: "Guns don't kill people; people kill people." It is the concept of instrumental cause as proposed by Thomas to explain the role of ministers and the sacraments that subtends the present-day notion of technology as a separate and neutral instrument.[10]

For Thomas, both the clergy and the sacraments are instruments. As such, they are mediators that simultaneously receive and transmit the action of God upon the world. To grasp the specific status of the instrument in Thomas, it is useful to contrast his account with that of Aristotle. In the *Eudemian Ethics*, Aristotle argues that:

> the relations between soul and body, artisan and tool, and master and slave are similar, between the two terms of each of these pairs there is no partnership (*koinonia*); for they are not two, but the former is one and the latter is part of that one, not one itself; nor is the good divisible between them, but that of both belongs to the one for whose sake they exist. For the body is the soul's tool (*organon*) born with it, a slave is as it were a member or tool (*organon*) of his master, a tool (*organon*) is a sort of inanimate slave. (*Eudemian Ethics*, 1241b)

This passage underpins a key point. The contemporary understanding of technology as the realm of neutral tools used to carry out the intentions of its human users presupposes an intelligible separation between user and tool. In Aristotle, the relation between the craftsman and tool is akin to that between soul and body. Just as there is no relation of exteriority between soul and body insofar as the latter is in the soul as an integral part of it, so also the tool is integrally enmeshed with the tool user. For Aristotle, the potter is called the *efficient cause* of a pot because he is the agent that transforms clay into a pot. No separate category is required to classify the potter's wheel because, for Aristotle, it is present in the potter himself. As incredible as such a notion is for us today, for Aristotle, no partnership or "community" is possible between potter and potter's wheel because the latter is an indivisible part of the former. It is for this reason that Aristotle can indiscriminately use the term *organon* when referring to the body, the tool, and the slave in relation to the soul, the artisan, and the master.

Thomas decisively explodes the Aristotelian schema in his effort to conceptualize the work of the minister in the articulation of divine salvation. Unlike Aristotle, Thomas distinguishes a conjoined instrument (*instrumentum coniunctum*) such as the potter's hand from separated instruments (*instrumenta separata*) such as the potter's wheel. With the concept of *instrumenta separata* Thomas expressed a belief widely held today concerning technology we have come to take for granted: that the instrument is separate from the user or, as Ivan Illich argued, that there is an intelligible distality between the tool and the user (2005, 72). Moreover, unlike the *organon* of Aristotle, instruments that separate the efficient cause and its effects call for explanations of the way the former produces the latter. To use the previous example, the separation of the potter from the wheel invites explanations of how the potter's wheel carries or conveys the potter's actions to the pot. It is in the elaboration of this consideration that a second widely held belief concerning technology was given form: that the tool is a neutral transmitter of human purposes. Accordingly, the foundation for the two present-day assumptions about technology—that it is separate and neutral—was set by Thomas's conceptualization of sacraments and ministers as instrumental causes (*causae instrumentales*).

To explain how God acts in the world, Thomas had to connect the principal efficient cause to its effects. To connect God to the world, Thomas first instrumentalizes divinity itself. He affirms the Trinitarian argument that Christ, the savior of the world, is a single person in whom two natures are united. Thomas, however, argues that the humanity of Christ is a conjoint instrument of his divinity (*humanitas instrumentum divinitatis*), just as the body is to the soul. By construing Christ's humanity as an instrument of his divinity, Thomas tacks on instrumental cause as a subspecies of the efficient cause. Since they are conjoined, Christ's humanity cannot distort or otherwise mar the efficacy of Christ's divinity. It is this concept of an undistorted mediator that furnishes Thomas a model to understand the instrumental efficacy of sacraments and the ministers.

Recall however that both the sacraments and ministers are defined as *instrumenta separata*, though in contrast to the sacrament, the minister is a "living instrument" (*instrumentum animatum*).[11] Consider as an example the sacrament of baptism in the Catholic Church. The pouring of water in combination with the utterance of a specific set of words conveys to the baptized six supernatural graces, including that of being cleansed of sin. The material qualities of water—color, smell, purity—are irrelevant to its capacity to receive and transmit divine grace. Polluted waters can wash away sins with the same efficacy as clean water because "the power to bestow salvation flow[s] from the divinity of Christ through his humanity into the sacraments themselves" (Thomas Aquinas, quoted in Heron 2018, 106). It is precisely its status as an instrument that allows water to cleanse regardless of its material qualities. If water can cleanse because it is an instrument of God, so should the minister, but with one difference. Thomas had to contend with the difference between the inanimate and animate instrument.

Unlike water which must be moved in order to move, men can move autonomously, they are auto-mobile. Unlike inanimate water which cannot not receive and transmit the saving grace of God, the animate instrument can disobey, rebel, or refuse. Thomas confronts this possibility by yoking the will of the minister to that of God. Accordingly, he stipulates that the minister follows God's will by acting in a ministerial way (*per modum ministerii*). A minister acts in a ministerial way insofar as "he intends to do what Christ and the Church do."[12] In order to do as the Christ does, the minister must wholly subject himself to that principal agent (*se subjiciat principali agenti*). To be wholly subject to the principal agent requires the minister to be an unobstructed conduit or medium of Christ's activity. To be a pure medium of the efficient cause, the minister must function as an instrument of God. It is to the diminution of human potentiality into an instrument that Thomas's notion of *intentio* refers when he says the minister must "intend to do what Christ and the Church do."

Crucially, *intentio* "names nothing subjective, and admits of no qualification whatsoever. Neither can it be considered right or wrong" (Heron 2018, 91). For Thomas, someone merely going through the motions expresses *intentio*. Yet, for us, habitual behavior and reflex actions would not be considered intentional in the ordinary meaning of that word. It is for this reason that Illich described instrumental cause as a "cause without an intention" (2005, 71–79). *Intentio* only implicates the work done and not the worker. Just as both clean and dirty water can wash away sins when used in the baptismal rite, so also a wicked priest or layman can successfully baptize by merely uttering the right words. The sacraments work because their efficacy depends only on the work done (*ex opere operato*) and not on the effort of the worker (*opus operantis*). Just as inanimate water cannot avoid being an agent of the efficient cause, so also the minister *qua* living instrument cannot but transmit the sacraments. It is this philosophical notion of instrumental cause that constitutes the paradigm of the widely held belief that technology is the domain of separable and neutral instruments entirely subjected to the efficient cause. Once, the priest was the exemplary instrument through which God worked. Today, technology is the instrument *par excellence* through which human beings work.

There is, however, a deeper consequence that spills over from construing the minister as a living instrument of God. In separating the work done from the worker, Thomas also sets the stage for a character called the *official* who is reciprocally defined by his office. When Thomas formulates the ministerial mode of action, he separates what the minister does (action) from who the minister is (being). The sacramental effects follow the ministerial acts precisely because they are conducted by anyone going through the motions, by somebody doing the prescribed work. The Greek word *leitourgia* (public or religious service) was translated into Latin as office (*officium*). Today, whether it be the office of a bishop or office of the president, *office* refers to the prescribed actions, the authorized tasks, the appropriate functions, that the occupant must engage in. Weber argued that such externalization of human actions into official acts was the decisive event in the formation of the Church. The "depersonalization of charisma," he wrote, describes the "the separation of charisma from the *person* and its linkage with the . . . *office*" (1978, v. 2, 1164, emphasis in original).

In the Greek Scriptures, *charism* refers to "activities gifted by the Holy Spirit" which included wise speech, healing, prophesizing, and speaking in different languages. The transformation of freely given gifts into offices such as bishops, priests, and deacons delineates the form of an institutional church. Church offices not only routinize the gifts of the spirit, as for example transforming the gift of wise speech into a canned sermon, but also distribute the abundance of grace into controlled flows of administered salvation. The *officium* is thus a diagram of how any office transforms human actions into official acts and instrumentalizes living beings.

When functioning as an instrument, one is acting upon another at the bidding of a third. As such, what one does is separated from who one is. As Agamben points out, such separation of action from being implies that "action becomes indifferent to the subject who carries it out" (2013, 54). This is precisely the attitude of the official who is doing his job—more so when he does it well. The official embodies the office just as the office is enacted by the official. An official cannot not embody the office because when he does not, he is fired or removed from the position. This is why, for example, it is comprehensible to us that a sitting US president can besmirch the office of the presidency. The thoroughgoing separation of who one is and what one does is expressed both in the fact that living instruments occupy offices and in the fact that all official actions are conducted in the name of someone else. It is this institutionalized division within humans that underlies Agamben's surprising claim that *officium* inaugurated a change in Western ethics and politics which has not yet been fully acknowledged.

The liturgical order refers to the essentially vicarious nature of official action. The minister acts upon the laity on behalf of God. To be an effective agent of God, the minister must function as an instrument of God. As an instrument of God, the minister cannot avoid carrying out his office. It is this idea of a living instrument of God that constitutes the theological paradigm for the contemporary idea of technology as a neutral and separate tool that transmits human intention. Moreover, it also foreshadows all those who do the job they are assigned, whether as president or as postman. If the hierarchical order describes the connected layers of rank-ordered offices, then it is the liturgical order that describes the instruments that populate these hierarchies, each one just doing their job. The following section describes the purpose for which this vast army of ever-growing functionaries is assembled.

THE PASTORAL ORDER: UNIVERSALITY

Recall that Weber characterized the institutional church as the consequence of the depersonalization of charisma. When Foucault gave his annual lecture in 1978, he made a remarkable claim that broadened the significance of Weber's insight:

> The Church is a religion that lays claim to the daily government of men in their real life on the grounds of their salvation and on the scale of humanity, and we have no other example of this in the history of societies. With this institutionalization of a religion as a Church . . . an apparatus was formed of a kind of power not found anywhere else and [this] . . . pastoral power, in its typology, organization, and mode of functioning, pastoral power exercised as power, is doubtless something from which we have still not freed ourselves. (Foucault 2007, 148 and passim)

Foucault argued the church is a unique institution for three reasons: it is organized to take hold of every aspect of quotidian life, and does so, in principle, for all humanity. There is no historical precedent for this comprehensive aim. Moreover, the Church is also unique for its manner of "conducting, directing, leading . . . [and] taking charge of men collectively and individually throughout their life." The Church undertakes its mission of universal salvation as a pastorate, that is, on the model of a shepherd's care of his flock. The pastoral style of governing souls is distinct from that of politics, child-rearing, or rhetoric, each of which, in different ways, seek to shape and influence men.

In contrast, says Foucault, the logic of pastoral caretaking infuses the Western Catholic church from "top to bottom": in its distribution of authority, the specification of tasks, and organization of offices. Not only the whole flock but each single sheep falls within the ambit of pastoral caretaking, which constitutes the historical paradigm of the scope and style of management or government today, of our bureaucratic order. Take as an exemplar the current situation of the global response to the COVID-19 pandemic. All the peoples of the world are subjected to public health management as is every individual, in ambition if not in fact. Whatever the merits of this specific manner of management, it is obvious that the contemporary art of governing aims at both the collective and the individual, both the population and the person. Biopolitics was the name that Foucault gave to this modern mentality of government, which takes for its object the whole life of all and of each (*omnes et singulatim*).

Whereas Foucault outlined the skeleton of the historical matrix of biopolitics, Agamben recently exposed its core. He finds the paradigm of biopolitics not first in the institutionalization of the Church as a pastorate but instead in Christology. Early Christian theologians had to resolve an apparent contradiction that derived from scripture, which referred to both the oneness of God and to the divinity of Jesus. If they asserted the divinity of Christ, then they risked being thought polytheists, but if they did not accept the divinity of Christ then they contradicted scripture—the word of God. These theologians also faced a second dilemma concerning the relationship of the divine with the world. One view, held by Plato and the Stoics, asserted that gods took care of all things great and small, whereas the contrary view, held by Aristotle and the Epicureans, held that the divine was transcendent and cared not for what happened to humans. If Christian theologians accepted that God involved himself with every detail of creation then that would diminish the grandeur of God, whereas to admit that God was unconcerned with men would be to abandon them. The Trinitarian logic elaborated by Christian theologians sought to address these dilemmas by dissolving their force. In doing so they would insert *oikonomia* (meaning management, administration, or government) into the very core of the Deity.

Agamben presents a filigreed study on the development of the Trinitarian argument that was accepted as doctrine at the Council of Nicaea in 325. He discusses how the notion of a Triune God allowed theologians to sidestep the horns of the dilemmas above. As stated by Tertullian, "setting forth Father and Son and Spirit as three, three however not in condition but in degree, not in substance but in form, not in power but in species" avoids the charges of polytheism (Agamben 2011, 40). The stipulation of three persons of one substance solved the first dilemma but did not address the second tension between an overactive and unconcerned God. It is in explaining the relation between the three persons of the Triune God that "management" would be inserted into God.

The term by which the unity of God was harmonized with the plurality of divine persons was *oikonomia*. The most fitting meaning of *oikonomia*, which was translated as government (*gubernatio*) in later Latin texts, is management or administration. Accordingly, God is an undivided being in his substance and only triple in terms of management or government of the world. The Greek word *oikonomia* had a largely domestic and nonpolitical sense, referring overwhelmingly to the practical art of running a household. The model of the Greek household was used to clarify the logic of the triune God. Just as a Greek patriarch's authority was not diluted by delegating the tasks to his son, so also God the father who created the world can delegate the task of its providential care to Jesus, the son, without thereby dividing the substantial unity of God. This "economic" argument entails that "God reigns over the world but does not govern it," and neatly solves the dilemma of an overly interfering or uncaring God.

However, argues Agamben, the Trinitarian argument preserved his unity only at the cost of dividing God's being from his action (2011, 53–67). That there is no necessary relation between God's being and his action entails a fracture between being and action. All that could be said of God's management of the world is that it was rooted in his free will. Rooting God's praxis on His willfulness or whim meant God's action in the world was arbitrary and hence mysterious.[13] *Oikonomia* both separates and connects the God who wills from the Son who acts, obscuring how providence works in the world. Through his painstaking examination of the sources, Agamben shows there is no fundamental tension between these two senses of management (*oikonomia*)—between "the articulation of persons within the deity" and "the incarnation and revelation of God in time." Both "are nothing but two aspects of the same single activity of 'economic' administration of divine life, which extends from the heavenly house to its earthly manifestations" (37). In this way, not only does economy refer to the mysterious structure of the triune God but also to the mysterious relationship between God and his creatures. It is perhaps not wrong to say that the Trinity is the economic articulation of a mysterious God's mysterious management of his creatures.

Another feature of interest embedded in the Trinitarian argument elaborated by the Church Fathers. This concerns the precise relation between God the Father and God the Son. Agamben suggests the controversy over Arianism most clearly lays out the stakes at issue. While both sides to the controversy agreed that the Son is generated or begotten by the Father, they disagreed about whether the "Son—which is to say the word and praxis of God—is founded in the father or whether he is, like him, without principle, *anarchos*, that is, ungrounded." Arius insisted that the only uncreated creator was God the Father from whom all creation, including the Son, emanated: "God being the cause of all things, is anarchic and altogether Sole but the Son being begotten . . . derived only being from the Father." In contrast, the bishops who were opposed to the Arian position, won the debate by stipulating the Nicene thesis whereby "The Son 'reigns together with the Father absolutely, anarchically, and infinitely'" (58). Christ is the divine savior of the world. The assertion that he is ungrounded and unprincipled (*anarchos*) implied that his praxis has no foundation in being. But, argues Agamben, this recognition of free action cannot be divorced from the Trinitarian *oikonomia* or government. After all, it is only when actions are free that the issue of its management emerges. Whether it is called management, administration, or government, the conduct of the actions and thoughts of persons is predicated on the presumption of their freedom. It is for this reason that Agamben writes "the praxis [which is] free and 'anarchic,' opens in fact, at the same time, the possibility and necessity of its government" (66).

The possibilities that are buried in the Trinitarian *oikonomia* now come into dim light. According to Trinitarian theology, the relation of the Father to the Son takes the form of hierarchical domestic management, akin to the delegation of tasks from the Greek patriarch to his son. The providential care of the world by God the Son is done in the name of God the Father and is accordingly liturgical. Such providential care is universal in that it includes all things great and small. In these respects, the Trinity is hierarchical in structure, vicarious in action, and universal in scope.

With the Trinitarian economy as an archetype, a circle that began to be traced starting with bureaucratic order is almost closed. Recall the bureaucratic order is hierarchical, instrumental, and universal. The celestial and ecclesiastical hierarchies, the office of the minister as living instrument, and the pastoral caretaking of the Church are paradigms of the bureaucratic order. Despite the evident structural similarity between the pastoral and bureaucratic orders, the circle cannot be closed because the pastoral order was elaborated with a view to salvation from this world into the next.

In contrast, the bureaucratic order can only promise salvation in this world—for example, through health, wealth, and wisdom. Yet, health is understood as the product of health care, wealth as the consequence of wealth management, and wisdom as the result of life-long learning. As such, the promised state of health requires remaining a patient, that of wealth requires remaining an economic producer whether as active worker or passive earner, and that of wisdom requires remaining a student. As these examples suggest, the promise of salvation proffered by the bureaucratic order requires remaining within it. Accordingly, there is no salvation from the bureaucratic order.

EPILOGUE

The management of increasing swathes of human life seems a practical issue, one that is purely technical and rational. After all, who doubts that organization by impersonal powers is more effective than the disorganization of personal relations? But what the genealogists of management have begun to reveal are, to quote Karl Marx (1976, 163), "the metaphysical subtleties and theological niceties" that subtend our bureaucratic order. The persistence and ubiquity of the management of human life cannot be understood without its religious underpinnings. It is the Trinitarian economy and its hierarchical and liturgical elaborations into the pastoral modes of care of all and each that constitutes the historical archetype to which our bureaucratic order must be referred. Bureaucracies are everywhere, they are hierarchies, and they are considered instruments of human purposes. Management captures and converts human actions into official acts; structures these official acts in a hierarchical order; and justifies these hierarchical structures by some mission. Hierarchically arranged offices populated by functionaries constitute the template of modern impersonal power. For instance, it is the office of the president and not a specific president that is authorized to take this or that action. By so making the office the seat and source of power, the bureaucratic order sacralizes power itself. It is a frequent complaint of bureaucratic power that it is faceless in two ways: that it treats all its subjects with equal disdain, which is to say, as "cases," and it also shapes bureaucrats into impersonal functionaries. Attention to the theological paradigms of our bureaucratic order shows that these features are not malign or accidental consequences of bad managers but rather constitutive features of management. The office presupposes the separation of the actor from the action and thereby necessarily transforms persons into functional instruments. Accordingly, no one can be held responsible for doing their job and no one can be found responsible for prescribing the job.

Consider Facebook as an exemplary instance of the bureaucratic order. Its organization chart depicts the articulation of offices occupied by its fifty thousand or so salaried employees. Each employee does her job which specifies the tasks to be accomplished. The chief operating officer's official acts differ from those of the programmer. Moreover, each is a functionary insofar as what they do is on behalf of someone else. Programmers work for product managers, who work for the chief of the apps division, who works for the chief executive, Mark Zuckerberg, who in turns works for someone else, whether shareholder or customer . . . with each acting as a more or less effective instrument of another. That vicarious action of functionaries constitutes an organization is, by now, familiar. What I want to focus on is Facebook *qua* technology, on the corporation itself and not its functionaries as instruments.

The officials at Facebook—whether programmer or president—collectively embody the stated corporate mission "to give people the power to build community and bring the world closer together." To that end, Facebook is a screen that mediates the relation of the world to itself. Unlike other corporations, Facebook's profits are not made by transforming the productive effort of employed workers into saleable commodities that consumers pay for. Instead, unpaid Facebook producers collectively generate the content they freely use. As the popular slogan says, "If the product is free, then you are the product." It is obvious that bureaucratic power goes hand in hand with the fabrication of official identities. Workers, administrators, and CEOs are no less official identities than are customers, clients, and citizens. The annual expenses of human resource training, consumer advertising, and political marketing costs is proof of the effort expended to fabricate the varieties of corporate identities. However, precisely because Facebook does not have sellers and buyers in the usual sense, it offers a telling instance of the depth of instrumentalization required by our bureaucratic order.

Instead of hiring workers and finding customers, Facebook mediates the relationship between the producer and consumer, which in the limit case, coincide. To accomplish its mission to bring the world closer together, Facebook mediates the relationship between the world and itself by encouraging everybody to become a Facebook user. To become a user requires only that one display oneself. For that very reason, however, this minimal requirement is the maximal degree of alienation. Those who use Facebook are free to display as much or as little of their lives as they want. But none, by definition, can avoid displaying themselves. To be on display means to curate oneself. To curate oneself is to manage oneself as museum curator would an exhibition. In posing for a selfie, the Facebook user is seduced into confusing an official act for a human action. Those who become a Facebook user forget that they are, *qua* users, necessarily and only poseurs. By the requirement of their office, Facebook users coincide entirely with their official profiles. In this respect, Facebook users starkly reveal what every official—whether president, professor, or postman—in the bureaucratic drama is: the display of a living instrument. It is in this sense that Ivan Illich remains profoundly insightful: the bureaucratic order is the ongoing inversion of an order not for this world.

NOTES

1. Tom Peters, "A Sweeping 'People' Agenda," https://tompeters.com/columns/a-sweeping-people-agenda/, accessed May 8, 2022.

2. For the following sections, the most important resources are Agamben (2011), Agamben (2013), and Heron (2018), in that order.

3. Fayol ([1916] 1949) is a locus classicus of the "principles of management" genre.

4. Pseudo-Dionysius refers to the unknown author of Greek texts dated to between fifth and sixth centuries CE. In this paper, I refer to *The Celestial Hierarchy* (henceforth *CH*) and *The Ecclesiastical Hierarchy* (henceforth *EH*) both published in *Pseudo-Dionysius: The Complete Works* (1987).

5. Pseudo-Dionysius is aware of the potential confusion arising from using the term *angels* to refer both to all created beings of the celestial hierarchy and to the group of beings occupying the lowest rung of the layered celestial hierarchy. All heavenly beings are angels because "all the heavenly powers hold as a common possession an inferior or superior capacity to conform to the divine." The least of these are also

named angels because this rank of beings does not participate in the more lustrous powers of the superior ranks. Accordingly, the least rank is named for the degree zero of angelic power which is to be as an angel, that is, a "messenger" or "envoy" of God.

6. Pseudo-Dionysius appears inconsistent with the structure of the rank of those seeking or being initiated. These range between three and five classes. For example, in *EH*, ch. 3 (432C) there are three: the catechumens, the possessed, and the penitents, while in *EH*, ch. 6 (532A), there appears to be two additional subclasses of those being initiated and beneath the ring of the sacred people.

7. *CH*, 301B, where both the examples of light and heat are used to explain the relative capacity of celestial and human beings to being gathered to God.

8. Cf. *CH*, 141A: "God is in no way like the things that have being, and we have no knowledge at all of his incomprehensible and ineffable transcendence and invisibility."

9. For a careful exposition of the terminological shifts leading up to the separation of laity from clergy, see Heron (2018, 54–64).

10. Mitcham noted that Philo of Alexandria, in the first century CE, explains that God created the world and reigns over it but does not govern it directly. Instead, "God works through secondary or instrumental causes (and) touches creation with the proverbial ten-foot pole of angels and other intermediaries." Mitcham thereby argued that the idea of the instrument derived from Christian reflections on the angelic mission to do God's work in the world. Mitcham suggested that "for the Christian, the paradigm of sacred technique can be found in the eucharist and other sacraments" (1984, 15).

11. More precisely, as an instrument with intelligence (*minister est sicut instrumentum intelligens*).

12. *Intendant facere quod facit Christus et Ecclesia* (Heron 2018, 90).

13. The crucial difference between Greco-Roman cosmology and Jewish-Christian theology is explained by Albrecht Dihle as, "This (Greco-Roman) philosophical theology or cosmology rests on a basic presupposition: the human mind has to be capable of perceiving and understanding the rational order of the universe and, consequently, the nature of the divine. . . . That presupposition is not invalidated by the fact that man, in his empirical condition, is not always capable of full appreciation of the cosmic order" (1982, 2–3). In contrast, "biblical cosmology was, however, completely different. There is no standard, no rule applicable to the creator and his creation alike. Creation results from the power and the pleasure or will of Yahveh and from nothing else" (4).

REFERENCES

Agamben, Giorgio. 2011. *The Kingdom and the Glory: For a Theological Genealogy of Economy and Government*. Stanford, CA: Stanford University Press.

———. 2013. *Opus Dei: An Archeology of Duty*. Stanford, CA: Stanford University Press.

Aristotle. 1935. *Eudemian Ethics,* Loeb Classical Library. Translated by H. Rackham. Cambridge, MA: Harvard University Press.

Dihle, Albrechte. 1982. *The Theory of Will in Classical Antiquity.* Berkeley: University of California Press.

Fayol, Henri. [1916] 1949. *General and Industrial Management.* London: Sir Isaac Pitman & Sons.

Foucault, Michel. 2007. *Security, Territory, Population.* London: Palgrave Macmillan.

Heron, Nicholas. 2018. *Liturgical Power: Between Economic and Political Theology.* New York: Fordham University Press.

Illich, Ivan. 2005. *Rivers North of the Future.* Toronto: Anansi Press.

Marx, Karl. 1976. *Das Capital.* Volume 1. Translated by Ben Fowkes. Harmondsworth, UK: Penguin.

Mitcham, Carl. 1984. "Technology as a Theological Problem in the Christian Tradition." In *Theology and Technology*, edited by Carl Mitcham and Jim Grote. Lanham, MD: University Press of America.

———. 2018. "Teaching with and Thinking after Illich on Tools." *The International Journal of Illich Studies* 6, no. 1: 172–78.

Pseudo-Dionysius: The Complete Works. 1987. Translated by Colm Luibheid and Paul Rorem. New Jersey: Paulist Press.

Weber, Max. 1978. *Economy and Society.* Berkeley: University of California Press.

Chapter 16

What Religion, What Technology?

A Wittgensteinian Approach

Andoni Alonso

It seems quite usual to think in terms of pairs: of religion and science, of science and technology, of religion and technology. The first term seems to oppose the second: using the last pair, technology comes from a rational milieu, religion has to do with faith; technology is true in its materiality, the most important truths of religion have to do with what is immaterial; technology frees us, religion demands submission; technology empowers us, religion limits us.

Is there a basic explanation on why these pairings exist? One approach is through using ideas developed in Ludwig Wittgenstein's philosophy of language.[1] While Wittgenstein never devoted a complete paper to science, technology, and religion, he made acute reflections in different places. I credit him with opening a complete and balanced new philosophical perspective on religion and science.

The idea of the disenchantment of the world is a good starting point for this exploration. Max Weber created the word *Entzauberung* to describe disenchanted modern life around 1919 (Ghosh 2014). The world ceased to have mystery; it became rationally predictable by science and nature and could be used and controlled. Over time, religion was progressively devalued as science and rationalization assumed the role of explaining the world and directing moral action. In a sense, technology is the materialization of science, its strong arm, because people deal more with devices and machines than with scientific theories. Wittgenstein was disenchanted with his times and considered religious belief as a sort of limit that reveals the trap of science and technology. Being is the wonder of the world, the marvel that reminds where enchantment lies. When he wrote his pieces on technique, Heidegger had already realized that the Western oblivion of being would end well (Heidegger 1977). Later in his life, Heidegger (1966) proposed that "only a god could" save us from the situation we are in due to science and technology. A similar assessment has been made by religious thinkers. For instance, Ivan Illich maintained that, to a great extent, contemporary technology is a degradation of Christian faith. Somehow, that degradation brings Jacques Ellul close to Illich to criticize the contemporary technology as a cage for human beings.

The above-mentioned authors shape the interpretative horizon that I will try to explain in what follows. I also introduce a few other questions related to the interplay of science, technology, and religion: What if technology becomes a religion in itself? Some extropianists, believers of the Omega Point, and even some defenders of the goddess Gaia seem to hold this belief. Is *technological superstition* possible? What about *technological idolatry*? Ultimately, I argue that variations between and within religions and their interplay with different kinds of technology require specific and nuanced investigation.

WITTGENSTEIN ON RELIGION AND SCIENCE

As it is well known, the *Tractatus* (Wittgenstein 2012) differentiates what we can know with sense from what cannot be expressed. The first includes facts and logical propositions; the last is the mystic, which does not belong to the senseless but is beyond sense. This is an important distinction: there are things that are simply absurd, and there are things that, regardless of our efforts, cannot be explained in words but yet often are the most important. This mysterious realm has often been neglected among logicians and analytical philosophers. The notion that there is something inexpressible was something to which Wittgenstein referred over and over during his life. Perhaps paradoxically, the supposed father of neo-positivism and analytical philosophy had a very sensitive and precise understanding of religion and its importance.

> My whole tendency, and I believe the tendency of all men who ever tried to write or talk Ethics or Religion was to run against the boundaries of language. . . . But it is a document of a tendency in the human mind which I personally cannot help respecting deeply and I would not for my life ridicule. (Wittgenstein 1980, 12)

The last sentence clarifies his position: what is important is to show this human unavoidable tendency to be aware of what is beyond expression and outside of rationality, something beyond facts, even if that seems impossible.

So logic, mathematics, physics, and all the scientific disciplines are not able to answer questions such as: Why does the world exist instead of being nothing? What is the good? What is the meaning of life?

Wittgenstein criticized science precisely because: "man has to awaken to wonder—and so perhaps do peoples. Science is a way of sending him to sleep again" (1980, 5). But somnolence is not the only side effect of science and technology:

> It may be that science & industry, & their progress, are the most enduring thing in the world today. That any guess at a coming collapse of science & industry were for now, & for a long time to come, simply a dream, & that science & industry after & with infinite misery will unite the world, I mean integrate it into a single empire, in which to be sure peace is the last thing that will then find a home. For science & industry do decide wars, or so it seems. (72)

Wittgenstein detected how a scientist's way of looking at the world would alter how we perceive the world in a significant and not always correct perspective.

What Wittgenstein had detected was a trend that would become stronger in the next years. At the same time religion becomes a more and more void "language game" thanks precisely to the rise of neo-positivism on one hand and the apparent incomprehensibility of the meaning of religious propositions on the other. His analysis of James George Frazer's *Golden Bough*, his remarks on religion, and his view on ethics point to the same target: religion is not a pseudo-logic (beliefs), a pseudo-science (myths), or a pseudo-technology (magic). It is not a surprise that William James had a great influence on Wittgenstein's ideas about religion (Goodman 2002). According to Wittgenstein, science and technology try to treat everything equally and this techno-scientific way of thinking, which has become preeminent, homogenizes all discourses. To integrate everything in a single rule. One could say, to transform techno-science into the true religion.

Again, in a broad sense religion reveals a key element: awareness of a general meaning or orientation. Having a set of religious beliefs implies one has a general view upon the world, as Wittgenstein would claim. From his point of view, for example, religion is a whole "form of life" that expresses a general sense to the world beyond logic and facts.[2] Religion speaks about the subject holding a belief, his or her passion and commitment, but not about how the world is explained. Religious beliefs are not true or false claims about the world or the relations of ideas but rather convey the existence of a comprehensive framework that extends beyond factual things to which one passionately adheres. A religion is more than a set of particular beliefs that could be tested and discarded if false.

Technological thinking rarely includes a similar awareness; at least it does not seem to happen in a conscious way. One can use technological devices, can be immersed into technological networks, and so on, without being aware of the "deep" or "complex" meaning of his or her actions. Use conveys some meaning automatically, but it is very limited. Günther Anders (1980) claimed that it is impossible to understand, assume, and feel the impact that is caused by contemporary technological uses. It is impossible to fully know what is fully implied in the use of a car, a mobile, a computer, or a washing machine, that is, all the social, ecological, political, and economic effects attached to daily technological use exceeds most people's knowledge. It requires a deep understanding of a multifaceted reality. Even more, being aware of some ethical difficulties may not deter one from continuing to use those technologies, even if doing so may open one to charges of hypocrisy.

To understand religious belief, Wittgenstein takes an indirect path focusing on miracles as statements close to religion. He asks "What is a miracle?" According to Wittgenstein, a miracle is not unexplainable phenomena, something that violates the laws of physics or medicine. It is a way of *looking at the world* (Wittgenstein 1965, 11).

> Take the case that one of you suddenly grew a lion's head and began to roar. Certainly that would be as extraordinary a thing as I can imagine. Now whenever we should have recovered from our surprise, what I would suggest would be to fetch a doctor and have the case scientifically investigated and if it were not for hurting him I would have him vivisected. And where would the miracle have got to? For it is clear that when we look at it in this way everything miraculous has disappeared; unless what we mean by this term is merely that a fact has not yet been explained by science which again means that we have hitherto failed to group this fact with others in a scientific system. This shows that it is absurd to say "Science has proved that there are no miracles." (10)

If no one still possesses a way of looking at the world that does not depend wholly on science, miracles no longer exist. Does anyone still have that way of looking? A miracle is also a very specific kind of surprise, similar to the surprise generated by pondering the mere existence of the world (why there is anything at all instead of nothing, as Heidegger or Wittgenstein would remind us), or even the mere existence of language.[3] So miracles, myths, or religious beliefs are not necessarily detached from our daily experience.

Wittgenstein provides another notion that is useful: superstition. In ordinary language, superstition often refers to "false belief," something that is proved false. Some rationalist defenders of science would say that religion in itself is no more than, again, a set of superstitions (false beliefs) (see Cooper 2017; Pandey 2009). What is it? In the *Tractatus,* Wittgenstein defines superstition as the belief in the causal chain. Logically there is no reason for the validity of *inference*. But later, in a different context he would say: "Religious faith & superstition are quite different. The one springs from fear & is a sort of false science. The other is a trusting" (1980, 48). There is a connection between the two: *inference* is a way of doing false science and false science is superstition. What other ways of doing false science exist? At least the very idea that science can explain more than what it is able to do. But there are more: progress is infinite, technology is able to solve any problem, everything can be quantified and transformed into data. But reversing the question may also be of interest: what happens when Religion surrenders and tries to transform itself in science? What happens when religion tries to adopt the methods and language of science? What happens when religious beliefs pretend to represent the *truth* of facts, the unquestioned representation of a certain part of the world? One of the moments that surprised Wittgenstein during the First World War was the practice of carrying the sacred form in the trenches in a stainless-steel cage, in order to avoid being blown out by bombshells. How could a bombshell *destroy* the body of Christ? This would be classified as superstition, according to him. In different writings, Wittgenstein notes that superstition is not limited just to religion. For instance, Frazer represents the pedestrian mentality of a contemporary English person.[4] His attitude of understanding myths and ritual as pseudo-science could be shared.

EXTENDING WITTGENSTEIN TO TECHNOLOGY

The preceding reflections open up new avenues to consider how religion relates to technology. One is that they have nothing in common; in fact, they are opposites. A general and somewhat crude argument is that technology depends on or is intertwined with science, science often contradicts religion, ergo technology goes against religion. Neo-positivists, understood in a very basic sense, would seem to accept this syllogism as valid and sound. Religion has to do with the immaterial, technology with worldly things. And it is relatively common to find some stances to prove the "religious irreverence" of technologists: many ask, for instance, why not ignore "divinely given limits" and radically improve the human body through technology? Why not treat the brain as nothing more than an extremely complicated corporeal computer? Why not reshape nature completely via biotechnology? If something can be built, shouldn't it? There are many areas where, prima facie, religion should oppose technology because this represents the main trend to de-spiritualizing. Different religious creeds and beliefs all around the world have already answered with a resounding "no" to those questions.

Yet it is not so clear or simple to assume a basic opposition between religion and technology. Various religions adopt different positions on the relationship: understanding in some Protestant branches, partial acceptance in Catholicism, disinterest in Buddhism, and so on. It does not make much sense to reject technique as a whole because that belongs to how human beings behave and live: I know of no creed that does. What are rejected are particular technologies. Some biotechnological techniques such as artificial insemination or certain uses of stem cells have been completely rejected by the Catholic Church for theological and moral reasons. Amish groups reject many technological devices because they will undermine communitarian and spiritual bonds. That rejection does not mean a complete refusal of any technology; they look for technologies reinforcing humility and equality (Wetmore 2007). Again, the key for making sense of a rejection of technology by religions is to pay attention to the rationale as well as the differences between them.

While John William Draper's old and often cited book (1998) on the inevitable conflict between religion—especially Catholicism—and science has influenced what is now a common opinion, that contemporary science and technology oppose religion as liberating forces against Middle Ages spirit, it is quite easy to find scientists and technologists who claim a spiritual and religious point of view for their work. Once the prototype of modern science, Isaac Newton was a devout alchemist, theologian, kabbalist, and historian of the Church (see Yates 1972). Many contemporary scientists and technologist see a tight bond between religious beliefs and scientific or technological achievements.[5] Some religions view technology as a gift from above that should be developed and used: is it not the task entrusted by Divinity to human beings? Some radical Islamic sects do not hesitate to use Western technology as a way to confront Western civilization as a whole, a means to gain a strategic advantage to fight *pagans*, to fight precisely that arrogant technological culture. There has been a sustained effort from religious groups to harmonize scientific and technological results to accommodate spirituality in contemporary times. For instance, Ian Barbour (2000) defends the congruence between contemporary science theories and religions, and Alvin Plantinga (2011), from an analytical point of view, sustains that it is not religion but naturalism—according to his definition, the idea that there is no such person as God or anything like God—that contradicts science because naturalists are necessarily atheists, whereas scientists can be religious people. Balance between science, technology, and spirituality would imply a real advance in moral life. How successful such efforts really were, however, is a quite different question.

Of the thousands of papers dealing with theology and technology, spirituality and religion, science and technology and religious beliefs, many highlight the challenges and questions that technologies pose to religions having to do with genetics, digital technologies including nanotechnologies, neurosciences, weaponry, and so on. Therefore, a narrowing—or maybe a generalization—is required to respond to this complexity. Religion appears at a certain historical moment, but what usually defines a religious position is precisely the need to explain the end of history, when the sacred enters into the world, when the spiritual takes over reality. While religious beliefs are constant and mostly the same,[6] technology responds to the movements of history. In fact, the distinction between "technics," made prior to modern science, and technology, which arises after Galileo, the Royal Society and so on, tries to draw a historical border of some sort. Contemporary technological development is unintelligible without science and science is unintelligible without religion.

Even more, religion arose before technology in every culture. The late arrival of the latter was greeted with a fight, criticism, or at least maladjustment. In Carl Mitcham's assessment, "Prior to the initial stirrings of distinctly modern attitudes, most religious philosophies were at least minimally wary of what is now called technology. The argument was fundamentally quite simple: that the pursuit and practice of technics distracts from higher things. This idea can be found in the Jewish–Christian scriptures as well as in Daoist and Buddhist teachings" (2009, 467).

But a related line of reasoning leads to a different question: has technology itself become a religion? If so, how does it occupy the traditional space given to spirituality? The hypothesis has become popular because it explains many phenomena and attitudes. Is *transhumanism* a philosophical or a religious trend? Is extropianism a scientific theory or a hidden religious belief?[7] What about trans-humanism? Do they have the promise of eternal live and the perfection of body, following the lead of conventional religions? It is precisely this understanding that Wittgenstein, Ellul, and Illich have criticized. If technology becomes *the* contemporary religion, its consequences must be analyzed.[8]

ON TECHNOLOGICAL IDOLATRY

With the popularization of computing and communication technologies at the end of the twentieth century, the insight that religion had a lot of to do with technology, at least to a certain point, was widely accepted. Information became a metaphysical, pseudo-religious concept. One book reflected that fact very early: *Omens of the Millennium* by literary critic Harold Bloom (1996). Bloom proposed that informational discourse was an actualization and reenacting of hermetic beliefs. Maybe without knowing that fact, the idea of guardian angels (avatars), a single language used by everyone (Adamic language), the transfiguration of the body (consciousness downloading), and so forth, points to those hermetic or gnostic discourses. Information becomes, according to Bloom, a sort of divine substance, the real essence of everything.[9] Hermetic tradition explains the culture where information gurus and digerati extract many of their bizarre forecasts and proposals (see Alonso and Arzoz 2004). Gnosticism is the theology for the contemporary theory of information. Somehow, in spite of this assessment, Bloom indicated enthusiasm for the approaching millennium, the glorious time for technology to satisfy hope and utopia. This is how the technological begins to become an idol in our times.

Two trenchant critiques of technological idolatry can be found in the work of Ellul and Illich. Ellul has articulated one of the most accurate and complete criticism on technology. He extends a critical understanding of technology that began with José Ortega y Gasset (2014) and includes Heidegger (1977),[10] but also opens up a new way of thinking. Ortega first wrote on technology in 1933, as it began to occupy a central place in Western philosophy. His notion of the society of masses takes technology as one of its distinctive features. He identified technology as a cultural feature and showed how the mass-man becomes unreflexively adapted to the use of technological systems and devices, knowledge is siloed into hyper-specialized disciplines (an expert knows everything about a tiny fragment and ignores everything else), and society loses its structure (*vertebración*) becoming a subject demanding rights but accepting no duties at all.[11] Heidegger considered technology as the fate of his society, how the Being reveals itself (*aletheia*) in that historical moment and puts humanity at the brink of disappearing, for instance, with the atomic bomb (see Heidegger 1977).

Ellul approaches technology in a richer and more radical way; technology has everything to do with power, politics, culture, and society; everywhere and every realm has to be measured against the technological imperative.[12] Ellul's work offers a notion that captures this superstitious element of the new religion: technological efficiency. Going against efficiency is often considered a mortal sin today; to go against technological efficiency is a direct attack not on a feature of the technological realm but directly against the very essence of that technological system.[13] Everything is calculated and rationalized based on efficiency and it is impossible to escape its demands. Efficiency represents the power to make or to transform, and rejecting its power, its use, becomes perhaps the only possible act of revolution against the system. In a sense, Christianity and anarchism coincide in a free renunciation to the power of efficiency (Ellul 2011).[14]

Illich's investigation showed that education, health, and transportation had all become subject to the all-embracing technological system, which possess a distinct feature: it transforms certain key social services into sacramental rituals. Education, health, transportation, and energy are no longer goods for individuals; they are rituals to which individuals have to submit. Needs are no longer what individuals have to fulfill but what institutions and states dictate to be adequate and to be administered, close to the notion of being *administered* in laic sacraments. The basic idea of proportionality about what is needed disappears. That is the mark of progress: there is no balance anymore between needs, society, and environment.

Ellul[15] and Illich shared insights about modern technology. In a meeting held in Bordeaux,[16] Illich, one of the more lucid critics of *industrialism*, confronts modern technology as a sort of heresy, a *corruptio optimi*, as he used to say. Illich stated that Ellul and he agreed on some important ideas:

> First, it is impossible to compare modern technique and its malevolent consequences with the material culture of any other society whatever. Second, it is necessary to see that this "historical extravagance" is the result of a subversion of the Gospel—its transformation into an ideology called Christianity. (Ellul and Illich 1995, 234)

Modern technique has no equal in humankind history precisely because its fundament is not technological but *theological*. Heidegger used to say that the essence of technology is not technological. In reality, this theological deviance is the result of a *subversion*, transforming that *episteme* into *praxis*. Maybe here we can go a little bit further; maybe here a possible explanation is that this essence is a sort of religious thinking that is heretical and, to a certain degree, superstitious in Wittgenstein's sense. Technology promises to reward faith in it by making possible the creation of Paradise on Earth through bureaucratic management. Ellul criticized this conception of science as religion, as he was worried about how society was becoming a sort of automaton driven by technology.

Illich summarized the problems encountered in this transmogrification of Christianity into modern technology. Contemporary post-industrial society secularizes Christian institutions such as compulsory schools (Illich 1971) or health care systems (Illich 1995). Technology, especially communication technologies, empowers this trend:

> It is not possible to account for this regime if one does not understand its genesis as growing out of Christianity. Its principal traits owe their existence to the subversion I just mentioned. Among the distinctive and decisive characteristics of our age, many are incomprehensible if one does not recognize a pattern: An evangelical invitation to each person has been twisted historically into an institutionalized, standardized, and managed social objective. (Ellul and Illich 1995, 234)

Illich devoted much of his effort to bring to light how organization and institutionalization is the final result for those needs and services, which corresponds to the organizational technology in Ellul's thought.

> Finally, one cannot correctly analyze this "regime of technique" with the usual concepts that suffice for the study of other societies. A new set of analytic concepts is necessary to discuss the hexis (the habitual state) and praxis of the epoch in which we live under the aegis of *la technique*. (234)

The preceding reflections reveal a completely unique characteristic of Western technology: its demand for a faith in a perfect future, despite all the disasters that accumulate around it, a faith extracted from the religious context but converted into the worldly context, and therefore into idolatry in the strict sense. There are no other such cases in the histories of other civilizations. The imperative of efficiency that Ellul criticizes is a form of contemporary idolatry; the conversion of society into a system ruled by technique that Illich denounces speaks of the same idolatry. The disenchantment of the world that Heidegger affirms follows similar paths: it is not possible to see our surroundings except as a source of "resources." The perplexity that arises is that, with technology being imbued with pseudo-religious characteristics, it is very difficult to criticize. For the ordinary individual, criticism is incomprehensible because there is no "outside" of the technical-scientific. Criticism becomes heresy rather than analysis in the eyes of religious followers.

THE VARIETIES OF RELIGIOUS AND TECHNOLOGICAL EXPERIENCE

One year later after Bloom's *Omens*, Noble published *The Religion of Technology* ([1997] 2013). His thesis was bold: contemporary technology is simply the result of Christian theology from the Middle Ages up to the present. Far from its contrary, religion is the very mother of the technological society, and the history of theology shows that fact. Christianity and technology go hand in hand through history. Humanity should restore the conditions in Paradise through technology, according to some Medieval theological schools.

This idea of restoration could be linked to certain gnostic notions, as Bloom reminds us: restoration from the Fall thanks to labor and technology can be used to explain many later developments in European history. Francis Bacon, for instance, tried to free human beings from the constraints of Nature, seeking to transform the world through science and experiment, and to search for immortality. This neo-Gnosticism is characterized by Bloom as despising the human body and aiming for its artificial reconstruction, proposing progress through technology to reach the "city of God," pursuing the task to unify humanity in a single creed (now in technological terms). In this argument, technology and Christianity are different sides of the same coin. "Technologies of transcendence"—computers, biotechnologies, cryogenics, and so on—promise to take humans beyond natural limits, to achieve immortality and complete and unlimited freedom.

To be fair, there are some Christian currents that express exactly the opposite ideas, a quite reflexive austerity and respect for the natural world. For instance, Francis of Assisi, patron saint of ecologists, declared that all creatures in this world, including animals and plants, are equally brothers and cannot be abused or destroyed. Other Franciscans, for instance, Hugh of Saint Victor, tried to build the first known philosophy of technology in Western culture in a proportionate way, not just as a quest for power (Illich 1981). Quietism and other forms of mysticism reject industrial and technological activity, as do the Amish and some variations of Protestantism, as mentioned before.

It is true that much of technological gibberish has adopted millennial and theological discourses on transcendence.[17] When Kevin Warwick (2000) implanted a chip in his arm, his claim was to go beyond the limits that nature forces upon our body. It is true, also, that the ideology of progress comes from Christian theology. But again, that does not mean a direct and simple translation from one realm to the other. This neo-gnosticism—gnostic thinking in a technological fashion—could be interpreted as some sort of superstition. Its blend of scientism and half-understood ancient religious beliefs, though, makes it a second-class religion.

There are some crucial ideas held in common by Wittgenstein, Illich, and Ellul. Technics is one thing, technology another. Humans have different techniques in each historical moment and in each civilization. Tools have accompanied human beings as far back as historians can go. But technology is far more than just a collection of tools and techniques, far more than just more powerful technics. Technology is a unique product of Western culture, a product of a particular civilization in a particular historical moment. Problems arise when one and just one model—technology—appears as the true way to do things, when diversity disappears. So, rejection or critique of all-embracing technology does not imply a rejection of all technics. The precise, disenchanting technology is what Heidegger, Ortega, Wittgenstein, and Illich deal with. Ellul's "technological system," Illich's institutionalization of human needs, Heidegger's *Gestell* ("enframing") of nature belong to a particular culture, one that has now spread around the globe, colonizing and absorbing many other cultures.

There are different societies with different techniques and tools and, also, there are many different religions around the world. It cannot be otherwise because human experience is diverse. Paraphrasing Wittgenstein, the problem is not why religion exists but whether it would be possible to be human without religious beliefs. The world, our consciousness, according to him, has a mysterious condition for us, and dealing with that mystery—in many different ways—is part of what we are. For that reason Wittgenstein's *Tractatus* had two parts, the first written by Wittgenstein, the second, the most important one, impossible to be written, would deal with beauty, the good, the meaning of life and so on. These ideas are impossible to explain fully. We cannot give a complete account of what surrounds us and never will: we cannot give a complete account of ourselves and never will. Trying to give these accounts results in many and varied responses. That variety reflects not only in the number of different religions but also differences within each religion. There are many currents within Islam, Christianity, and Judaism. So, placing Christianity as the root of malevolent scientific and technological movements in Western society lacks precision and ignores crucial differences. It is quite clear that Roger Bacon, Hugh of Saint Victor, and Francis of Assisi are far from of one mind on this matter. Each one adopted a very different and sometimes opposing position. To equate Francis of Assisi with Draper would be absurd.

Taking Wittgenstein's difference between religious belief and superstition, certain attitudes toward technology could be considered as superstition. Superstition consists in transforming amazement about a mysterious reality, the wonder of existence, into something to be managed, something devoid of mystery. Ellul and Illich state this fact plainly: the perversion of Religion results in the technological system. But the stress should be placed into the word *perversion*.

To intelligently make sense of this situation, distinctions matter, as Aristotle would say, and one trustworthy guide in making these important distinctions is Carl Mitcham. Mitcham has traced how Christianity has related to technology and how technology is structured, as well as how they are understood from different cultural perspectives. Here I will follow Mitcham's paper "Playing Technology in Religious-Philosophical Perspective: A Dialogue among Traditions" (2010). I begin with the latter concept. Modern technology, according to him, is better understood in terms of design:

> What is most distinctive of and central to modern technology, which itself is most highly manifest in modern engineering, is the activity of engineering design. The designing process as such is not to be found in premodern technology and constitutes a distinctive way of turning making into thinking,

engendering at one and the same time a special kind of making and a unique form of technical thinking. (12)

That design can be considered a divide between technics and technology—what Ortega, Heidegger, and Ellul might consider the pre-technological culture, though they did not explain it fully or in such a precise way. Analysis of modern design methods explains many aspects of contemporary production, such as the division of labor, mass production, and consumption. It is this kind of technology that today relates to religion and that opens up contemporary problems as to how they should harmonize if it is even possible.

There is not a single answer but a gamut of possible responses. One appealing notion has been the *alienation* of contemporary technology, the separation of humans from the products of their labor and from each other. Coming from Karl Marx and restated many times but with slightly different considerations, it seems to be the perfect answer in many cases. If consumerism is the other side of labor and abstract work, this single answer should suffice. Alienation would explain any anti-technological, pro-religion movements. But surprisingly, there are those that are pro-religion and pro-technology, who do not consider alienation as a problem at all.

Considering that there are a variety of religions and a variety of technologies, perhaps there is no sharp binary relationship: there may be gray zones. In addition, if we consider the "religious experience" in the sense William James (2009) gave, it follows that religion would be something above or different from particular churches or established creeds. The notion of considering varieties of religions in connection with varieties of technologies is a crucial insight and, to my knowledge, Mitcham is a pioneer in this space. In what I see as a very Wittgensteinian attitude, Mitcham proposes the existence of that variety as enriching rather than forcing a reductionist answer.

Acknowledging diversity also seems to correspond better with history. Is Christianity essentially pro-environment? If one reads Francis of Assisi, yes. But other Christian movements would interpret respect and love for nature as something pagan. A better response is to find a *typology* rather than searching for the *essence* of the problem. Different attitudes can be found. It is possible to understand Christianism as opposed to technology, for instance those currents that despise the world, wealth, and comfort. But also there is Christianism that can find technology as a way to ascend to God. Still, they are many differences inside each of those movements. Soren Kierkegaard and Leo Tolstoy opposed technology but in different ways. Friedrich Schleiermacher was a religious but in a different way than how certain neo-gnostic computer-mediated religious believers think of themselves.

Other religions are similarly diverse. Take Buddhism, for instance. The influential Western philosopher Arthur Schopenhauer drew some of his philosophical key ideas from Buddhism. He influenced Wittgenstein, and some scholars claim that there is a clear connection between Japanese Buddhism and Wittgensteinian logic, ethics, and his ideas on Religion (see Gudmunsen 1977; Read 2009; Canfield 1975; Richards 1978). With that in mind, it is necessary to rethink how technology would be part of Buddhist society. Of course this cannot be explained fully here. And also there is the reticence of speaking about a religious culture quite remote and difficult to understand from our Western mentality. Still, this effort is worth undertaking. When thinking about a Buddhist response to technology, Mitcham offers a hint about how activism could be understood:

> Buddhism does lack a certain kind of determination to do things that seems fundamental to Christianity. I also made some small counter comment to the effect that in a world so full of determinations to take action with regard to many things, perhaps there is some reasonable place for a complementary criticism of activism. (2010, 32)

A closer study would complicate matters further. It may be frustrating not finding a common thread able to explain everything with a single answer but, as Wittgenstein used to say, one has to explain the world "as I found it" (*Tractatus*). This does not imply a sort of relativism but simply an acknowledgment of the plurality of human experience. It is clear that modern technology produces what we can call "cultural stress." Technology is part of culture, as is religion or art. It is important to keep in mind the difference between technique and technology, as it has been pointed out. Societies and cultures have their own technique, but only the West has technology—that is, a way of producing tools and processes that claims to be universal. Perhaps the irruption of engineering, as described by Mitcham, helps to clarify this universal character. It leads to a domination of other cultures and the consequent destruction of their way of understanding the world. That is why its possibility of provoking social changes today is overwhelming all around the world. Understanding how to accommodate, metabolize, or cope with those changes is a task for philosophy. For sure, technology has become one of the central issues for philosophy in the twenty-first century. Religious, political, and ethical issues are pressing matters when considering technology because, in a way, technology accepts no limits to the transformations it causes, no proportion to action. The lack of proportion, which Illich denounced, is precisely what ethics and politics must support: there must be proportionality in human action, as Aristotle pointed out. No doubt that reflecting on how to accommodate religion and technology in these times is a useful enterprise.

To understand the relationships between religion and technology, it is necessary to have an idea of what could be called the Western techno-scientific trend. Philosophy could contribute from its point of view. There have been many answers, and, among them, Wittgenstein offered a quite useful and clarifying one. As human beings, we adopt a form of life, Wittgenstein stated, and our language opens different "language games." But just practicing those different games does not mean automatically that we are playing correctly. Sometimes we try to find definite answers by simplification: there should be a "language game" able to unite all that diversity. That aim for unification aims to simplify the studied object. The superstition of using the scientific and technological point of view to explain everything is, Wittgenstein would state, what leads us to the present situation. *Describing* instead of *explaining* or *valuing* is the privileged method, then. When we accept a *faith in the endless progress of technology,* then we are superstitious. The chain of reasoning is simple: since historically we progressed to a certain point, we will keep on progressing. From a logical point of view this does not follow.

Again, what Wittgenstein seems to point out, both in the *Tractatus* and in later works, is precisely that limit: what we can and cannot know in a techno-scientific way. To think that science and technology is capable of providing an unequivocal answer to any question is a superstitious leap of faith; logic does not accompany us in this enterprise. The ability of technology to produce instruments, procedures, and systems makes us think that truth is simply effectiveness, the ability to extend our power over reality. However, we are at a moment where such power seems to have turned against us. To understand progress as a historical law is a crass error because history, like society and culture, has no laws.

Indeed, a search for unification comes at the cost of an adequate analysis of variations. Being aware of those similitudes and differences makes us confront an unavoidable feature of human beings: difference and variety are part of being humans. We adopt a certain Eurocentric bias if we reduce all of this problem in terms of Christianity and technology. There are other possibilities, such as Buddhism, where technology could have different and completely new sense, and similarly, Islam, Judaism, and so on. A fine-grained gaze upon different realities could give a more accurate picture of what is going on with religion and technology. In this sense, Mitcham's contribution is extraordinarily valuable because it brings to the fore the need to understand from different cultural perspectives, from different societies that also give different answers to how to relate to the world. Therefore, instead of understanding religion—in this case basically Christianity—as the explanatory source of Western technology, Mitcham opens a much wider and diverse field of interpretation and this is, undoubtedly, an important contribution to the current discussion on technology and its cultural implications.

NOTES

1. My approach is inspired by Carl Mitcham and Robert Mackey's approach to philosophy of engineering. As they adroitly point out, "Linguistic philosophy takes the analysis of language as its starting point and, as in the later Ludwig Wittgenstein, suggests that clarification of language use can enable us to see through certain conundrums that have accumulated in both popular and professional philosophical thought" (2010, 52).

2. Wittgenstein (1980) wrote,

> It appears to me as though a religious belief could only be (something like) passionately committing oneself to a system of coordinates. Hence although it's belief, it is really a way of living, or a way of judging life. Passionately taking up this interpretation. And so instructing in a religious belief would have to be portraying, describing that system of reference & at the same time appealing to the conscience. And these together would have to result finally in the one under instruction himself, of his own accord, passionately taking up that system of reference. It would be as though someone were on the one hand to let me see my hopeless situation, on the other depict the rescue-anchor, until of my own accord, or at any rate noticed led by the hand by the instructor, I were to rush up & seize it. (45)

3. "Let yourself be struck by the existence of such a thing as our language-game of: confessing the motive of my action" (Wittgenstein 1976, 224).

4. "What narrowness of spiritual life we find in Frazer! And as a result: how impossible for him to understand a different way of life from the English one of his time! Frazer cannot imagine a priest who is not basically an English person of our times with all his stupidity and feebleness" (Wittgenstein 1965, 5).

5. For instance, Kelly says: "In a new axial age, it is possible the greatest technological works will be considered a portrait of God rather than of us" (2010, 153). Kelly calls himself a "technological theologian."

6. Sacred texts, obviously, are subject to different interpretations along time; there is a historicity in that. Primitive communitarian Christianism differs its interpretation from, for instance, the Inquisition; the Medieval Koran was an example of tolerance and peaceful coexistence, and this is not the case for some today. Still, there should exist something as an "eternal meaning."

7. Extropianism derives its name from the physical concept of "extropy," the opposite of entropy. Basically, this movement preaches the possibility of immortality for human beings based on the continuous improvement of their conditions thanks to technology. See More (2003).

8. In 1971, the French group *Survivre et Vivre* claimed:

> Science has created its own ideology having many features of new religion that we can denominate as "scientism." . . . Scientism has supplanted by far any traditional religion. . . . The power of the word "science" upon the spirit of the majority of the public has a certainly irrational mystical essence. [*La science a créé son idéologie propre, ayant plusieurs des caractéristiques d'une nouvelle religion, que nous pouvons appeler le scientisme. . . . Il a, de loin, supplant toutes les religions traditionnelles. . . . Le pouvoir du*

mot "science" sur l'esprit du grand public est-il d'essence quasi mystique et certainement irrationnel.] (Pessis 2014, 145–46)

9. Bloom (1996) wrote,

> When Newt Gingrich tells us that our national economic future depends completely upon information, then I recall that the ancient Gnostics denied both matter and energy, and opted instead for information above all else. Gnostic information has two primary awarenesses: first, the estrangement, even the alienation of God, who has abandoned this cosmos, and second, the location of a residuum of divinity in the Gnostic's own inmost self. That deepest self is no part of nature, or of history: it is devoid of matter or energy, and so is not part of the Creation–Fall, which for a Gnostic constitutes one and the same event. (27)

10. Ellul never quoted Heidegger. Mitcham's book (1994) is one of the best summaries of how philosophy of technology evolved. Mitcham's sections on Ortega, Heidegger, and Ellul synthetizes this change of perception.

11. See Esquirol (2011). Some authors suggest an ambiguity in Heidegger's valuation of modern technology. It is true that some ecological movements have relied on Heidegger's ideas, but another interpretation is that we must reach the end of this technological-calculative step and try to find salvation confronting that abyss.

12. Mitcham devoted many writings to Ellul's thought from the beginning of his work: "Jacques Ellul and his Contribution to Theology" (1985), "Propaganda" (1967), and with Robert Mackey, "Jacques Ellul and the Technological Society" (1971), among others. Mitcham was working closely with Illich for decades and contributed several works such as *The Challenges of Ivan Illich: A Collective Reflection* with Lee Hoinacki (2002), "En memoria de Iván Illich, un anarquista entre nosotros" (2002), and, coedited with Jerónimo and Garcia, *Jacques Ellul and the Technological Society in the 21st Century* (2013). In a short piece, Mitcham reviews Ellul's influence (2012) and explains his distance from both Ellul and Illich because his perspective has shifted from Christianism to Buddhism in relation to technology.

13. By technological system, Ellul identifies the coupling of machines among themselves but also with a general system embracing organizations, propaganda, economy, and institutions based on control and efficiency.

14. This is another coincidence with Illich because both despised the effort of the Church to gain power.

15. Almazán (2016) wrote,

> The Technique of organization, which applies to great masses and equally to large commercial or industrial businesses (and, therefore, depends on the field of economics) as it does for States and for administrative or police life. . . . Today everything that belongs to the field of Law is tributary to the Technique of organization. [*La Técnica de la organización, que se refiere a las grandes masas y se aplica igual a los grandes negocios comerciales o industriales (y, por tanto, depende del campo económico) que a los Estados y a la vida administrativa o policial. . . . Hoy todo lo que pertenece al campo jurídico es tributario de la Técnica de organización.*] (69)

16. "Technique and Society in the Work of Jacques Ellul" held at the University of Bordeaux, November 11–13, 1993.

17. Mitcham's (2017) work on transcendence offers a good taxonomy of how technology relates to religion in different creeds, not only applied to the Christian faith. What is clear is the elemental and degraded Christian background of those Western techno-hermetic discourses.

REFERENCES

Almazán, Adrián. 2016. "El Sistema Técnico en la obra de Jacques Ellul." *Papeles de relaciones ecosociales y cambio global* 133: 65–81.
Alonso, Andoni, and Iñaki Arzoz. 2002. *La nueva ciudad de Dios: Un juego cibercultural sobre el tecno-hermetismo.* Madrid: Siruela.
Alonso, Andoni, and Carl Mitcham. 2004. "Semblanza de un pensador crítico. En torno a Ivan Illich." *Telos: Cuadernos de comunicación e innovación*, no. 60: 16–26.
Anders, Günthers. 1980. *The Obsolescence of Man*, Volume II: *On the Destruction of Life in the Epoch of the Third Industrial Revolution.* https://libcom.org/article/obsolescence-man-volume-2-gunther-anders.
Barbour, Ian G. 2000. *When Science Meets Religion: Enemies, Strangers or Partners?* New York: HarperCollins.
Bloom, Harold. 1996. *Omens of Millennium: The Gnosis of Angels, Dreams, and Resurrection.* New York: Riverhead Books.
Canfield, John V. 1975. "Wittgenstein and Zen." *Philosophy* 50, no. 194: 383–408.
Cooper, David. E. 2017. "Superstition, Science, and Life." In *Wittgenstein and Scientism,* edited by Jonathan Beale and Ian James Kidd, 28–38. Oxford: Routledge.
Draper, John William. 1998. *History of the Conflict between Religion and Science.* https:/www.gutenberg.org/files/1185/1185-h/1185-h.htm.
Ellul, Jacques. 2011. *Anarchy and Christianity.* Eugene, OR: Wipf and Stock.
Ellul, Jacques, and Ivan Illich. 1995. "Statements by Jacques Ellul and Ivan Illich." *Technology in Society* 17, no. 2: 231–38.
Esquirol, Josep M. 2011. *Los filósofos contemporáneos y la técnica: De Ortega a Sloterdijk.* Barcelona: Gedisa.
Ghosh, Peter. 2014. *Max Weber and "The Protestant Ethic": Twin Histories.* Oxford: Oxford University Press.
Goodman, Russell B. 2002. *Wittgenstein and William James.* Cambridge: Cambridge University Press.
Gudmunsen, Chris. 1977. *Wittgenstein and Buddhism.* Dordrecht: Springer.
Heidegger, Martin. 1966. "Only a God Can Save Us: The Spiegel Interview," https://dasein.foundation/s/Only-a-God-Can-Save-Us.pdf
———. 1977. *The Question Concerning Technology.* Translated by William Lovitt. New York: Garland.
Hoinacki, Lee, and Carl Mitcham, eds. 2002. *The Challenges of Ivan Illich: A Collective Reflection.* New York: SUNY Press.

Illich, Ivan. 1971. *Deschooling Society.* London: Marion Boyars.
———. 1981. *Shadow Work.* Boston: Marion Boyars.
———. 1995. *Medical Nemesis: The Expropriation of Health.* London: Marion Boyars.
James, William. 2009. *The Varieties of Religious Experience.* https://csrs.nd.edu/assets/59930/williams_1902.pdf.
Jerónimo, Helena M., José Luís Garcia, and Carl Mitcham. 2013. *Jacques Ellul and the Technological Society in the 21st Century.* Dordrecht: Springer.
Kelly, Kevin. 2010. *What Technology Wants.* Sidney: Penguin.
Mitcham, Carl. 1967. "Propaganda." *CrossCurrents* 17, no. 1: 117–20.
———. 1985. "Jacques Ellul and his Contribution to Theology." *CrossCurrents* 35: 1–8.
———. 1994. *Thinking through Technology: The Path between Engineering and Philosophy.* Chicago: University of Chicago Press.
———. 2002. "En memoria de Iván Illich, un anarquista entre nosotros." *El País*, December 10, 2002. https://elpais.com/diario/2002/12/10/agenda/1039474807_850215.html.
———. 2009. "Religion and Technology." In *A Companion to the Philosophy of Technology*, edited by Friis, Jan Kyrre Berg Olsen, Stig Andur Pedersen, and Vincent F. Hendricks, 466–73. Oxford, Blackwell.
———. 2010. "Placing Technology in Religious-Philosophical Perspective: A Dialogue among Traditions." *Philosophia Reformata* 75, no. 1: 10–35.
———. 2012. "Connecting with Ellul: An Episodic Engagement." *The Ellul Forum* 50 (Fall): 16.
———. 2017. "Religious Transcendence." In *Spaces for the Future: A Companion to Philosophy of Technology,* edited by Joseph Pitt and Ashley Shew, 92–97. London: Routledge.
Mitcham, Carl, and Robert Mackey. 1971. "Jacques Ellul and the Technological Society." *Philosophy Today* 15, no. 2: 102–21.
———. 2010. "Comparing Approaches to the Philosophy of Engineering, Including the Linguistic Philosophical Approach." In *Philosophy and Engineering: An Emerging Agenda*, edited by Ibo van de Poel and David E. Goldberg, 49–59. Dordrech: Springer.
More, Max. 2003. "Principles of Extropy." https://web.archive.org/web/20131015142449/http://extropy.org/principles.htm.
Noble, David. F. [1997] 2013. *The Religion of Technology: The Divinity of Man and the Spirit of Invention.* New York: Knopf.
Ortega y Gasset, José. 2014. "Meditación de la técnica." *SCIO: Revista de filosofía* 10: 187–91. https://revistas.ucv.es/index.php/scio/article/download/641/612.
Pandey, K. C., 2009. *Religious Beliefs, Superstitions and Wittgenstein.* New Delhi: Readworthy Publications.
Pessis, Céline. 2014. *Survivre et vivre: Critique de la science, naissance de l'écologie.* Paris: Échappée.
Plantinga, Alvin. 2011. *Where the Conflict Really Lies: Science, Religion, and Naturalism.* New York: Oxford University Press

Read, Rupert. 2009. "Wittgenstein and Zen Buddhism: One Practice, No Dogma." In *Pointing at the Moon: Buddhism, Logic, Analytic Philosophy*, edited by Mario D'Amato, Jay L. Garfield, and Tom J. F. Tillemans, 13–26. Oxford: Oxford University Press.

Richards, Glynn. 1978. "Conceptions of the Self in Wittgenstein, Hume, and Buddhism: An Analysis and Comparison." *The Monist* 61, no. 1 (January): 42–55.

Warwick, Kevin. 2000. "Cyborg 1.0." *Wired* 8, no. 2: 144–51.

Wetmore, Jameson M. 2007. "Amish Technology: Reinforcing Values and Building Community." *IEEE Technology and Society Magazine* 26, no. 2: 10–21.

Wittgenstein, Ludwig. 1965. "A Lecture on Ethics." *The Philosophical Review* 74, no. 1: 3–12.

———. 1976. *Philosophical Investigations*. Translated by G. E. M. Anscombe. Oxford: Blackwell.

———. 1980. *Culture and Value*. Edited by G. H. von Wright and H. Hyman. Oxford: Blackwell.

———. 2012. *Tractatus Logico Philosophicus*. New York: Simon and Schuster.

Yates, Frances. 2003. *The Rosicrucian Enlightenment*. London: Routledge.

Chapter 17

Bioethics, Philosophy, and Religious Wisdom

A Critical Assessment of Leon Kass's Thought

Larry Arnhart

Do modern science and technology give us greater understanding of and greater power over the world? Or should we recognize the limits of modern scientific knowledge? Do we need a premodern kind of wisdom that goes beyond modern science? Should we worry that modern science and technology do not provide us moral guidance for their proper uses? Can we develop a philosophic or religious bioethics that can guide us in regulating biotechnology so that it promotes human dignity rather than human degradation?

Leon Kass has thought deeply about such questions concerning science, technology, and ethics. His thinking has influenced the debates over bioethics through his published writings, his work as Chair of the US President's Council on Bioethics (2001–2005), and through his influence on other scholars, such as those who write for the journal *The New Atlantis* (Kass 2003b).

"I esteem scientific discovery, and I treasure medical advance," Kass told the *Chicago Tribune*. "But it's very clear that the powers we are now acquiring to alter the human body and mind also pose a certain threat to the long-term future of the things that make us human" (Jeremy Manier and Ron Grossman, "Bush's Guardian of Bioethics," August 12, 2001, https://www.chicagotribune.com/news/ct-xpm-2001-08-12-0108120380-story.html). He has contended that we need a bioethics—what he calls a "richer bioethics"—rooted in a wise understanding of nature and human nature that teaches us how to live a worthy human life, so that we can defend those conditions of human dignity against the threat of dehumanization by modern science and the technological manipulation of nature (Kass 2005). Kass has said that we can find that humanizing wisdom by reading the "Great Books" of philosophy, literature, science, and religion (particularly, the Bible).

His argument has been weakened, however, by a couple of major problems. First, it is ambiguous in that while he recognizes that unaided natural reason may at least seemingly contradict supernatural revelation, which he identifies as the choice between Athens and Jerusalem, he shifts back and forth on the question of which of these positions should be definitive when they conflict. Moreover, while his philosophical defense of an Aristotelian and Darwinian ethical naturalism is plausible, his religious defense of biblical revelation suggests an atheistic religiosity that is self-contradictory and self-deceptive.

The problems raised by his atheistic religiosity are compounded by his later interpretation of modern scientific knowledge as limited to a crudely reductionistic and mechanistic view of nature. These problems may be resolved by adopting a Darwinian science of human nature that provides the intellectual understanding and the moral standards that allow us to respond properly to the modern scientific and technological project. This proposal is what Kass himself proposed in his early writings as "a more natural science."

I argue that this Darwinian science of nature and human nature is best studied as part of a Darwinian liberal education that unifies the natural sciences, the social sciences, and the humanities within the framework of an evolutionary science of life within the natural order, which would include the evolutionary history of the universe from the Big Bang to the present. I also suggest that this Darwinian liberal education helps us to see the natural limits to any attempt to change human nature through biotechnology: the limited means provided by biotechnology and the limited ends set by natural human desires.

MUST WE CHOOSE ATHENS OR JERUSALEM?

The ambiguity of Kass's position in the reason/revelation debate became evident in how different readers saw conflicting messages in Kass's first book on the Hebrew Bible—*The Beginning of Wisdom: Reading Genesis* (2003a). Richard Sherlock (a Christian believer) thought the book showed that Kass was "a person of faith" (Sherlock 2005). Alan Jacobs (also a Christian believer) said that despite Kass's claim that his book was "addressed to believers and nonbelievers alike," this was not really a book for believers like himself; and so nonbelievers would be more comfortable with Kass's "philosophic reading" of the Bible. Kass summarized his interpretation of Genesis in one sentence: "The book of Genesis is mainly concerned with this question: is it possible to find, institute, and preserve a way of life that accords with man's true standing in the world and that serves to perfect his godlike possibilities?" (2003a, 661). Jacobs responded: "It seems to me that not a single significant word in this sentence accords with what the book of Genesis is about. Genesis, and the culture from which it emerges, doesn't seem to me to give a damn about our 'true standing in the world' and our 'godlike possibilities'; rather, as far as I can tell, it is about God and what He has done, and is doing, to repair what His rebellious and arrogant creatures have broken: our relations with ourselves, with one another, with the creation, and with God Himself" (2003, 34).

While Sherlock put Kass on the side of Jerusalem, and Jacobs put him on the side of Athens, Hayyim Angel (an Orthodox rabbi) placed him somewhere in between the two poles, because Kass showed "an unorthodox step toward revelation" or "a step toward a faith commitment" (Angel 2012).

Readers find mixed messages in Kass's Genesis book. For Sherlock's identification of Kass as "a man of faith," the crucial passage was this:

> There are truths that I think I have discovered only with the Bible's help, and I know that my sympathies have shifted toward the biblical pole of the age-old tension between Athens and Jerusalem. I am no longer confident of the sufficiency of unaided human reason. I find congenial the moral sensibilities and demands of the Torah, though I must confess that my practice is still wanting. And I am frankly filled with wonder at the fact that I have been led to this spiritual point, God knows how. (2003a, xiv)

This is what Sherlock saw as Kass's profession of faith. But notice that Kass speaks only of his "sympathies" and "moral sensibilities" as shifting toward the biblical pole. He does not affirm his faith in the existence of God or in the Bible as His revelation. Notice also that Kass says "my practice is still wanting." A few pages earlier, he says that he is *not* "religiously observant" (xii).

Kass does not believe any of the theological doctrines of biblical religion. He says that he has deliberately avoided "any specific doctrine." Most importantly, he denies the doctrines of the immortality of the soul in an afterlife with rewards for the saved in Heaven and punishments for the lost in Hell (2017, 21, 35).

Why then did Kass's "sympathies" shift toward the Hebrew Bible? In his new biblical book—*Founding God's Nation: Reading Exodus*—Kass says that after the birth of his first child, he and his wife joined a Conservative synagogue in 1967. He says they were "preparing ourselves to offer our children an experience of Jewish tradition that they could later embrace or reject as they wished. Better, we thought, to be something rather than nothing, and *our* something was nothing to be ashamed of" (2021, x–xi). Notably, he does not say that this "experience of Jewish tradition" led him to become a pious Jewish believer.

And yet, Angel, a leader of Orthodox Judaism, a rabbi and biblical scholar, says that Kass's "greatest moment" in the Genesis book is this passage: "If we allow ourselves to travel its narrative journey, the book may reward our openness and gain our trust. Who knows, we may even learn who (or Who) is speaking to us, and why" (Kass 2003a, 17). Angel says this shows how a secularized reading of the Torah can lead to "a step toward a faith commitment" (2012, 70).

Yet Kass has never made more than a first step: he says that he does not read the Bible in the manner of "those fundamentalist Protestants and Orthodox Jews who approach the text piously and who study it reverently." Instead, he will offer a "philosophic reading" of the Bible—reading it in the same way he reads Homer's *Iliad*, Plato's *Republic*, or Aristotle's *Nicomachean Ethics* (2003a, 1–2).

Kass admits that this approach contradicts what he says about the opposition of reason and revelation or Athens and Jerusalem:

> The Bible, I freely acknowledge, is not a work of philosophy, ordinarily understood. Neither its manner nor its manifest purposes are philosophical. Indeed, there is even good reason for saying that they are *anti*philosophical, and deliberately so. Religion and piety are one thing, philosophy and inquiry another. The latter seek wisdom looking to nature and relying on unaided human reason; the former offer wisdom based on divine revelation and relying on prophecy. There is, I readily admit, a reason to be suspicious of a philosophical approach to the Bible. (3)

Sometimes Kass stresses the tension between reason and revelation, which forces us to choose one side or the other. At other times, however, he suggests overcoming this tension by finding some middle ground between the two, and that middle ground must be *nature* (44). But then Kass must contradict himself because he says repeatedly that the philosopher's appeal to nature contradicts the pious man's appeal to revelation.

As Kass indicates, the Hebrew Bible has no word for "nature," which has led some people to believe the very idea of nature is absent from the Bible. If everything is created by God, then one might think that everything exists not by any natural order but only by the contingent will of God. If there is no natural order, then philosophy or science as the inquiry into the causal regularity of the universe is futile. The only wisdom would be unquestioning obedience to the seemingly arbitrary contingencies of God's inscrutable will.

Kass admits: "We run the risk of distorting the biblical teaching by referring anachronistically to the Bible's view of 'nature,' or indeed by using the term at all in this volume. Nevertheless, we shall do so, albeit nervously, in order to bring our study of the biblical text into conversation with other wisdom-seeking activities. We shall, no doubt, have later occasions to visit this question of nature. For now, let the reader beware" (44).

Kass never resolves this fundamental contradiction in both affirming and denying the opposition between natural reason and divine revelation, which could be done two ways. One way would be to admit that he was mistaken in turning away from the Aristotelian and Darwinian naturalism of his first book—*Towards a More Natural Science* (1985)—and moving toward biblical revelation. If he were to do this, he could still read the Bible for whatever philosophical wisdom it might contain, but without any faith commitment to the truth of revelation. He would also have to correct the Bible to conform to a philosophical conception of natural morality and natural understanding. The other way to resolve the contradiction would be for him to profess his faith in revealed religion as superior to natural reason.

In at least some parts of his later book on Exodus, Kass seems close to taking this second way—choosing revelation over reason. He says that while he has "no single epiphany to report" from his years of reading the Bible, his reading of Exodus has had a profound effect on him. "I have lived *with* the book and allowed it to work on me," he reports, "and it has changed me" (2021, xiii). The biggest change has come from his reading of the last third of Exodus, which is devoted to the construction of the portable Tabernacle that the people of Israel will carry with them as they wander for forty years in the Sinai desert.

Kass thinks the ritual enactments in the Tabernacle "speak to the human soul's deep longings for transcendence and that—quite mysteriously—can bring a numinous Presence into the daily lives of ordinary human beings" (2021, xv). "Having witnessed the Tabernacle's raising," Kass says, "I try to imagine it occupied, myself among the assembled," and thus "we bear collective witness to His awesome Presence": "When performing the prescribed rituals or raising our voices in worship and song, we may on occasion be lifted up to otherworldly states of feeling and awareness, sensing for a moment that attachment to God is the core and peak of existence. Could this be what is meant by knowing His Spirit and feeling His Presence?" (604)

This may sound like religious experience, but what Kass says about the Divine Presence in the Tabernacle suggests an atheistic religiosity—religious *feelings* of transcendence, of being in touch with God, but without believing any religious *doctrines* about the real existence of God. God "exists" only in the minds and actions of people who feel awe and reverence in their experience of "the ecstatic passions of Dionysus" elicited by religious ceremony (2021, 430).

God requires daily sacrifices in the Tabernacle. At the beginning and end of each day, a young lamb is to be burned on the altar (Ex 29:38–42). These sacrifices are imitations of a human meal, but the meal is for God. Why? Kass explains: "Surely He has no ordinary need for nourishment. Are the offerings then solely for our sake, to remind us daily—when we rise up and when we lie down—of what we owe for our existence, given us not for our merit but as an act of grace? Are the offerings of gratitude intended to introduce a similar gracious disposition into our souls?" (2021, 499) Yes, but Kass sees more here than that:

> The sacrifices are not only for the human beings; they are important also for Him. Strange though it is to say, the Lord needs the sacrifices, not to eat, but analogously to our need for food: in order to live in our world. He 'needs' for human beings to recognize His presence in order to be Himself fully present in His world. The purpose of the daily sacrifices, He comes close to saying, is to keep the association alive: [if] you bring the daily sacrifices to the door of the Tent of Meeting before the Lord, [then] "*I will meet with you and speak unto you there.*" If there are no sacrifices, there can be no meeting. The Lord will go into eclipse—not as an act of will or as punishment to us, but as an unavoidable consequence of being ignored. If God's Presence is unnoticed, unknown, or unacknowledged, He is not Present. Not to be known is, in a very real sense, to cease to be. I-Will-Be-What-I-Will-Be depends on His creatures for "Being-What-He-Is." (2021, 500)

God exists only in the religious thoughts and actions of the human beings who know or acknowledge Him. If He were not recognized by those who believe in Him, He would "go into eclipse," and He would "cease to be."

Kass thinks this point is clear when God says that He needs the ritual sacrifices in the Tabernacle "that I might dwell among them" (Ex 29:43–46). According to Kass, this states the "ultimate purpose" of God and the purpose of the whole Torah (see 2021, 500–503, 598, 603). When God dwells in the religious life of Israel, there is a *mutual* benefit: it benefits Israel that they come to know God, but it also benefits God to exist as part of Israel's life forever. If Israel were to stop worshiping God, then God would be dead. As Kass says, "The Lord God of Israel needs the recognition of His children for His living Presence in the world" (689).

Kass draws a similar conclusion from the first chapter of Genesis, particularly Gn 1:27: "And God created the human being in His image, in the image of God He created him, male and female He created them." Kass sees this verse as providing the biblical basis for seeing man as the most godlike of the animals, and thus supporting the moral equality of all human beings in their human dignity. He also sees this equal human dignity as the fundamental principle for religious bioethics, which claims that we ought to prohibit any biotechnological alteration of the human body or mind that would violate that equal human dignity.

How can Kass interpret the creation story in Genesis 1 so that it shows us that this is a *truth*, even a self-evident truth—that God created human beings in His image? Kass states his interpretation first in his Genesis book and then repeats it in almost the same words in other writings (2003a, 36–40; 2017, 310–14; 2021, 591–93). I need to quote some of this at length:

> In the course of recounting His creation, Genesis 1 introduces us to God's *activities and powers*: God speaks, commands, names, blesses, and hallows; God makes, and makes freely; God looks at and beholds the world; God is concerned with the goodness and or perfection of things; God addresses solicitously other living creatures and provides for their sustenance. In short, God exercises speech and reason, freedom in doing and making, and the powers of contemplation, judgment, and care.
>
> Doubters may wonder whether this is truly the case about God—after all, it is only on biblical authority that we regard God as possessing these powers and activities. But it is indubitably clear, even to atheists, that we human beings have them, and that they lift us above the plane of a merely animal existence. Human beings, alone among the creatures, speak, plan, create, contemplate, and judge. Human beings, alone among the creatures, can articulate a future goal and use that articulation to guide them in bringing it into being by their own purposive conduct. Human beings, alone among the creatures, can think about the whole, marvel at its many-splendored forms and articulated order, wonder about its beginning, and feel awe in beholding its grandeur and in pondering the mystery of its source.
>
> Note well: these self-evident truths do *not* rest on biblical authority. Rather, the biblical text enables us to confirm them by an act of self-reflection. Our reading of this text, addressable and intelligible only to us human beings, and our responses to it, possible only to us human beings, provide all the proof we need to confirm the text's assertion of our special being. . . .
>
> In addition to holding up a mirror in which we see reflected our special standing in the world, Genesis 1 teaches truly the bounty of the universe and its hospitality in supporting terrestrial life. (2017, 312–13)

Notice that we confirm the truth of Genesis 1 "by an act of self-reflection," because in reading the text we are "holding up a mirror in which we see reflected our special standing in the world."

Notice also that the teaching of Genesis 1 should be clear "even to atheists," and it should "inspire awe and wonder, even in atheists" (313). Atheists like Kass? Does Kass not suggest that God's mental powers exist only as an anthropomorphic projection or mirror of human mental powers? Doesn't Kass express the same idea in his Exodus book in saying that God needs to dwell in the minds of His believers who acknowledge Him, because without that human acknowledgment, God would "in a very real sense . . . cease to be"? Does Kass not thus suggest that the meaning of God creating man in His image is that man has created God in his image?

This is Kass's atheistic religiosity: he recognizes that human beings have a natural longing for God that can be satisfied through religious feelings, but he cannot himself affirm any doctrinal faith in God's existence, because God does not exist outside of those human religious feelings. This atheistic religiosity is incoherent self-deception. It's incoherent in trying to both affirm and deny the existence of God. On the one hand, Kass presents the people of Israel as affirming God's existence in their religious ceremonies. On the other hand, Kass tells us that God exists *only* in these ceremonies, and that without these ceremonies, God would "cease to be." This atheistic religiosity is self-deception, because it's a fake religiosity that doesn't work if we *know* it's fake. A genuine religiosity requires not just religious feelings but also a doctrinal faith in God's real existence outside of the human mind.

Kass's philosophic reading of the Bible leads him to affirm the natural human desire for religious belief, but it cannot affirm the doctrinal truth of that religious belief. He chooses Athens over Jerusalem, while still being *open* in some manner to Jerusalem. He could do that by adopting the natural evolutionary science of religious belief.

Since the evolutionary science of religion can recognize that the natural desires of evolved human nature include both a natural desire for *intellectual* understanding and a natural desire for *religious* understanding, that science can leave the choice between Athens (intellectual understanding) and Jerusalem (religious understanding) as an open question for human beings. And yet even in recognizing that the desire for religious understanding is natural, that science remains neutral about the truth of revelation, because it can neither confirm nor deny the divine as a supernatural reality beyond nature. Remarkably, among those evolutionary psychologists who argue for religious belief as a natural propensity of evolved human nature, some (such as Justin Barrett) are theists, while others (such as Jesse Bering) are atheists (Barrett 2004; Bering 2011; Guthrie 1993).

THE NATURAL HUMAN DESIRES IN
THE BIOTECHNOLOGY DEBATE

Kass did not introduce his biblical atheistic religiosity into public bioethics debates. That became evident in his chairing of the President's Council on Bioethics. He never introduced Bible-reading into the Council's meetings (Arnhart 2005b; Briggle 2010). In *Beyond Therapy: Biotechnology and the Pursuit of Happiness*, the best of the Council's reports, there are no references to the Bible and only a few vague references to "souls with longings for the eternal" (President's Council 2003a, 200, 206, 288, 299).

In his writings on the Bible, Kass has often tried to show how the moral teaching of the Bible provides religious reasons for limiting biotechnology (Kass 2003a, 4–9, 20–21, 242–43; 2021, 284–85, 604–605, 702). But although the Bible can reinforce our natural moral sense, the Bible cannot stand alone as a moral guide, because it often lacks moral authority, moral clarity, and moral reliability. It lacks moral authority because many people doubt that it is truly a revelation from God. It lacks moral clarity because its moral teaching is often too vague to give us precise moral instruction. And it lacks moral reliability because some of its teachings are immoral and therefore need to be corrected by our natural moral sense (Arnhart 2009).

In a section of *Beyond Therapy* that considers the "appreciation of the giftedness of life," it is said that "although it is in part a religious sensibility, its resonance reaches beyond religion." There is no attempt to identify God as the giver of life. Instead, nature and human nature are identified as "the naturally given," which has arisen as "wondrous products of evolutionary selection" (President's Council 2003a, 287–90). The report appeals repeatedly to a purely naturalistic ethics rooted in evolved human nature and the natural human pursuit of happiness as the complete and comprehensive satisfaction of natural human desires (205, 235, 260, 265, 270). The Kass Council implicitly took the side of Athens over Jerusalem.

In doing this, the Kass Council returned to the ethical naturalism that Kass had first proposed in his early writings. Kass later rejected this ethical naturalism in his biblical writings. But then when he began chairing meetings of the President's Council in 2002, he had to revive that ethical naturalism, because it provided the only rationally defensible grounds for a "richer bioethics" rooted in an evolutionary science of human nature that would not require religious faith.

In *Toward a More Natural Science*, Kass criticized modern natural science as "quite deliberately, most *un*natural." But he proposed that modern science could become a "more natural science" that would be both Aristotelian and Darwinian in its comprehensive understanding of nature. Such a science could move "from nature to ethics." "A more natural science might be useful for ethics," because it would show how ethics is "part of nature," and so "the natural, rightly understood, might even provide some guidance for how we are to live" (1985, xi, 346–48).

Kass's more natural science of ethics was based on an evolutionary biology and psychology of the natural human desires. "What are the things human beings by nature desire: the pleasant, the beautiful, or participation in the eternal; children, freedom, distinction, or understanding? Answers to these questions might inform not the prescription of rules but an ordering of lives, according to a full standard of human flourishing." This natural biological ethics could be applied to bioethical disputes because it would "allow us to recognize and discourage certain dehumanizing attitudes and practices" in biomedical technology that would frustrate those natural human desires that constitute the flourishing of our nature (1985, 346–48). In 1994, Kass's second book—*The Hungry Soul*—continued his search for "a more natural and richer biology and anthropology, one that does justice to our lived experience of ourselves as psychophysical unities—enlivened, purposive, and open to and in converse with the larger world" (1994, 9).

But then, in 2002, with the publication of his *Life, Liberty, and the Defense of Dignity*, Kass announced that he was reversing his position. In his chapter on "The Permanent Limitations of Biology," he explicitly turned away from his "more natural science" of biological ethics. He spoke of "the insufficiency of nature for ethics" and "the difficulty in looking to biology—even a more natural science more true to life—for very much help in answering the questions about how we are to live." Instead of a rational study of nature, he advised, we should look to "insights mysteriously received from sources not under strict human command," and we should "acknowledge and affirm the mysteries of the soul and the mysterious source of life, truth, and goodness" (2002, 296–97). In 2003, with the publication of his Genesis book, Kass provided further confirmation of his turn away from natural human reason to religious mystery, even though, as we have seen, his message was confusing in its ambiguity.

It's remarkable, therefore, that once Kass began leading the discussions of the President's Council in January of 2002, he had to return to his earlier project of developing a "more natural science" that could move "from nature to ethics." Instead of asking the Council to look to "insights mysteriously received from sources not under strict human command," Kass organized their ethical deliberations around the natural human desires of evolved human nature. In *Beyond Therapy*, he explained: "We have structured our inquiry around the desires and goals of human beings, rather than around the technologies they employ, the better to keep the important ethical questions before us." Those desires included the desires for better children, superior performance, ageless bodies, and happy souls. The report offered "reasons to wonder whether life will really be better if we turn to biotechnology to fulfill our deepest human desires" (President's Council 2003a, xvi).

In *Beyond Therapy*, "a richer bioethics" was said to depend on judging biotechnology by how well it achieved the satisfaction of human desires (21). It was said that the root of all human activity is the natural desires (149). The natural human desires were identified as natural adaptations of the human mind as shaped by evolution (89, 200, 219, 226, 246, 260, 287). The natural desire for happiness was seen as the comprehensive desire that motivates all human action. Aristotle, John Locke, and Thomas Jefferson were right about the pursuit of happiness as the ground of all ethics (205, 210–11, 235, 260, 265, 270).

The report's primary objection to a thoughtless reliance on biotechnology to satisfy our desires was the "Midas problem": when biotechnology gives us what we think we desire, we might discover this is not truly desirable for us (xvi–xvii, 85, 149, 155, 183, 234, 279). "To avoid such outcomes, our native human desires need to be educated against both excess and error" (300). Thus, the naturalistic ethics of the *Beyond Therapy* report is an ethics of *informed desire*: all human action is motivated by desire, but desire needs to be informed by rational deliberation, by prudent judgment, about what is truly desirable for us.

Summarizing the report's general bioethical assessment of biotechnology, Kass explained:

> We want better children—but not by turning procreation into manufacture or by altering their brains to gain them an edge over their peers. We want to perform better in the activities of life—but not by becoming mere creatures of our chemists or by turning ourselves into tools designed to win or achieve in inhuman ways. We want longer lives—but not at the cost of living carelessly or shallowly with diminished aspiration for living well, and not by becoming people so obsessed with our own longevity that we care little about the next generation. We want to be happy—but not because of a drug that gives us happy feelings without the real loves, attachments, and achievements that are essential for true human flourishing. (President's Council 2003a, xvii)

This is what a "richer bioethics" grounded in evolutionary ethical naturalism can teach us. Ultimately, this bioethics must appeal to the natural moral sense of our evolved human nature based on the principle that the good is the desirable (Arnhart 1998, 2005a; Bloom 2013; Greene 2013; Haidt 2012).

THE EVOLUTION OF NATURAL KINDS AND NATURAL ENDS

In Kass's recent writing, however, he has presented a false caricature of modern science that would render a scientific ethical naturalism impossible. He has claimed that modern science must be crudely reductionist and antiteleological, saying that it is "[indifferent] to questions of being, cause, purpose, inwardness, hierarchy, and the goodness or badness of things, scientific knowledge included" (2017, 300). But his earlier writings (particularly in *Towards a More Natural Science*) presented a truer and richer account of modern science as capable of recognizing emergent complexity and natural teleology, which would support the scientific ethical naturalism and the "richer bioethics" of *Beyond Therapy* (1985, 249–75, 346–48).

An ethical naturalism rooted in evolutionary science must assume that human beings exist as a distinct species or kind of animal with a characteristic set of traits. It must also assume that these natural traits of the species include natural desires that incline human beings to certain ends. These assumptions allow one to argue that whatever frustrates the natural ends of the human species is contrary to human nature, and whatever fulfills those ends is according to human nature. One must be able to affirm the reality of natural kinds and natural ends. It is crucially important, therefore, that Darwinian biology provides a scientific explanation for why living beings emerge in the world as distinct kinds with distinct ends.

We might think that this contradicts the Darwinian principle of evolutionary continuity—that all differences between species are only differences in degree and not in kind. In *The Descent of Man*, Darwin declared: "the difference in mind between man and the higher animals, great as it is, certainly is one of degree and not of kind. We have seen that the senses and intuitions, the various emotions and faculties, such as love, memory, attention, curiosity, imitation, reason, etc., of which man boasts, may be found in an incipient, or even sometimes in a well-developed condition, in the lower animals" (2004, 151).

In Kass's early writings, however, he saw that despite Darwin's *explicit* statement that humans differ only in degree, not in kind, from other animals, he *implicitly* recognized human differences in kind. That is, Darwin saw that human beings have some moral and mental traits that other animals do not have at all.

In *The Descent of Man*, Darwin noted that self-consciousness is uniquely human: "It may be freely admitted that no animal is self-conscious, if by this term it is implied, that he reflects on such points, as whence he comes, or whither he will go, or what is life and death, and so forth." Morality is also uniquely human: "A moral being is one who is capable of comparing his past and future actions or motives, and of approving or disapproving of them. We have no reason to suppose that any of the lower animals have this capacity. . . . Man . . . alone can with certainty be ranked as a moral being." And language is uniquely human: "The habitual use of articulate language is . . . peculiar to man" (2004, 105, 107, 135).

Darwin was thrown into self-contradiction—both affirming and denying that humans are different in kind from other animals—because he failed to see how he could affirm *emergent* differences in kind without affirming any *radical* differences in kind. Emergent differences in kind can be explained by evolutionary science as differences in kind that naturally evolve from differences in degree that pass over a critical threshold of complexity. So, for example, we can see the uniquely human capacities for self-consciousness, morality, and language as emerging from the evolutionary development of the primate brain, so that at some critical point in the evolution of our hominid ancestors, the size and complexity of the brain (perhaps particularly in the frontal cortex) reached a point where distinctively human cognitive capacities emerged at higher levels of brain evolution that are not found in other primates. With such emergent differences in kind, there is an underlying unbroken continuity between human beings and their primate ancestors, so there is no need to posit some supernatural intervention in nature—the divine creation of the human soul—that would create a radical difference in kind in which there is a gap with no underlying continuity of natural causes (Morowitz 2002).

Simona Ginsburg and Eva Jablonka have shown how an evolutionary neuroscience can explain this emergence of the human mind as passing through the three levels of mind identified by Aristotle in *De Anima*. The basic nutritive and reproductive soul belongs to all living things—plants and animals. The second level, the sensitive soul, belongs to all animals. The third level, the rational or symbolizing soul, is specific to humans. The crucial evolutionary transition marker of the rational soul is language. The rational soul gives humans the capacity for grasping and sharing the abstract symbolic values of the good and the just that make human morality and politics unique. Ginsburg and Jablonka (2019) show that all three levels can be explained by evolutionary biology, thus confirming Aristotle's biological psychology.

Darwinian science recognizes not only natural kinds but also natural ends. In Kass's early writings, he saw how Darwin's evolutionary science was open to Aristotelian natural teleology. Kass saw that while Darwin's science did not recognize an *external* or cosmic teleology by which all of nature is directed by design to some cosmic end, Darwin did recognize the *internal* or immanent teleology evident in the purposiveness—the directedness toward a goal—of individual organisms. Natural selection is a purposeless process with purposeful products. Therefore, Darwin's friend Asa Gray was right in declaring: "Let us recognize Darwin's great service to Natural Science in bringing back to it teleology." Moreover, this Darwinian immanent teleology supports a Darwinian ethical naturalism. Kass wrote: *"The end is a standard as well as a goal.* Teleological analysis will be concerned both to identify the end and to evaluate how well or badly it is achieved." Biological science can recognize "the tacit ethical dimension of animal life," and thus the "natural, animal bases for the content of an ethical life" (Kass 1985, 252–64; see also 1994, 12, 14, 39, 59–63, 76–79). Even if morality cannot be grounded in the *cosmic* teleology of nature, it can still be grounded in the *immanent* teleology of human nature (Arnhart 1998, 238–48; Gotthelf 2012). This makes it possible for Kass's richer bioethics to be grounded in natural science.

FOR A THICKER BIOETHICS

Kass agrees with the argument of John Evans that the public bioethical debate has become too "thin," and that it needs to become "thicker" (Evans 2002). Kass's "richer" bioethics is what Evans would call a "thicker" bioethics. For bioethics to become richer or thicker, it must become part of a Darwinian liberal education, which became clear in the work of the President's Council as led by Kass (Briggle 2010).

Evans uses the metaphor of thick and thin to distinguish the *substantive* rationality of the bioethical debate from the late 1960s to the early 1970s and the *formal* rationality that came to dominate bioethics beginning in the late 1970s. In the earlier period, the *substantive* rationality of the debates over biotechnology required that people argue about whether some technology such as human genetic engineering was consistent with ultimate values or ends, which required that people argue about those ultimate ends. So, for example, a scientist promoting human genetic engineering might argue that this was the best means for achieving the ultimate end of human control of nature and human nature, so that human beings could engineer the perfection of their species by eliminating genetic defects and pursuing genetic enhancements. But then a theologian might argue that this was "playing God," in that man was trying to become his own self-creator and thus take the place of God the Creator, in violation of God's ends. There would then be a debate over which ultimate end should be higher—striving for a human God-like power over nature for human self-perfection or a humble and reverent acceptance of a God-given human nature with all its imperfections. Since many conflicting ends could be considered in such debates, reaching agreement on which end should predominate was difficult if not impossible, so that the debates could become endless.

By contrast, beginning in the late 1970s, bioethics became dominated by professional bioethicists who developed bioethics as a specialized field of study promoting an argumentation of *formal* rationality. According to this view, any biotechnological means that maximized ends was ethical; and the ends were predetermined by consensus of the experts to be limited in number. The ends were eventually reduced to four principles: personal autonomy (informed consent), beneficence (benefits greater than costs), nonmaleficence (avoiding harm), and justice (fair distribution of benefits, costs, and risks) (Beauchamp and Childress 2001). It was assumed that these were universal ends to which all human beings could agree. All other ends for which there was no universal agreement were excluded from the bioethical debates. Once they had agreed to their four ends, bioethicists would only debate about calculating the best means to these four ends, and they saw no need to debate about ends other than these four. In this way, the formal rationality of their debates was *thin*.

Kass's thicker or richer bioethics accepts the four principles adopted by the professional bioethicists—autonomy, beneficence, nonmaleficence, and justice. But in his writings and in his work with the President's Council, Kass has shown that a deep deliberation about bioethical issues must consider many important moral ends beyond these four principles; and because these many moral ends often conflict with one another, different people will come to different conclusions about how to weigh these ends. The moral deliberation about these ends will often not reach consensus.

For example, consider the debate over whether parents should be free to use stimulant drugs—such as Ritalin (methylphenidate) or Adderall (amphetamine)—to modify the behavior of children who are inattentive, impulsive, or hyperactive. For *thin* bioethics, there might be only two principles for this debate—autonomy and nonmaleficence. Are parents exercising their autonomy in giving these drugs to their children? Are these drugs safe for the children? If the answer to both questions is yes, then we should allow parents to use these psychotropic drugs in helping them to rear their children.

But for the *thick* bioethics of the *Beyond Therapy* report, there are other important moral considerations (President's Council 2003a, 87–92). For example, one crucial part of parental rearing of children is the moral education of children through shaping their moral character, so that they are capable of self-control and behaving appropriately in society. Will behavior-modifying drugs interfere with this moral education? Will this teach children that good behavior is caused by chemistry, and that the responsibility for their conduct belongs not to themselves but to their pills? Will this diminish their sense of moral agency? The *thin* bioethics of the professional bioethicists does not ask such questions.

Furthermore, while thin bioethics requires only a narrowly specialized training—learning the four moral ends and how biotechnological means can maximize those ends—thick bioethics requires a broadly interdisciplinary liberal education that seeks wisdom about the conditions for a flourishing human life and how biotechnology might impede or promote that human flourishing. That liberal education is best pursued, Kass believes, through reading and discussing the Great Books of the Western intellectual tradition. One can see that in the anthology of ninety-two selected texts published by the President's Council—*Being Human: Readings from the President's Council on Bioethics* (President's Council 2003b). The authors of the texts include scientists (such as Edward O. Wilson, Richard Feynman, and James Watson), physicians (such as Hippocrates and Richard Selzer), philosophers (such as Plato, Aristotle, Lucretius, and Thomas Hobbes), poets (such as Homer, Shakespeare, and Walt Whitman), and novelists (such as Leo Tolstoy, George Eliot, and Willa Cather).

In its devotion to a bioethics rooted in liberal education, the President's Council was different from the other five general federal bioethics commissions preceding it. Those other commissions were concerned mostly with developing specific public policy proposals. By contrast, the Kass Council made few policy proposals. Kass has admitted that the Council had "no demonstrable effect" on "specific policy issues." He conducted the Council's meetings and supervised the Council's reports to promote seminar-like discussions that allowed open debate about the moral and intellectual questions raised by bioethical disputes without ever reaching consensus. He said that *Beyond Therapy* was "a purely educational work, with no policy recommendations." Pursuing this educational goal, he hoped that the published discussions and reports of the Council would be adopted as readings for college seminar courses on bioethics or for groups of ordinary citizens who wanted to discuss deep questions about the implications of biotechnology for human life (Kass 2005, 229, 240–41, 244–47).

We can see here that Kass's educational goal is to revive the tradition of liberal education as a unification of all knowledge, an education that embraces all the arts and sciences. As suggested in the Council's *Being Human*, we need such an education in the quest for wisdom about the meaning of our humanity, of our human nature within the natural order of the whole. For example, as we have seen, the Council's report *Beyond Therapy* is organized around natural human desires—"desires for longer life, stronger bodies, sharper minds, better performance, happier souls, better children" (Kass 2005, 235). These natural desires are the "essential sources of concern" that set the standards for any moral assessment of biotechnology. They constitute "what is naturally human," "what is naturally and dignifiedly human." They are "naturally given" to us as inherent in our human nature. If we seek the source of this gift, we find that our natural desires are "wondrous products of evolutionary selection," because "the human body and mind" are "highly complex and delicately balanced as a result of eons of gradual and exacting evolution" (President's Council 2003a, 286–87).

This unification of knowledge founded on evolutionary science suggests something like what Edward O. Wilson (1998) called "consilience." Wilson argued that the natural human desire to understand the world as an orderly whole was a quest for the fundamental unity of all knowledge. This longing for a comprehensive knowledge of the whole began with ancient philosophers such as Thales and Aristotle. It was renewed by the Enlightenment of the seventeenth and eighteenth centuries. Now, Wilson claimed, the progress in modern science has created a realistic prospect for satisfying this ancient longing by developing a web of causal explanations that would combine all the intellectual disciplines. Crucial to this unification of knowledge is its foundation in evolutionary biology as explaining the nature of human beings and their place in the natural whole, including the evolution of the universe from the Big Bang to the present, which some historians now call Big History (Christian 2004; Christian, Brown, and Benjamin 2014). Darwinian liberal education must encompass all of this (see Arnhart 2006, 2012; Buss 2016; Carroll 2004; Gottschall 2012; Richards 2019; Slingerland 2008).

THE NATURAL LIMITS OF BIOTECHNOLOGY

One objection to this pursuit of a Darwinian liberal education that studies an evolved human nature is that this assumes the stability of that human nature, even though we know now that biotechnology is giving us the power to change and even abolish that human nature. After all, isn't Kass's bioethics driven by his fear that biotechnology can lead to what C. S. Lewis called "the abolition of man"?

The problem with this objection, however, is that it exaggerates the power of biotechnology for changing human nature (see Arnhart 2003). The most fervent advocates of biotechnology welcome the prospect of using it to transform our nature to make us superhuman. The most fervent critics of biotechnology warn us that its power for transforming our nature will seduce us into a Faustian bargain that will dehumanize us. Both sides agree that biotechnology is leading us to a "posthuman future" (Fukuyama 2002).

This is a mistake. It ignores how evolution has shaped the adaptive complexity of our human nature—our bodies, our brains, and our desires—in ways that resist technological manipulation. A Darwinian view of human nature, one truer to the facts of human biology and human experience, reveals the limits of biotechnology, so that we can reject both the redemptive hopes of its optimistic advocates and the apocalyptic fears of its pessimistic critics.

Biotechnology will always be limited both in its technical means and in its moral ends. It will be limited in its technical means because complex behavioral traits are rooted in the intricate interplay of many genes, which interact with developmental contingencies and unique life histories to form brains that respond flexibly to changing circumstances. Consequently, precise technological manipulation of human nature to enhance desirable traits while avoiding undesirable side effects will be very difficult if not impossible. Biotechnology will also be limited in its moral ends because the motivation for biotechnological manipulations will come from the same natural desires that have always characterized human nature.

In *Beyond Therapy*, Kass and his Council recognized both these natural limits on biotechnology. For example, they noted the technical limits to any attempt to use genetic engineering to design "better children": "Growing recognition of the complexity of gene interactions, the importance of epigenetic and other environmental influences on gene expression, and the impact of stochastic events is producing a strong challenge to strict genetic determinism. Straightforward genetic engineering of better children may prove impossible, not only in practice but even in principle." Consequently, "genetically engineered 'designer babies' are not in the offing" (President's Council 2003a, 38, 276). They also recognized that biotechnology would be limited in its moral ends as set by the natural desires of evolved human nature, including those desires around which the whole discussion in *Beyond Therapy* is organized.

CONCLUSION

I have argued that Kass's richer "more natural science" of his early writings supports a richer public bioethics as part of a Darwinian liberal education that unifies all intellectual disciplines—the natural sciences, the social sciences, and the humanities—within the unifying framework of Darwinian evolutionary science. The aim of liberal education is to probe all the fields of intellectual inquiry to understand how the complex interaction of natural propensities, cultural traditions, and individual choices shapes the course of human experience within the cosmic order of nature. Darwinian science provides a general conceptual framework for such liberal learning grounded in the scientific study of the evolution of life within the evolution of the universe. Kass's primary contribution to this Darwinian liberal education has been in developing what he has called "a richer bioethics, one that recognizes and tries to do justice to the deep issues of our humanity raised by the age of biotechnology" (2005, 221).

Darwinian liberal education can help us understand our human place in nature. We are neither mindless machines nor disembodied spirits. We are animals. As animals, we display the animate powers of nature for movement, desire, and awareness. We move to satisfy our desires in the light of our awareness of the world. We are a unique kind of animal, but our distinctively human traits—such as symbolic speech, moral deliberation, and conceptual thought—are emergent elaborations of powers shared in some form with other animals. So even if the natural world was not made for us, we were made for it, because we are adapted to live in it. We have come from nature. It is our home.

REFERENCES

Angel, Hayyim. 2012. "An Unorthodox Step toward Revelation: Leon Kass on Genesis Revisited." *Tradition: A Journal of Orthodox Jewish Thought* 45: 61–70.
Arnhart, Larry. 1998. *Darwinian Natural Right: The Biological Ethics of Human Nature*. Albany: State University of New York Press.
———. 2003. "Human Nature Is Here to Stay." *The New Atlantis*, no. 2 (Summer): 65–78.
———. 2005a. "Evolutionary Ethics." In *Encyclopedia of Science, Technology, and Ethics*, edited by Carl Mitcham, 715–20. Farmington Hills, MI: Macmillan.
———. 2005b. "President's Council on Bioethics." In *Encyclopedia of Science, Technology, and Ethics*, edited by Carl Mitcham, 1482–86. Farmington Hills, MI: Macmillan.
———. 2006. "Darwinian Liberal Education." *Academic Questions* 16, no. 4 (Fall): 6–18.
———. 2009. "The Bible and Biotechnology." In *Biotechnology: Our Future as Human Beings and Citizens*, edited by Sean D. Sutton, 123–57. Albany: State University of New York Press.
———. 2012. "Biopolitical Science." In *Evolution and Morality*, edited by James E. Fleming and Sanford Levinson, 221–65. New York: NYU Press.
Barrett, Justin. 2004. *Why Would Anyone Believe in God?* Plymouth, UK: AltaMira Press.
Beauchamp, Tom L., and James F. Childress. 2001. *Principles of Bioethics*, fifth edition. New York: Oxford University Press.
Bering, Jesse. 2011. *The Belief Instinct: The Psychology of Souls, Destiny, and the Meaning of Life*. New York: Norton.
Bloom, Paul. 2013. *Just Babies: The Origins of Good and Evil*. New York: Crown.
Briggle, Adam. 2010. *A Rich Bioethics: Public Policy, Biotechnology, and the Kass Council*. Notre Dame, IN: University of Notre Dame Press.
Buss, David M., ed. 2016. *The Handbook of Evolutionary Psychology*, second edition. 2 vols. Hoboken, NJ: John Wiley and Sons.
Carroll, Joseph. 2004. *Literary Darwinism: Evolution, Human Nature, and Literature*. New York: Routledge.
Christian, David. 2004. *Maps of Time: An Introduction to Big History*. Berkeley: University of California Press.
Christian, David, Cynthia Stokes Brown, and Craig Benjamin. 2014. *Big History: Between Nothing and Everything*. New York: McGraw-Hill.
Darwin, Charles. 2004. *The Descent of Man, and Selection in Relation to Sex*, second edition. New York: Penguin.
Evans, John H. 2002. *Playing God? Human Genetic Engineering and the Rationalization of Public Bioethical Debate*. Chicago: University of Chicago Press.

Fukuyama, Francis. 2002. *Our Posthuman Future: Consequences of the Biotechnology Revolution*. New York: Farrar, Straus, and Giroux.

Gottschall, Jonathan. 2012. *The Storytelling Animal: How Stories Make Us Human*. Boston: Houghton Mifflin Harcourt.

Ginsburg, Simona, and Eva Jablonka. 2019. *The Evolution of the Sensitive Soul: Learning and the Origins of Consciousness*. Cambridge, MA: MIT Press.

Gotthelf, Allan. 2012. *Teleology: First Principles, and Scientific Method in Aristotle's Biology*. Oxford: Oxford University Press.

Greene, Joshua. 2013. *Moral Tribes: Emotion, Reason, and the Gap Between Us and Them*. New York: Penguin.

Guthrie, Elliott. 1993. *Faces in the Clouds: A New Theory of Religion*. New York: Oxford University Press.

Haidt, Jonathan. 2012. *The Righteous Mind: Why Good People Are Divided by Politics and Religion*. New York: Pantheon.

Jacobs, Alan. 2003. "Leon Kass and the Genesis of Wisdom." Review of *The Beginning of Wisdom: Reading Genesis*, by Leon R. Kass. *First Things*, no. 134 (June): 30–35.

Kass, Leon R. 1985. *Toward a More Natural Science: Biology and Human Affairs*. New York: The Free Press.

———. 1994. *The Hungry Soul: Eating and the Perfecting of Our Nature*. New York: The Free Press.

———. 2002. *Life, Liberty, and the Defense of Dignity*. San Francisco, CA: Encounter Books.

———. 2003a. *The Beginning of Wisdom: Reading Genesis*. New York: The Free Press.

———. 2003b. "Ageless Bodies, Happy Souls: Biotechnology and the Pursuit of Happiness." *The New Atlantis*, no. 1 (Spring): 9–28.

———. 2005. "Reflections on Public Bioethics: A View from the Trenches." *Kennedy Institute of Ethics Journal* 15: 221–50.

———. 2017. *Leading a Worthy Life: Finding Meaning in Modern Times*. New York: Encounter Books.

———. 2021. *Founding God's Nation: Reading Exodus*. New Haven, CT: Yale University Press.

Morowitz, Harold. 2002. *The Emergence of Everything: How the World Became Complex*. Oxford: Oxford University Press.

President's Council on Bioethics. 2003a. *Beyond Therapy: Biotechnology and the Pursuit of Happiness*. Washington, DC: President's Council on Bioethics.

President's Council on Bioethics. 2003b. *Being Human: Readings from the President's Council on Bioethics*. Washington, DC: President's Council on Bioethics.

Richards, Richard. 2019. *The Biology of Art*. Cambridge: Cambridge University Press.

Sherlock, Richard. 2005. "Jerusalem and Athens." *Modern Age* 47, no. 1: 64–68.

Slingerland, Edward. 2008. *What Science Offers the Humanities: Integrating Body and Culture*. Cambridge: Cambridge University Press.

Wilson, Edward O. 1998. *Consilience: The Unity of Knowledge*. New York: Knopf.

PART IV

Science and Technology Studies

Chapter 18

Ethics and the Search for Scientific Knowledge
The Whole Truth and Nothing but the Truth?

Carlos Verdugo-Serna

I have to confess that discussing some questions about the relationships between ethics, truth, and pure or basic science, including the issue of the existence of forbidden, dangerous, or discouraged knowledge, makes me feel rather hesitant and trembling. It is very difficult to add something new and fruitful on these subject matters. Nevertheless, I hope that my discussion, at least, can help to renew interest mainly on the problem of forbidden knowledge or truth and to attempt to say something relevant to current discussions that are still and surely will continue to be very much alive.

Of course, there are other very important issues concerning the relations between science and ethics. Thus, science understood as an activity or process aiming to obtain knowledge, gives rise to some serious moral problems and objections not only regarding its goals, but also considering its means, conditions, and consequences. There is also the need to critically reexamine those relations, at this time, concerning the constant growing power of science, often connected with new technological possibilities that may harm society and our planet, the entanglement of scientific research with financial and political interests, and the use of humans and animals in research, among other related trends.

Since it is common wisdom to claim that surely one of the main objectives of science is the search for truth or for true knowledge, I will start by examining the connections between scientists, science, and truth. This will be very important for assessing or rather, as I will argue, for rejecting the claim that it is in principle morally acceptable to know everything or to know every possible truth and that, after all, there seems to be, as some philosophers have argued, no morally prohibited truths.

As the title of this work indicates, my goal is to reconsider the ethical issues that have arisen about the uses of truth in scientific research. I will show that the search for truth or pure knowledge can be and has been used for two purposes:

1. as a sort of criterion of demarcation between pure or basic science and applied science and technology
2. as a means to immunize science from moral objections

I want to argue that (1) can be accepted for establishing a working distinction between two important human activities—that is, science and technology. In fact, this criterion has been supported by some philosophers and by certain scientists. But I will also show that there are compelling reasons to reject (2).

TRUTH, PURE AND APPLIED SCIENCE, AND TECHNOLOGY

The idea that science can be distinguished from technology according to differences in their aims, goals, methods, and objectives has been adopted by scientists working in very diverse areas of scientific research. For example, the winner of the 1963 Nobel Prize in Physiology, Sir John Eccles, claims that the distinction lies in the difference between objectives of science and those of technology. According to Eccles, on the one hand, the scientist "tries to understand or comprehend the natural world as he experienced it. On the other hand, the technologist utilizes the knowledge about the natural world for practical purposes" (1970, 137).

In the realm of physics, a very similar view concerning the pure intellectual aim of science is found. Thus, Stephen Hawking asserts that "the eventual goal of science is to provide a single theory that describes the whole universe" (1988, 10). Even some critics of the distinction between pure science and technology, such as Hans Jonas, have acknowledged that there are some branches of science or, at least, some scientific research whose results or products—that is, knowledge—can be considered "pure" or "disinterested" because its discoveries are very far from or actually devoid of technical applicability. As he remarks, one clear example is cosmology. Its main subjects are the expansion of the universe, the evolution of galaxies, the Big Bang, and black holes, and Jonas points out that "these are matters for knowing only and for no possible doing on our part" (1982, 599).

Additional examples of areas of research in physics in which the motivations and aims of scientists were never related to some practical applications were pointed out by the physicist Melvin Schwartz:

> I am sure that Einstein, when he thought of special relativity at the age of twenty-three . . . had no idea that it would end up having any practical value whatsoever. Applicability was just not a factor in asking the questions at that time, and it is still not a factor in asking the questions. The basic reason that we ask the questions in our field, and I think this is true in very many of the other pure research fields, is that we would like to have a deep understanding of the fundamental relationship of physical objects to each other. (1977, 81)

In sum, it is still possible to defend the thesis that there is a clear distinction between pure science and applied science or technology, which can be based on the different goals, aims, or objectives of those human activities.

Thus, although we have to accept that, as many authors have remarked (e.g., Hottois 1979; Latour 1987), the boundary between pure science and technological applications has become rather blurred or should be eliminated, it can be argued that a distinction between them still holds.

Nevertheless, accepting that the search for truth or knowledge can be considered as the objective or aim of science and that this objective can provide an adequate criterion for distinguishing pure science from its possible applications, does not imply that the search for truth, knowledge, or understanding gives science moral immunity.

Before examining and criticizing the use of the search for truth as a means to immunize science from moral objection or any ethical culpability, the assertion that the aim of scientists is or should be the search for truth must be clarified.

THE WHOLE TRUTH AND NOTHING BUT THE TRUTH?

If the expression "To search for 'the whole truth'" is understood as an attempt to find and collect all logically and empirically true statements, it seems quite obvious that no scientist *qua* scientist would accept this as a reasonable and possible task to carry out, much less a duty.

As Jan Łukasiewicz remarked, not even the sage as described by Aristotle aimed to know everything or to be omniscient—that is, to know all the particular facts, but only some general truths:

How different is Aristotle's idea of perfect knowledge! He, too, thinks that a sage knows everything; *yet he does not know detailed facts*, and has only knowledge of *the general*. And as he knows the general, in a way he knows all the details falling under the general. Thus *potentially* he knows everything that can be known. But potentially only: actual omniscience is not the Stagirite's ideal. (1970, 2)

Today many scientists would agree that the most important general or universal truths worthy to search for and to find are those needed to explain and predict empirical facts about the world. This view has been supported by many influential philosophers of science. One clear example is Karl Popper, who accepts the idea that the aims of science are to search for truth and to develop satisfactory scientific explanations. But he also said that what we actually want and need is more than mere truths—that is, we should look for interesting and relevant truths with a high degree of explanatory power. As he remarked:

It is very important that we try to conjecture true theories, but truth is not the only important property of our conjectural theories; for we are not particular interested in proposing trivialities or tautologies. "All tables are tables" is certainly true—it is more certainly true than Newton's and Einstein's theories of gravitation—but it is intellectually unexciting: it is not what we are after in science. . . . In other words, we are not simply looking for truth, we are after interesting and enlightening truth, theories that offer solutions to interesting problems. (1972, 54–55)

So, it is clear that, in principle, scientists have no obligation to try to know everything that it is possible to know or to search for the whole truth. Even more, it can be argued that, in fact, as Michael Boylan points out, we have to reject the claim that "Whatever can be known about the physical world should be known." Or, if we talk also about the social world we should reject the more general statement claiming that "What can be known should be known" (Boylan 2005, 3).

Carl Hempel makes an even stronger argument: to construe the goal of science or scientific research as the search of knowledge of the truth, the whole truth, and nothing but the truth about the natural and the social world turns out to be an impossible goal. For, if by a total or complete knowledge about the world we understand a set of sentences describing not only particular events in past, present, and future but also the laws of nature connecting them, then we have to accept that this impossibility is of a logical nature:

But it may be of interest to note that the ideal of total knowledge as just characterized is unattainable for purely logical reasons, and no being can achieve omniscience in this sense. For the sentences expressing such a total knowledge would have to be formulated in some suitable language: but no matter how rich a language may be, there are always facts that cannot be expressed in it. (Hempel 2001, 362)

Finally, as I will indicate below, there are some important arguments to support the view that, contrary to a common and old idea, the search for knowledge or truth is not always good or acceptable and, therefore, that we have to accept that this search must somehow be limited by moral and social constraints. This issue is closely related to the problem of forbidden, dangerous, or discouraged knowledge.

THE SEARCH FOR TRUTH OR PURE KNOWLEDGE AS AN IMMUNIZATION TO MORAL OBJECTIONS

The demarcation between pure, fundamental, or basic science, on the one hand, and applied science or technology, on the other, is also commonly used in many discussions on the moral responsibility of the scientist or on the possible limits of scientific research, including the issue concerning the existence of "forbidden" knowledge or truths.

They play a very important role: they protect scientists devoted to pure or basic science or to the search for pure or disinterested knowledge from claims of responsibility for any unethical applications developed using knowledge produced by their research activities. Therefore, if it is accepted that, by definition, the aim of basic scientists is the search for truth and nothing but the truth, then the culprits must be in the next house, as it were—that is, where applied scientists and technologists live.

The view that pure science or basic scientific research is morally innocent has been strongly defended by some philosophers of science such as the late Argentinian philosopher Mario Bunge: "The natural scientist wants to find new laws of nature, and the social scientist wishes to describe and explain society. Basic scientists, in sum, wish to understand reality, not to dominate: they are after knowledge, not power" (1991, 96).

Bunge argues that from the point of view of its goals, basic science is either innocent and does not raise moral problems or is immune to ethical criticisms because basic scientists "have no opportunities for doing harm, except through simulation, theft, or sloth—and even so the damage they can do is limited" (97). He also claims that applied scientists, since they seek knowledge or truth having potential use, have plenty opportunities for mischief: "In sum, basic science is innocent, for it seeks only knowledge of what there is, was, or may be. . . . Applied science or technology can be either good or evil, according whether they promote life or, on the contrary, endanger it" (105).

He asserts that the results or products of pure research are ethically neutral because they can be used for good or bad purposes. Nevertheless, Bunge is also ready to admit that research processes are not ethically neutral.

A similar view has been argued by Evandro Agazzi who, after indicating that pure and applied sciences can be considered as efforts to provide knowledge, states:

> In the case of pure science the goal of this knowledge is (to put it briefly) the discovery of *truth*, in the sense of establishing "how thing are," while in applied science this goal is the realization of some action or practical *result*. Admitting the specific aim of pure science to be the search for truth, it is clearly immune from moral objections in itself (i.e., it constitutes a perfectly legitimate value). (1989, 52)

Having said that, Agazzi also acknowledges that in the process of the acquisition of pure knowledge the means used for this purpose can raise moral objections.

In short, all of these quotations show clearly that the search for truth or pure knowledge as the goal or objective of pure science or knowledge has been used not only for demarcating pure science from applied science or technology, but also to immunize pure science from moral objections.

THE PROBLEM OF FORBIDDEN TRUTH OR KNOWLEDGE

Many authors writing about the relations between science and ethics, including the possibility of forbidden, discouraged, or dangerous knowledge especially in connection with scientific knowledge, have made reference to the story about Adam and Eve who were punished for violating the deity's order not to eat from a specific tree. According to the story in the book of Genesis, this was the tree of the knowledge of good and evil. Actually they were punished and expelled from the Garden of Eden not only for disobeying that prohibition but, also, to prevent them from eating from the tree of life and thus living forever.

If we wish to use this biblical story as one of the first registered cases of forbidden knowledge we have to consider that it was related to moral knowledge. Descriptive knowledge was not subject to that prohibition. In his very important article "Problems of Forbidden and Discouraged Knowledge:

Intrinsic and Extrinsic Constraints," Richard Rudner offers a rather different interpretation of the biblical account:

> Indeed, if we equate partaking of the fruit of the tree of knowledge with the acquisition of science (say, with the acquisition of scientific method) . . . everything that is the case is open to science. I would argue that there is nothing in the nature of science, hence *no intrinsic constraint* upon science that closes any subject to it. (1977, 30)

Later, I will analyze Rudner's article in full, especially his proposal to identify science with a certain method or logic of inquiry. It is also very important to examine some consequences of his proposal and to evaluate his claim that there are *extrinsic constraints* on science or scientists, most of which are based on societal and ethical values.

While for many people the biblical source is only a mythical tale similar to the Prometheus story, the issue of the relationship between ethics and theoretical or practical knowledge was similarly raised in ancient philosophy where it was recognized that knowledge is like a double-edged sword that can be used to do good or evil.

Thus in Plato's *Republic* we find Socrates asking: "It is not true that he who knows how to guard against disease is also most able to infect with it and escape detection?" (333e) This seems to be one the first and clear formulations of the problem of the moral responsibility of people possessing some kind of knowledge or expertise. In Greece this problem was solved in part by requiring that the students of medicine and practicing physicians take the Hippocratic Oath. promising that they will use medical treatment to help the sick, and never to injure or wrong them.

Of course, it is very difficult to deny that knowledge is one of the most important human goods in itself and a basic and necessary tool for human actions and aims. This has been accepted both in Western and Eastern philosophy. For example, Dharmakīrti, the Indian Buddhist philosopher (d. 660 CE) in chapter one of his *Nyāya-Bindu* (*A Short Treatise of Logic*) states: "All successful human action is preceded by right knowledge" (1962, 1).

Nevertheless, today more than ever before in history, human knowledge, especially scientific knowledge, has become a tremendous power that can be used for the benefit or detriment of human beings. Thus, few people would deny the need to establish some limits, constraints, or regulations on pure or applied science. After all, if knowledge is power that continuously and increasingly is opening new unimaginable possibilities for doing good and evil, surely this power cannot enjoy unlimited freedom.

Certainly, we are very aware that these are issues have been and continue to be the focus of many discussions by scientists and philosophers concerning the relationship between science and ethics, including some moral problems of scientific research, for example, the risks of using some revolutionary techniques such as recombinant DNA or the development of genetic engineering. Besides the important International Asilomar Conference in 1975 organized to analyze these risks and possible biohazards, the United States National Academy of Sciences convened a Forum on "Research with Recombinant DNA" in 1977. The proceedings of the Forum led to the publication of the book *Research with Recombinant DNA* (National Academy of Sciences 1977a). The papers of this volume were described as exploring and discussing "the scientific, legal, and moral issues that have been raised by the recombinant DNA technology." The impact and importance of this debate was described by Stephen P. Stich in the following terms:

> The debate over recombinant DNA research is a unique event, perhaps a turning point, in the history of science. For the first time in modern history there has been a widespread public discussion about whether and how a promising though potentially dangerous line of research shall be pursued. At root the debate is a moral debate. (1982, 590)

There was a vigorous debate in the 1977 Forum on issues such as the existence of a "free inquiry principle" and, on the contrary, that the recombinant DNA research and its necessary restrictions have nothing to do with the question of freedom of inquiry, but, as Jonathan King says: "This is a question of freedom of manufacture, of modifying the environment, of modifying living organisms, not of asking questions about them, but in the route which you take in getting the answer" (1977, 39–40). Other questions raised included the importance of criticizing the assumption that truth can be considered a good and the societal benefits of scientists learning truth. A strong rejection of a similar assumption about the search for knowledge was put forward by Kurt Mislow: "I do not agree that increased human knowledge is of paramount importance. I do not agree that the real enemy is ignorance. . . . I can think of lots of examples where knowledge is extremely dangerous" (1977, 277–8). Finally, another important issue raised was the need to revise the nature of the social contract between science, technology, and society. Thus, David Baltimore emphasizes that "the responsibility for controlling the fruits of science falls on the total society" (1977, 240).

Similar concerns about the new possibilities opened by scientific advances have been emphasized by philosophers of science, of technology, and of engineering. A recent and illuminating discussion about these issues can be found in Adam Briggle and Carl Mitcham's book *Ethics and Science* (2012). They underline that among some main trends in contemporary science that raise ethical concerns are the increasing power of science that is intimately related to new and increasing threats and risk:

> As scientific research grows, it continues to influence our lives in surprising ways—sometimes hopeful, sometimes frightening, often ambivalent or unclear. There is thus an ongoing need for critical assessment of the relationship between science and society.... What ideals should inform the practice of science?... Who should exercise authority or responsibility in and over scientific practice?... How can progress in science be defined and measured in the context of broader ethical norms and social goals?... Is it sufficient to appeal to the pursuit of truth or curiosity as motives or to an 'invisible guiding hand' that turns independent research programs to the common good? (2012, 11–12)

Unfortunately, the book does not deal with certain important questions that are formulated only for further research and discussion. These are: "Is there any scientific research that should be restricted or even banned on principle? Why or why not? More generally, are there any societal values that justifiably limit the quest for knowledge" (Briggle and Mitcham 2012, 21). The next section examines some answers to these very pressing questions.

PROBLEMS OF FORBIDDEN, DANGEROUS, AND DISCOURAGED KNOWLEDGE

Are there subject matters or things we should not know? Should there be forbidden questions or certain topics or truths in science? I argue that there are some very good reasons to answer both questions in the affirmative. Thus, I think we should reject the claim that "it is in principle morally acceptable to know everything, and there are no morally prohibited truths" (Agazzi 1989, 53). A clear example of a clash between, on the one hand, epistemic values related to scientific research and the quest for new knowledge, and, on the other, certain societal values that are basic for the existence and preservation of important social institutions, is the 1954 Wichita case.

In a University of Chicago Jury Project, some jury proceedings were recorded as a part of a research. These recordings were done with the knowledge of the trial judges but without consent of the jury. In 1955 this method of research became public. This led to a strong rejection by the Attorney General of the United States and provoked many angry hearings, for example, before the US Senate. In one of these federal hearings Senator James A. Eastland said to one of the researchers, "Now, do you not realize that to snoop on a jury, and record what they say, does violence to every reason for which we have secret deliberations?" (Warwick 1982, 112)

As a result of these hearings legislation was passed expressly prohibiting the recording of federal petit or grand juries for any purpose. Not only was jury taping prohibited but also direct observation of jury deliberation. Thus, to the question: Are there things we should not know? there is a clear and justified answer: we should not obtain knowledge of the secret deliberations of a jury, because it harms a society value that is essential to administer a needed and proper justice. In "Types of Harm in Social Research" published in *Ethical Issues in Social Science Research* (1982), Donald P. Warwick claims that the Wichita case can be considered as a clear example of doing harm to society by undermining the function and legitimacy of a much needed social institution. Of course, we need to accept, as Warwick also remarks, that you can harm and benefit not only individuals but also society as a whole or some subgroups, and government.

The subject matter connected to the jury case, and two additional areas that should be forbidden for social research, have been remarked by Alasdair MacIntyre as follows:

> It is wrong for anyone, and therefore for both journalist and social scientists, to intrude upon the grief of a recently bereaved person or family; and it is wrong for me or any other stranger to read without permission other people's diaries or letters; and it is wrong to violate the integrity of the jury process. . . . We need sanctuaries, we need to be able to protect ourselves from illegitimate pressure, we need places of confession, and we need to disclose ourselves in different degrees to people to whom we stand in different degrees of relationship. Intimacy cannot exist where everything is disclosed, sanctuaries cannot be sought where no place is inviolate, integrity cannot be seen to be maintained—and therefore cannot in certain cases be maintained—without protection from illegitimate pressures. (MacIntyre 1982, 188)

I fully agree with MacIntyre. Moreover, I want to emphasize that the problem here is not concerned with the morality of certain types of techniques or methods of research for studying these sanctuaries: what is morally reprehensible is the scientific aim or goal to acquire knowledge or some truths about them. These are evident examples of forbidden knowledge. It is also an evident clash among values: epistemic versus personal intimacy or privacy. Here I quote Robert L. Sinsheimer, who has been one of the first scientists to formulate the question: Can there be "forbidden" or "inopportune" knowledge? He remarks:

> If one believes that the highest purpose available to humanity is the acquisition of knowledge (and in particular of scientific knowledge, knowledge of the natural universe) then one will regard any attempt to limit or direct the search of knowledge as deplorable—or worse. If however, one believes that there may

be other values to be held even higher than the acquisition of knowledge—for instance general human welfare—and that science and possibly other modes of knowledge acquisition should subserve these higher values, then one is willing to (indeed, one must) consider such issues as . . . the selection of certain areas of scientific research as more or less appropriate for that social context. . . . In short, if one does not regard the acquisition of knowledge as an unquestioned ultimate good, one is willing to consider its disciplined direction. (1978, 23)

A similar view has been argued by Nicholas Rescher in *Forbidden Knowledge and Other Essays on the Philosophy of Cognition*:

One should never lose sight of the fact that knowledge is only one of human goods among others. . . . While knowledge represents an important aspect of the good, it is by no means a sovereignly governing factor that is always and everywhere predominant. It is only one component in a wider framework of human purposes and interests. . . . And the competing interests must be weighed here—the value of knowledge vs. the welfare of people. . . . Freedom of inquiry is unquestionably a great good but it is not an absolute one. . . . At many junctures of life we face a situation of conflict where we must balance goods, claims and rights against each other. No Moses has come down from the mountain with tablets graven on high to the effect that the quest for knowledge is an overriding priority—that freedom to investigate and to inquire, to disseminate and to teach automatically outweighs any other interest with which it can come into conflict. (1987, 10–14)

But, then, are we not here violating some inalienable rights belonging to scientists, such as the "right to know" or the value of "freedom to search for truth"? Or as Charles Fried has asked in his article "Problems of Consent in Sex Research: Legal and Ethical Considerations":

What then of the claim of freedom of inquiry and the pursuit of pure knowledge? I am a rigid Kantian believer in the right to pursue knowledge freely and without restrain. But that is the researcher's right, and obviously one cannot exercise this right by deliberately and intentionally violating the rights of others. For instance, one cannot pursue this right by kidnapping research subjects. (1980, 38)

In sum, the freedom and the right to know and to pursue knowledge or to search for truth cannot be considered as absolute and without limitations or constraints. In other words, it should be clear that science, understood as a human activity or a process whose main goal is the search for truth or knowledge, has to admit as any other human activity, certain moral limits. Thereby, all scientists engaged in this activity have certain moral obligations mainly to do no harm.

Finally, I want to examine Rudner's claim mentioned above that if we understand that the term "science" can also be understood as referring to certain specific method—that is, the so-called "scientific method"—then there would be *compelling* reasons to claim that nothing in principle should be forbidden to science and there would be no intrinsic limitations on science. In other words, there are no subject matters beyond the capacities of the scientific method. By *scientific method* he understands the rationale or logic of inquiry that is used to give a systematic and organized account of the way natural and social reality is. As Rudner states:

> Everything that is the case is open to Science. Of anything that may be the case, an inquiry into whether it is the case is intrinsically open to Science. I would argue that there is nothing in the nature of science, hence *no intrinsic constraint* upon science that closes any subject to it.... What I am suggesting, then, in saying that there are no intrinsic subject matter constraints on science, is that there is no sound argument that implies that *any* subject matter is *closed* to a method of inquiry thus characterized. (1977, 30–31)

But, at the same time, he emphasized that to accept the claim that science as a method is free from intrinsic constraints does not imply that there are no *extrinsic constraints*, for example, moral ones, upon some particular procedures or *techniques* used in scientific research. Thus, according to Rudner if we identify science with a method, then all of the limitations or constraints "on how or where the method is to be applied, turn out to be *extrinsic* constraints" (30).

Among some of the most important conclusions that Rudner derived from his claim that science can be identified with the method or logic described above, are: (a) science proscribes no inquiry into what may be the case, nor proscribes or commands any subject matter, but also (b) neither does science prescribe any subject matter. Indeed he states that the openness of any subjects to the scientific method "is quite compatible with our never investigating some of them." But more important, that there is also nothing *intrinsic* to science as a method that dictates or prescribes the use of any specific technique of research. This is one of the main reasons for Rudner to reject absolutely that the abominable practices and techniques used by Nazi doctors could be morally justified by appealing to "science."

I highlight again that the openness here is about scientific method. The method by itself has no aims, it can be used by scientists for different purposes, among them to describe, explain, or predict certain events or to test scientific hypotheses or theories. Thus, the decision to investigate (or not) some subjects are made by scientists and this decision can, in principle, be subjected to extrinsic limitations, especially of moral or social nature.

It seems to me that after all, we should reject Agazzi's claim that "it is in principle morally acceptable to *know* everything, and there are *no morally prohibited truths*" (1989, 53).

I close my analysis of Rudner's contribution to the problem of forbidden and discourage knowledge by quoting in full one of his final conclusions:

> The second, and final, vexing question I want to propound is whether the intrinsic entry into Science of even so "cool" a set of values as the epistemic ones may nevertheless have the effect of closing off some subject matter to Science. Perhaps the sharpest way of pointing up the issue is to consider whether the decision *to* inquire into a subject matter at all is more likely to lead, causally, to the end of human or indeed sentient life than is the decision not to inquire into that subject matter. If such a circumstance *is* a possible one, then it seems necessary to conclude that some subject matter could be closed to scientific inquiry. (1977, 39)

CONCLUSION

I have argued that it is still acceptable for scientists and philosophers of science to claim that the main (or at least one of the main) objectives of science is the search for truth or true knowledge. Of course, this should not be interpreted as asserting that scientists should attempt to find and collect all logical and empirically true statements. As Popper has remarked, what scientists are looking for are not trivialities or tautologies but rather interesting and relevant truths with a high degree of explanatory power. Hempel also has emphasized that the ideal of total knowledge turns out to be logically impossible.

I have also shown that the search for truth or pure or disinterested knowledge can and has been used as a sort of criterion of demarcation between pure or basic science and applied science or technology. This use does not seem to raise strong rejections.

Nevertheless, what is more difficult to accept and what ultimately must be rejected is any attempt to use the search for truth, pure knowledge, or understanding as an argument for the claim that science is free from moral problems or from ethical constraints or social limitations. I am not referring here to the widely held recognition that science as a process or as a human activity is subjected to moral constraints in certain areas, for example, research on human or animal subjects and subjects that are potentially dangerous such as recombinant DNA. Rather, the rejection is related to the claim that if the aim or goal of pure science is the search for truth, then it is totally immune from moral objections in itself. This claim is based on the questionable, and I believe, wrong assumption that knowledge or truth is the most important human value. The arguments put forward by Sinsheimer and Rescher clearly manifest that this is not the case. Both authors agree that knowledge has to be regarded instead as only one good among other goods, such as general human welfare, privacy, or intimacy.

In addition, the Wichita case and the observations made by MacIntyre about the human need for sanctuaries and intimacy are important arguments for establishing the fact that there are some subject matters that scientists should not know because they clearly do harm to society and to personal or individual values, and research in these areas should be forbidden. These limits also apply to any claim that scientists have some sort of inalienable rights to know or to search for truth. As Fried states, scientists cannot exercise their right to pursue knowledge by deliberately violating the rights of others.

In the end, even if we agree with Rudner's proposal to identify science with a method or logic of inquiry that in principle is free from ethical and social intrinsic constraints so that we have to accept the openness of all subject matters to science, this fact does not mean that there are not extrinsic constraints upon scientists. Thus, to questions formulated by Briggle and Mitcham (2012, 21), "Is there any scientific research that should be restricted or even banned on principle? More generally, are there any societal values that justifiably limit the quest for knowledge?" I answer that they should be answered in the affirmative.

ACKNOWLEDGMENT

I would like to thank Dr. Rodrigo Lopez, David Wayne Memmott, Raul Carrasco, and Patricio Varas for their assistance with this manuscript.

REFERENCES

Agazzi, Evandro. 1989. "Ethics and Science." In *Logic, Methodology and Philosophy of Science VIII*, edited by Jens E. Fenstad, Ivan T. Frolov and Risto Hilpinen, 49–61. Amsterdam: North-Holland.

Baltimore, David. 1977. "Potential Uses." In National Academy of Sciences, 1977a, 237–40.

Boylan, Michael. 2005. "The Ethical Limitations on Scientific Research." *Journal of Philosophical Research* 30, Issue Supplement, "Ethical Issues for the Twenty-First Century": 15–26.

Briggle, Adam, and Carl Mitcham. 2012. *Ethics and Science: An Introduction.* Cambridge: Cambridge University Press.

Bunge, Mario. 1991. "Basic Science Is Innocent; Applied Science and Technology Can Be Guilty." In *Nature and Scientific Method*, edited by Daniel O. Dahlstrom, 95–106. Washington, DC: Catholic University of America Press.

Dharmakīrti. 1962. *A Short Treatise of Logic, Nyaya-Bindu*. In *Buddhist Logic*, edited by Th. Stcherbatsky, vol. 2, 1–153. New York: Dover.

Eccles, John C. 1970. *Facing Reality: Philosophical Adventures by a Brain Scientist*. New York: Longman, Springer-Verlag.

Fried, Charles. 1980. "Problems of Consent in Sex Research: Legal and Ethical Considerations." In *Ethical Issues in Sex Therapy and Research*, edited by William H. Masters, Virginia E. Johnson, Robert C. Colony, and Sarah M. Weems, vol. 2, 21–41. Boston: Little, Brown.

Hawking, Stephen W. 1988. *A Brief History of Time: From the Big Bang to Black Holes*. New York: Bantam Books.

Hempel, Carl G. 2001. *The Philosophy of Carl G. Hempel: Studies in Science, Explanation, and Rationality*, edited by James H. Fetzer. New York: Oxford University Press.

Hottois, Gilbert. 1979. *L'Inflation du langage dans la philosophie contemporaine*. Brussels: Editions de L'Université de Bruxelles.

Jonas, Hans. 1982. "Freedom of Scientific Inquiry and the Public Interest." In *Contemporary Issues in Bioethics*, edited by Tom L. Beauchamp and LeRoy Walters, 598–601. Belmont, CA: Wadsworth.

King, Jonathan. 1977. "Discussion to 'The Involvement of the Public' by Daniel Callahan." In National Academy of Sciences 1977a, 38–41.

Latour, Bruno. 1987. *Science in Action: How to Follow Scientists and Engineers through Society*. Cambridge, MA: Harvard University Press.

Łukasiewicz, Jan. 1970. "Creative Elements in Science." In *Selected Works*, edited by L. Borkowski, 1–15. Amsterdam: North-Holland Publishing.

MacIntyre, Alasdair. 1982. "Risk, Harm, and Benefit Assessments as Instruments of Moral Evaluation." In *Ethical Issues in Social Science Research*, edited by Tom L. Beauchamp, Ruth R. Faden, R. Jay Wallace, Jr. and Le Roy Walters, 175–89. Baltimore: Johns Hopkins University Press.

Mislow, Kurt. 1977. "Discussion to 'The Economic Implications of Regulations by Expertise: The Guidelines for Recombinant DNA Research,' by Roger G. Noll and Paul A. Thomas." In National Academy of Sciences 1977a, 277–78.

National Academy of Sciences. 1977a. *Research with Recombinant DNA: An Academic Forum, March 7–9*. Washington, DC: The National Academies Press. https://doi.org/10.17226/20351.

———. 1977b. *Science: An American Bicentennial View*. Washington, DC: The National Academies Press.

Popper, Karl R. 1972. *Objective Knowledge: An Evolutionary Approach*. Oxford: Clarendon Press.

Rescher, Nicholas. 1987. *Forbidden Knowledge and Other Essays on the Philosophy of Cognition*. Dordrecht: D. Reidel.

Rudner, Richard. 1977. "Problems of Forbidden and Discouraged Knowledge: Intrinsic and Extrinsic Constraints on Science." In *New Dimensions in the*

Humanities and the Social Sciences, edited by Harry R. Garvin, 30–40. London: Associated University Presses.

Sinsheimer, Robert L. 1978. "The Presumptions of Science." *Daedalus* 107 (Spring): 23–35.

Stich, Stephen P. 1982. "The Recombinant DNA Debate." In *Contemporary Issues in Bioethics*, edited by Tom L. Beauchamp, and LeRoy Walters, 590–98. Belmont, CA: Wadsworth.

Schwartz, Melvin. 1977. "Forum III. The Use of Knowledge. Frontiers Expansion or Inward Development." In National Academy of Sciences 1977b, 81–84.

Warwick, Donald P. 1982. "Types of Harm in Social Research." In *Ethical Issues in Social Science Research,* edited by Tom L. Beauchamp, Ruth R. Faden, R. Jay Wallace, and LeRoy Walters, 101–24. Baltimore: Johns Hopkins University Press.

Chapter 19

A Short History of Science, Truth, and Politics in the United States, 1945–2021

Daniel Sarewitz

The idea that we live in a post-truth era appears to be widely held by opinion leaders today. The Oxford Dictionaries named *post-truth* their 2016 "word of the year,"[1] and defined it as "relating to or denoting circumstances in which objective facts are less influential in shaping public opinion than appeals to emotion and personal belief." Post-truth thus highlights the widespread conviction that a social preference for truthfulness as a foundation for dialogue and action in the public sphere has waned. Charlatans, demagogues, and simple ignoramuses are on the rise; experts, scientists, and rational, truth-telling politicians are in retreat. As the British journalist Matthew d'Ancona writes in his elegant tract *Post-Truth: The New War on Truth and How to Fight Back*, "It is a battle between two ways of perceiving the world, two fundamentally different approaches to reality.... Are you content for the central value of the Enlightenment, of free societies and of democratic discourse, to be trashed by charlatans—or not?" (d'Ancona 2017, 5).

And what, exactly, is this thing called truth? Those announcing the advent of the post-truth era are not clear on this point (presumably finding it self-evident), but invocations of objectivity and the Enlightenment provide some hint of what they seem to have in mind. Truth is won from disciplined inquiry and empirical validation; as a cultural ideal, it gets its cachet from its direct link to the fact-generating power of science. Truth is thus a tool for guiding rational action for human betterment. Perhaps most importantly, truth can be clearly separated from non-truth. This is the Enlightenment project, and the credo that d'Ancona invokes. The rejection of truth as the sine qua non for rational action is thus an assault on the Enlightenment, on science, on modernity, and progress. It opens the door to nihilistic opportunism, wielded wantonly for the acquisition and exercise of power, and acceded to by the masses for their emotional gratification. The post-truth perspective tells us that many of the difficulties of our day can be traced to the rejection of truth in favor of mere belief, emotion, and even lies.

If we are now in a post-truth era, we must be leaving behind a time of truth. As observed in the introduction to a collection of scholarly essays on the decline of reason in democratic society, "A trend has been gathering momentum in modern culture away from science as a means to think about human affairs and an approach to the truth" (Thompson and Smulewicz-Zucker 2018, 7). Such declarations of post-truth convey nostalgia for the day when respect for expertise and science in democratic societies was a political and social norm. When was that day?

I nominate August 6, 1945, when, in a single instant, the power of scientific truth to translate human aspiration into action became apparent to all. On that day, Einstein's recognition of the equivalence of mass and energy, having been made palpable through the coordinated labor of thousands of scientists and engineers in the building of the first nuclear weapons, became a fearsome demonstration of the power truth. Science, made visible and worldly in technology, brought a horrific war to an end through the sudden wielding of a godlike might against Hiroshima.

Mass-energy equivalence: this is truth with a capital T. But what of it? From the moment that this truth was made known to all, its value, meaning, and implications were contested. Physicists had now acquired a mythic cultural mantle as discoverers and purveyors of truth, which brought with it great political influence, even as physicists and other scientific truth-creators bitterly debated among themselves the moral and practical import of these truths. The scientific truth-nugget at the core of this debate fades into near invisibility when surrounded by the contexts of the human world.

Thus, a second moment: February 20, 1958. By this date, the United States had accrued more than seven thousand nuclear weapons; the Soviet Union's stockpile was nearing nine hundred (Norris and Kristensen 2010). Two leading scientists, Linus Pauling and Edward Teller, debated on national television the ethical, public health, and policy aspects of atmospheric nuclear weapons testing and the proper management of the nation's growing nuclear arsenal. These guys were capital E experts. Pauling had won the 1954 Nobel Prize in Chemistry; in 1962 he would win the Nobel Peace Prize for his work on global nuclear disarmament. Teller was a renowned physicist who made important contributions to quantum theory and also participated centrally in the development of nuclear weapons technologies, most notable the hydrogen bomb.

This debate between scientific experts is extraordinary for the near absence of factual statements that might be regarded in some narrow sense as "truth." In arguing against atmospheric nuclear tests, Pauling (quoting a petition that he coauthored), says that "'Each amount of radiation [from a test] causes damage to the health of human beings all over the world, and causes damage to the pool of human germ plasm such as to lead to an increase in the number of seriously defective children that will be born in future generations.' This statement *is* true" ("Fallout and Disarmament," 150, emphasis in original). Teller counters that "this alleged damage which the small radioactivity is causing—supposedly cancer and leukemia—has not been proved, to the best of my knowledge, by any kind of decent and clear statistics" (155–56). Such mutually negating exchanges between experts are today utterly commonplace in the public sphere, but I want to make two additional points. First, in 1958 the scientific basis for saying anything at all that might amount to a "truth" about the public health and the genetic effects of a single atmospheric weapons test was extraordinarily thin. Second, the Pauling-Teller debate was not really about the truth of these matters at all; it was rather a sweeping and passionate philosophical, moral, and ideological confrontation about what role nuclear weapons can and should play in America's national security. It pitted the pacifist sensibilities of Pauling against Teller's conviction that the nation could not protect itself against Soviet aggression without an overwhelming nuclear capability.

Where did that confrontation of values and beliefs leave viewers of the debate? Here's how the moderator wrapped things up: "I'm sure both our guests would agree that [the issue's] ultimate solution rests in our hands, that each of us bears the moral obligation to examine the evidence, draw conclusions from this evidence, and act upon our convictions" (163).

During what one supposes must have been the heart of the era of truth, the moderator was calmly telling viewers that these two world-renowned scientific experts were not, after all, much help to them and that it was up to each citizen to make sense of the competing assertions of truth, interpreted through one's individual sensibilities, in arriving at a judgment about these momentous, indeed existential, matters.

A third moment: January 10, 1961. In his famous farewell speech, President Eisenhower warned that "in holding scientific research and discovery in respect, as we should, we must also be alert to the equal and opposite danger that public policy could itself become the captive of a scientific-technological elite."[2] Eisenhower had in effect translated the final charge of the moderator in the Pauling-Teller debate into a general warning for democracies. It was an extraordinarily acute prophecy. My argument here is that the political conditions leading to what many say is a post-truth world is a terrifying culmination of this prophecy. Yet were a latter-day Eisenhower to make such a statement today he'd be dismissed as a know-nothing, a charlatan, a fool, an avatar of post-truth itself.

I only have space to provide the briefest thumbnail sketch of what happened in the following sixty years, but the perspective I want to share is essentially a political economic one. Starting in the late 1950s the United States began to ramp up investments in scientific research, and in the training of more scientists, and this growth has continued more-or-less unabated ever since.[3] The concerns of publicly funded science itself began to expand from a fairly limited portfolio related mostly to matters of national defense and natural resources to an ever-broadening engagement across all traditional scientific disciplines and emerging cross-disciplinary problems. In parallel, the development of new tools for scientific research led to incredible growth in the types and amounts of data being collected, the scales of observation, the methods available for analyzing data, theories devised to help make sense of the analyses, and the volume of publications and other products reporting on all this new knowledge.[4] In the process, science gradually became engaged with an expanding array of problems of direct concern to society, both applying new research tools to existing problems (such as reducing poverty, predicting weather, and improving the management and exploitation of natural resources) and identifying new problems (such as the cancer-causing potential of synthetic chemicals and the risks of nuclear energy) that demanded policy attention.

These new problems increasingly could be characterized as those of the "Risk Society" (see Beck 1992). At the time of Eisenhower's farewell speech, technological risk created by nuclear weapons proliferation was the central place where experts (such as Pauling and Teller) engaged in the science and politics of risk, but the publication of Rachel Carson's *Silent Spring* a year later eloquently signaled a rapid increase in scientific and political focus on technological risks to environment, health, and social well-being, from nuclear energy to pesticides to industrial chemicals. In the "Risk Society," an expanding class of experts with knowledge bearing on every conceivable subject was matched by an expanding demand from policy makers and politicians looking for solutions to mounting problems they were facing; a demand met by new research funding programs and new government agencies;[5] the growth of expert advisory committees to federal agencies; and the growth and proliferation of think tanks, philanthropic foundations, and advocacy organizations devoted to applying science to the development of policies for the solution of society's problems.

But—and here's the rub—risk is a subjective, cultural phenomenon, mediated within complex social and natural systems (e.g., see Douglas and Wildavsky 1983). One person's risk is another person's opportunity. Or way of life. Scientists tried to domesticate risk, defining it in simple quantitative terms as the magnitude of an event multiplied by its probability, but as the Pauling-Teller debate should have made clear from this start, this was, at best, a quixotic conceit. Reducing the risk of, say, a nuclear reactor meltdown, a low-dose toxic chemical exposure, climate change, or pandemics to single numbers demanded simplifying assumptions that were direct invitations to being contradicted by other, equally fanciful (or plausible) numbers (e.g., Porter 1996; van der Sluijs 2016). The science quickly became politicized, and simultaneously the politics became scientized, conditions that, once established, seemed to become permanent, as in cases ranging from nuclear power to genetically modified foods to the cancer-causing potential of various chemicals to climate change. In political economic terms, this state of affairs has been good for science, because it led to continued demands, and funding, for more research and more expertise, and it has been good for politics, because it has allowed action to be deferred until the day when more and better science would provide more reliable guidance for effective action.

Risk-related policy debates thus became sites for dueling experts wielding truths backed by competing scientific assessments of risks, benefits, and potential or actual damage to individuals and communities. Often, but by no means always, these disputes mapped in a fairly straightforward way onto political divisions, with industrial interests typically aligning with conservative politics to assert low levels of risk and excessive costs of action, and interests advocating environmental protection aligning with more liberal regimes for which the proper role of government included regulation of industry to reduce risks, even uncertain ones, to public health and well-being. Under these conditions scientific claims of truth became proxies for the values and interests of those who wielded them. What should have been carried out as political debates about competing values and interests were waged in the language of science and competing claims of truth. After sixty years of such debates, the notion of science and politics as distinct enterprises came to be ridiculous.

When people talk about post-truth, the idea of truth they have in mind seems to derive its legitimacy and power in the political world from science's power to derive facts that are context-free—true in every situation where they are asserted and applied. Yet the most certain and immutable of scientific truth nuggets are actually those that are most cut off from the dilemmas of our complex social and natural worlds. This is a definitional point. Derived from experiments under highly controlled settings that intentionally shield the phenomenon being studied from the complexities of the world, truths can be isolated and validated precisely because they operate outside the contexts and contingencies of the real world. These are the truths inside the A-bomb and the flat-panel display. It is the truth of an object falling in a vacuum undergoing a known and invariant acceleration due to gravity. But in the real world, where vacuums are abhorred, that same object, say a dollar bill or a plastic bag, under the influence of air currents might actually move upward rather than downward (see Cartwright 1983).

As scientists have worked to understand complex systems whose behavior could never be subject to controlled experiments, they moved farther from the canonical ideal of discovering and empirically validating context-free scientific truths. Often the best they can do is represent complexity by simplifying it through mathematical models. As the scope and scale of both science and the risk society have grown, so have the ambitions of complex system models. In the early 1960s, researchers sought to model how best the United States might avoid and survive a nuclear war in the next decade (e.g., Dalkey 1965); by the 1980s and 1990s, earth systems modelers sought to simulate the behavior of nuclear waste storage sites for thousands of years into the future, and to predict the behavior of the earth-atmosphere-biosphere-social system for the coming centuries. Today scientists are proposing to model the "exposome," to capture the "diversity and range of [human] exposures to synthetic chemicals, dietary constituents, psychosocial stressors, and physical factors, as well as their corresponding biological response" (e.g., Vermeulen et al. 2020). Such models may generate (and require as inputs) crisp numbers, often precise to several decimal places, for phenomena that are abstractions, that cannot be measured, controlled, confirmed, or even observed, such as the ratio of costs to benefits of a large dam, the number of excess cancers that will occur in the population as a result of a certain concentration of a chemical in drinking water, the dollar value of the global ecosystem, the number of species going extinct each year, or the precise global average atmospheric temperature that must be maintained to prevent dangerous climate change (Saltelli et al. 2020).

Where is the truth in all this? The power of the idea of a post-truth world requires truth on its pedestal, the magnificent truth of August 6, 1945. But what counts as "truth" when our understanding of the basic systems we study and seek to act on is so frail, so incomplete, and the problems to which we are applying our imperfect and contestable knowledge are so thoroughly steeped in and even defined by competing values, interests, worldviews, and aspirations?

Notwithstanding such difficulties, by the early 2000s, the Democratic Party had discovered that invoking science to support its political agendas, especially around issues of environmental and health risk, was good politics, and it sought to position itself as the party of science (and thus of truth and rationality). Republican President George W. Bush was persistently and productively attacked by liberals for ignoring science in his political decisions, and his successor, Barak Obama, wasted no time in proclaiming, in his first inaugural address, that he would "restore science to its rightful place."[6] These political dynamics—catalyzed by a sixty-year commitment to the pursuit truth-on-a-pedestal—reached their apotheosis in the confluence of the Trump presidency and the COVID-19 pandemic. Trump was certainly a shameless liar, but the fact that tens of millions of voters didn't really mind was a consequence not of a post-truth world but of a political environment that had fetishized truth and the scientific-technological elite as a convenient short-cut to dealing with the hard work of democratic politics. Scientific truth claims had become such an obvious proxy for politics that Trump supporters could easily and not unreasonably fail to see much difference between Trump's bluster about climate change being a hoax, and a claim that, say, civilization would come to an end if humans do not stop burning carbon by the year 2050 in order to modulate the average temperature of the Earth's atmosphere with a precision of tenths of a degree Celsius.

Then, late in the Trump administration, COVID-19 came along to reveal that the experts were nearly powerless and truthless in the face of the pandemic. Policies for handling international travel, business closures, mask wearing, and school operations looked like a crap shoot as different countries and regions adopted radically different approaches, and scientists continually reversed course while disagreeing with one another about the meaning of data and the best ways forward. Trump enthusiastically and publicly stoked conflict with mainstream scientists, perhaps to help burnish his populist appeal. Meanwhile, open scientific questions—Do paper masks protect against transmission? Does hydroxychloroquine have any therapeutic value?—became political litmus tests. When one reasonably robust and complex truth did emerge—the new vaccines work pretty well—the damage had already been done, and overselling vaccine efficacy merely exacerbated the contempt with which many people, on all sides of the political spectrum, held such messy truths. The newly elected Biden administration in turn sought to justify a consistent set of policies around testing, social distancing, mask wearing, travel, and vaccination through appeals to scientific evidence—but simultaneously tried to suppress dissenting views from scientists, medical practitioners, and political analysts,[7] who for their part viewed administration policies as both scientifically and politically unjustified.[8]

The challenges of postindustrial society are simply not amenable to being characterized as authoritative agglomerations of concrete, invariant scientific facts. As COVID-19 has made completely clear (and as Pauling and Teller made clear decades earlier), the science associated with these challenges typically is permissive enough to support a wide range of distinctive, and sometimes even incommensurable, policy pathways. Eisenhower warned of the dangers of yielding democratic judgment over fundamental value choices to a narrow class of scientific and technical experts. He was insisting, presciently, that the dilemmas of modernity could never be reduced to technocratic prescriptions. What he did not anticipate was that the overweening ambitions of a growing science enterprise to tell the truth about *everything* would be the foundation not for political monopoly, but for epistemic chaos. The internet and social media have amplified these conditions by making it easy for anyone with a computer or smartphone to find whatever truth nuggets they need to bolster their own view of how the world looks and what ought to be done about this problem or that. These recent developments were made possible by the entrenched political economics of science and technical expertise, enabled by a puerile yet politically irresistible notion of truth-on-a-pedestal, and its decisive role in guiding democratic action. To call the outcome "ironic" would be an understatement. Far from living in a post-truth world, we live in a miasma of competing truths, available to all.

Is it still possible to restore some dignity to the ideal of truth in public affairs? We might start by returning to lessons of the Pauling-Teller debate of 1958—in the era before anyone was worrying about post-truth. We might establish, say, a series of high-profile national events where well-regarded scientists who disagree with each other on technical, philosophical, and ideological grounds debate the contentious socio-technical dilemmas of our day. Such public performance could help make newly apparent that worldly truth-making is a messy affair, a core function of democratic governance in which, as the man on the TV said, "each of us bears the moral obligation to examine the evidence, draw conclusions from this evidence, and act upon our convictions." We are going to have to demythologize truth before we can reembark on establishing its proper place in our world.

NOTES

1. See https://languages.oup.com/word-of-the-year/2016/. Accessed Dec. 7, 2021.

2. Available at: https://www.archives.gov/milestone-documents/president-dwight-d-eisenhowers-farewell-address. Accessed August 19, 2022.

3. For example, federal non-defense research and development expenditures increased about fifteenfold between 1958 and 2021 in constant dollars (from about $5 billion in FY 1958 to about $75 billion in FY 2022; see https://www.whitehouse.gov/omb/historical-tables/, table 9.7, accessed December 7, 2021); the US science and engineering workforce has grown more than sevenfold during this same period (see https://ncses.nsf.gov/pubs/nsb20198/u-s-s-e-workforce-definition-size-and-growth, accessed December 7, 2021).

4. It has long been recognized that published scientific output grows exponentially (Price 1963); one recent study estimates a doubling time of about nine years (Bornmann and Mutz 2015).

5. For example, the National Oceanic and Atmospheric Administration and Environmental Protection Agency were created in 1970, and the Department of Energy in 1977.
6. For a fuller treatment of these events, see Sarewitz (2009).
7. For example, see Vinay Prasad, "At a Time When the U.S. Needed Covid-19 Dialogue between Scientists, Francis Collins Moved to Shut It Down," *STAT News*, December 23, 21. Accessed June 23, 2022. https://www.statnews.com/2021/12/23/at-a-time-when-the-u-s-needed-covid-19-dialogue-between-scientists-francis-collins-moved-to-shut-it-down/.
8. For example, see https://brownstone.org/ for a sampling of anti-mainstream perspectives on COVID-19 policies by highly credentialed experts.

REFERENCES

Beck, Ulrich. 1992. *Risk Society: Towards a New Modernity*. London: Sage.
Bornmann, Lutz, and Rüdiger Mutz. 2015. "Growth Rates of Modern Science: A Bibliometric Analysis based on the Number of Publications and Cited References." *Journal of the Association for Information Science and Technology* 66, no. 11: 2215–22.
Cartwright, Nancy. 1983. *How the Laws of Physics Lie*. Oxford: Oxford University Press.
Dalkey, Norman C. 1965. "Solvable Nuclear War Models." *Management Science* 11, no. 9: 783–91.
d'Ancona, Matthew. 2017. *Post-Truth: The New War on Truth and How to Fight Back*. New York: Random House.
Douglas, Mary, and Aaron Wildavsky. 1983. *Risk and Culture: An Essay on the Selection of Technical and Environmental Dangers*. Berkeley: University of California Press.
"Fallout and Disarmament: A Debate between Linus Pauling and Edward Teller." 1958. *Daedalus* 87, no. 2: 147–63. https://www.jstor.org/stable/20026443.
Norris, Robert S., and Hans M. Kristensen. 2010. "Global Nuclear Weapons Inventories, 1945–2010." *Bulletin of the Atomic Scientists* 66, no. 4: 77–83. https://journals.sagepub.com/doi/pdf/10.2968/066004008.
Price, Derek J. de Solla. 1963. *Little Science, Big Science*. New York: Columbia University Press.
Porter, Theodore M. 1996. *Trust in Numbers*. Princeton, NJ: Princeton University Press.
Saltelli, Andrea, Gabriele Bammer, Isabelle Bruno, Erica Charters, Monica Di Fiore, Emmanuel Didier, . . . and Paolo Vineis. 2020. "Five Ways to Ensure That Models Serve Society: A Manifesto." *Nature* 582, 482–84.
Sarewitz, Daniel. 2009. "The Rightful Place of Science." *Issues in Science and Technology* 25, no. 4 (Summer), 89–94.

Thompson, Michael J., and Gregory R. Smulewicz-Zucker, eds. 2018. *Anti-Science and the Assault on Democracy: Defending Reason in a Free Society*. Amherst, NY: Prometheus Books.

van der Sluijs, Jeroen P. 2016. "Numbers Running Wild." In *The Rightful Place of Science: Science on the Verge*, 151–87. Tempe, AZ: Consortium for Science, Policy & Outcomes.

Vermeulen, Roel, Emma L. Schymanski, Albert-László Barabási, and Gary W. Miller. 2020. "The Exposome and Health: Where Chemistry Meets Biology." *Science* 367, no. 6476: 392–96.

Chapter 20

Moral Narratives of Technological Change in the Early Green Revolution

Suzanne Moon

The moral dimensions of technological change have drawn attention from many scholars working in science and technology studies (STS). Philosophers, anthropologists, sociologists, historians, and others have examined how moral beliefs, moral agency, and technology become intertwined over time. Yet only a few of these studies theorize morality as carefully or explicitly as they theorize power, for example. In the history of technology, some engagement with moral theory usually appears in studies of engineering or professional ethics (Kranakis 2004; McDonald 2009; Knowles 2003), although Jameson Wetmore (2015) and Edward Jones-Imhotep (2016) are two notable examples in which authors tackle the topic of morality directly. Otherwise, moral behavior or moral dimensions of larger histories are often treated more anecdotally. In this chapter, I attempt a stronger interdisciplinarity, using both history and philosophy of technology to better understand the role of everyday moral reasoning in technological change. This approach draws attention to the vital place of moral reasoning in everyday sociotechnical life and suggests the important role moral narratives and negotiation play in adapting to fast-changing realities.

Bringing history and philosophy together in ways that might satisfy practitioners of both disciplines is famously difficult. Philosophers may find historical investigation overly descriptive and only partially helpful in working through complicated moral questions. Historians reject the idea that their work is "merely descriptive," as serious historical scholarship goes well beyond simple narration. Yet they may characterize philosophical investigation as arid theorizing, unable to offer a meaningful perspective on the messy rough and tumble of real social worlds. In doing so, they may overlook the powerful lucidity offered by philosophers' systematic development of (among other things) moral reasoning and the vital questions that such work raises about the place of morality in sociotechnical orders.

Nevertheless there are ways that the two disciplines can and do speak to each other, as has been especially evident in studies of engineering and technology ethics, where the historical and the philosophical both are crucial to the analysis of ethics in professional life (Mitcham 2020; Winner 1986). If the normative focus of philosophy puts it at odds with the tendency of scholarly history to eschew prescription, historical projects often have an implicit (and carefully constrained) normativity that can serve interdisciplinary conversation. For example, histories that explore the emergence and perpetuation of injustice are often implicitly interested in showing the historicity and thus mutability of such forms of social order, empowering action rather than resignation. Philosophers, for their part, may find a fully textured history too messy to allow focus on central questions they wish to explore. But philosophical investigations are nevertheless critically informed by actual moral lives, and in the case of the philosophy of technology, technologies actually at work in a complex world. There is ample common ground to inform an interdisciplinary discussion.

This chapter draws on both disciplines to explore the moral narratives that circulated during the early years of the Green Revolution, roughly 1965–1975, with specific attention to Indonesia. The Green Revolution is the name given to the wide use of a suite of technologies developed initially in the 1950s, including dwarf hybrid grains, fertilizers, and pesticides, which, under proper conditions, could dramatically increase crop yields for a few staple crops, including wheat and rice. Exploring the early years of the Green Revolution offers a window into the role of publicly circulating moral narratives that tried to make sense of this newly emerging technological system. Over time the Green Revolution had dramatic consequences that touched individuals well outside the relatively small numbers of experts and farmers who used or studied the technologies involved. Yet outside that group of users and experts, most people would have learned of this technology primarily through the media. Journalists and commentators offered both information about the Green Revolution technologies, and moral narratives to explain the deeper social meanings of this dramatic shift in agricultural practice. Although later debates focused mainly on whether Green Revolution technologies reduced hunger or poverty as advocates frequently claimed (Moon 2014), this chapter shows that the moral narratives of the early period were never solely focused on problems of hunger.

Taking a closer look at the moral inflections of public discourse is not only the start of better understanding the diverse concerns raised in this technological system, but also a departure point to look for deeper moral concerns that inform early and later debates. What aspects of Green Revolution technologies did historical actors identify as relevant to moral questions? What was it about these technologies that convinced journalists and writers to engage in moral negotiation about the technology? What patterns of moral responsibility do their narratives about the Green Revolution construct? In what ways do particular moral debates justify compulsion or otherwise structure forms of moral agency? How is the relationship between technological and moral agency implicated in their arrangements? How do these proposed patterns of responsibility reinforce or overwrite existing expectations? A strongly visible common denominator in the moral narratives that emerged around the Green Revolution was a questioning of the proper moral relationship between governments and citizens.

Margaret Urban Walker (2007, 9) suggests that morality should be thought of as "a socially embodied medium of mutual understanding and negotiation between people over their responsibility for things open to human care and response." Although historians have often explored morally inflected debates about technology, especially in work bent on unearthing the ways that technology is embedded in struggles for social and political power, such studies often lack careful attention to the character of moral reasoning in play, failing to question more deeply the nature of the technomoral world that is being constructed in such struggles. Historical studies of engineering ethics by contrast unite the best features of contextualized history and rigorous ethical analysis. Such histories provide ethical insights that can be applied to improve professional practice or inform awareness of ethical issues among policymakers, experts, and others responsible for technical projects (Pariso 2015; Tang and Nieusma 2017). They also provide essential historical context that helps explain why certain moral or ethical arrangements gain priority under particular circumstances.

I argue that historians should look beyond the limits of engineering ethics and its focus on professional technologies to rigorously explore other contexts in which technological activities prompt some form of moral debate about technology, and the historical actors who involve themselves in these discussions. The tools offered by moral and political philosophy can help guide this historical investigation. Studies of "everyday" moral reasoning of the kind that might appear in public outlets, authored by journalists, politicians, activists, and others, provides a broader picture of how technology is integrated into moral lives and how the character of this integration changes over time. Of those who have theorized the relationship between technology and morality in "everyday" ways, the most prominent is STS scholar Bruno

Latour (1992). He has argued that humans routinely attempt to delegate morality to technology when humans are themselves morally unreliable. His example of the "sleeping policeman" ("speed bump" in American parlance), an intentionally constructed hump in a roadway meant to slow down traffic, highlighted how technology could help fix human failures to fulfill accepted moral obligations or duties. Drivers are understood to be obligated to show due care for pedestrians in return for the privilege to drive on a street. If drivers routinely fail in that duty, adding a speed bump will enforce the desired moral behavior. This action of "delegating to the object" enrolls technology, in this case the speed bump, in the work of maintaining moral order. Yet, as Wetmore (2015) has demonstrated in his history of safety belts, delegating moral obligations to an object can be a fraught and difficult process, exposing deep disconnects in presumptively "shared" understandings of what counts as moral behavior, and prompting serious moral negotiations between businesses, consumers, activists, and others who have some relationship to the technology in question. Negotiating the (re)distribution of responsibility between people and objects may raise serious issues involving what moral obligations ought to exist, who they pertain to, and whether technological means are themselves a morally acceptable approach to enforcement, among others. Recognizing that moral life may have material elements is a vital insight. But it is only the start of the analytical task. We should further seek to understand the nature of the moral negotiation that arises in response to technological change, and its connections to deeper moral rifts or dissatisfactions in society. Such investigations can reveal how technology and social or political morality can be co-constituted with technological change.

This chapter explores everyday moral narratives about the early Green Revolution that appeared in newspapers and magazines. Such accounts are interesting because although the Green Revolution became a large enough project to materially affect the lives of billions of people, actual direct engagement with the technology was limited to a relatively small percentage of those people. If it would be incorrect to see the technology as fully invisible to the wider public, the general population of non-farmer consumers would probably have gained most of their knowledge through media reporting. I do not claim that news reports "reflected" public opinion. Still, I do argue that reporting could provide people who otherwise had no experience with this technology moral frameworks to reason with. Since it would take considerably more historical research to determine what effect this reporting had on policymaking, practice, or even public opinion, the modest goal of this chapter is to analyze and contextualize the moral narratives offered to make sense of this technology and consider the role of technical knowledge in these accounts. What was at stake, morally speaking, in adopting Green Revolution technologies from the perspective of historical actors? What were the starting points of subsequent moral negotiations about the meanings of these technologies?

To engage with this history, I draw on Walker's work (2007), which develops what she terms an "expressive-collaborative" theory of morality. She highlights the fundamentally interpersonal character of morality as a set of practices that "show what is valued by making people accountable to each other for it" within complex social orders (15–16). She points out (as many feminist philosophers do) that not all people in a given society have the full range of opportunities to speak or act on moral concerns and that the distributions of moral responsibilities are directly affected by this unevenness in the moral landscape (10). To put it briefly, she calls for philosophical investigation to be more sharply attuned to the actual functioning of morality in human societies, the deeply social process of negotiating how and why to assign, assume, and deflect responsibility.

Walker's call for a detailed understanding of the practices of responsibility resonates with the temporally grounded aims of historical analysis. Moreover, the expressive-collaborative perspective suggests how to interrogate the public debates about Green Revolution technologies, especially with respect to the ways different positions asserted particular patterns of moral responsibility. As we will see, there were a range of different—often mutually exclusive—interpretations of how the technology would or should change or solidify moral responsibilities. Exploring the character of moral claims made at the time highlights both shared moral concerns and points of significant disconnect. How these concerns are or are not ultimately addressed tells us much about the (very likely uneven) moral orders produced by implementing Green Revolution technologies. The public character of these debates also helps us see how certain sociotechnical imaginaries—shared ideas about the desirable or harmful futures that may emerge as a result of a particular set of sociotechnical arrangements—get taken up well beyond communities of technical experts (Jasanoff and Kim 2015).

THE GREEN REVOLUTION IN INDONESIA

In the late 1960s and early 1970s, Indonesia, like many other countries, started experimenting with the combination of hybrid rice varieties, fertilizers, and pesticides that made up the technologies of the Green Revolution. Promising to dramatically reduce (or, by some, to eliminate) imports of rice, these varieties ultimately became a staple of Indonesian agriculture. Although Indonesia's experience has not received full historical treatment, many scholars have investigated the history of the Green Revolution in other contexts (Baranski 2020; Cullather 2010; Olsson 2017). Hybrid dwarf wheat varieties, developed by a team of scientists in Mexico, could produce substantial increases in yields—in some cases as high as 300–400 percent over conventional varieties—provided they received considerable inputs of fertilizers, water, and pesticides. Their unprecedented response to inputs made this technology a risky one and challenging to implement in some places (Cullather 2010). Farmers knew that conventional varieties would die if dosed with so much fertilizer, making education about the new varieties vital. The ability to obtain adequate water might be beyond any given farmer's control. Such risks meant that interested officials or advocates needed to persuade (and sometimes compel) large numbers of farmers to experiment with this new suite of technologies.

The Green Revolution came to Indonesia shortly after the bloody political transition between Indonesia's first president, Sukarno, and army general, and later president Suharto in 1965 (Ricklefs 2008, 273–344; Elson 2001). Faced with political trauma and inflation, Indonesia was on the brink of economic collapse. Among other steps that Suharto's government took to put Indonesia's economics on a new footing was their participation in trials for new high-yielding hybrid rice developed at the International Rice Research Institute (IRRI) in the Philippines. One of the IRRI's early successes, the variety dubbed IR8, was tested on a large scale in the Philippines, Pakistan, India, and Indonesia, and was found by many to be a definitive proof of concept that rice yield increases on a large scale were within reach. For Indonesia, yield increases offered the promise of food self-sufficiency or at least less dependence on expensive international markets. And they were not wrong to be concerned about food. Serious mistakes in food purchasing, which coincided with weather-related crop failures in the late 1960s and early 1970s, precipitated conditions of real and desperate hunger during the period. In the context of 1960s inflation and the debt, Suharto's government found the prospect of reducing the strain that food imports imposed on the national budget appealing.

Green Revolution varieties were systematically tested in Indonesia starting in the late 1960s, building on a history of rice improvement through increased fertilizer use pursued since the late 1950s. The nature of these projects has some bearing on the moral discussions that emerged later. Suharto's New Order Green Revolution programs were ostensibly built on one of the most successful rice improvement projects, the so-called "Demonstrasi Massal" or "mass demonstration" project begun by the Agricultural University at Bogor in 1963 (Hansen 1971; Rieffel 1969). Student extension workers moved to farming villages to live full-time and offer courses on topics like fertilizer use. The mode of instruction was critical. The mix of formal teaching with informal daily interaction helped build trust between farmers and students; farmers were therefore more willing to try the recommended improvements.

The key to their success, according to reports, was the combination of effective technology (in this case, fertilizers) and functional, trusting relationships. The Demonstrasi Massal project was intended to serve as a pilot program for improvement projects that could be replicated across the country.

After the disruptions of 1965–1966, the new Indonesian authorities hoped for large yield increases by scaling up the Demonstrasi Massal approach. The BIMAS program (Bimbingan Massal, which translates to "mass guidance") focused primarily on introducing fertilizers but later included tests of Green Revolution varieties (Collier and Sajogyo 1970). The critical element in the original project's success, the combination of technology and close engagement with farmers, however, proved to be both too time and labor-intensive for the government (Hansen 1971). Organizers neglected the trust-building aspects of the original project and instead focused on fertilizer distribution and simpler, faster modes of instruction. Predictably, trust declined as a result. The BIMAS programs used coercive methods to enroll farmers in the tests, for example, by requiring participation of all farmers in a given area. They eventually spread over more than 100,000 acres of rice land. Those who advocated this approach used the "demonstration effect" of agriculture to justify the use of coercion. In theory, farmers only needed to see yield improvements to immediately adopt the technology, making coercion only a transient requirement. Coercion would provide compliance, and the demonstration created by compliance would produce a willingness to continue without further coercion. Harun Zain, the governor of West Sumatra, noted at a conference: "It is truly the case that farmers will change their habits as soon as they understand the benefits" (*Tempo*, 1973d). Although seeing good results could indeed be encouraging, adoption rarely went this quickly or easily.

A good example of the problems they faced appeared in the so-called *gotong-royong* program for distributing seed and fertilizer. The Indonesian term *gotong-royong* refers to a moral economy of mutual aid. The BIMAS *gotong-royong* program involved enrolling a major non-state, non-Indonesian participant, the Swiss company Ciba, to compensate for the limitations of Indonesia's over-stretched bureaucracy. Rather than distributing seeds and fertilizer through the government, Ciba distributed seed and fertilizer packages in predetermined amounts on the state's behalf. They also provided aerial pesticide spraying to compensate for farmers' unwillingness to use pesticides in the unprecedented quantities the new technologies required (Arndt 1968; Mears 1970; Lansing 1991). The package approach attempted a technological fix to make up for the loss of farmer education and the intensive informal engagement that would otherwise have been needed to gain their cooperation in trying this unusual new technology. By prepackaging seeds and fertilizer in the required amounts and forcing pesticide spraying, they expected farmers would simply follow the procedure implied by the contents of the packages. The project proved a dramatic and costly failure. Ciba did better at distributing fertilizer than previous programs, but the packages of inputs were not necessarily suitable to all conditions. Pesticide spraying was largely ineffective because it was done without regard for the different timing of rice grown in different areas. Meanwhile, farmers refused to view the packages as required inputs and often sold what they saw as "excess" fertilizer to local agricultural estates that grew sugar, tobacco, or other export crops (Arndt 1969; Hansen 1972).

The rural sociologist often regarded as the father the Indonesian Green Revolution, Sajogyo (who, like many Indonesians, went by one name), noted that because the parameters of use of these new varieties were so different from any that farmers had experienced, it was difficult to convince them that such advice, coming from a distant authority, could possibly be reliable (Collier and Sajogyo 1970; Palmer 1976). And the fact that these packages were poorly adapted to some local conditions drove that conclusion home. Collier and Sajogyo reported in 1970 that fields planted with IR5 and IR8 yielded no more than 60 percent of what was expected, a finding confirmed by others (Kolff 1971). After the first year, when poor yields caused farmers to default on their loans, the Indonesian government did not give up. In 1971, officials suggested that the program required coercive methods, although, as they put it, not "overtly" coercive (Hansen 1972). Sajogyo noted that successful demonstrations—when they happened—tended to be most effective in spreading the technology (Collier and Sajogyo 1970).

Although it is easy to read farmer pushback as resistance, what happened is more complicated. Was farmer skepticism related to the suspicion that state actors or distant foreign companies were not competent? Were they frustrated more by methods that compelled their participation or by the unexpected behavior of these new technologies? Collier and Sajogyo's study of the BIMAS areas suggests that these programs were not wholly uninteresting to farmers. They reported in 1970 that farmers in BIMAS areas did use more fertilizer than their non-BIMAS compatriots, and areas planted to Green Revolution varieties grew steadily throughout the 1970s. If it was not the instant success that bureaucrats hoped for, it ultimately opened the door for wider use of Green Revolution technologies around the region. Such successes, however, never fully resolved a central moral issue raised repeatedly in the reporting on the Green Revolution, the question of whether compulsion was justified or justifiable.

REPORTING IN THE INTERNATIONAL ANGLOPHONE MEDIA

This background on the Indonesian experience with the Green Revolution is helpful context for the moral debates that arose in newspapers and journals from the late 1960s to the 1970s. I will focus on three different kinds of publications, each of which offered a distinct moral interpretation of Green Revolution technologies at this time. The first category includes high-profile publications with an international readership, including the *Economist* (United Kingdom), the *Far Eastern Economic Review* (Hong Kong), and the *New York Times* (United States). Although other non-anglophone publications would reward attention, I chose these three as widely read news outlets that drew educated readers interested in international business and politics for this case study. The second category contains just one publication, the hybrid public affairs/scholarly journal *Bulletin of Indonesian Economic Studies* (henceforth *BIES*). Published in Australia and featuring Australian and Indonesian economists and public intellectuals, it focused on expertly written, broadly accessible analysis and commentary on Indonesia's economic development. Finally, I explore reporting in two Indonesian outlets, including the news magazine *Tempo*, which originally modeled itself on *Time* magazine and remains a highly respected news magazine to this day, and *Prisma*, a public affairs journal featuring the work of public intellectuals. The readership of both journals consisted mainly of well-educated Indonesians (although *Tempo* had the wider circulation of the two). In all three categories, contributors were professional journalists and experts in fields such as economics and anthropology writing for sophisticated general audiences.

The *New York Times*, the *Economist*, and the *Far Eastern Economic Review* tended to link Green Revolution technologies to population growth and hunger. In the late 1960s, many experts in the developed world were making pessimistic arguments about world hunger, including in Paul Ehrlich's 1968 best seller *The Population Bomb*. They argued that any progress toward economic development in the developing world was being outpaced by population growth (Halfon 2006). Development agencies that followed this line of reasoning had a dual focus on increasing food production and reducing the average number of births, mainly via birth control. As a World Bank report put it (as quoted in the *Economist* 1970), at stake was nothing less than providing a "morally acceptable level of existence" for all people. The idea that this technology can reduce hunger, thus providing a fairer and more humane world, has been one of the primary moral claims backed by advocates of the Green Revolution from its earliest days.

But the moral claims visible in these publications went well beyond framing the Green Revolution as a humanitarian tool. They also outlined specific distributions of moral obligations and even the assertion of arguably new moral relationships that should accompany Green Revolution technologies. Most reporting of this sort portrayed the Green Revolution as a sound technological answer to an urgent global problem. Any failures or hesitancy in the implementation produced a cascade of blame. One strain of analysis assigned blame for such failures partly or mostly to farmers. Journalists variously describe farmers as backward, superstitious, and resistant to scientific expertise or modernity, and therefore major obstacles, not just to their own country's aims, but to global human well-being (*Economist* 1968; *New York Times* 1967a; Etienne 1966; Goodstadt 1966). A parallel to books like Ehrlich's *Population Bomb* portraying population growth and food inadequacy as a threat to the entire world, some journalists asserted that Asian farmers had a moral obligation to the entire human race to adopt this presumptively world-saving technology.

It is worth pausing to consider the magnitude and unprecedented nature of that assumption. Normally, anxieties about the food supplies of other countries would be limited to the implications for certain geopolitical relationships or for providing humanitarian aid to stave off warfare. Criticisms of food policies across borders were nothing new, but the consequences were rarely portrayed as a serious threat to food-secure countries. That a farmer in, for example, Indonesia, owed it to humanity to change their methods of rice production (and their habits of reproduction) was a startlingly broad moral claim.

It is no surprise that "blame the farmer" reporting echoed the civilizing mission narratives of earlier times. Their analysis emphasized the newness of the technology—the term "revolution" was repeatedly picked up. However, they varied in their assessment of how to get Asia's farmers to take these technologies on board. They uniformly neglected to mention the sorts of problems farmers experienced with these technologies, as exemplified by Indonesia's experiences discussed above. Some baldly recommended coercive techniques to get farmers to adopt the new seeds. They justified their arguments based on an assumption of an insurmountable cultural backwardness among farmers and the perceived urgency of food supply and population growth. For example, the 1966 article "Asia's Needless Hunger" that appeared in the *Far Eastern Economic Review* argued that decisions about how to farm were far too important to be left to farmers, thus making coercion a practically and morally necessary action. The *Economist* article "Miracle, Maybe" (1968) (after offering a decidedly muddled and incorrect account of how rice agriculture works) suggested that agricultural extension agents should trick farmers by downplaying the newness of the varieties, an approach that resembled in its disrespect for farmers' skills the Ciba project in Indonesia. Embracing a moral logic that made Asian farmers responsible for the survival of the human race, these narratives made a case for a significant level of social coercion, a practice generally regarded as anathema within many Western democracies that funded Green Revolution work. Notably, even in later years, when the context of population growth and hunger no longer commanded the attention of journalists, and coercion was more frowned on, one article in the *Economist* entitled "The 35,000 Villages That Know That Growth Works" (1979)—the author actually visited thirty-five villages—still characterized "good farmers" as those who had adopted Green Revolution technologies. Even without explicit coercion, praise still accrued primarily to those who complied with Green Revolution thinking.

Some reporters, whether or not they criticized farmers, also directed blame to Asian governments for a number of reasons. They criticized Asian governments for failing to provide adequate incentives and resources to farmers or for corruption or incompetence in the administration of Green Revolution programs. Notably, few complained about agricultural extension services or commented on how they should proceed. Instead, they focused on decision-making and bureaucratic or political bungling at higher levels (*Economist* 1971; *New York Times* 1970, 1967a,1967b). Governments were criticized for failing to provide enough water, pesticides, or fertilizers to farmers who needed them to succeed with the new varieties. The author of "Asia's Needless Hunger" (Goodstadt 1966) also criticized land redistribution policies that created "uneconomic" parcels of land in the interest of politics rather than a scientific distribution of land in favor of food supplies. Donors too were criticized if they did not support food aid or provided insufficient help to push these projects along. In pieces like this, the need to provide knowledge (or coercion) to implement the Green Revolution technologies vanishes, but the embeddedness of these technologies in wider networks emerges. Here the question of responsibility emphasizes the responsibility of governments and donors to competently go about the work of implementation.

With its focus on high-level government activities (and that of some international agencies), this narrative is more explicit in asserting the existence of reciprocal moral obligations between Asian governments and the "donor nations" who had supplied the technology in the first place. If donor nations did their part by supplying funds, seeds, fertilizers, or pesticides, recipients were obligated to use those donations for Green Revolution work in the ways intended by the donors (Cullather 2010; *Far Eastern Economic Review* 1966a, 1966b; Goodstadt 1966). The pairing of these obligations only makes sense if Asian nations accept a priori the moral reasoning reporters attached to the technologies—that the moral obligation to accept these technologies came from Asia's outsized contribution to the problem.

It is hardly novel to notice that technical aid often came with strings attached. Yet, it is still useful to observe that the obligations reporters asserted took for granted that a failure to live up to these obligations was a failure for all of humanity, not just a failure of good relations between countries. Most reporting assumed that the only morally reasonable response to the existence of Green Revolution technologies was to use them. This way of thinking subsumes all of grain-producing Asia into a global moral community whose composition Asians played no part in constituting, and within which Asians were given both special obligations and a preestablished definition of moral behavior relating to Green Revolution technology. In this narrative, donor nations assumed a right to define the character of the moral obligations attached to this technology in part because of their instrumental role in making the technology possible. The framework offers Asian nations only limited options for moral action: comply or fail the human race.

The power of these assumptions is most highlighted by an article that calls them into question. In an op-ed in the *New York Times* (1968), Takeshi Watanabe, the first director of the Asian Development Bank from 1966 to 1972, turned the argument around, suggesting that Asians had held up their end by making significant yet unacknowledged strides. He argued that the real problem was unrealistic expectations on the part of donors based on donor ignorance of conditions on the ground in much of Asia. Calling attention to donor ignorance is a powerful moral claim; it demands that blame cannot properly be assigned without clearly understanding the Green Revolution technologies under the particular circumstances that pertained to Asia. It subtly rejects the idea that providers of Green Revolution technologies have an unquestionable right to assert the moral parameters of judgment. Rather than seeing this system as an indispensable technical fix that needed only competence, diligence, and finance to succeed, it asserts the importance of local context, context that Asian nations themselves needed to provide. Without rejecting the urgent claims about the global good that would accrue from resolving problems of population growth or food supply, Watanabe points out that farmers and Asian governments were not the only ones who needed additional education to uphold the moral balance of the global community. Rejecting the stark "adopt or fail the world" logic of other narratives, Watanabe intertwined and reinforced both the technical and moral agency of Asian nations.

REPORTING IN THE *BULLETIN OF INDONESIAN ECONOMIC STUDIES*

Reporting that appeared in the *Bulletin of Indonesian Economic Studies* contrasts starkly with that in the news outlets mentioned above. Contributors to *BIES* paid little heed to issues like global hunger or the dangers of population growth, although they did address Indonesia's periods of food crisis in 1968 and 1974. Although the new hybrid varieties were interesting enough in and of themselves to be described in some detail, most contributors focused not on whether or not the varieties should be used but instead on who should decide whether or not they should be used. Commenting specifically on the BIMAS projects described earlier, authors like H. W. Arndt (1969) and Peter McCawley (1973) roundly condemned the project because they provided more fertilizer and pesticides than farmers typically wanted, in their view, wasting money. The position they took was that governments were neither knowledgeable enough nor adequately responsive to farmers' needs to intervene successfully in the complex business of farming. Incompetent administration and large-scale government intervention were, for them, the underlying problems. They saw the deal with Ciba as similarly flawed: just bringing in a private partner would not solve the overriding problem. Their skepticism, widely shared among contributors, was rooted in more than just Green Revolution technologies. During Indonesia's food crises, they criticized the Indonesian government's food logistics organization BULOG (Badan Urusan Logistik) for failing to obtain adequate imports and for their inaccurate knowledge about yields (Panglaykim, Penny, and Thalib 1968, Penny and Thalib 1969).

Despite some similarities with the critiques of government offered in the publications above, the underlying assumptions about desirable moral relationships are strikingly different. Although critics publishing in other outlets also highlighted incompetent or corrupt governments, reports in *BIES* rarely if ever criticized farmers. Authors like Arndt and McCawley, for example, argued that farmers should have the freedom to apply their farming expertise as they saw fit and make choices about inputs without compulsion from state officials. They offered practical reasons to justify this, pointing, for example, to black market sales of excess fertilizer in the BIMAS project. Clearly, farmers would follow their own thinking whatever the government did. And indeed, as most contributors were economists, the explicit focus of most of their critiques was economic efficiency.

Yet underpinning the pragmatic suggestions were some important moral assumptions. The most evident of these was the idea that the Indonesian government (or probably any government) had any right to demand compliance from farmers, no matter how well intended or effective the new technology was. Farmers, in this view, had limited moral obligations to their governments, much less to international donors or all of humanity. Therefore, they were not appropriate targets of blame. They rejected economic coercion as both unjustifiable and unwise. It is worth wondering whether coercion that could be shown to serve economic efficiency would have gotten a hearing, but I do not have sufficient evidence to speculate. The moral obligations in the relationship between farmers and governments were remarkably one-sided. Governments were obligated to tread lightly, if at all, on the private business of its people. Mild financial incentives were acceptable, but forcing material practices on farmers was deemed an unacceptable intrusion into their freedom to operate as they saw fit. In return for a governing environment that allowed them to operate with minimal intrusion in the business of farming, the only implicit reciprocal obligation of farmers was simply to produce effectively for the market.

The economists' position advocated for the efficiency and shared benefits of free markets, a key element of broader neoliberal frameworks. The moral good that emerges is the freedom of farmers to decide whether to use the varieties as intended or at all. This freedom was predicated on an interpretation of farmers as smart, skilled, and knowledgeable—just the opposite of what was evident in the international reporting mentioned above. Yet, just as with the case above, reporting in *BIES* often failed to acknowledge this system's unusual technical qualities, which might have complicated their thinking. As already mentioned, the significant fertilizer requirements and the vulnerability of Green Revolution varieties to pests ran counter to long-standing farmer experiences. The only way any farmer could determine how much fertilizer was too much was either to be educated about the new varieties, preferably in relation to their specific local conditions, or to go through a protracted and probably risky process of trial and error. And although economists criticized government intervention as too coercive, they paid little attention to the real ecological and financial considerations that would limit a given farmer's choices. For example, farmers with poor access to water could not make a go of Green Revolution varieties. In this case government intervention, perhaps in the form of improved irrigation works, would be essential to giving the farmer real freedom of choice.

The issue of intervention became central to the introduction of Green Revolution technologies worldwide. The practical necessity of education or information for farmers to make what economists would consider a "rational" decision is given no attention. Although it does not seem that any of these critics would quarrel with noncompulsory participation in demonstrations, the nature of the technology meant that farmers could not make decisions that were both free and informed without some intervention that these critics found morally problematic.

REPORTING IN INDONESIA

Mentions of Green Revolution varieties by Indonesian contributors in journals *Tempo* and *Prisma* in this period are surprisingly fleeting, given how central these new varieties became. I found few articles that focused in detail on them. Yet by paying attention to wider discussions of government agricultural programs, it is possible to gain some insight into the moral reasoning at work in reporting on food, agriculture, and hunger. Like the contributors to the *BIES*, those reporting in these Indonesian outlets gave little or no attention to problems of world hunger and global population growth (although attention to Indonesian hunger is central). Indonesians were more interested in places within the country that had notably low population densities, sometimes as potential sites of opportunity (*Tempo* 1973b). When Suharto's government introduced a birth control–focused program, it was understood less in terms of overpopulation and more in terms of the economic and quality-of-life benefits it would provide for individual families (Niehof and Lubis 2003). Whether journalists implicitly rejected the moral framing that appeared in the international anglophone press is impossible to say with the sources at hand. At the very least, however, we can speculate that those moral claims (of which Indonesian journalists would certainly have been aware) did not carry enough weight or interest to find their way into news articles that targeted Indonesian audiences.

Indonesian commentators focused their critique on government actors rather than farmers, but their critical moral logic was unique, especially during years of serious rice shortage in 1968 and 1973. Rather than being concerned with freedom, as contributors to *BIES* were, they emphasized failures on both sides to fulfill duties, and resulting dysfunctions in reciprocity, especially where government failed to provide the competent handling of agricultural issues that was seen as their duty to the people of Indonesia. Logistics rather than issues of production or yield drew heavy attention from Indonesian journalists in *Tempo* and *Prisma*. Indonesian journalists repeatedly called BULOG's technical competence into question in terms very nearly as harsh as those applied to Asian farmers in the pages of the *Economist* or the *New York Times*. The failings of BULOG in properly estimating the rice supply mentioned earlier were taken up in *Tempo*, just as they had been in *BIES*. Reporting also considered problems in the quality and safety of the rice provided and the government's failure to protect farmers from rapacious middlemen—an old but ongoing problem (*Tempo* 1972, 1971, 1973c, 1973e). Although critics certainly leveled accusations of incompetence, the strongest critique emphasized the government's failure in their duty to care for the people of the nation, a duty they implicitly asserted as the foundation of legitimate governance, even in country under authoritarian rule (*Tempo* 1971, 1972). That this problem was an issue of dysfunctional reciprocity becomes clear when reporters interview or quote government sources. BULOG officials blamed farmers for a dishonest self-reporting of yields, which left BULOG with bad information and harmed the nation. They cite this as a failure of farmers to live up to their obligations as citizens. Farmers were motivated to underreport yields to relieve tax burdens or loan payments, an act met with sympathy by many reporters. Reporters for their part described such farmer "corruption" of the agricultural systems with no subtext of criticism or blame—quite the opposite. They tended to portray subversion by farmers as either inevitable or as a reasonable response given the government's failure to live up to their basic obligations. Goenawan Mohamed, *Tempo*'s editor in these years, was a frequent critic of the authoritarian New Order government, pointing out the myriad ways that it failed the Indonesian people. It is no surprise therefore that *Tempo*'s journalists emphasized the failures of the state (Steele 2005). Still, their careful reporting of opposing perspectives highlighted not just government failures but also a dysfunctional moral relationship rooted in the problems of authoritarianism that underpinned the practical problems of food availability in Indonesia.

BULOG officials also (indirectly) criticized Suharto's practice of siphoning off the best quality rice to the military. Reporters gave special attention to the failures of the state (especially in the context of growing authoritarianism) as the starting point. If the state failed to deal honestly and fairly with its people, did citizens owe them honesty and fair dealing in return? This question was posed by Indonesian writers in both Indonesian outlets and in *BIES*. (Panglaykim, Penny, and Thalib 1968; Penny and Thalib 1969; *Tempo* 1973c, 1972). It was already clear by this time that Suharto's government reinforced its power at the expense of its citizens. Complaints about government interference in agricultural cooperatives, too, suggest deep unhappiness with the relationship between the state and citizens. Cooperatives were constitutionally affirmed institutions in the Indonesian economy, designed to encourage bottom-up, self-driven development. Suharto's leadership interfered in the operation of cooperatives to ensure political loyalty and to shift the focus of cooperatives to match his administration's economic goals. One *Tempo* article (1973a) used the phrase *Belanda minta tanah* to refer to these takeovers—a phrase from the colonial era that roughly translates to "the Dutch always want more and more," or "give them an inch, and they'll take a mile" (Arifin and Nasution 1981; LP3ES 1981). Deploying a phrase that calls to mind the oppressive colonial past could hardly be anything but criticism of government overstepping its boundaries and acting in ways that violated the very moral foundation of state authority. Notably, this bad behavior was not merely a result of the state's size or habits of intervention: it was the context of authoritarianism that made this a problem.

To understand the contrast between perspectives in the Indonesian press and those from *BIES*, both critical of government interventions, consider their respective critiques of fertilizer distribution for Green Revolution projects. *BIES* focused on the government forcing more fertilizer on farmers than they wanted. *Tempo,* however, gave more attention to routine problems of access. Because distributors operated only out of regional government centers, farmers who wanted or needed fertilizer had to incur travel costs, sometimes considerable, given Indonesia's weak rural transport infrastructure at that time. Travel increased the real price of fertilizer to the point that it was no longer an economical choice for many farmers, undercutting the incentive. The failure of the government to live up to their obligations put cooperation, and therefore the functioning of the state, in jeopardy. *Tempo* argued for more accessible distribution, sensitive to farmers' limitations and needs. Therefore, rather than seeing the problem as one of "bloated" government, *Tempo* suggested that there needed to be more government actors, in more places, that were better able to provide the services that their citizens needed. Only then could one fairly expect farmers to take up the new production technologies that the government wanted them to. In other aspects of food supply, they make similar requests for more and better government activity, including computerized tracking of yields, better transport, and more expert administrators. They emphasized the real-world consequences of failure to take these actions: long lines for rice rations, malnutrition, and death (*Tempo* 1972, 1973c).

Addressing the problem of hunger within the Indonesian context was neither about systematically increasing yields nor was it about shrinking government. Instead, it was about resolving the underlying dysfunction that undercut the expected reciprocity between the state and its citizens, a dysfunction reporters asserted as rooted first and foremost in the state's failings. Whether these failures resulted from corruption or incompetence, the solution was not less government, nor yet a technological fix coming from outside the country, but instead the reinvigoration of a properly reciprocal relationship between the government and its people.

CONCLUSION

This exploration of moral reasoning in early reporting about the Green Revolution offers useful insight into the narratives offered to interpret the meaning and value of this technological system. One striking result shows that the moral stakes of this technology, as interpreted by non-experts, were never solely centered on its value for reducing hunger. The claim about the moral necessity of Green revolution technologies (and, assuming the technologies were taken up, their moral value) was only sometimes and for some commentators focused on its relation to hunger. The more widely shared moral concern was less about hunger per se than it was about sorting out how these technologies affected the moral obligations and rights of governments and their farmer-citizens. In particular, these accounts betray serious disputes about the place of coercion in connection with a new technology whose value can only be seen if deployed on a large scale. Outlets like the *New York Times* and the *Economist* frequently argued that a greater good justified coercion of farmers. Contributors to *BIES* flatly rejected any form of coercion and posed a minimal set of obligations between governments and farmers as the only justifiable way to introduce new technology. Indonesian accounts subtly suggest that projects implemented without due attention to reciprocity between the government and farmers amounted to coercion—but that a functional relationship, in which governments live up to their obligations, would make cooperation likely and intervention in farmers' work justifiable.

Although more research is needed, these moral positions offer us a useful context for understanding the later moral critiques of Green Revolution technologies as they emerged. In later years, one significant concern with the Green Revolution was whether the benefits of the technology were evenly distributed. As it became clear that not all farmers could benefit from this technology, and even that some farmers would be harmed, for example, by being unable to compete with larger landowners or through redirection of resources to other farmers, the question of whether Green Revolution technologies really answered problems of hunger became more urgent. Coercion seemed more thinkable when reporters treated Green Revolution technologies as an undoubted common good, necessary for humanity. In this way, the efficacy of technology is bound up with the permissibility of coercion. If the strident claims of *New York Times* reporters that coercion was essential to save the world lost credence, notions of minimal government involvement turned out in practice to be tantamount to rejecting the technology altogether. Introducing the Green Revolution always involved top-down involvement of one kind or another, and thus the thorny moral question of coercion has never ultimately disappeared from these debates.

In making this analysis, I was inspired by Walker's framing of morality as expressive-collaborative, emergent from social negotiation, a standpoint that I find especially valuable when considering relatively new technologies. The reporting explored here does not offer a full historical picture of that negotiation. Still, it does provide a sense of the diverse ways that Green revolution technologies (and debates about technologies to provide food and prevent hunger) were enmeshed in a particular set of moral narratives that were not in themselves limited to questions about hunger. Each perspective is partial, betraying significant limits in knowledge about local circumstances that affected implementation or about the technological system itself. Notably absent in each discussion are the voices of farmers themselves, who are never interviewed and instead are spoken for, although to be sure the agency of farmers is given a more prominent position in *BIES*, *Tempo*, and *Prisma*, than in the *New York Times*, the *Economist*, or the *Far Eastern Economic Review*. These rhetorical choices—witting or not—worked to constrain or steer the moral conversation. Each suggests a very different form of moral order; each turns on the unresolved problem of coercion. Although this chapter cannot demonstrate that moral perspectives offered in magazines, newspapers, and other media outlets directly shape the outcome of the Green Revolution, a historically interesting question, the mere presence of such accounts arguably shaped the narratives that wider publics used to think about these technologies. For those with no personal experience with the technology, it might be their only source of information at all. Such everyday moral reasoning may emerge within specific historical contexts but may continue to shape understandings of technology even when circumstances have changed. They may also reflect deep-seated unresolved moral conflicts whose foundation goes beyond the specific technology in question. Paying attention to the process of moral negotiation may therefore reward both historians intent on understanding the ways that sociotechnical orders come into being and philosophers who may find the temporal context valuable as they consider how best to intervene in public moral reasoning about technological change.

REFERENCES

Arifin, and Muslimin Nasution. 1981. "Dialogue: Cooperatives and Government Participation." *Prisma* 23: 22–26 (English language edition).
Arndt, H. W. 1968. "Survey of Recent Developments." *Bulletin of Indonesian Economic Studies* 4: 1819.
———. 1969. "Survey of Recent Developments." *Bulletin of Indonesian Economic Studies* 5:12–13.
Baranski, Marcus. 2020. *Globalizing Wheat: Success and Failure of the Green Revolution*. West Lafayette, IN: Purdue University Press.
Collier, William L., and Sajogyo. 1970. *Preliminary Analysis of Rice Farmers in 37 Villages*. Bogor: Agro Economic Survey, Dept. of Agriculture.
Cullather, Nicholas. 2010. *The Hungry World: America's Cold War Battle against Poverty in Asia*. Cambridge, MA: Harvard University Press.
Economist, The. 1968. "Miracle, Maybe." October 19, 1968, 54.

———. 1971. "Miracle Rice." March 27, 1971, 85.

———. 1979. "The 35,000 Villages That Know That Growth Works." July 14, 1979, 48.

Ehrlich, Paul R. 1968. *The Population Bomb*. New York: Ballantine Books.

Elson, R. E. 2001. *Suharto: A Political Biography*. Cambridge: Cambridge University Press.

Etienne, Gilbert. 1966. "Scratching a Living." *Far Eastern Economic Review*. April 7, 1966, 17–19.

Far Eastern Economic Review. 1966a. "Secret Conclave." September 15, 1966.

———. 1966b. "Strings Needed." February 24, 1966.

Goodstadt, L. F. 1966. "Asia's Needless Hunger." *Far Eastern Economic Review*. March 24, 1966, 559–60.

Halfon, Saul. 2006. *The Cairo Consensus: Demographic Surveys, Women's Empowerment, and Regime Change in Population Policy*. London: Rowman & Littlefield.

Hansen, Gary E. 1971. *Episodes in Rural Modernization: Problems in the Bimas Program*. Honolulu, HI: East-West Technology and Development Institute.

———. 1972. "Indonesia's Green Revolution: The Abandonment of a Non-Market Strategy toward Change." *Asian Survey* 12 (November 1, 972): 932–46. https://doi.org/10.2307/2643114.

Jasanoff, Sheila, and Sang-Hyun Kim. 2015. *Dreamscapes of Modernity: Sociotechnical Imaginaries and the Fabrication of Power*. Chicago: University of Chicago Press.

Jones-Imhotep, Edward. 2016. "Malleability and Machines: Glenn Gould and the Technological Self." *Technology and Culture* 57, no. 2: 287–321.

Knowles, Scott. 2003. "Lessons in the Rubble: The World Trade Center and the History of Disaster Investigations in the United States." *History and Technology* 19, no. 1: 9–28.

Kolff, John. 1971. "The Distribution of Fertiliser." *Bulletin of Indonesian Economic Studies* 7: 56–77.

Kranakis, Eda. 2004. "Fixing the Blame: The Quebec Bridge Collapse." *Technology and Culture* 45, no. 3: 487–518.

Lansing, John Stephen. 1991. *Priests and Programmers: Technologies of Power in the Engineered Landscape of Bali*. Princeton, NJ: Princeton University Press.

Latour, Bruno. 1992. "Where Are the Missing Masses? The Sociology of a Few Mundane Artifacts." In *Shaping Technology/Building Society: Studies in Sociotechnical Change*, edited by Wiebe Bijker and John Law, 225–258. Cambridge, MA: MIT Press.

LP3ES (Institute for Economic and Social Research, Education and Information). 1981. "Cooperatives: Search for Self-Reliance, a Field Report." *Prisma* 23: 27–63.

McCawley, Peter. 1973. "Survey of Recent Developments." *Bulletin of Indonesian Economic Studies* 9: 1–27.

McDonald, Allen J. 2009. *Truth, Lies, and O-Rings: Inside the Space Shuttle Challenger Disaster*. Miami: University of Florida Press.

Mears, Leon A. 1970. "A New Approach to Rice Intensification." *Bulletin of Indonesian Economic Studies* 6: 106–11.
Mitcham, Carl. 2020. *Steps toward a Philosophy of Engineering: Historico-Philosophical and Critical Essays*. Lanham, MD: Rowman & Littlefield.
Moon, Suzanne. 2014. "The Green Revolution." In *Ethics, Science, Technology and Engineering: A Global Resource*, edited by J. Britt Holbrook, second edition, 413–16. New York: Macmillan.
New York Times. 1967a. "Food Crisis Engulfs Billion and a Half Rice Eaters." January 22, 1967.
———. 1967b. "The Failure of Philippine Rice: A Bumper Crop of Paper Plans." January 20, 1967.
———. 1970. "A Decade of Disappointment for Asia." January 19, 1970.
Niehof, Anke, and Firman Lubis, eds. 2003. *Two Is Enough: Family Planning in Indonesia under the New Order (1968–1998)*. Leiden: KITLV Press.
Olsson, Tore. 2017. *Agrarian Crossings: Reformers and the Remaking of the US and Mexican Countryside*. Princeton, NJ: Princeton University Press.
Palmer, Ingrid. 1976. *The New Rice in Indonesia*. Geneva: United Nations Research Institute for Social Development.
Panglaykim, J., D. H. Penny, and Dahlan Thalib. 1968. "Survey of Recent Developments." *Bulletin of Indonesian Economic Studies* 4: 1–34.
Pariso, Christopher. 2015. "Bhopal and Engineering Ethics: Who Is Responsible for Preventing Disasters?" *Business & Professional Ethics Journal* 34, no. 3: 353–76.
Penny, D. H., and Dahlan Thalib. 1969. "Survey of Recent Developments." *Bulletin of Indonesian Economic Studies* 5: 1–33.
Ricklefs, M. C. 2008. *A History of Modern Indonesia since c. 1200*. Stanford, CA: Stanford University Press.
Rieffel, Alexis. 1969. "The Bimas Program for Self-Sufficiency in Rice Production." *Indonesia* 8: 103–33.
Steele, Janet. 2005. *Wars within the Story of Tempo an Independent Magazine in Soeharto's Indonesia*. Singapore: Institute of Southeast Asian Studies.
Tang, Xiaofeng, and Dean Nieusma. 2017. "Contextualizing the Code: Ethical Support and Professional Interests in the Creation and Institutionalization of the 1974 IEEE Code of Ethics." *Engineering Studies* 9, no. 3: 166–94.
Tempo. 1971. "Perang Melawan Bubuk." March 6, 1971, 39.
———. 1972. "Jerawat-Jerawat Sang Ratu." December 16, 1972, 5–10.
———. 1973a. "Bak Belanda Minta Tanah." June 16, 1973, 4–6.
———. 1973b. "Masalah Penduduk di Propinsi Basah." May 12, 1973, 26–27.
———. 1973c. "Menekan Beban Impor." June 23, 1973, 40.
———. 1973d. "Sumatera Di Bukittingi." June 16, 1973, 18–19.
———. 1973e. "Tengkulak Kakap Sampai Teri." June 16, 1973, 38–39.

Walker, Margaret Urban. 2007. *Moral Understandings: A Feminist Study in Ethics*, 2nd edition. Oxford: Oxford University Press.
Watanabe, Takeshi. 1968. "Asia Needs Help to Keep Hope Alive." Op-ed in the *New York Times*. January 19, 1968.
Wetmore, Jameson. 2015. "Delegating to the Automobile: Experimenting with Automotive Restraints in the 1970s." *Technology and Culture* 56, no. 2: 440–63.
Winner, Langdon. 1986. *The Whale and Reactor: A Search for Limits in the Age of High Technology*. Chicago: University of Chicago Press.

Chapter 21

Momentum, Interrupted

Developing Habits of Discernment in Engineering and Beyond

Jen Schneider

Every ten years, the Western States Communication Association convenes a panel of prominent rhetoric scholars tasked with writing essays on "the status of rhetorical criticism." It is a chance for the field to reflect on itself, on new directions, and on existing problems and questions in rhetoric. As the respondent for the 2020 panel, rhetorician of science Leah Ceccarelli performed a "metacriticism" of these top essays, a rhetorical analysis of the rhetoric of rhetoricians. Seeking to understand the "rhetoric of inquiry" invoked by these fellow scholars, she examined their modes of argumentation, the turns of phrase they used to persuade or connect to the audience, and the patterns of language meant to evoke affect and response (Ceccarelli 2020, 365). In particular, she sussed out their own "terministic screens"—a concept from Kenneth Burke (1968)—in which "when some views are selected, others are deflected from our attention" (367). The terministic screens explored in Ceccarelli's analysis reveal how rhetorical scholars themselves intentionally (or not) adopt, develop, and circulate language that in turn shapes the field of rhetoric and its rhetorical practices.

This metacriticism has stuck with me, partly because Ceccarelli's approach reveals interesting things about how rhetoric as a discipline has changed over time, where some of its major disagreements or fissures may lie, and where it may be headed. I am not formally trained as a rhetorician, but frequently collaborate with those who are, and this analysis provided useful insights to me as a friendly outsider. But perhaps just as important, what resonated about Ceccarelli's approach was that, by paying exquisite attention to the modes of argumentation themselves—the figurative language, the careful verb choice, the turns of phrase—she also pulled back the curtain on the field's current preoccupations, revealing how scholars' rhetorical and discursive choices reflect our concerns, values, and commitments, even when we may not explicitly state them.

In this essay, I borrow from Ceccarelli's approach in seeking to understand some of the motifs that recur in Carl Mitcham's significant body of work in the philosophy of engineering, which, given his influential role, serves as a kind of proxy for the field itself. By attending to the cultural aspects of communication practices, particularly in science and engineering spaces, one uncovers the patterns and modes of argumentation that inform scholarly debates and values in these areas (e.g., see Janse 2017). This essay thus endeavors to use some of the tools of rhetoric and communication to reveal throughlines in Mitcham's philosophy of engineering work, as they may also be throughlines present in the philosophy of engineering writ large. This essay explores a few of the philosophical and practical tensions he grapples with, perhaps leaves unresolved, and then returns to across many essays and subjects of analysis—he jokes that his is a "stumbling" toward a philosophy of engineering (Mitcham 2019, 347).[1] But the stumbles themselves may be revealing of the important questions at the heart of the philosophy of engineering.

There is a slippage in Mitcham's essays between engineering as a profession and its associated technical activities, on the one hand, and engineering as a metaphor for life, on the other. I re-create that slippage in this essay. A metacritical approach to Mitcham's writings offers not just insight into his own thinking as a critic of engineering, but proposes opportunities to reflect more broadly on challenges posed by living with and within major sociotechnical systems—including information systems and energy systems, both of which receive some attention below. In this exploration I rely heavily on the work of another thinker—someone who is not a philosopher or in direct conversation with Mitcham, but who might be considered a kindred spirit of his—the artist Jenny Odell. Odell has written a compelling contemporary treatise on our relationship to information technologies and the natural world, topics that have received Mitcham's attention, though less than I would have expected. My hope is that an analysis of the recurring metaphors or other language forms in Mitcham's work, supplemented by interdisciplinary insights from scholars such as Odell, reveals some accompanying conceptual ruptures or openings that point to important philosophical questions many of us are preoccupied with, even though we are not all philosophers of engineering. They offer a helpful starting point for meditating on the interplay between our built, digital, and natural environments and our attempts to shape, navigate, and reshape those environments. Above all, as I reflect on my own language here, I see that I often use cartographic, wayfinding, habitation, or landscape metaphors to connote efforts to locate how one moves with and through these ideas.

Below I examine three types of metaphors frequently used by Mitcham, which can be grouped into dialectical pairings. One pairing is speed/attention. Speed refers to a key characteristic of engineered systems that constantly entice us to direct our attention elsewhere and "elsewhen"—out of the present moment and into virtual environments and fabricated time-spaces. Adherence to speed drives us out of embeddedness in the present moment, and largely disrupts habits of attention and reflection. Attention, on the other hand, refers to the ability to discern when it is important to stay present, focused on the tasks at hand, but also attentive to larger contexts and movements in the culture, politics, or society. A second pair of metaphors present throughout Mitcham's work is that of alienation/bridging. The former evokes distance, isolation, and atomization, and can be contrasted with the latter, which refers to the ways that we reconnect, return, or rejoin. The third pairing is momentum/interruption: momentum is most frequently expressed by Mitcham in terms of movement, emergence, promotion, and directionality (sometimes echoing progress narratives), while interruption refers to slowing, stopping, turning, and reversing. These three pairings often work together or overlap as a means of navigating and exploring—there are those wayfinding metaphors—what I believe is Mitcham's larger preoccupation, which has to do with how we develop the habits and skills of discernment appropriate to this engineered age.

Discernment involves the ability to notice, evaluate or judge, and then act; various thinkers have equated it with wisdom (e.g., Sayrak 2019) or prudence or *phrónesis* (Aristotle 2002, 1140a25–1140b30; Flyvberg 2001). It is a useful concept for describing the challenges of connecting attention, presence, and action across micro and macro scales. At times, Mitcham's work calls for us to pay much closer attention to the micro—habits of thinking, reflection, design, play, constraining the excesses of the self, and so on. At other times, his work asks us to think historically, sociologically, longitudinally, even politically, and to consider the work of the self in context.

Above all, Mitcham's work invites us to think about how one develops habits of discernment that tell us when to focus on individual behaviors, values, and commitments, when to organize or identify with our particular collective (such as with a profession), and when to think more broadly about our place in politics and society. Discernment also involves developing the skill of judgment: how is action constrained or enabled by cultural, political, organizational, and historical forces? Discernment is the ability to know when to perceive and then to act through a microethics lens, and when it is more appropriate to consider macroethics (Herkert 2005; Mitcham 2019, 304–7). Moving between the two, synthesizing, or bridging them requires discernment. Ignoring the small leads to poor engineering, as in the case of the *Challenger* disaster;[2] ignoring the large can lead to other forms of catastrophe in which engineering is implicated, such as climate change. Individual action matters, sometimes a great deal, and yet it is inadequate to focus solely on individual ethics in the face of the problems confronting us today. This is true for engineers and for many others.

The sections below therefore explore some of the "terministic screens" in Mitcham's work—screens created through dialectical pairings described above. These screens suggest some of the challenges inherent in developing an ethical discernment in relation to technosocial complexity—challenges that are not always neatly resolved. Throughout, I take some liberties to extrapolate from Mitcham's writing about engineering to dilemmas each of us may face as we find our way in and through large technosocial systems. As he puts it, "We are all to some extent engineers, insofar as we design, construct, and operate in the microworlds of our lives" (66).

SPEED/ATTENTION

Speed and attention make for an interesting pairing. Attention metaphors frequently have to do with sight—reflection, insight, focus—and therefore imply an embodiment, a perspectival mode of thinking and being informed by the environment in which an individual perceives, feels, and acts. Speed metaphors generally conjure the machine, external or runaway forces, and technological systems in which it may be difficult to exert agency. Speed can make careful attention seem impossible, or at best partial. The speed of technological change is often in tension with our need to understand what technologies do; what they are for; how they change us, our humanity, and the natural world.; and, of course, whether we want or need them. Technologies appear, often seemingly out of nowhere (though historians know otherwise), and they seemingly disappear from our attention just as quickly, often through their ubiquity and embeddedness. Engineers and engineering may function similarly.

It makes sense, then, that metaphors of recognition and concealment abound in Mitcham's histories of engineering, perhaps as an effort to rectify the tendency of engineers, and engineering, to fade into the background of our histories, philosophies, and cultural analyses. Engineering is frequently "occluded" (2). The term "engineer" appears and disappears in the historical record and cultural discourse (17). Engineering itself, as it has been historically practiced, may be disappearing or dying (218–19). Engineering is "everywhere but not everywhere recognized" (22). Philosophy has paid it "inadequate attention" (53). It requires repeated efforts at "promotion," another common noun (or verb, in the case of "promote") throughout his essays (34, 36, 74, 78, 227, etc.).

The call to see engineers more clearly, and to acknowledge their agency and contributions, and for engineers to also see better—to reflect—are coupled. Mitcham calls on engineers to develop habits of reflection, introspection, and seeing, and argues that those who study technoscience must also pay more attention to engineering, to render it visible. At times, it needs promoting; at other times, it must be held at a critical distance, kept at arm's length for inspection and interrogation—forms of keen attention. Engineering must be held up for recognition, but also held responsible. In the essay "Ethics into Design," Mitcham warns, "The human practice of designing simply as designing can be said to deepen the tendency inherent in all play by exhibiting a marked inclination to distance the designer from self-examination or social responsibility" (49). All play and no (reflective) work dulls the engineer's ethical senses.

The call to pay keener attention to engineering is complicated by the problematics of speed, which is inherent in modern engineering, and in some understandings of modernity itself (e.g., Virilio 1977; see also Wajcman 2015). Speed sweeps up into the machine of "progress" both the engineer and the critic. Speed is often the enemy of reflection; it privileges acceleration and forward momentum. This is clearly articulated in an essay reflecting on the constant churn of software development and the replacement of simple machines. In the essay "Convivial Software," Mitcham draws on the work of one of his great influences, Ivan Illich, to propose three criteria by which we might judge software in terms of its conviviality, the capacity it has to bring us home to ourselves, reconnect us to the conditions of production, and resituate us in community. The criteria are stability, transparency, and simplifiability (134–35). The "con" in "conviviality" suggests a "with"-ness, unlike computerized machines which have taken us on a "trajectory toward living more *away from*" (132).

The typewriter, a trustworthy and reliable machine, offers one example of a convivial technology. Mitcham recalls, "I bought a typewriter when I was in high school and used it continuously, with only minor repairs and cleaning, for 30 years" (134). The shift to electronic word processing offered an improvement upon the typewriter, he argues, but the periodic software upgrades after that did not. In fact, such upgrades required he master new forms of interaction and functionality that were largely unnecessary and destabilizing, offering little by way of improved performance; they were, in essence, alienating.[3] Another key difference between typewriter and word processing software is that one can see how a typewriter works by observing the machine. Ostensibly, with enough time and tinkering, one could repair a typewriter; the average user cannot do so with word processing software, which is "black boxed" (Stahl 1995; O'Neil 2016).[4] And it is difficult to "simplify" a word processing program such that it might mimic the typewriter; the typewriter is an inherently simple machine, and is a good example of a technology in which there is a "right to repair" (Hanley et al. 2020). The word processor is not easily customized, made simpler, if one prefers, or repaired. Presets on many types of software are rigid and may not be removed or easily adjusted, options ostensibly lost in service to the promise of ease and convenience. Mitcham ends this essay by reflecting on the "speed trap" inherent in most modern electronic technologies. Of the frequent "updating" of software he asks, "Who wants to be going so fast that it becomes impossible to live in the present? Why can't updating and innovating be practiced with dedication to stability instead of creative destruction?" (137)

It would be easy to dismiss these reflections on typewriters and the speed of change as those of a Luddite shaking his fist at software engineers. But remember that Mitcham is not anti-technology, nor is he anti-engineering. In fact, Mitcham's "Convivial Software" critique—published in 2009—presages much of what Jenny Odell argues in her bestselling book *How to Do Nothing: Resisting the Attention Economy* (2019). In this book, she addresses the question of how we are to live in a digital society, which has proven to be deeply alienating, both from lived, embodied social connection and from how we experience the natural world. But she is not a Luddite, either. Living and working in the heart of Silicon Valley, Odell is an astute analyst of technological "progress"—she teaches at Stanford University and has been an artist-in-residence at both the San Francisco dump and at Facebook, sites where one is able to observe the waste technology creates. *How to Do Nothing* is a series of essays reflecting on reclaiming one's attention from the many technologies and algorithms created by "Big Tech." These technologies take us *away from*, and create a sense of existence outside of time and space. They can alienate us from others and the physical world, often via attention traps, such as the infinite scroll (Collins 2020) and pull-to-refresh (Hanin 2021). These attention traps also encourage speed and discourage reflection. Like Mitcham, in other words, Odell is critical of technologies that emphasize speed, opacity, and alienation.

Also like Mitcham, Odell discerns between technologies that separate and those that connect. She writes, "I am not anti-technology. . . . Rather, I am opposed to the way that corporate platforms buy and sell our attention, as well as to designs and uses of technology that enshrine a narrow definition of productivity and ignore the local, the carnal, and the poetic" (Odell 2019, xii). This statement of interest shares much in common with the notion of "conviviality"—Mitcham quotes Illich as valuing "autonomous creativity in relation to intercourse among persons, and the intercourse of persons with their environment" (131). Or as Mitcham puts it elsewhere, the "fundamentally ethical impulse . . . to remain connected" (65) persists, even in the face of our "turbocapitalist charged technological present" (102). Illich has clearly and deeply shaped Mitcham's critique of speed and alienation here, and the work of Illich, Mitcham, and Odell all calls us back to our bodies, the relationships those bodies have to other bodies, and to the physical world they inhabit. They invite us to consider how it is we want to live, outside of the exigencies of speed and technological "innovation." They are suspicious of speed, the turbo-charged, and the surface.

Even more than Mitcham, Odell explicitly locates possibilities for conviviality through one's relationship with the natural world. For Odell, a commitment to "bioregionalism" is an appropriate corrective to the alienation encouraged by modern social media technologies. She writes of getting to know plants and animals in her area, of paying particular and sustained attention to them. This is an example of the micro. But it is through the micro that one also starts to understand one's place in the macro. In one instance, Odell follows a creek to a stream to a river, which leads to an understanding of and relationship with the watershed itself. Similarly, she collects rainwater in a jar, and this leads her to learn about atmospheric rivers, rivers that connect the rain falling in her backyard to the Philippines, where her mother is from. Bioregionalism as Odell imagines it shares much in common with conviviality as Illich and Mitcham imagine it—as a commitment to presence, situational awareness, relationship to our tools and environment, responsibility to one another. Odell simply makes explicit the centrality of the natural world in heightening our sense of relationship to the earth and one another.

For these thinkers, a response to the narrow, obsessive habits of speed encouraged by today's information systems is to widen one's view, to step back and pay attention to one's context and relationships. A bioregionalist perspective, like a convivial one, requires focused attention away from the "penumbra" or shadow (202), and toward an intentional shining of light on more wholesome relationships and environs. Odell puts it this way: "It is with acts of attention that we decide who to hear, who to see, and who in our world has agency. In this way, attention forms the ground not just for love, but for ethics" (Odell 2019, 154). Mitcham, for his part, prefers to refer to duty *plus respicere* when he writes about engineering ethics: "The moral dimension of taking more into account is realized when it links engineering practice into general considerations of and reflection on the good" (177). To consider duty *plus respicere* may look like "radical reflection" (98) or "ethical reflection and criticism" (103)—vision metaphors that connote our ability to see things that may have been formerly invisible to us or that are occluded by social relations. Avoid technologies or social arrangements that interfere with our ability to make these connections, to connect with place and the other beings in that place, and to pay attention, these authors advise. Avoid the lure of the penumbra and its tendency to occlude. Adjusting our sights at different times, and adjusting the speed with which we move or are moved, brings different things into relief, or alternatively lets other things blur or be lost. The terministic screens of speed/attention reveal how such dynamics function.

ALIENATION/BRIDGING

A commitment to speed and a sustained lack of attention can often lead one to a sense of profound alienation. Like machine metaphors, rhetorics of alienation have a long history and are common in literature and in modes of speech (Prosser 1968). Such rhetorics emphasize communication errors or breakdowns, estrangement, isolation, loneliness, and insulation. Alienation is a frequent theme or concept in Marxist criticism, modern literature, and science fiction—it is so ubiquitous as to be considered essential for understanding the human experience. But alienation manifests in particular ways for engineers and engineering, because of the design of many engineering programs that de-emphasize the social and emphasize abstract puzzle solving, but also because of engineering's embeddedness in and centrality to capitalist systems. It is notable that even engineering practice itself is frequently compartmentalized, fragmented, or depoliticized.

It should not be surprising, then, that alienation is a frequent motif in Mitcham's philosophy of engineering. Mitcham argues that some of the activities most common in engineering involve "dis-embedded"-ness (77). Many types of engineering require a deep focus on abstraction, representation, miniaturization, modeling, and compartmentalization—what Mitcham refers to as a "world of living artifice" (69). A mythical or stereotypical view of the engineer as lone inventor, asocial and obsessively problem-solving—tinkering, playing, getting lost in the "flow" of design work—certainly persists in the larger culture; this stereotype reinforces the perception of alienation that surrounds engineering. Pragmatically speaking, we know that much engineering education invites students to spend hours on dis-embedded problem-sets, to "plug and chug" (see Downey 2005; Lanning and Roberts 2019) and to resist design work as not being "real" engineering (Downey and Lucena 2003). Both the stereotype and some persistent practices in engineering education suggest a particular kind of alienation from the end products, users, and contexts of engineering, in spite of the fact the engineering requires awareness of context.

Industrial processes, too, produce alienation through compartmentalization. The design of the Manhattan Project is an illustration of this—groups of scientists and engineers were assigned to work on different parts of the bomb, without any one group knowing what another group was working on.

Compartmentalization was a feature of the project, meant to ensure security but also efficiency via hyper-focus, free from ethical concerns about how and whether the bomb might be used, or even, in some cases, knowledge that one was working on a bomb. In a discussion of the ethical reckoning scientists and engineers underwent following the dropping of the bomb, Mitcham notes that "the experience of Einstein and others that there is a temptation within science to forget human beings and their fate is also surely a criticism of technoscience itself" (102; see also Thorpe 2017).

When we become too engrossed in certain types of model-play or rote habits of work, or too caught up in gadget-making and gadget-making cultures, or when our engineering is shaped by industrial cultures which separate us from the end-uses of our artifacts, we risk a deepening distancing from "self-examination or social responsibility" (49). Or, as Mitcham puts it in a pithy note of caution: "Do not let miniature making become so miniature that it ceases to reflect and engage the real world" (49). Miniature making, or what we might think of more commonly as modeling, is integral to engineering, but the miniature does not perfectly simulate the world. We must keep both the model and the thing it represents in our sights.

The key is to develop awareness across scales. This skill is frequently represented in Mitcham's essays through the metaphor of the bridge. In physical terms, bridges (which are frequently designed by engineers!) are structures meant to connect places that are divided, usually by bodies of water but also by other natural or human-made divisions such as crevasses and those caused by busy thoroughfares. They thus symbolize connection, ease, and efficiency. But bridges are also marked by their difference: I am on land on one side of a bridge. As I cross the bridge, I am suspended between worlds, so to speak. Once I arrive at the other side, I am back on land. The bridge therefore literalizes a sense of suspension or disorientation, which may be resolved once the bridge is crossed. Similarly, the "bridge" in music provides a moment of contrast between parts of a song that are otherwise similar; the difference is pleasing because we are taken away from the familiar for a moment, knowing we will be returned to it on the other side (though we may be changed as a result of crossing over). We are invited to notice the contrast. Figuratively speaking, bridge metaphors may encourage identification—the phrase "bridging the gap" comes to mind—or may emphasize division (that is a "bridge too far"). Bridges can represent obstacles or challenges that must be overcome, and which can be overcome, often with the help of technology; they can also connect us to places that were heretofore inaccessible. The "bridge" is, in short, a remarkable, flexible, and powerful metaphor.

Engineers (and the rest of us) need bridges to move between sustained, focused attention and the larger contexts in which we live and work. Bridges help us to connect specific or compartmentalized design work, for example, to the larger context in which that design will eventually be embedded. We also need conceptual bridges that allow us to reflect on the ethical dimensions of that design and its larger context. Engineers must find ways to "remain connected to social bonds and the limitations of the human connection . . . [and] to what is pragmatically known about the world" (65). Mitcham alternately argues that bridges between engineering and philosophy are needed on the one hand, and perhaps are futile on the other. As the creators of "microworlds," it "makes sense for [engineers] to build bridges to philosophers (who already to some degree practice engineering)" (66). Later Mitcham asks whether "philosophy is important to engineering in a way that would make it worthwhile to construct a bridge between the two" (105). He notes that such interdisciplinary efforts at bridging are difficult (254) and at last asks, "Why should engineers be philosophers?" (259). Perhaps this is an example of Mitcham "stumbling" around with the philosophy of engineering, seeking solid ground of his own.

Nonetheless, we need bridges for the respite they provide us from our familiar ways of seeing, and for the new perspectives they grant us; again, it should be no surprise that attention metaphors emphasizing reflection and introspection (metaphors of sight and seeing) are common in the work of a philosopher focused on engineering ethics. Reflection granted by bridging permits us to see connections and may promote relationships. Divisions created by technological development, such as between producers and consumers, "can only be bridged or reunited through the systematic development of ways to relate the different worlds" (124). It is through the cultivation of ethical practice, such as committing to a "public good," that such alienation between engineering and users, or technology and citizens, can be overcome (129). Bridges provide new vantage points, connect disparate worlds and worldviews; they alter perspectives. All are necessary for engineering *plus respicere*.

It occurs to me that like our information systems, our energy systems, also, promote alienation that must be overcome with bridging. On a research trip to Europe many years ago I attended an art installation whose aim was to make us more aware of (to see, literally) our habits of energy consumption by making the invisible, visible. For example, one designer had created a power cord that visibly showed when current was flowing through it. This small change—making the cord casing transparent—invited us to encounter our typical habits of making energy production, distribution, consumption, and pollution invisible. We could see that even when an appliance was turned off, it still drew current from the cord—a "vampire device." In the United States, at least, our power cords are almost always opaque, and so we are not confronted with the ways we passively consume, and therefore waste, electricity. Electricity consumption as vampire takes a bite out of us in the form of increased costs, decreased efficiency, and a lot of pollution. Many of us remain disconnected or alienated from these externalities in a number of ways.

This is a minor example. But it points to the many ways in which our energy systems—like most of our sociotechnical systems—are designed to disappear into the background such that we do not interact with them in any conscious way, except when they break (Bowker and Star 1999). The products of engineering are thus meant to be naturalized, to fade into the background because they join seamlessly with our everyday functioning. They come to seem natural because they are integrated so smoothly with us and our behaviors, and as such, they functionally disappear, even though they shape our behaviors, just as we shape them. "Because engineering does not so much produce knowledge as materials, artifacts, and physical processes that merge into the fabric of material culture and social order," according to Mitcham, "its cultural influences tend to be more hidden, even while its physical impact is ultimately more pronounced" (21). This seamlessness can have unintended consequences, such that we cease to question whether our technologies really are necessary or beneficial. Sometimes bridge moments—chances to occupy new and unfamiliar vantage points—encourage us to see things differently, to observe anew. Bridges need not be used just for smoothing things over. Technological or design choices that create opportunities for conceptual bridging between action and consequence can promote ethical reflection. Such deliberate pauses for reflection and suspension might bring into sight systemic effects, such as the increasingly threatening climate change threats caused when we burn fossil fuels.

Furthermore, it is only for some that energy infrastructure and byproducts have remained invisible. Work by scholars such as Gabrielle Hecht (2012), Winona LaDuke (2002), and Gwen Ottinger (2013) make clear that many suffer if they are unlucky enough to live near, say, a uranium mine or an oil refinery. For these frontline communities, energy production is the opposite of invisible: it is hyper-present. Indeed, until very recently, American energy production (and industrial production in general) has been concentrated in "sacrifice zones" that have had uneven and unjust impacts on communities, impacting low-income areas and people of color disproportionately (Lerner 2012).[5] Wealthy Americans are often able to block the development of large-scale energy production facilities near their living and recreation spaces, whether they are oil refineries or offshore wind farms. A real challenge for environmental and energy justice advocates has been to make the means of production (and distribution, consumption, and waste disposal or pollution) visible to those who are otherwise alienated from that production. Recognizing the need for bridges between the micro and the macro, and knowing when and how to build and cross them, is essential to cultivating discernment.

MOMENTUM/INTERRUPTION

This last section returns to this essay's beginning, with a set of metaphors that toggle between the organic and the mechanical. Mitcham's historical essays in particular are rife with language describing collective or emergent momentum, which is primarily interrupted or slowed through active reflection and intentional objection. His momentum metaphors alternate between biological metaphors invoking "emergence" or evolution, and mechanical metaphors, including the "machine" and the "dynamo."

On the one hand, verbs such as "emerge" and "arise" appear dozens of times throughout Mitcham's historicizing of engineering, engineers, engineering projects, and engineering organizations (148, 156, 167, 222, etc.). "Emergence" language implies a natural arising or occurrence, one that occurs at the systems level, and that is directional, aiming toward more complexity, implying "progress," and defying easy narrativization. According to McAllister, "In biological systems, emergence occurs in species evolution [. . . and] in earth history and climate processes central to Anthropocene conditions, in which collective, micro-scale, human behaviors produce macro-scale, long-term planetary effects" (2020, 81). Emergence implies movement, momentum, and directionality, but defies our ability to locate agency and narrative arcs—where does emergence begin, exactly? When does it end? For example, in the essay "The Importance of Philosophy to Engineering," Mitcham writes, "Engineers, it will be suggested, are the unacknowledged philosophers of an *emergent* techno-lifeworld" (53, italics added) and "Engineering also emerged as a recognized human activity at a particular point in history—namely the seventeenth and eighteenth centuries" (61). Engineering ethics also emerges at various points in history (e.g., 167 and 207), and at other times it "arises" (145) or is on a "trajectory" (208).

Conversely, engine metaphors such as that of the "dynamo" refer to human technologies rather than natural processes of evolution. Indeed, it is not surprising that "engine" and "engineer" share some linguistic roots.[6] Machine metaphors are so common as to be nearly universal and can be found in language reaching back to the Middle Ages (Glebkin 2013), and the "machine" is both a literal and figurative concern of many science and technology studies (STS) scholars. Machines are handy conceptual figures because they connote something that moves reliably, that has component parts that can be understood, and which are transparent and perhaps even "fixable" (Janse 2017). But like emergence metaphors, engines are machines that seem to have an agency all their own. They are machines that are difficult to slow or stop once they are up and running—automation plus power. Dynamos, in a purely engineering sense, are just generators. But figuratively speaking, they connote a mindless unstoppability—constant forward motion. They have no capacity for reflection or self-awareness; they exist to move, to do their work.

More specifically, dynamos are also "self-exciting"; in metaphorical terms, they do not need human intervention to proceed once begun and seem animated with their own life force and momentum. Once they get going, they become actors in context, in networks, in landscapes, in society. Engineering is itself a dynamo in some of Mitcham's essays; it engages in a *"dynamic disintegrating of premodern ways of making and building"* (18, italics added). Again, the Manhattan Project provides an apt case study for understanding this dynamism. The entire apparatus devoted to building the "gadget" took on dynamo-like characteristics such that the enterprise would have been very difficult to interrupt or knock off-course once begun. For Mitcham, engineering is often like this, an "engine that designs and produces," and which is at the "dynamic core of technological creation" (1). Yet it is engineers who are doing the designing and producing, and engineering is the " 'dominant but masked character' at the heart of technological and social change" (2). The dynamo/momentum suggests that engineering itself is a black box, historically speaking, and is frequently seen as being an "engine of progress"— unstoppable and mighty, yet void of intention and reflection (228).

The usefulness of figurative language that references "emergence" and the "dynamo" is that these metaphors help us describe systems-level change that occurs rapidly and is collective and complex, thus making genealogical work and historical specificity challenging; these are the metaphors of technological determinism. Such metaphors are useful for describing, at a high-level, phenomena such as the fracking boom (also tellingly called the "fracking revolution") in the United States, for example. Since the early 2000s the growth in shale gas production has completely changed the energy profile of the United States. There are debates as to whether the fracking boom has busted, but most utilities are shifting or have shifted away from building new coal-fired and nuclear power plants—the latter largely because of cost—and many are investing instead in natural gas plants and renewables. The systemic impacts of the boom have been enormous and wide ranging.

In material terms, the physical infrastructure of the fracking boom seemingly grew up overnight.[7] I remember driving over the front range of Colorado one year without seeing very many fracking rigs at all, even on the Western Slope, a place accustomed to oil and gas exploration. It seemed like the very next year, frack rigs were everywhere, and along the front range, too, their long goosenecks pecking at the scrub brush, metronomic in their movement, tiny dynamos themselves, generating all that future electricity.

Over the course of the next several years, natural gas production would come to displace coal as a source of the nation's electricity supply, finally dethroning "King Coal" at the top of the heap, a dynamic few could have predicted only years before (Rapier 2017). Who created this sudden and dramatic change in our energy and environmental landscape? Engineers? Oil and gas companies? Environmentalists? The George W. Bush administration? The Halliburton loophole? Petroleum engineering programs? American consumers? Regulators? Politicians? The answer, of course, is yes—all of these, and more. How else to describe such a dramatic transformation of the built environment and its ripple effects into our energy economy than as one that "emerged" due to a complex confluence of factors, entanglements of individual and collective choices, knotted together to create a "dynamic momentum" pushing us toward a natural gas "revolution"?

For a variety of reasons (lock-in, path dependency), interrupting momentum in fossil fuel development has been particularly difficult. Yet we see in the fracking example how the terministic screens examined in the other sections come into play as well. Coalitions in favor of natural gas development effectively argued early on that it was an apt "bridge fuel" to a decarbonized economy, largely because the bridge metaphor has enough strategic ambiguity that it can encompass a variety of competing visions for energy futures (Delborne et al. 2020). These coalitions also had momentum, emergence, and dynamism on their side, and we live with the results of that today, even as the boom has slowed from its most breakneck pace. Fracking-as-dynamo, too, insists that we not pause too long to reflect on its negative impacts but instead focus on its benefits, such as jobs, or cheap electricity. As Mitcham and Jessica Smith put it, again referencing Illich, "The problem with advanced forms of energy production is that they progressively depend on expertise and the alienation of a majority—turning citizens into consumers" (332).

But the ghosts of our alienation leave traces: the landscape in many parts of the West, especially, is littered with fracking wells, some in production, some abandoned. Natural gas is a significant and formidable aspect of our current energy system, and its physical infrastructure is significant, though placed in landscapes not often visible to most energy consumers. Its ubiquity and diffuse-ness also make accountability and answerability for its externalities difficult. Water supplies continue to be periodically threatened, and methane leaks pose serious economic and environmental threats.[8] Areas that were formerly primarily agricultural or residential are now home to essentially what looks like heavy industry. Decommissioning has come to look more like abandonment; momentum carries us forward only in the direction of building and construction; deconstruction does not hold our interest perhaps because it only invokes costs, not pay-offs. These places where the "boom" went off were rapidly and dramatically altered, creating both location-specific problems (e.g., water pollution) and collective and diffuse ones, like climate change (see Thompson 2021).

The changes wrought by the fracking boom represent a "speed trap" of their own. As was the case with the Manhattan Project, there has been such an immense and complex web of social, economic, and political investment in rapid expansion of the technology that it is difficult to locate where individual and even collective resistance could have intervened—offered meaningful opportunities for interruption—on spatial and time scales that could have slowed or stopped the boom writ large; there is also disagreement about whether interrupting the boom would have been desirable. Moreover, it is becoming less and less clear who will be responsible for the industrial detritus and pollution left behind. To return to our atomic bomb example: careful historical work shows that some scientists involved in the development of the atomic bomb did resist, and after the war they organized against atomic warfare (Spencer 1995). Similarly, anti-fracking activists did organize, and continue to organize, against the deleterious impacts of the practice (e.g., Kinchy 2017). Impacts of protest have been uneven—achieving moratoriums or injunctions here and there. COVID-19 led to a slow-down in production, certainly. And yet natural gas production and fracking continue on a large scale, and natural gas has proven rhetorically fungible as well, at times miraculously becoming no longer a fossil fuel or even a bridge fuel, but, in the case of the EU, a "green" and "sustainable" one.[9]

To return to our metaphorical analysis, what might interruption of such momentum look like? Mitcham writes of interruptions in terms of the rise of "nodal events," and "turns" away from existing paths. "Nodal events," "nodal points," or "inflection points" (215, 224) allow opportunities to punctuate narratives of emergence with examples of individual or collective agency and resistance. Disasters happen, whistleblowers blow whistles, engineering ethics "missionaries" go on the stump, organizations revise codes of ethics, new curriculum "turns" toward awareness of political power and its contexts (214–16). A whistle being blown is a particularly evocative metaphor for a punctuation, or interruption. Such interruptions—whistleblowing, monkey-wrenching—can cause the "engine of progress" to run off the tracks. It is possible that such "strategic gestures" may add up over time, eventually accumulating into a movement with real political and strategic power (Bsumek et al. 2019).

Like Mitcham, I wonder if these analyses are not just informative for engineers and engineering, but for all of us caught up in using, building, and sustaining large technological systems, some of which have become monsters demanding our love. The example of how so many contribute labor, data, and information freely to social media companies comes to mind. These questions are not easily answered, as we waver between seeing history as a collection of emergent phenomena on the one hand and as a sequence of events driven or punctuated by human agency on the other. We create historical narratives and meaning out of both. Yet developing a theory of social action that connects individual agency and reflection with the necessity for collective action is pressing, given the scale of the challenges we face.

CONCLUSION

At the heart of developing discernment is the ability to be present at times to one's specific task and at other times to the context that engendered the task itself. This paradox may explain how Mitcham's texts appear at times in thrall to the innovations wrought by engineering, and at other times focused on the ethical failings of engineering writ large. The language he uses to write about his own approach to engineering as a subject invokes that of a pendulum swinging on the one hand, and a heart divided on the other—one a mechanical metaphor, the other a bio-emotional one. In the preface to the collection of his essays, *Steps Toward a Philosophy of Engineering* (2019), he writes, "Although I have been thinking about engineering and technology for many years, I remain unsure of a final judgment and swing between trying to think with and think against engineering" (xiii). Or, as he puts it in "Professional Idealism among Scientists and Engineers," "the divide between optimistic promotion and critical assessment of technoscience is one that runs through all of us, often separating our own hearts and minds" (90). Mitcham's work maps an engineering ethics that requires one to skillfully move between the engineering problem or task, and the problematics of engineering in society. This requires both embrace and distance, promotion and critique. A destabilized or off-balance posture to engineering is a requirement for doing this sort of work. This may be because engineering is itself "schizophrenic" (Mitcham's word) in that it is continually caught in a dialectic between corporate duty and public responsibility (170).

We are invited to grapple with this divide. Mitcham articulates the dilemma as he both reflects on his own experience as a user of technology who is often sympathetic to the engineer as historical figure, and as one who critically interprets the histories, cultures, and failures of engineering. Throughout his work he returns again and again to the value of play, fun, design, games, construction, tool use, miniature or model construction, and world-making (49, 77, 79, 146, 263, etc.)—all of which are essential to engineering. In addition to being a philosopher and academic, Mitcham is also a tinkerer, a builder, and a designer, and he finds delight in the practices of construction and of playing with materials (here is perhaps another connection he shares with Jenny Odell, an artist who tinkers with trash/refuse, turning it into art). This is true of him intellectually, as well, as he moves from disciplinary sandbox to disciplinary sandbox—here history, there philosophy, here policy, there art. His work evinces a sideways delight in high-level technological accomplishments, such as those achieved via space exploration. In my many interactions with him over the years, Mitcham has often been protective of engineering and dismissed any easy jabs that might be made by the humanities and social scientists against engineers. He is unhappy with those who "seem to have mounted cannons on their areas of the philosophy island in order to fire away at selected domains of the engineering world" (54). Engineers, he notes, "are among the most self-reflective of professionals" and should be recognized as such (217).

In other words, an analysis of Mitcham's metaphors suggests that we must be able to *pay attention*, to be able to expand or contract our attention appropriately and at the right time, without becoming overly preoccupied with just one mode of observation. Knowing how and when to do this requires discernment, which takes practice and effort. To put it another way, his work leads us to ask: what habits of mind, what conditions must we cultivate, in order to better pay attention? How might paying attention lead to developing discernment, which is key for guiding ethical action? And in particular, what might it mean to develop discernment as an engineer, one who operates across scales and is charged with constructing our material reality? What prevents us, or prevents engineers, from developing that discernment? In exploring these questions, I hope to build—though perhaps not in an explicitly, canonically philosophical way—on the work of philosophers of engineering who articulate the importance of discernment within the context of virtue ethics (e.g., Harris 2008; Miller 2020; Schmidt 2014). But I also seek to make connections to larger cultural preoccupations with how we develop discernment as humans functioning within large, fast-paced technological and information systems.

The terministic pairings explored in this essay provide some direction. Speed is countered with attention; alienation with bridging; and momentum with interruption. Through attention, bridging, and interruption, we might dull the deleterious aspects of engineering and instead emphasize its pro-social aspects. A more reflective engineering practice counters the impulse to focus primarily on plugging and chugging, the miniature, and the model. It calls us to pay attention to our own agency as engineers and to skillfully move back and forth between the micro and the macro, adjusting our perspectives as we go. Engineering as bridge-building offers another corrective. In material terms, bridging emphasizes contributions to the built environment that facilitate ingenious crossings and movement across obstacles. In symbolic terms, bridge-building valorizes perspective-taking and interdisciplinary connection-making. Bridging enables us to stop and evaluate from new vantage points; it provides opportunities for constructive conceptual destabilization, as bridges allow us to remain suspended between worlds. We can focus not just on making sure engineers know how to build the *thing*, whatever it is (bridge or otherwise), but also on encouraging connections and collaborations between ways of thinking and being, and on developing comfort with being temporarily suspended in space, not having our feet on solid ground.

Finally, there is the importance of creating habits of interruption. This requires that engineers understand the extent and limits of their own agency and develop the capacity to critically understand how organizations and other social structures shape our lives, enabling or constraining action. It requires that engineers understand the many ways they might be seen as "cogs in the machine"—dynamos of the state, or capitalism—but also how resistance, questioning, and interruption are always possibilities, especially in the collective form. There are always opportunities to ask what is engineering *for*, who it serves, what world we are building together.

Wayfinding is not always easy, whether we are engineers or otherwise; we are deeply embedded in and contributing to technosocial systems. It can be disheartening to even try to find our way. The macro view is often overwhelming and discouraging; the micro view, in contrast, is enticing in its simplicity. It seems to offer safe haven in a seemingly apolitical world of model-building, while the macro is messy, politicized, and sometimes frightening. The problems of the world are immense; it is easy to focus on the hopelessness of things. In the conclusion to *How to Do Nothing*, Odell argues, "If you become interested in the health of the place where you are, whether that's cultural or biological or both, I have a warning: you will see more destruction than progress" (2019, 186). Living in almost any city in America, one can see the destructive traces of technological momentum: polluted air, congested freeways, homeless encampments. In rural areas, one sees main streets emptied out, farms in bankruptcy, and technological and social service deserts. In both, one sees more destruction than progress.

Yet there are creative opportunities in that immense destruction, in those macro-level challenges, if we focus on how the micro might be engineered differently. Instead of simply lamenting Mitcham's "expansive disintegration"—where engineering destroys as it builds, Odell calls us to explore what she calls "manifest dismantling," or the intentional deconstruction of what has been built to move us toward a healthier social and environmental future. Her examples include dam removals intent on restoring ecosystems, farmlands painstakingly restored through thoughtful "do-nothing" approaches, and wetlands restored through community partnerships. It is no mistake that "manifest dismantling" is a play on the term "manifest destiny," with all of its implied colonialist ideologies. Manifest destiny—the ultimate terministic screen focusing on speed, alienation, momentum as the United States "developed" the West—was constructive for some and destructive and even genocidal for others. Manifest dismantling, on the other hand, points to the ways in which un-building, or conscious deconstruction, might be used to engineer healthier and more robust social and environmental relationships; to undo some of the damage manifest destiny wrought, not by building more and faster, but by intentional deconstruction and simplifiability. This kind of conviviality is what Odell means by "doing nothing." Engineering, paradoxically, is essential to the success of such efforts. But it must include capacities for reflection, bridging, and interruption if it is to contribute meaningfully.

Perhaps it is enough to say that what Mitcham's work—and Odell's, for that matter—invites us to do is to refuse any too-easy assumptions about what engineering is and what it can do. Engineering is not to be uncritically celebrated, nor unnecessarily damned. We are called instead to explore how power functions in and through engineering, how its emergence and momentum are made to seem natural or inherent, and the ways in which it shapes or deconstructs our social and environmental worlds in both welcome and unwelcome ways. In "The True Grand Challenge for Engineering: Self Knowledge," Mitcham argues that engineers often function as the "unacknowledged legislators of the world" (279). It is hard to imagine a more transformative force in modern human experience than engineering, yet "neither engineers nor politicians deliberate seriously on the role of engineering in transforming our world" (279). Instead, he argues, we should be examining the push to engineer, the push to innovate, the push to build. We should be developing additional habits of discernment: all of us should be having more discussions about the "relationship between engineering and the good life" (285). Such conversations require space, time, and connection. But they are clearly worth the effort.

NOTES

1. All citations to Mitcham's work in this paper refer to *Steps toward a Philosophy of Engineering: Historico-Philosophical and Critical Essays* (2019). It is a compilation of papers and presentations, most of which at least slightly revised, that he has given or composed over the last twenty-five years. Given that the object of this paper is to analyze his *corpus*, and because this book is the easiest text for readers to reference, in-text citations refer to the book, rather than individual essays, and usually only include the page number, unless they are part of a parenthetical citation that includes another author.

2. The *Challenger* was an American space shuttle that exploded shortly after take-off in 1986. The *Challenger* disaster is a well-known case study in engineering ethics because the explosion was caused by the failure of two small "O-rings," a flaw

that had been known to engineers before the launch; their concerns were not heeded. NASA's organizational culture—including the pressure to launch despite warnings about the O-rings—has largely been blamed for the explosion and the deaths of the six crew members aboard.

3. Illich explores the concept of "counterproductivity" throughout his body of work, but especially in *Tools for Conviviality* (1973). In the case of technology, counterproductivity refers to the point at which a technical process or product, or the pursuit of one, works against its original goals. A word processor, to use Mitcham's example, seeks to make document production faster, easier, and simpler. But by adding so *much* functionality, it actually complicates what could be a very straightforward task.

4. See also Borgmann's (1984) discussion of "devices" as contrasted with "focal things" and "focal practices."

5. See also Lylla Younes, Ava Kofman, Al Shaw, and Lisa Song, "Poison in the Air," *ProPublica*, November 2, 2021, https://www.propublica.org/article/toxmap-poison-in-the-air.

6. Christoper McFadden, "The Origin of the Word 'Engineering,' " *Interesting Engineering*, September 5, 2017, https://interestingengineering.com/the-origin-of-the-word-engineering.

7. Bobby Magill, "Fracking the USA: New Map Shows 1 Million Oil, Gas Wells," *Climate Central*, March 27, 2014, https://www.climatecentral.org/news/fracking-the-usa-maps-show-americas-1.1-million-oil-and-gas-wells-17226.

8. See, for example, Dino Grandoni and Steven Mufson, "Biden Unveils New Rules to Curb Methane, a Potent Greenhouse Gas, from Oil and Gas Operations," *Washington Post*, November 2, 2021, https://www.washingtonpost.com/climate-environment/2021/11/02/biden-methane-rule-epa/, and Naveena Sadasivam, "Study: Toxic Fracking Waste Is Leaking into California Groundwater," *Grist*, October 26, 2021, https://grist.org/accountability/fracking-waste-california-aqueduct-section-29-facility/.

9. See Shashi K. Yadav, "Natural Gas is a Fossil Fuel, but the EU Will Count It as a Green Investment—Here's Why," *The Conversation*, February 22, 2022, https://theconversation.com/natural-gas-is-a-fossil-fuel-but-the-eu-will-count-it-as-a-green-investment-heres-why-175867.

REFERENCES

Aristotle. 2002. *Nicomachean Ethics*. Translated by Joe Sachs. Newburyport, MA: Focus.

Borgmann, Albert. 1984. *Technology and the Character of Contemporary Life: A Philosophical Inquiry*. Chicago: University of Chicago Press.

Bowker, Geoffrey C., and Susan Leigh Star. 1999. *Sorting Things Out: Classification and Its Consequences*. Cambridge, MA: MIT Press.

Bsumek, Peter K., Steven Schwarze, Jennifer Peeples, and Jen Schneider. 2019. "Strategic Gestures in Bill McKibben's Climate Change Rhetoric." *Frontiers in*

Communication, August 19, 2019. https://www.frontiersin.org/articles/10.3389/fcomm.2019.00040/full.

Burke, Kenneth. 1968. *Language as Symbolic Action: Essays on Life, Literature, and Method*. Berkeley: University of California Press.

Ceccarelli, Leah. 2020. "The Rhetoric of Rhetorical Inquiry." *Western Journal of Communication* 84, no. 3: 365–78.

Collins, Grant. 2020. "Why the Infinite Scroll Is So Addictive." *UX Collective*. December 10, 2020. https://uxdesign.cc/why-the-infinite-scroll-is-so-addictive-9928367019c5.

Delborne, Jason A., Dresden Hasala, Aubrey Wigner, and Abby Kinchy. 2020. "Dueling Metaphors, Fueling Futures: 'Bridge Fuel' Visions of Coal and Natural Gas in the United States." *Energy Research & Social Science* 61 (March). https://doi.org/10.1016/j.erss.2019.101350.

Downey, Gary. 2005. "Are Engineers Losing Control of Technology? From 'Problem Solving' to 'Problem Definition and Solution' in Engineering Education." *Chemical Engineering Research and Design* 83, no. 6 (June): 583–95.

Downey, Gary and Juan Lucena. 2003. "When Students Resist: Ethnography of a Senior Design Experience in Engineering Education." *International Journal of Engineering Education* 19, no. 1: 168–76.

Flyvberg, Bent. 2001. *Making Social Science Matter: Why Social Inquiry Fails and How It Can Succeed Again*. Cambridge: Cambridge University Press.

Glebkin, Vladimir. 2013. "A Socio-Cultural History of the Machine Metaphor." *Review of Cognitive Linguistics* 11, no. 1: 145–62.

Hanin, Mark L. 2021. "Theorizing Digital Distraction." *Philosophy & Technology* 34: 395–406.

Hanley, Daniel A., Claire Kelloway, and Sandeep Vaheesan. 2020. *Fixing America: Breaking Manufacturers' Aftermarket Monopoly and Restoring Consumers' Right to Repair*. Open Markets Institute. https://www.openmarketsinstitute.org/publications/fixing-america-breaking-manufacturers-aftermarket-monopoly-restoring-consumers-right-repair.

Harris, Jr., Charles E. 2008. "The Good Engineer: Giving Virtue Its Due in Engineering Ethics." *Science and Engineering Ethics* 14, no. 2: 153–64.

Hecht, Gabrielle. 2012. *Being Nuclear: Africans and the Global Uranium Trade*. Cambridge, MA: MIT Press.

Herkert, Joseph R. 2005. "Ways of Thinking about and Teaching Ethical Problem Solving: Microethics and Macroethics in Engineering." *Science and Engineering Ethics* 11, no. 3: 373–85.

Illich, Ivan. 1973. *Tools for Conviviality*. London: Marion Boyars.

Janse, Maartje. 2017. "'Association Is a Mighty Engine': Mass Organization and the Machine Metaphor, 1825–1840." In *Organizing Democracy: Reflections on the Rise of Political Organizations in the Nineteenth Century*, edited by Hank te Velde and Maartje J. Janse, 19–42. Basingstoke, UK: Palgrave Macmillan.

Kinchy, Abby. 2017. "Citizen Science and Democracy: Participatory Water Monitoring in the Marcellus Shale Fracking Boom." *Science as Culture* 26, no. 1: 88–110.

LaDuke, Winona. 2002. *The Winona LaDuke Reader: A Collection of Essential Writings*. Penticton, BC: Theytus Books.

Lanning, Joel, and Matthew W. Roberts. 2019. "Fighting 'Plug and Chug' Structural Design through Effective and Experiential Demonstrations." Paper presented at the 2019 ASEE Annual Conference & Exposition, Tampa, FL. https://peer.asee.org/32839.

Lerner, Steve. 2012. *Sacrifice Zones: The Front Lines of Toxic Chemical Exposure in the United States.* Cambridge, MA: MIT Press.

McAllister, Brian J. 2020. "The Rhetoric of Emergence in Narrative." *Diegesis* 9, no. 2: 80–95. https://www.diegesis.uni-wuppertal.de/index.php/diegesis/article/view/388.

Miller, Glen. 2020. "Western Philosophical Approaches and Engineering." In *The Routledge Handbook of the Philosophy of Engineering*, edited by Diane P. Michelfelder and Neelke Doorn, 38–49. New York: Routledge.

Mitcham, Carl. 2019. *Steps toward a Philosophy of Engineering: Historico-Philosophical and Critical Essays*. Lanham, MD: Rowman & Littlefield.

Odell, Jenny. 2019. *How to Do Nothing: Resisting the Attention Economy*. Brooklyn: Melville House.

O'Neil, Cathy. 2016. *Weapons of Math Destruction: How Big Data Increases Inequality and Threatens Democracy*. New York: Crown.

Ottinger, Gwen. 2013. *Refining Expertise: How Responsible Engineer Subvert Environmental Justice Challenges*. New York: NYU Press.

Prosser, Michael H. 1968. "A Rhetoric of Alienation as Reflected in the Works of Nathaniel Hawthorne." *Quarterly Journal of Speech* 54: 22–28.

Rapier, Robert. 2017. "How the Shale Boom Turned the World Upside Down." *Forbes*, April 21, 2017. https://www.forbes.com/sites/rrapier/2017/04/21/how-the-shale-boom-turned-the-world-upside-down/.

Sayrak, Inci O. 2019. "Mindfulness beyond Self-Help. The Context of Virtue, Concentration, and Wisdom." *Journal of Communication and Religion* 42, no. 4 (winter): 28–38.

Schmidt, Jon A. 2014. "Changing the Paradigm for Engineering Ethics." *Science and Engineering Ethics* 20, no. 4: 985–1010.

Spencer, Metta. 1995. " 'Political' Scientists." *Bulletin of the Atomic Scientists* 51, no. 4 (July): 62–68.

Stahl, William A. 1995. "Venerating the Black Box: Magic in Media Discourse on Technology." *Science, Technology, & Human Values* 20, no. 2 (spring): 234–58.

Thompson, Jonathan. 2021. "Why Reducing Methane Emissions Matters." *High Country News*, October 29, 2021. https://www.hcn.org/issues/53.11/infographic-energy-industry-why-reducing-methane-emissions-matters.

Thorpe, Charles. 2017. "The Political Economy of the Manhattan Project." In *The Routledge Handbook of the Political Economy of Science*, edited by David Tyfield, Rebecca Lave, Samuel Randalls, Charles Thorpe, 43–56. New York: Routledge.
Virilio, Paul. 1977. *Speed and Politics*. Translated by Marc Polizzotti. Los Angeles: Semiotext(e).
Wajcman, Judy. 2015. *Pressed for Time: The Acceleration of Life in Digital Capitalism*. Chicago: University of Chicago Press.

Chapter 22

Innovation Policy Driven by the Market
The Second Great Disembeddedness

José Luís Garcia

This essay presents a critical reflection on the dominant form of innovation policy implemented since the end of the 1980s. This criticism intends to prove that innovation policy has consisted of neoliberal techno-economic orientation, institutionalized and developed by States, national agencies, and supranational entities, in the service of the expansion of private control over the economy and of reliance on market mechanisms in more domains of human and natural life. My analysis clarifies the commitment of current innovation policy to the constant creation of inventions, particularly ones that are technological or have a technological component, that can be converted into highly profitable goods in the world economy, which has depended on the cooptation of scientific and technoscientific research and other forms of knowledge and information.

The argument is based on the exploration of two perspectives. The first one argues that the innovation project is a new manifestation of historicism, specifically of both technological and marketological historicism. It is claimed that, from a fallacious representation of history as the fulfillment of an evolutionary and ascending process of freedom and rationalization of the relationships among human beings and of humans with nature, goods produced by innovation always emerge as superior to their predecessors and the market system appears as a culminating system. The second perspective breaks from the common tendency to overemphasize social influence in the market and to underestimate the influence of the market in society. Instead, I develop Karl Polanyi's idea of the "embeddedness" and the "disembeddedness," key in his classic study *The Great Transformation* ([1944] 1957), to show how the construction of national markets in the eighteenth and nineteenth centuries disconnected economic production from other social contexts, to understand how innovation, implemented since the 1980s, has generated a second great movement that further disembeds the economy from society.

The argument is developed in four steps. The first section describes the context of the mobilization of innovation, its antecedents, and the emergence of the notion of technological innovation for commercialization purposes. The second discusses the transition from a negative perspective of "the new," common prior to modernity, to a positive one, which became dominant during the twentieth century and which serves as the basis for the contemporary historicist view of progress, which I describe as "techno-liberal historicism." In the third section, I propose that Polanyi's thesis, that society had been subordinated by the force and magnitude of the market mechanism, can be extended to be the basis for understanding a new great disembedding driven by neoliberal economics powered by technoscience that has occurred over the past four decades. The fourth section analyzes and exemplifies this movement by extending the idea of the abundant production of "fictitious commodities" into socially and ecologically crucial spheres, such as those of science, knowledge, human communication, sociability, biological life, health, and food.

MOBILIZING FOR INNOVATION

Innovation began to be embraced as a basic concept in economic policy in the market liberalism doctrine known as Thatcherism, Reaganism, neoliberalism, and the "Washington consensus," which achieved a dominant position at the end of the Cold War in the early 1990s as a more interconnected global market developed. Over the last thirty years, countries, whether developed or developing, have regularly supported innovation programs by issuing guidelines, training promoters, and investing funds. They have given an extraordinary amount of attention to technological achievements, the role of innovation in global economic competition, the advantages and gains that companies can achieve, and technology markets. Take three illustrative cases from three different continents. The US Department of State's dedicated Innovation Policy document proclaims and informs: "The United States is the most innovative economy in the world. American companies drive global innovation and the development of advanced and emerging technologies. The State Department is committed to removing barriers abroad, protecting intellectual property, and maintaining America's technological edge." The European Union's website for stimulating innovation states: "Investing in research and innovation is investing in Europe's future. It helps us compete globally and preserve our unique social model. EU support for research and innovation adds value by encouraging cooperation between research teams across countries and disciplines that is vital to make breakthrough discoveries." In the same vein, the Indian Union in 2019 announced the formation of a National Research Foundation for innovation and R&D, and in 2021–22 approved a budget that for the first time included a pillar to ensure its funding.[1]

A new activism has used campaigns, fairs, contests, and advertising, to encourage the creation of innovative items. Avant-garde slogans— "Empowered by Innovation," "Innovating for a Safer World," "Living Innovation," "Powered by Innovation"—are frequently evoked to curry favor for innovation programs and the technologies, industries, employment forms, and markets that accompany them. As part of this innovation crusade, universities formed the new field of "innovation studies," which is accompanied by a vast academic bibliography that includes numerous handbooks expounding models and theories of innovation that emphasize their technical and economic relevance, and, in rare cases, the importance of social and cultural conditions, but almost always with little critical thinking. The literature on innovation revealing the bet of political, economic, and university entities has grown in parallel with the incentive given to it as a guideline for societies.[2]

After the Second World War, governmental guidelines and laws began to favor innovation as an economic instrument that promised productivity and profit. It is an essential element of market capitalism, in which the invention of new technologies is expected to produce more goods at a greater profit with a smaller workforce. As early as 1911, Austrian economist Joseph Schumpeter, in his classic analysis of capitalist economy, had reformulated the development theory based on the idea that capitalism is a dynamic system in which innovation happens constantly and disruptively, leading to new businesses and ways of doing business that replace existing ones—a process he later called "creative destruction." For Schumpeter, growths and downturns are inevitable and cannot be eliminated or rectified without obstructing the creation of new wealth through innovation (Schumpeter 1911). "The eternal storm of creative destruction," he wrote in 1942, "is the essential fact of capitalism, with the innovative entrepreneur as the central actor" (Schumpeter [1942] 1976, 87). According to Schumpeter, the innovative entrepreneur is responsible for bringing new products to the market through more efficient combinations of production factors. Entrepreneurs' innovations are seen as drivers of long-term economic growth, in which new products destroy old companies and business models. They lead to new products, new production methods, innovative scientific discoveries, new sources of raw materials, new ways of organizing industries, and new markets.[3]

While Schumpeter introduced innovation into economic theory, he did not coin the term *technological innovation*, or even analyze it in detail. Instead, this credit goes to Rupert W. Maclaurin, an economic historian at the Massachusetts Institute of Technology, who, in the 1950s became the first theoretician of *technological innovation*. Although he is commonly ignored by studies concerning its origins, not only was Maclaurin one of the first to use that term, but he also developed the idea as a sequential process that began with scientific research and concluded with commercialization. In a matter of decades, technological innovation became the predominant representation of innovation (Godin 2008).

At the end of the twentieth century, science and technology indicators were redesigned as innovation indicators. The language of research and development (R&D) emerged to position the technoscientific complex in the service of commercial innovation; the complex is now seen as a lever to institutionalize innovation in all possible areas in the search for new industrial, commercialization, and labor-saving fronts. Subsequently, innovation agencies were set up to promote innovation policy; funding priorities were established to favor research with greater possibilities of generating innovations; economic incentives were established for research carried out in a business context; and funding agencies and even universities have made generating patents a mandatory job requirement for researchers. In all these efforts, governments, supranational entities such as the European Union, and global entities such as the Organization for Economic Cooperation and Development (OECD) were supported by academic sectors, including consultants whose innovation models framed, guided, and justified policies.[4]

In the late 1970s and early 1980s, the richest countries invested heavily in a general restructuring of the technological and productive economic network. This process became known as the Third Industrial Revolution and involved substantial investments in computers, microelectronics, computer networks, biotechnologies, artificial intelligence, and new materials. The goal was to launch a new cycle of increasing material wealth through the constant creation of new products and services, especially technological or technology-based ones, that could be converted into highly profitable new commodities in an economy that was rapidly globalizing. The meaning of the new was fundamentally changed to refer primarily to the technological new. Despite the many negative anticipated and realized human and social effects of this restructuring, especially on the most vulnerable groups, its implementation was generally welcomed. How should this decision be understood?

INNOVATION AS A NEW HISTORICISM: TECHNOLOGICAL HISTORICISM

The positive connotation and support for innovation nowadays stands in stark contrast to the suspicion it engendered in the past. It was seen as disturbing the social order and those who sought novelty were considered unwelcome. In a comprehensive study of the historical representation of innovation, Godin (2015) argues that this notion is now so closely linked to an economic ideology that we tend to forget that it was, for centuries, a contested political notion.

In his work, Godin distinguishes two "epistemes": a first, which extends from the end of the Middle Ages in Europe to the nineteenth century, when innovation was essentially rejected, and a second, in the twentieth century, when innovation acquired a character of nobility (2015). The epistemes are dominant but not absolute, as new various literary and cultural movements, technological innovations as the printing press, and technicians and pioneers were highly valued from the Renaissance until the late eighteenth century.[5] In the second half of the twentieth century, innovation became a genuine watchword, bringing together a diversity of other terms and concepts. The term "technological innovation" came to encompass almost all of what novelty came to mean (2015).[6] It is this attraction to the new—neolatry—that integrates the modern structure of thought, in which the triumph of a new notion of historical time turned toward the future and open to change was decisive.

In pagan antiquity the appreciation of the past generally prevailed, accompanied by the idea of a decadent present. In the Middle Ages the present was enclosed between the burden of the past and the hope for an eschatological future. Since the Renaissance, a new concept of time has emerged. Time is no longer cyclical, following natural processes, but progressive, tracking the "perfection" of human reason. Maritime explorations and the formation of scientific knowledge in the fifteenth, sixteenth, and seventeenth centuries, later combined with the first Industrial Revolution, the process of mass production, and economic expansion, supported the assertion of the superiority of modern (over the ancient) and the foundation of a historical consciousness directed toward becoming (over being). In the eighteenth and nineteenth centuries, certain expressions of social development and a new set of formative philosophies and beliefs allowed humans to expand their horizons and thus reject the fatalism in the relations among men and between men and nature and God that were typical of the European collective mentality. The idea that human beings were free to create the course of their own lives, the confidence in the intelligibility of reality and in the autonomous faculty of reason to disclose it, and the belief in humanity's predisposition to perfectibility contributed to the notion of Progress that ensued in Europe and later in the United States.

As Reinhart Koselleck (2002) clarifies, the way is cleared for a genuinely historical time, in which the future is open and will be determined by human acts in pursuit of the goals that humans set. Francis Bacon put the metaphor of aging in the background; Blaise Pascal put the human progress of *raison* in opposition to the aging of the world; and Gottfried Wilhelm Leibniz formulated a dynamic concept of time, able to account for the temporality that is inherent to progress in which the best of all worlds is only the best when it constantly improves. But above all, Kant conceived an idea of progress as a temporal mode of history that had never previously been conceptualized. Progress becomes a transcendental category in which the conditions of knowledge converge with the conditions of acting and deed. It is clear, Koselleck notes, that from here on there is a path that leads to Hegel and Marx.

The prior senses of history had to be replaced by an understanding that it was the coming-to-be of a sequenced, grand, benevolent plan whose purpose was mankind's aggrandizement. In the influential Hegelian system and its idea of the conquest of historical time by reason, history is governed by an ultimate plan of continuous improvements inscribed in an ascending dynamic of rationality that goes from the past into the present. Comte de Saint-Simon and the ideological-technological movement inspired by his thought drove engineering projects for railroads, waterways, and interoceanic canals, in line with the conviction that industrialization and technological change were the desirable means to achieve the end of prosperity that would be the culmination of historical evolution. Marx believed that the development of the new capitalist mode of production—of monstrous cruelty as he described it in *Capital*—had to proceed to its logical historical end, which was socialist society. For liberal apologists of the nascent industrial capitalist economy, the consequences of the destruction of traditional structures were the cost to be paid for the increased wealth brought about by the increment of productive capacity. For critics of capitalism like Marx, it was the final chapter in the prehistory of human society and the possibility of achieving the (eschatological) goal of all history. Finally, the true History would emerge from the potentialities inscribed in the new productive forces and from the class struggle inscribed in the antagonism that they gave rise to between the bourgeoisie and the proletariat. For some and others alike, the calamities brought about by the industrial capitalist economy were mere damage in the way of the future.

The understanding of the historical process as following a special sequence or direction, concomitant with rationality, is known as "historicism." An early understanding of historicism emerged from the middle of the nineteenth century from German thinkers such as Ernst Troeltsch and Wilhelm Dilthey. They argued that knowledge of society and the human condition has an intrinsically historical character, and therefore a philosophical approach to historical knowledge is necessary. This conception of historicism sought to validate history as a science but proposed replacing scientific models of knowledge with historical models, rooted in concrete, perspectivist contexts, always in need of interpretation, a shift applicable not only to history, but also to economics, law, political theory, and philosophy.

There are many twentieth-century critics of historicism and a large consensus on the rejection of any view of the course of history in which it appears as the fulfillment of a plan or program. Liberal thinkers such as Karl Popper (1957) and Isaiah Berlin (1954) are often referred to as critics of historicism, which they link to socialist and communist progressivism far from the political mainstream. It is true that the socialist and communist mainstreams throughout the twentieth century have espoused an unwavering faith in humanity's progress, with few exceptions. This tendency has probably contributed to widespread neglect of the compelling historicism found in much liberal progressivism of the nineteenth and twentieth centuries. It also leads to ignoring relevant intellectual works that can be included in the critical thinking about historicism, in which Leo Strauss (1953), Koselleck ([1979] 2004), Ernest Gellner (1992), Fernando Catroga (2009), and, in the field of Marxism itself, Henri Lefebvre (1970), are examplars.

Perhaps unsurprisingly given its content, the term *historicism* has varied meanings, some of which are even irreconcilable. In this text, it means a representation of deterministic historical time in which all events in history appear as obeying more or less rigid laws or trends. These laws could be discovered, and their discovery would enable the future to arrive, to reach the ends of history, even as it allowed individuals or groups freedom in determining their political and social action. In broad strokes, the kind of historicism professed by many modern views tends to conceive history as an evolutionary, cumulative, ascending, and teleological thread of a rational pattern whose laws irresistibly drag humanity forward. It is a historicism that values the present as superior to the past, justifies it as a harbinger of the future, and celebrates novelty or supposed self-created novelty as a step toward achieving the preordained established perspective. What has been lost deserves no regret, and future progress is not to be awaited, but rather brought into reality.

While historicism has receded in the face of criticism of historiography, the current view of innovation follows its structure of thought: it has assumed its mantle. Innovation has become understood as rationality's irrevocable dynamic, which conveys the impression that it is taking us into the future; as beneficial since it is the result of a historical trend of progress, in particular because it is attached to technological progress; and, finally, of change, for it appears as a force that can be guided by human action toward humankind's greatness and liberation. Innovation is impregnated with the historicist temperament, whereby the consecration of the new is now in conjunction not with history but with technology and with the market. And so it is an appropriate concept to understand the phenomenon of technological innovation, as well as its current attraction, as "technological historicism." This expression was suggested by Hermínio Martins (2011, 88) specifically to describe Marx's perspective of the process of man's self-production. I use it to conceptualize the emergence of technological novelties as an expression of a rational and inexorable path in the evolution of reason—understood as technical reason—which leads to their adoption without hesitation. Technological achievements, then, appear as the force that activates human beings' realization of their aspirations. In this view, the technologies of the present are always credited with overcoming the many human difficulties of the past.[7] Technological voluntarism, the doctrine that holds that technological entrepreneurship is solely based on a will that is conceived as *potentia absoluta*,[8] takes hold and technological change is exalted.

Technological historicism, associated with Hegelian teleology transposed to Marxism-Leninism, is first found in the USSR in the ideas of Nikolai Bukharin, who, despite being a victim of Stalin's authoritarianism, exercised an undeniable influence on its economic, scientific, and technological policy. This important leader of the Communist Party of the USSR formulated a socialist technological historicism that exalted technological advances as the engine of the exponential growth of the industrial productive forces, believing that such growth tended to be related to the transition to communism.[9] But, considering that today's technological historicism corresponds to an understanding of technology as a commodity and the market as a result of an increasing evolution of reason in terms of the productive system and circulation of goods, it is appropriate to speak of a techno-liberal historicism, a historicism that is both technological and marketological. Technology and the market would be history's new motors, its new drivers. Commercial technological innovation thus becomes a new historical inevitability.

To resist the forces promoting this new historicism, it is necessary to adopt a reflexive and critical attitude of current innovation politics and the parallel processes of change in the scientific and technological spheres, which nowadays are completely intertwined with the "industrial revolutions," announced and unannounced, and the expansion of the market economy.

THE FIRST GREAT DISEMBEDDEDNESS

The openness to the new and the modern idea of progress is integrated in the same historical complex of the Industrial Revolution and the introduction of machines that began to enable large-scale production and to imply a whole different context in terms of investment, production factors, raw materials, and circulation and acquisition of manufactured products. It is in this background that a new economy is established in the nineteenth century, at the center of which lies the modern market. Only by recognizing the assumptions of techno-liberal historicism can the political meaning of the systematic introduction of new technologies be unveiled.

A good introduction to this process can be found in Karl Polanyi's celebrated work *The Great Transformation* ([1944] 1957), which is devoted to the formation of the market economy, its implications, and developments. This study is an early yet accurate interpretation of how society came to be subjected to the pervasive influence of an economic sphere which then distinctly emerged from the rest of social life. It also yields an insight that has received little or no attention concerning the relationship between the introduction of new means of production and the dynamics of a productive process that becomes organized in the form of buying and selling that is shaped by the market system. Polanyi's argument is extremely pertinent for understanding the astonishing change that has been occurring since the late twentieth century under the incitement of innovation in very sensitive domains of social, political, cultural, and natural life. Changes in science, knowledge, communication, and sociability are stimulated by computer science, information technologies, and artificial intelligence; biology, health, and food are enhanced by biotechnosciences.

Focusing on England's transition from the preindustrial period to industrialization, Polanyi demonstrates how state intervention, influenced by economic liberalism, engaged in a movement to replace past isolated and regulated markets into a self-regulating market system, that is, one based only on competitive exchange and profit. The old markets, of ancestral existence, were part of a specific community and had been shaped by its moral, labor, and religious norms. The construction of national and international markets resulted not from the "natural" dynamics of markets but from the deliberate action of the state. The new markets were set up as self-regulating mechanisms according to the principle that the fields of production and distribution of goods should be controlled, regulated, and directed only by market prices. Consequently, even if usually the economic order is merely a function of the social, in which it is contained, in societies subject to a market economy, a trend has emerged in which the social order became conditioned by the logic, magnitude, and power of the economic system ([1944] 1957, 71). This shift is seen as goods such as land, human labor, and money, at first components only included in social and moral orders, had been converted into elements of the market exchange before they were integrated into the market. For Polanyi, separating the economic sphere from social and moral contexts is a decisive difference from earlier societies.

There is, thus, the problem of the relationship between economy and society, which is thought by Polanyi to operate through the theory of the "embeddedness" and the "disembeddedness." He states: "The control of the economic system by the market . . . means no less than the running of society as an adjunct to the market. Instead of being embedded in social relations, social relations are embedded in the economic system. The vital importance of the economic factor to the existence of society precludes any other result" (57). *Embeddedness* means the quality of being firmly and deeply rooted or united in something; *disembeddedness* indicates that something has been separated or extracted from where it was fixed. What Polanyi seeks to clarify with the ideas of "embeddedness" and "disembeddedness" is that all previous human economies were coupled to familial, social, religious, and other contexts and obligations. Polanyi's research shows that during the nineteenth century, economic liberalism led, through state action, to a decontextualization of the economic system from social interactions and norms, while at the same time subordinating other societal values to the pursuit of wealth, embracing the market as its fundamental institution. As a consequence of the market's control of the economic-productive apparatus, it has increasingly exerted more control or power over nature's resources and over human beings in their daily activities.

Three years after *The Great Transformation*, Polanyi published "Our Belief in Economic Determinism" (1947), in which he emphasized that in the nineteenth century the constitution of the market system meant not a mere influence of the market on society, but a determination:

> By making labour and land into commodities, man and nature had been subjected to the supply-demand-price mechanism. This meant the subordinating of the whole of society to the institution of the market. . . . Instead of incomes being determined by rank and position, rank and position were now determined by incomes. The relationship of status and contractus was reversed—the latter took everywhere the place of the former. To speak merely of an "influence" exerted by the economic factor on social stratification was a grave understatement. . . . The working of a capitalist society was not merely "influenced" by the market mechanism, it was determined by it. The social classes were now identical with "supply" and "demand" on the market for labour, land, capital, and so on. Moreover, since no human community can exist without a functioning productive apparatus, all institutions in society must conform to the requirements of that apparatus. . . . Here was "economic society"! Here it could truly be said that society was determined by economics. Most significant of all, our views of man and society were violently adjusted to this most artificial of all social settings. (100)

At the epistemological level, the notion of "embeddedness" establishes that the economy, including the market, cannot avoid being under the constant influence of sociability (cultural socialization, exchanges of reciprocity, formal and informal status systems, social networks, trust), power (hierarchies, possession of capital), and contingency (unforeseen consequences of rational action, uncertainty). In terms of historical and social analysis, disembeddedness refers to the specific situation of societies that are subject to the powerful conditioning of the market economy. This understanding of the reasoning of embeddedness and disembeddedness is contradicted by the prevailing position of the "new economic sociology." This field of study emphasizes that the market system is a social entity and subject to sociality. But this understanding is done in such a way that the market system is completely subsumed in society. According to this, nothing else counts for understanding and explaining the social world but social relations, sociability, or social influence. This is an epistemological perspective that, in the most extreme versions, devalues modes of causal interference in social reality that are not strictly social.[10]

One needs to consider the social interactions that take place in the market and the networks created by their expectations, affinities, animosities, and conflicts. However, one also needs to consider that relations between people and between people and nature change when they take the form of market relations, and to understand that market value can suppress moral and social values on goods and human and social activities. On the one hand, the theory of the "embeddedness" and the "disembeddedness" indicates that it is society that gives economy a meaning and an end, not the other way around; on the other, it indicates that the market has become a powerful social structuring factor whose mechanism intensely alters the substance of society itself and relations with nature.

Polanyi's idea that the project of a fully self-regulating market economy is nothing more than a utopian design of liberal economic and political thinking becomes therefore apparent. The guidelines that seek to develop greater market autonomy ultimately lead to increased social tension and the need for regulatory intervention by governments. An economy disembedded from society is not a real possibility, even if it has justified policies of market expansion, and generally causes devastating consequences for society and nature. Polanyi's historical look at the constitution of market society reveals that a social order permanently threatened by a free or deregulated market economy came to require constant intervention by political institutions to regulate its social effects. His analysis assumes that the movement of disembeddedness always generates a political response that prevents the totalization of the disembeddedness of the economy in order to prevent social chaos. In the twentieth century, the New Deal in the United States and the Welfare State in Europe were two examples of a substantive response.

Polanyi argued that mechanical inventions of the Industrial Revolution and their great production capacities, operating in a modern market economy, were the decisive conditions that led to disembeddedness (Polanyi [1944] 1957, 40). For Polanyi, the decisive transformation that constituted the establishment of the market economy and the nature of this institution can only be understood by taking into account "the impact of the machine," the "use of specialized machines," "elaborate machines," in a society—nineteenth-century England—that was already involved in the expansion of trade, both nationally and internationally (40–41). This happened since, he argues, "the extension of the market mechanism to the elements of industry—labor, land, and money—was the inevitable consequence of the introduction of the factory system in a commercial society" (75). Prior to modern markets, most goods were not produced in large quantities with the intention of being purchased or sold, that is, they were not regarded as commodities. As they were indexed to the market, they began to have its value dictated by the mercantile exchange and became what Polanyi calls "fictitious commodities" (72–73).

It is therefore not only the market economy that is a driver of disembedding. In a text that builds on Polanyi's theory, Adam Briggle and Carl Mitcham note that modern science, knowledge, and technology have developed autonomous dynamics that collide with other social spheres. Like neo-liberal economics,

> science is detached from its social context and conceptualized as without value or purpose. . . . Modern technology can be understood as the disembedding of pre-modern techniques. . . . For example, pre-modern artisans constructed a house only when someone needed a house and generally worked with materials present in a given location, with attention paid to particular materials and the users for whom they built. The wood for houses built in/near a deciduous forest was different from the wood for houses in/near an evergreen forest, with joists and rafters of each type sized differently to accommodate the differing strengths of each tree species, based on assessments that were passed from master to apprentice. By contrast, a modern engineered house is disembedded from any context and designed for materials that can come from all over; some materials, such as drywall, are even designed for non-specific construction uses. Engineers determine what load a beam can carry by using standardized tables and mathematical formulas instead of a personal assessment of the growth and grain of some particular piece of timber. Embedded houses reflected and are limited by their context. When disembedded construction is limited, it is because of the availability of money. (376–77)

If we accept Polanyi's thesis, the market mechanism and its link with technoscientific and industrialized production result in more and more elements that can enter the productive and mercantile chain to increase the quantity and diversity of goods and profits seeking to save labor costs, and, at the same time, introduce new and more powerful technological means of production. The elements that integrated human, social, and natural life are transmuted into factors of production, subject to purchasing and selling and, therefore, they acquire the form of commodity, destined to be consumed and to generate more resources than were invested. Under the ideals of economic liberalism embracing an economic system committed to mercantile expansion, this process of production logically tends to be a progressive plunder of nature, human life, technological development itself, and anything that can be transformed into mere means of producing profit through the consumption of goods and services. To foment consumption, it is vital that the commodity fiction can also cover money itself through credit, hence the importance that financial markets and financial capitalism have had.

As a convulsion of the economic sphere is linked to a convulsion of the political sphere, at the end (part 4, ch. 20) of *The Great Transformation*, Polanyi correlated the role that fascism came to play from the 1930s onward with the general crisis of the market economy, which would have been at the origin of the joint decomposition of the planet's political system and economic system and led from the first great world war conflict to World War II. In the essay already mentioned, published in the aftermath of World War II, he argued that it was possible that industrial civilization would destroy human beings (Polanyi 1947). He described the difficulties of this civilization as follows: "The present condition of man can be described in simple terms. The Industrial Revolution, some 150 years ago, introduced a civilization of a technological type. Mankind may not survive the departure; the machine may yet destroy man; no-one is able to gauge whether, in the long run, man and the machine are compatible. But since industrial civilization cannot and will not be willingly discarded, the task of adapting it to the requirements of human existence *must be* solved, if mankind shall continue on earth" (96).

THE SECOND GREAT DISEMBEDDEDNESS

Against the backdrop of the market mechanism that has been described, innovation is an orientation toward industrial reinvention and all sorts of services that seek in science and technoscience the provision of a wealth of knowledge and means for the reproduction and extension of the mechanism itself. What differentiated the innovation agenda promoted since the 1980s was its institutionalization as an economic policy and its aim to reconfigure the scientific and technological fields so they can be completely deployed toward the creation of economic value and in market competition. There are many indicators, in areas from computer science and media to biological sciences and health, of the annexation of these research domains by industrial objectives and business and commercial rationale. There have also been important changes to the status and mandate of researchers, who are increasingly living under the imposition of becoming producers of goods rather than being producers of knowledge.[11]

Innovation policy is also combined with financial reinvention as a fundamental agent in the speculation of capital itself. Innovation is an instrument of economic policy that activates the invention and introduction of novelties in the economic field in both existing and new areas, always according to the principle of their acceptance by the markets. Economic gain is the main objective of this innovation, an innovation that continues the logic of subordination of society to the economy, but now in a systematic and exacerbated way.

Polanyi's thought remains highly relevant for understanding the neo-liberal impetus, the permanent technological innovation, and the spread of the market and meritocratic mentality that has occurred since the last two decades of the twentieth century. Domains that had been assumed as pillars of the welfare state, such as social security, health, and education systems, have been privatized, and areas previously free of capitalist exploitation through the emergence of new technological industries—especially computer and genetic technologies—have been taken over by mercantile powers under the guidance of innovation and technological fetishism. This conjuncture can be considered as *a second great disembeddedness* of the economy in relation to the social structure; in other words, a new and enormous process that, through the market economy, extends the control of the economic-productive system, to which society and more natural elements are but an appendix.

The transformation I am describing has been interpreted in the light of the transition to "cognitive capitalism" (e.g., Fumagalli and Lucarelli 2007; Boutang 2011; Lazzarato 2014; Pasquinelli 2015), given the increasing importance of cognitive work involved in production, distribution, and reception processes, as well as in products, services, and even sources of supply, or even into a "digital capitalism" (e.g., Schiller 1999; Picciano and Spring 2012; Fuchs and Mosco 2015), since the cognitive element is expressed in the digital and in a capitalist economic system. It should be noted, however, that highlighting the cognitive factor should neither neglect materiality—as shown with women's reproductive labor (Federici 2004) and semi-slave labor, often found in Africa, that extracts minerals for computers (Parikka 2015; Bratton 2016)—nor restrict it to scientific and technological knowledge alone, for it involves others such as design, aesthetics, marketing, market and consumer studies, and the communication and creative resources of the workforce.

An extremely negative consequence of innovation policies is their marginalization of research areas that are less susceptible to innovation potential according to a technological and marketological rationale. Examples include basic research, the humanities, many environmental problems, uncertainties associated with technologies, alternative forms of agriculture, and the improvement of living conditions and health of people most seriously in need. They also accelerate the invention of technical-scientific devices and systems capable of continuously increasing production in the most diverse areas and with greater savings in labor costs. Technological change, particularly that associated with computers, information technology, and digitalization, has been accompanied by both the mirage of the end of work and the fear of massive technological unemployment, but brought neither. According to several studies on automation, the so-called intelligent machines, and the future of work, the implications of the second great disembeddedness are the corrosion of the division between work and leisure; the subsumption of all actions in the same technological space; the expansion of the dimension of the activities to which workers are subjected; the creation of new forms of labor precariousness; and the permanent control via social networks of work processes.[12] Once again, technological change in industrial and market civilization has not led to the emancipation of workers, but rather to the increase of their subjection.

However, the most critical factor of innovation policy concerns the opening of new areas to the productive system and the process of commodification. Otherwise, it would be impossible to include such areas in this mechanism due to their lack of scientific-technological knowledge and legitimization. This is the case of the ongoing transformations that are leading to the constitution of new industrial and commercial areas. This expansion is seen especially in areas that are based on the notion of information (computers, digitalization, internet, but also biotechnology, pharmaceuticals, food, etc.). The new branches of industry operate on a global scale and combine technoscience, appropriation of the reproductive function, and strict control over distribution.

During the second great disembeddedness, innovations that often "cluster" as multiple research areas are integrated and redeployed, notably computer science, information technology (e.g., software, internet, smartphones, data, tablets, iPads, gaming, new media, artificial intelligence), and biotechnologies (e.g., genetically modified food, plant breeding, forest genomics, genomic medicine, vaccines, artificial breeding, genetic diagnosis and counseling, tissue bioengineering, pharmacogenomics). Technology, information, data, imaging markets, and biomarkets have clearly become essential in today's market economy. These are just a few examples of the many forms of knowledge, technology, vital faculties of human beings, and biological life forms that have been subject to a disembedding of their social and natural contexts.

An exemplary case of what can be called "disembedded innovation," much discussed at political, sociological, and philosophical levels and the object of serious social conflicts, is that of transgenic seeds and their dissemination in plantations and marketing. Traditionally, seeds were renewable and regenerative biological entities that were part of ecosystems that generated products predominantly focused on local needs, in line with culture and social organization and selected according to knowledge shared by farmers over centuries. Transgenic seeds, on the contrary, in which genetic engineering breaks the seed's unity as the source for a crop and the next generation of seeds, are commodities produced by capital-intensive companies that have their patents and whose aim is to profit on the world food market.[13] Just as happened with the land, labor, and money, analyzed by Polanyi, scientific knowledge, technologies, genes, genomes, and biological entities, through increased technoscientific capacity and the extension of intellectual property rights and patents, have now been converted into "fictitious commodities" of emerging industries that drive new markets.

The role played by computer science and digital technologies in this whole situation is decisive. In technical terms, transnational and national agents were only able to move through the world market due to the existence of a constantly progressing instrumental base, which guarantees an increase in the capacity to collect, reproduce, process, and transmit information of various types and profoundly changes the forms of production and distribution. The technological infrastructures that support the pooling of innovations depend on computing processing capabilities and internet use. Therefore, the technological change must be considered not as a mere extension of previous technologies, but as a caesural change.

Nevertheless, the current orientation toward innovation has an element that is as attractive as it is fallacious. In many cases, what is intended to be fostered is better designated "pseudo-innovation." This term captures the tendency to over-invest in the aesthetic dimension of products, combined with marketing and advertising, to confer the value of new and original as a product differentiation strategy. Much of the production of technological devices by big companies is guided by the desire to add features that mobilize taste and sensibility for consumption, not to improve function. This aesthetic inflection of industrial production indicates that companies invest in the incessant creation of new goods to compete with existing ones, using them to stimulate consumption. In this sense, much innovation is not even intended to be incremental, but merely superficial.

This is a logic linked to the superimposition of exchange value over use value, a well-known characteristic, dubbed commodity fetishism, of market capitalism. The concept of pseudo-innovation seeks to remove the camouflage that conceals the orientation toward massification methods, assembly operations, and market calculations that are built into the product and aim at the industrial mobilization of taste and purchase. These methods are overshadowed by the communicative or cultural dimension incorporated into the products as a strategy to obtain an innovative reputation. The goal is to develop use devices that capture the deepest recesses of emotions, affections, and culture, and use them as an economic engine. The current capitalist economy, in which information, knowledge and aesthetics play a crucial role, is engaged in a trend of culturalization of industrial production.[14] The world of artistic creation and production will have anticipated the way for the new industrial processes. If the flag of creativity has been assimilated by technological innovation, then there seems to be evidence that twentieth-century industrialization and commercialization follow in the footsteps of the industrialization of culture.

Innovation has been essentially an economic policy aimed at the relentless pursuit of productivity and profit by extending the market mechanism to new spheres of the natural world and human life. The role of institutionalizing scientific and technoscientific research is part of pro-market economic policy, which has been at the service of profound transformations in broad areas of possible innovation, encompassing not only products and processes, but also knowledge and sources of raw materials and market structures. As a corollary, new forms of dispossessing nature and human beings have been triggered, resulting in a vertiginous increase and concentration of knowledge and power.

I argue that the seduction of the fallacious idea that the trajectory of technology and the evolution of markets are correlated to the direct and linear development of rationality is a manifestation of techno-liberal historicism where technological commodities appear both as promise and justification for each novelty. Thus, the technological new is conceived in a precise and effective manner always confirming the bets that are being made. Consequently, the interests, the economic calculations, the power logics, the possible alternatives, and the conflicts inherent to the innovation projects and the technological options remain obscured. The innovation driven by technoscientific neoliberalism is driving the continuous collapse of entire productive sectors generating more labor exploitation and inequality, inventing new forms of despoliation of nature, and exerting constant pressure on democratic values. All this is happening at the same time as the discovery of new resources, the application of methods of production, and the constitution, distribution, and consumption of fictitious commodities. The unexpected consequences of all kinds, the catastrophic social effects, environmental crisis, and the moral confusion that accompany innovation are neglected. Since the 1980s, we have been experiencing a second major disembeddedness.

CONCLUSION

The ideological orientation that has conditioned the direction of societies since the dawn of the twenty-first century seeks to impose technological change and market relations not because they are expected to increase freedom, equity, and ecological protection. The ruling elites treat people as though they are economic and mercantile beings by nature, denying the primacy of social and symbolic being. They fervently aim to increase the capacity of machines and to pursue the project of unlimited domination of nature and the search for economic wealth. The general mood of the current historical period is increasingly techno-economic, with a steady trend toward market-technology-science fusions, located far from communities, dominated by experts and companies that deliver the water supply, electricity, means of transport, communication, food, and so on, on which the majority of humanity depends, and on whose failure results in anxiety and suffering.

Becoming aware of the consequences of the techno-economic sphere for the acceleration of the path in which contemporary societies are embroiled does not have to mean acceptance of it. On the contrary, it can be a way of insisting on the defense of an idea of the human being and of society that grants primacy to the sphere of social relations, to the sense of community and to contradicting a perspective based on the primacy of technology and the economic system. Admitting in descriptive and interpretive terms that in contemporary societies techno-economic change operates as a motor that revolutionizes the social, political, and legal structure, the world of the arts, beliefs, customs, and scales of values, does not imply the defense, epistemological or political, of economic and technological determinism. Acknowledging that the techno-market sphere has extensively conditioned everything else is a necessary step in constructing a world that is more socially just, more respectful in its relations with nature, and less reckless with the power of technical devices.

Societies now face dilemmas of enormous magnitude generated by the tendencies described, including those inherent to this technological civilization, such as the global ecological crisis, the specter of nuclear wars, and the possibility of the biological control of the human being—and also the commodification of life forms, from the simplest to the most complex—through genetic engineering. At the same time, the contemporary world retains serious problems of scarcity and illness that it had at the entrance of industrial societies and has added the specter of destruction with the permanent threat of nuclear war and climate change. Industry, science, and technology, whose successes have undeniably played a role in improving the conditions of human beings, have become a source of difficulties and uncertainties in a system that is currently under the impulse of economic neoliberalism and deregulated technological innovation. The regime of permanent innovation as an engine of economic growth, the construction of future markets in the biological and other fields, and as a means of discoveries in the service of power, violence, and war, raises new and difficult moral and political questions, above all an unprecedented horizon of programmed obsolescence, threats, and dangers that arise from the action of human beings themselves. Moreover, all this occurs in a circumstance in which citizens begin from a position of weakness ("irrationality") when it comes to debating and challenging this new regime.

The oligarchies that rule the world have sought to mold society to fit the prevailing techno-economic system in order to preserve it. To intervene in this system consciously and responsibly, to open it up to public discussion and put it at the service of the common good—to re-embed it—is necessary to have a truly democratic society. Success in this task requires a radical shift, namely, to think of the human being and society in a way that is very different from the one imposed by those who believe in technology and the market economy as the ultimate ends of human life.

NOTES

1. For the United States, see https://www.state.gov/innovation-policy/; for the EU, see https://european-union.europa.eu/priorities-and-actions/actions-topic/research-and-innovation_en; and for the Indian Union, see https://www.psa.gov.in/article/innovation-and-rd-highlights-union-budget-2021–22/2529.

2. There are more than a dozen manuals published by prominent publishers. These include Antonelli (2022), Régnier et al. (2022), and Engel (2022).

3. Schumpeter distinguished invention from innovation. The former is the creation of a product. The latter refers to making or altering a product so that it is successful. The idea of innovation, central to this article, is problematic, because in today's "market society" success means consumption, which is not necessarily the best criterion. Ideally, drawing from the political register, the common good should be the criterion of "success."

4. Innovation literature is saturated with models. Before the 1960s the term *models* was rarely used; Godin (2017) suggests that the term model has both a scientific and rhetorical function. See also Barbosa (2011), Garcia (2012), Fernández-Esquinas (2012), Godin and Vinck (2017).

5. Regarding cultural and technical creation, Martins (2011, 85–86) refers to the theological-scientific-technological parallels that artists and artisans have evoked since the Renaissance to imply cooperation with the Deity in the framework of a Christian culture that emphasized the image of God as "creator of creators." These parallels made craftsmen, artists, and engineers demiurges, as it were, who could create, not ex nihilo, but over preexisting materials, thus expanding the power available to humankind and creating unprecedented wonders.

6. In this regard, it is also worth remembering the role of Futurism, which began before World War I with Italian writers and artists but extended its influences later through other contexts and was a cultural avant-garde that praised both the new and the technological new. For Italian futurists, the primordial *locus* of aesthetic values was fundamentally the future and not nature or the art of the past. This future was projected and sublimated as a world under a process of mechanization of existence. The Italian futurists were vehicles of neolatry in a technocratic version. Their glorification of the new was accompanied by the glorification of technology, including that linked to war. Thorough notes on this issue can be found in Marshall Berman (1982) and Martins (2011, 152–55).

7. My first incursions into the conceptualization of innovation as a (providentialist) historicist manifestation, although without naming the concept "technological historicism" and "techno-liberal historicism," are found in Garcia (2012).

8. My use of *potentia absoluta* (absolute potency) here refers to a will that is not ordered by a moral end or that is the sole cause *per se* of its acts. The will as *potentia absoluta* means that it is not the good that is willed, but to will that is good. On the explication of the notion of will as *potentia absoluta* in modern philosophy, see Muralt (2002).

9. The current of German engineers and philosophers called "reactionary modernists" by Jeffrey Herf (1984), which had a strong influence on science, technology, and industrial policy before and after the Nazis took power, also accepted modern technology, exalted the new, and pushed for scientific rationalization. Herf's study devoted to reactionary modernists shows how they, although illiberal and authoritarian, adopted a positive stance toward technological progress, which brings them closer to technological historicism. However, they rejected Enlightenment reason and embraced romanticism. Rather than being identified with technological historicism, reactionary modernists should be considered more strictly a stream of highly technological romanticism.

10. For a discussion on this topic, see, among others, Granovetter (1985), Swedberg (2003), Portes (2010), Mingione (2011), and Dale (2012). Also in this regard, it is worth noting that the winner of the Nobel Prize in Economic Sciences, Joseph Stiglitz, argues that Polanyi's concept of embeddedness indicated that he "saw the market economy not as an end in itself, but as means to more fundamental ends" (2012, 77). Fred Block (2003) shares the same type of position: for this author, the term *embeddedness* expresses the idea that the economy is not autonomous, as it must be in economic theory, but subordinated to politics, religion, and social relations.

11. In this regard, see, among others, Pestre (2003), Krimsky (2003), Nowotny et al. (2010).

12. See Benanav (2020) and Carbonnel (2022), among others.
13. See Kloppenburg (1988), Shiva (1997), Lacey (2005), and Garcia (2009), among others.
14. See Assouly (2008), and Alonso and Fernández (2018), among others.

REFERENCES

Alonso Benito, Luis Enrique, and Carlos J. Fernández Rodríguez. 2018. *Poder y sacrificio: Los nuevos discursos de la empresa.* Madrid: Siglo XXI.
Antonelli, Cristiano, ed. 2022. *Elgar Encyclopedia on the Economics of Knowledge and Innovation.* Cheltenham, UK: Edward Elgar.
Assouly, Olivier. 2008. *Le capitalisme esthétique: Essai sur l'industrialisation du goût.* Paris: Éd. du CERF.
Barbosa, Marcos. 2011. "Formas de autonomia da ciência." *Scientiae Studia* 9, no. 3: 527–61.
Benanav, Aaron. 2020. *Automation and the Future of Work.* London: Verso.
Berlin, Isaiah. 1954. *Historical Inevitability.* London: Oxford University Press.
Berman, Marshall. 1982. *All That Is Solid Melts into Air: The Experience of Modernity.* London: Verso.
Block, Fred. 2003. "Karl Polanyi and the Writing of 'The Great Transformation.'" *Theory and Society* 32, no. 3: 275–306.
Boutang, Yann Moulier. 2011. *Le capitalisme cognitif, comprendre la nouvelle grande transformation et ses enjeux.* Paris: Editions Amsterdam.
Bratton, Benjamin. 2016. *The Stack: On Software and Sovereignty.* Cambridge, MA: MIT Press.
Briggle, Adam, and Carl Mitcham. 2009. "Embedding and Networking: Conceptualizing Experience in a Technosociety." *Technology in Society* 31, no. 4: 374–83.
Carbonnel, Juan Sebastián. 2022. *Le futur du travail.* Paris: Editions Amsterdam/ Multitudes.
Catroga, Fernando. 2009. *Os passos do homem como restolho do tempo: Memória e fim do fim da história.* Coimbra: Almedina.
Dale, Gareth. 2012. "Double Movements and Pendular Forces: Polanyan Perspectives on the Neoliberal Age." *Current Sociology* 60, no. 1: 3–27.
Engel, Jerome S., ed. 2022. *Clusters of Innovation in the Age of Disruption.* Cheltenham, UK: Edward Elgar.
Federici, Silvia. 2004. *Caliban and the Witch: Women, the Body and Primitive Accumulation.* Brooklyn: Autonomedia.
Fernández-Esquinas, Manuel. 2012. "Hacia un programa de investigación en Sociologia de la Innovación." *Arbor,* "Sociologia de la Innovación" 188, no. 753: 5–18.
Fumagalli, Andrea, and Stefano Lucarelli. 2007. "A Model of Cognitive Capitalism: A Preliminary Analysis." *European Journal of Economic and Social Systems* 20, no. 1: 117–33.

Fuchs, Christian, and Vincent Mosco, eds. 2015. *Marx in the Age of Digital Capitalism*. Lam Ed. Leiden: Brill.
Garcia, José Luís. 2009. "Biocapital et nouvelle économie politique de la vie." *Revue de l'Institut de Sociologie*, 14: 7–38.
———. 2012. "El discurso de la innovación en tela de juicio: Tecnologia, mercado y bien estar humano." *Arbor*, "Sociologia de la Innovación" 188 (753): 19–30.
Gellner, Ernest. 1992. *Reason and Culture: The Historic Role of Rationality and Rationalism*. Oxford: Blackwell.
Godin, Benoît. 2008. "In the Shadow of Schumpeter: William Rupert Maclaurin and the Study of Technological Innovation." *Minerva* 46, no. 3: 343–60.
———. 2015. *Innovation Contested: The Idea of Innovation over the Centuries*. New York: Routledge.
———. 2017. *Models of Innovation: The History of an Idea*. Cambridge, MA: MIT Press.
Godin, Benoît, and Dominique Vinck, eds. 2017. *Critical Studies of Innovation: Alternative Approaches to the Pro-Innovation Bias*. Cheltenham, UK: Edward Elgar.
Granovetter, Marc. 1985. "Economic Action and Social Structure: The Problem of Embeddedness." *American Journal of Sociology* 91, no. 3: 481–510.
Herf, Jeffrey. 1984. *Reactionary Modernism, Technology, Culture, and Politics in Weimar and the Third Reich*. Cambridge: Cambridge University Press.
Kloppenburg Jr., Jack Ralph. 1988. *First the Seed: The Political Economy of Plant Biology, 1942–2000*. Cambridge: Cambridge University Press.
Koselleck, Reinhart. 2002. *The Practice of Conceptual History: Timing History, Spacing Concepts*. Translated by Todd Samuel Presner. Stanford, CA: Stanford University Press.
———. [1979] 2004. *Futures Past: On the Semantics of Historical Time*. New York: Columbia University Press.
Krimsky, Sheldon. 2003. *Science in the Private Interest: Has the Lure of Profits Corrupted Biomedical Research?* Oxford: Rowman & Littlefield.
Lacey, Hugh. 2005. *Values and Objectivity in Science: The Current Controversy about Transgenic Crops*. Lanham, MD: Lexington.
Lazzarato, Maurizio. 2014. *Signs and Machines: Capitalism and the Production of Subjectivity*. New York: Semiotext(e).
Lefebvre, Henri. 1970. *La fin de l'histoire*. Paris: Editions de Minuit.
Martins, Hermínio. 2011. *Experimentum humanum: Civilização tecnológica e condição humana*. Lisbon: Relógio D'Água.
Muralt, André de. 2002. *L'unité de la philosophie politique: De Scot, Occam et Suarez au libéralisme contemporain*. Paris: Vrin.
Mingione, Enzo. 2011. "Embeddedness." In *International Encyclopedia of Economic Sociology*, edited by Jens Beckert, and Milan Zafirovski, 231–36. London: Routledge.
Nowotny, Helga, Dominique Pestre, Helmuth Schulze-Fielitz, Eberhard Schmidt Assmann, and Hans-Heinrich Trute. 2010. *The Public Nature of Science under Assault*. Heidelberg: Springer.

Parikka, Jussi. 2015. *A Geology of Media*. Minneapolis: University of Minnesota Press.
Pasquinelli, Matteo. 2015. "Italian Operaismo and the Information Machine." *Theory, Culture & Society* 32, no. 3: 49–68.
Pestre, Dominique. 2003. *Science, argent et politique: Un essai d'interpretation*. Paris: INRA.
Picciano, Anthony G., and Joel Spring. 2012. *The Great American Education-Industrial Complex*. New York: Routledge.
Polanyi, Karl. [1944] 1957. *The Great Transformation: The Politics and Economic Origins of our Time*. Boston: Beacon Press.
———. 1947. "On Belief in Economic Determinism." *The Sociological Review* a39, no. 1: 96–102. doi:10.1111/j.1467-954X.1947.tb02267.x.
Popper, Karl. 1957. *The Poverty of Historicism*. New York: Routledge and Kegan Paul.
Portes, Alejandro. 2010. *Economic Sociology: A Systematic Inquiry*. Princeton, NJ: Princeton University Press.
Régnier, Philippe, Daniel Frey, Samuel Pierre, Koshy Varghese, and Pascal Wild, eds. 2022. *Handbook of Innovation and Appropriate Technologies for International Development*. Cheltenham, UK: Edward Elgar.
Schiller, Dan. 1999. *Digital Capitalism: Networking the Global Market System*. Cambridge, MA: MIT Press.
Schumpeter, Joseph A. 1911. *The Theory of Economic Development: An Inquiry into Profits, Capital, Credit, Interest, and the Business Cycle*. Translated by Redvers Opie. Cambridge, MA: Harvard University Press.
———. [1942] 1976. *Capitalism, Socialism and Democracy*. London: Allen and Unwin.
Shiva, Vandana. 1997. *Biopiracy: The Plunder of Nature and Knowledge*. Boston: South End Press.
Stiglitz, Joseph. 2012. Prefácio to *A grande transformação: As origens políticas e económicas do nosso tempo* by Karl Polanyi, 65–79. Lisbon: Edições 70.
Strauss, Leo. 1953. *Natural Right and History*. Chicago: University of Chicago Press.
Swedberg, Richard. 2003. *Principles of Economic Sociology*. Princeton, NJ: Princeton University Press.

PART V

Science and Technology Policy

Chapter 23

Irrational Energy Ethics

Adam Briggle

Imagine you are the energy czar. Starting from scratch, you have absolute power to distribute energy to the world's population of eight billion people. Your job is to achieve three goals. You need to make everyone happy and healthy. You need to be fair. And you need to protect the environment. Basically, you need to solve our biggest energy ethics dilemma: provide everyone with a decent standard of living while preventing dangerous climate change. What should you do?

You realize that energy is vital to life: around two thousand calories daily is the bare minimum to sustain human metabolism. Beyond that, though, energy is vital to the *good* life. One way to measure progress is by the share of total energy used and consumed by humans that comes from food calories (see Smil 2017). In the Roman Empire, food accounted for 45 percent of total energy. Even in the most developed economies of the 1820s, food was still about 30 percent of total energy. Today, food is less than 2 percent of the total energy supply for human civilization. The total energy supply is a measure of all the things that power civilization, including not just its human bodies (via food) but also its engines, generators, and other machines. So, this captures energy from oil, coal, natural gas, nuclear, wind, solar, hydro, and other supplies (the total energy supply in 2019 was about 273,000 terra-watt hours compared to 6,300 TWh in 1820). This means that more energy is being used to provide for security, freedom, health, happiness, play, travel, science, entertainment, art, education, and other aspects of culture that go beyond mere survival (that is, merely fueling the human body).

We might say, then, that energy humanizes us—allowing us to more fully flourish—or even that access to sufficient energy is a human right. Yet even if an infinite amount of energy were available, everyone cannot claim a right to an unlimited amount of energy, however, given the environmental consequences of the portfolio of fuels used today and for the foreseeable future. Further, there are diminishing rates of return when it comes to energy consumption and human well-being. After surveying a growing body of empirical literature, the energy analyst Vaclav Smil concludes that "no indicator of high quality of life—very low infant mortality, long average life expectancy, plentiful food, good housing, or ready access to all levels of education—shows any substantial gain" once the average energy consumption rises above 65 GJ per person per year (Smil 2010, 721). Beyond this level, there are no "important gains in physical quality of life or . . . greater security, probity, freedom, or happiness" (726). Smil puts it colorfully: "We could all be perfectly happy living at the level of consumption and income as Frenchmen in 1959" (in Voosen 2018, 1324).

This raises the idea of an energy "golden mean," a level that is neither too little nor too much. Carl Mitcham and Jessica Smith Rolston (2013) make this point by questioning the widely held assumption that there is a linear relationship between more energy consumption and greater well-being. Like Smil, they advance what they call a "Type II energy ethics" that challenges more orthodox views about energy growth and happiness. Similarly, the Greenhouse Development Rights framework uses the notion of a consumption threshold (in their case, based on income rather than energy) to establish the parameters for climate responsibility: those above the threshold should bear the burdens of emergency climate action (see Baer et al. 2009).

So, as energy czar, you should distribute roughly 2,000 watts or 65 GJ annually per capita, perhaps as an energy card or allowance of some sort (see also Jochem et al. 2002). There could be some room for differences, but inequalities could not grow so large that they become unjustifiable or unsustainable. This roughly equal and limited distribution scheme would ensure your three goals of well-being, fairness, and sustainability.

Curiously, global average energy consumption is currently about 65 GJ per capita annually. The problem is the extreme inequalities in distribution. According to UN data on its seventh Sustainable Development Goal (ensure access to affordable, reliable, and modern energy for all), roughly 780 million people globally lack access to electricity and nearly 2.8 billion people lack access to clean cooking fuels (UN 2020). Meanwhile, the 580 million residents of North America consume on average 330 GJ of energy per person per year, which is roughly five times the optimal amount.

Now modify our thought experiment. Imagine you are the energy czar in our present reality, and you can redistribute current energy consumption at will. Given that the average global consumption is already 65 GJ annually per capita, it would be a simple fix. You could cut from the more developed nations and add to the less developed nations to achieve the optimal energy allocation *without any net increase in energy production or consumption*.

IRRATIONALITY IS UNAVOIDABLE

It does not work that way in the real world. A North American's energy consumption is not a set of credits that can be moved to someone else's ledger. It is the reality of heating and cooling, driving, electricity, entertainment, food and water, and more. Much of the energy overshoot stems from sprawling, car-centered development patterns. Even if it were morally sound and politically feasible, it is not technically possible to shuffle energy in a redistributive scheme. There are material path dependencies. Four-fifths of the pipes and wires going to my home can't be redirected elsewhere. Take away my energy "credits" and I still need to get to work. We often *need* energy at higher levels than required for happiness. That is another way to describe the irrationalities of the high-energy world (see Illich 1974). Even a homeless American consumes about 130 GJ of energy annually, due to the energy used in providing social services such as police, roads, libraries, courts, and the military (Madrigal 2008). So, even if the energy czar made all Americans homeless, they would only be halfway to the target!

And, of course, redistribution in the name of sufficiency is not politically feasible. Hardly anyone would willingly submit to any significant reduction in their material standard of living. Witness, for example, the Yellow Vest Protests in Paris against a gasoline tax. People would resist enforced energy sufficiency even if we assured them that it was for their own good. No politician would gain power on this kind of platform. The ideas of rationing energy consumption, deliberately shrinking the economy, or capping total wealth are non-starters in countries with even a modicum of capitalist or democratic organizing principles.

It is just as politically unlikely and far more ethically dubious to suppose that developing nations would willingly submit to a 65 GJ ceiling on their consumption. Why should they limit themselves when millions of others have for so long enjoyed the fruits of energy super-abundance? Who would have the moral authority to impose such limits on them? They are just as entitled as anyone in the developed world to international tourism, streaming entertainment, and large homes even if it won't ultimately make them happy.

Let me be clear. Energy consumption beyond roughly 65 GJ annually per capita is indeed irrational: it is an ecologically damaging overshoot with no compensating benefits for human well-being. The energy rich get trapped in more needs and stuck on a hedonic treadmill, consuming more and more to sustain the same level of happiness. Those consuming far more energy than the 65 GJ golden mean are not happier, healthier, or smarter than those living lives at the golden mean. They are just more wasteful and destructive. It is irrational.

My position rests on a commonsense notion of instrumental rationality. Someone is instrumentally rational when they adopt suitable means to reach their ends. In this case, the irrationality consists in the energy means outstripping the requirements for the goal of human flourishing. People look for more happiness via more energy consumption—beyond the 65 GJ threshold—and they don't find it there. The means don't achieve the ends. This is not an argument based on an objective notion of what people ought to value. Even when subjective measures of well-being are used, there is no empirical evidence that more energy is related to greater happiness in developed nations (Okulicz-Kozaryn and Altman 2020).

And yet this irrationality is inextricable, at least on timelines that matter for sustainable development objectives. Any real-world energy ethics, unlike the imaginary energy czar, must work within our irrational system. This means taking preferences as given and seeking the best means for satisfying them. Philosophers may object to such irrationality—indeed they should—but it won't change the fact that this is our situation. Utopian schemes for energy sufficiency are fantasies. Real energy ethics assumes irrationality and makes the best of it.

The energy czar is an imaginary, all-powerful technocrat. Like Plato's philosopher king, the czar is an attempt to replace political action with making or fabrication, that is, to replace the inherently unpredictable acts born from human plurality with a single maker who conjures the idea that everyone else implements. The czar thought experiment depends on a division between one who knows (who perceives what really is good) and the many who do not know. These kinds of thought experiments, the thinker Hannah Arendt (1958) notes, constitute "escape plans" from politics. They are an attempt to give a solidity and rationality to human affairs that cannot be had. Political action is frail, futile, uncertain, and unpredictable.

Given the pressing environmental threats posed by high-energy civilization, it is tempting to seek an escape from the irrationality—to find a mountain from which to look down and rearrange the pieces of an insane world. Such a perspective can be illuminating, but it offers no entry point for an ethics that could gear into the world. An ethics of energy sufficiency is rational yet unworkable, because of the material, historical, institutional, and cultural givens of our age. Advanced industrial, nation-state capitalist society with its ruling ideals of production and consumption establishes the boundaries for allowable ethical discourse and action. Of course, people can and do float ideas outside of these boundaries (e.g., in academic philosophical discourse). Yet, there is an Overton Window or range of ideas deemed feasible, respectable, or allowable in mainstream discourse and channels of power. Given this reality, irrational energy overshoot is bound to happen. The trick is not to prevent it, but to make it less destructive.

This means that to achieve the goal of a decent standard of living for all, we are going to have to continue growing energy production and far overshoot the current ideal average of 65 GJ per capita. Indeed, most forecasts call for energy production and consumption to double by 2050. This is simultaneously crazy and necessary; it is irrationally just.

Now, it is possible that a climate disaster would change everything. Kim Stanley Robinson's climate fiction novel *The Ministry for the Future* (2020), for example, begins with a heatwave that kills twenty million people. This kind of event could radicalize, mobilize, organize, and energize social networks in completely new ways. A new kind of ethics might emerge. For example, the elite class of the carbon industrial complex (especially those funding climate change denial) may be imprisoned for crimes against humanity. (Of course, this still leaves path dependency problems of the high-energy built environment.) Global financial institutions might create a new carbon currency that drives projects to massively drawdown atmospheric carbon dioxide levels. Our modern capitalist and nation-state system with its legal structures, dominant discourse, and growth imperative had a beginning, and it will have an end. No socioeconomic system lasts forever. If such a moment does arise, then a new kind of energy ethics might prevail. An ethics of rational sufficiency could take root in a world rocked by climate catastrophe. An example of this would be the "donut economy" envisioned by Kate Raworth (2017), that conceives of the golden mean as a band of human well-being above a floor (access to sufficient food, water, education, etc.) and below an ecological ceiling (that would prevent destructive overshoots such as pollution, ocean acidification, species loss, etc.).

In the absence of such a shock, however, any viable energy ethic must work within the parameters of the irrational imperative of energy overshoot. To seek a more rational relationship with energy is a laudable private venture. Turn off the television, YouTube, Netflix, and your email, and buy less stuff! As John Stuart Mill ([1861] 2001) wrote, "all honour to those who can give up for themselves the personal enjoyment of life. . . . He may be a rousing proof of what men *can* do, but surely not an example of what they *should* do." Mill, as well as Garrett Hardin (1968), notes that self-restraint is not a form of *social* morality. There are just too many more-or-less thoughtless and careless people to trust that private virtue or rationality will save the day. And, as I've argued above, social or public policies to limit energy consumption are non-starters. The one exception is energy efficiency measures, but here we encounter the well-known "rebound effect" or Jevon's Paradox, wherein greater efficiency often leads to more, not less, energy consumption.

There is already a default energy ethics at work in the real world. It is not controlled by a puppet master czar, but rather it is implemented with varying degrees of conscious reflection by major energy actors, agencies, and institutions around the world (like the US Department of Energy, the International Energy Agency, BP, and hundreds of other energy players). In what follows, I trace the past, present, and future of this default irrational energy ethics. How did we get into this mess and how, absent a disaster, are we going to get out of it?

PAST: MODERN ENERGY ETHICS

Premodern ethical-political thinkers often espoused "perfectionist" political regimes that would guide citizens toward excellence and rationality. Nowadays, we would call this paternalism, though the ancients thought of it more in terms of self-sufficiency. An energy paternalism would begin, like our energy czar, with a sense of what is really best for people (i.e., roughly 65 GJ—not much less, not much more) and then implement rules and institutions to nudge or even coerce people toward those outcomes. This is the logic of virtue-based political systems that seek sufficiency, that is, the golden mean. In his proto-economic writings, for example, Aristotle uses the term *enough* in fairly objective ways, as in "retail trade is not a natural part of the art of getting wealth; had it been so, men would have ceased to exchange when they had enough" (1984, 1257a17–18).

Modern ethics-politics, beginning with Machiavelli, lowered the bar of expectations. Take people as they are, he advised, rather than worry about how to make them better. Thomas Hobbes gave this a different twist, insisting that there is no ultimate good toward which people could be steered anyway (for these interpretations of Machiavelli and Hobbes, see Strauss 1953). Later liberal democratic theorists like Ronald Dworkin built on this tradition. He argued that since citizens differ in their conceptions of the good life, the government only treats them as equals if it is neutral on matters of ultimate concern. For him, this is not a retreat from moral issues but is the "constitutive political morality" of liberalism (Dworkin 1978, 127). Our ruling morality is neutrality. That means leaving the question of the good life open—to be answered by individual, not collective, choice.

Liberal democracy, then, maintains that the state should provide rights plus (to varying degrees) the economic arrangements (including energy infrastructure) and civil liberties to realize those rights. But it should neither promote the good nor justify its actions by an appeal to any particular conception of the good. No czar gets to dictate the limits within which you must operate—even if they can prove that they are making better choices for you than you would make for yourself. The right to be irrational is a fundamental modern axiom.

Modern energy and technology appear to be the paradigms of neutrality. Insofar as liberal democratic states promote the extraction, production, and development of modern energy resources (like the pursuit of technological innovation more generally), they may only do so on the assumption that energy is neutral with respect to any conception of the good life. No matter what ideal of flourishing you have, the assumption runs, you will need modern energy services to pursue it.

Yet as philosopher Albert Borgmann (1984) noted, this means that modern technologies including energy services like road infrastructures, oil and gas pipelines, and electricity grids are not offered as choices for individuals but rather they are compelled on people as the *basis for choice*. Further, modern energy technologies do not leave the question of the good life open, but rather fill in an answer in terms of an inconspicuous but pervasive pattern that he calls the device paradigm. I prefer the more accessible term *convenience*.

The point is that our default modern energy ethics smuggles a certain ideal of the good life under the guise of neutrality. A paradigmatic example of convenience is the typical grocery store. Picture the aisles of foods from around the world. Imagine the energy it took to get those mini corn dogs in the freezer or that banana to Canada in January. Or picture the materials from around the world convened in the store itself. Modern energy technologies liberate us from any particular place or season. They stitch things together—convening, for example, Asian and North American tourists in throngs at the Vatican or plastic polymers from oil drilled in Texas in a catheter in Finland. This is the moral purpose of modern energy: to liberate and enrich the human condition by making commodities readily available, that is, by convening them at our fingertips.

This modern sense of convenience was first developed in the 1600s by John Locke, who wrote in the *Second Treatise* that "civil government is the proper remedy for the inconveniences of the state of nature" ([1689] 1994, ch. 2, § 13). The central theme of Locke's political philosophy is *increase*. To increase productivity is to provide security from the whims, threats, and limits of nature and allow us to enjoy "the conveniences of life." Locke notes that nature is miserly, refusing to grow in proportion to our desires. This results in a zero-sum game or the kind of static world of sufficiency assumed by the premoderns. Whatever one person gains is at the expense of taking from others. It is not within nature's power to extend her limits, so how can the human condition be rendered more fruitful and convenient?

Locke's answer is human labor. Through labor, we create value. Locke is sketching a political and economic system premised on private property and the technological mastery of nature in the name of convenience. That engineers would receive a royal charter is telling, because for Locke, the job of the public sphere is to manage, secure, and cultivate the growing economy by fostering industrial and technological labors. Locke argues that the Native Americans were "rich in land and poor in all the comforts of life . . . for want of improving it by labor have not one-hundreth part of the conveniences we enjoy" (ch. 5, § 41). Locke notes that a king in America is worse off than a day laborer in England, because the poorest members of a productive society reap material benefits unattainable by even the elite of unproductive societies.

Convenience has become our new public answer to the question of the good life, replacing the premodern paternalist schemes seeking excellence, righteousness, or rationality. Thomas Tredgold's 1828 definition of civil engineering captures this well as "the art of directing the great Sources of Power in Nature for the use and convenience of man." Engineers were chartered to direct the energies of nature in the making of a more convenient world. And this has been unofficially sanctioned as part of the modern Western ruling morality. In other words, the command of energy is seen as a neutral enabler of different visions of the good life and not itself a substantive vision of the good. As the UN Sustainable Development Goal cited above suggests, access to reliable, modern energy services (like electricity) is widely understood as a necessary condition for a decent human life.

So, instead of being guided by an energy czar aiming to implement an ideal of sufficiency, people are governed by networks of energy agents and agencies implementing an ideal of convenience. Unlike the sufficiency ideal (65 GJ), the convenience ideal has no governor and no golden mean, which is why we are stuck with irrational energy overshoot. There is no non-subjective standard for "enough" convenience (witness the private cabins now for first class on international flights). The absence of any such standards is in stark contrast to the way Aristotle wrote confidently about retail trade not being the "natural" way of getting wealth, because it does not stop when people have *enough*. The Latin *convenientia* means "meeting together, agreement, accord, harmony, conformity, suitableness, [or] fitness." In the premodern paradigm, something could be described as convenient if it was in accord with nature, or a morally appropriate fit. The Aristotelean virtues are about striking the right balance. But the mechanical worldview behind Tredgold's engineering project lent a radically new meaning to convenience as ease or the absence of trouble. Thomas Tierney argues that the different meaning has to do with how "suitable" has changed:

> Convenience is no longer a matter of the suitability of something to the facts, nature, or a moral code; suitability in the modern meaning of convenience refers back to the person, the self. Something is a convenience or convenient in the modern sense of these words if it is suitable for personal comfort or ease. (1993, 39)

There is no need for a limiting doctrine of the mean, because more experiences and commodities can always be made more comfortably available. There is no "objective" standard over and above individual preferences or, in economic parlance, demand.

Using technological labor (especially through engineering) to harness energy to provide convenience is the modern project. Modern energy ethics is a moral vision about the good life (commodities rendered conveniently available) and justice (extending this kind of life to more and more people). It is also, as we have noted, fundamentally irrational, because it has no governor to prevent the mindless and destructive excess that characterizes much of the developed world. As Henry David Thoreau noted, we set elaborate traps to catch convenience only to get our own legs caught in the trap!

PRESENT: ECOMODERN ENERGY ETHICS

Locke was writing when the global population was about five hundred million. Is the modern project of increase and convenience a recipe for disaster for a planet with eight billion people? After all, Earth is finite and the growth of a global high-energy civilization is already causing massive environmental problems even as billions of people languish in energy poverty and forecasts call for global energy consumption to double. Can the (irrational) ethics of convenience-for-all be sustained?

The energy analyst Vaclav Smil thinks not: "The benefits of high energy use that are enjoyed by affluent countries, that is by less than one-sixth of humanity consuming >150 GJ per capita, cannot be extended to the rest of the world" (2010, 728). He argues that attempts to keep growing energy production and consumption will not work. If the energy comes from fossil fuels, it will accelerate climate change. Renewable energy sources lack sufficient power density to do the job. And energy efficiency measures are good, but "without concurrent limits on consumption they become a part of the problem rather than an effective solution because they stimulate rather than reduce the overall energy use" (727). As an example, Smil offers the car: in the United States from 1965 to 2005, fuel economies (efficiencies) improved and the average miles driven per car increased. People used the money saved from more fuel-efficient cars to drive more.

Smil concludes that we must abandon the imperative of increase and endless convenience. The global energy problem is "a moral dilemma" that can only be solved, by putting "in place rational limits that guarantee a decent quality of life for an increasing proportion of humanity while preserving the integrity of the only biosphere our species will ever inhabit" (2010, 728). In other words, Smil thinks we need to pursue a sufficiency project of not just increasing energy access for those in poverty but also decreasing the irrational energy overshoot of the wealthy.

Smil is obviously right. Rational limits make sense—everyone could be happy and healthy on a more equitable and sustainable planet if all made do with 65 GJ. Yet, as noted above, Smil's call for rational sufficiency is itself irrational. It violates what the science policy scholar Roger Pielke (2010) calls the "iron law" of climate change: when economic growth collides with proposals for reductions in emissions or consumption, economic growth will win the political argument every time. This law holds even when further economic growth fails to bring improvements to human well-being, that is, even when it is irrational.

Pielke is a prominent voice in ecomodernist theory, which is the best explicit expression of the default energy ethics project already at work. Again, this is an "irrational" energy ethics, because it takes people's (often irrational) desires as immutable. Given that people are not going to be convinced to consume less, how do we achieve the goals of justice and sustainability? The default answer is to make the project of convenience-for-all sustainable: green growth. Smil may be right that this is foolish. Then again, it might work. In any case, it is worth spelling out the ethics of green growth, because it is the basic formula being followed by high-energy civilization.

As the name implies, ecomodern is an extension of modernity (see Symons 2019). It adopts Locke's insight that human labor, far more so than nature, creates wealth, resources, and convenience. Copper wire does not just pop out of the ground; it has to be made. The natural supply of copper is finite, but scarcity, on a pragmatic level, only means an increase in price. When prices go up, people can use their ingenuity to either find more of a resource or a replacement for it. Fiber optic cables can replace copper wire and PEX can replace copper pipes. After all, people do not want the copper, they want the streaming entertainment or hot water. More generally, people do not want energy per se; they want what energy allows them to do and to have, especially comfort and convenience.

The economist Julian Simon (1981) was one of the first to spell out this extended Lockean logic, calling the human mind the "ultimate resource," because it is what creates value. From Simon and his followers (e.g., Lomborg 2001; Shellenberger 2020), the basics of green growth can be distilled: (a) as we begin to use more energy, we initially harm the environment; but (b) we then pass peak impact and the environment actually improves (for example, London in 2015 had three times more GDP per capita and thirteen times less air pollution than London of 1952); (c) pessimists err in underestimating future technological capabilities and overestimating the human tolerance for regulations; (d) nature is almost always more dangerous than technology; (e) the best metric for human flourishing is economic growth; and (f) justice is best measured in absolute historical terms such that just results are achieved when the living standards of the poorest people are raised, even if extreme inequalities remain (recall, even a pauper in England is better than a king in America, because the pauper benefits from the conveniences of a productive society).

The ecomodern theorists Michael Shellenberger and Ted Nordhaus summarize the formula: "The solution to the unintended consequences of modernity is, and always has been, more modernity—just as the solution to the unintended consequences of our technologies has always been more technology" (2011). For example, we solved the ozone hole problem not by limiting consumption of refrigerators and air conditioners but through technological fixes. There is a dialectical logic to this ethics. We use technology to overcome serious problems like food spoilage and heat stress. This almost inevitably creates new unintended problems like the ozone hole, but these problems are better and more manageable than the ones we had before. We just keep refining solutions. Climate change is a good example as climate-related deaths have actually dropped precipitously due to the securities afforded by a fossil-fueled civilization (Epstein 2014). True, climate change is a new unintended problem, but it is more manageable and less immediate than the original problem of being exposed to the elements and the vicissitudes of nature.

Green growth is also known as decoupling, because the goal is to unhitch resource use from economic growth. As in the London case or the ozone hole, peak impact is reached, beyond which material prosperity continues to rise even as negative environmental impacts diminish. This default energy ethics pursues projects to decarbonize (but *not* shrink the economy). It is the grand pursuit of decoupling on a global scale. Again, it may be a fool's errand (see Parrique *et al.* 2019)—it is an *irrational* ethic after all! Nonetheless, it is the guiding logic and existential gamble of our high-energy civilization.

FUTURE: BEYOND EARTH, BEYOND BODIES

In 2015, the billionaire tech mogul and philanthropist Bill Gates launched the Breakthrough Energy initiative to invest over $1 billion "to make sure that everyone on the planet can enjoy a good standard of living, including basic electricity, healthy food, comfortable buildings, and convenient transportation, without contributing to climate change" (https://www.breakthroughenergy.org, accessed December 29, 2020). Their website notes that global energy consumption is forecasted to double by 2050: "And that's a great thing: The more access to energy people have, the larger our economies grow and the better our lives become."

Elsewhere, Gates writes, "If I had to sum up history in one sentence it would be: 'Life gets better.' . . . And the reason is energy" (Gates and Gates 2016). The command of energy allows us to power lights, refrigerators, air conditioners, and other machines, and it allows us to build the infrastructure that can deliver clean water and provide security. Over the next thirty years, humanity will use modern energy to build the equivalent of a new New York City *every month*. "Without access to energy," Gates continues, "the poor are stuck in the dark, denied all of these benefits and opportunities that come with power."

Gates is articulating our default energy ethic of green growth, that is, the ecomodern theory discussed above. He makes no mention of *reducing* consumption by the rich, just increasing consumption by the poor via greener technologies. Gates is a fan of Smil's scholarship, yet he is more optimistic that a technological *breakthrough* will allow us to have our cake and eat it too—to keep growing the economy without killing the planet. Smil and Gates offer two different technological imaginaries for the future.

The sufficiency ideal of Smil, like that of Mitcham and Rolston's Type II energy ethic (2013), is part of a worldview that might be called precautionary (see Briggle 2015). This is a view premised on limits and the dangers of humanity overstepping its proper reach. It is the view of those who see sustainability in terms of ecological footprints, limited carrying capacities, and upper bounds to energy consumption. From this view, we are heading for a wall and collapse unless we shrink or degrow (see Latouche 2010). This view assumes, as Smil did above, that we are forever Earth bound and must reverse the growth imperative if we are to survive as a species on a finite planet.

Writer Amitav Ghosh (2017) puts this view in poignant terms. It is true, he notes, that by any account of distributive justice, the poor of the Global South have been deprived are entitled to their share of the economy. Ghosh doesn't buy the optimistic logic of the ecomoderns, however. So, we are obligated to produce more energy to reduce poverty, but this moral obligation is a recipe for planetary suicide: "Our lives and our choices are enframed in a pattern of history that seems to leave us nowhere to turn but toward our self-annihilation" (Ghosh 2017, 111). Our irrational energy ethics will kill us in the name of a utopian ideal of convenience for all.

The green growth ideal is part of a far different imaginary, one that may well be so techno-fabulist that it proves to be an unrealizable fantasy. Then again, it may prove to be our future. We are, at least, on that trajectory. From this perspective, we are not running into a wall, but rather pushing through a bottleneck (see Karlsson 2015). We will (to use a favorite ecomodern term) *breakthrough* to the other side of peak impact where we can rewild nature, bring species back from extinction, and fine-tune the ppm of CO_2 in the atmosphere. If we extend our timeline, the present level of destruction may not seem so irrational after all. At least it need not be permanent. The Industrial Revolution was destructive, but in the long run the benefits will far outweigh the costs.

This techno-optimist imaginary is often criticized for being captured by short-sighted cycles of innovation and profit. But it is actually very good at seeing things from the long run. It sees a future, say, where lab-grown meat means that we can restore the prairies and ocean fisheries while still enjoying "steak" and "fish." On another long-term front, even though entropy will always exist, it may be technically possible to reach efficiency gains of such an extent that they eliminate the counterproductive rebound effect.

Rooted as it is in the modern vision of liberation via the technoscientific control of nature, this view is limitless. Humanity is not defined by its ape-like origins—artificial intelligences will join the "Republic of Humanity" (see Fuller 2019). Further, humanity is not destined to remain on Earth. Jeff Bezos, founder of Amazon and one of the world's wealthiest people, has said that his most important project is Blue Origin, an aerospace manufacturing company that is making rockets for extraterrestrial resource extraction and space colonization. Bezos thinks we will be able to zone Earth as residential only, because energy extraction and heavy industry will be able to be done on other planets and asteroids. As the tagline for Blue Origin reads, "Earth, in all its beauty, is just our starting place."

Of course, we still have to get through the present bottleneck before we get to Mars and we are not on track to reach climate targets. Yet there are hopeful signs about the near future. Consider just two. First, most developed nations are decarbonizing their economies. True, it is not fast enough, but there is *rational* hope that decarbonization rates will increase (see Drawdown 2020). Second, through 2040 "all of the growth in energy demand" will come from developing economies (BP 2019). Per capita energy consumption in wealthy nations has pretty much leveled out. Greater efficiencies almost guarantee that future Earthlings won't be guzzling more and more energy. So, there is an upper limit to consumption, one that does not require the heavy hand of an energy czar.

CONCLUSION

The energy czar thought experiment divided the current "energy pie" on the claim that everyone can flourish sustainably with an equal allotment of 65 GJ. Our default ecomodern energy ethic grows the pie rather than slicing it into equal shares. Maybe, though, given material path dependencies and value commitments to affluent lifestyles, this will work out as a more practical path to the same goals. In other words, perhaps the work of the energy czar is being accomplished through the ecomodernist project that we have been exploring.

Recall the czar's task: distribute energy in a way that (a) makes everyone happy and healthy; (b) is fair; and (c) reduces the risks and impacts of environmental problems like climate change. We can call these three goals the good life, justice, and sustainability. The ecomodernists argue that (a) global standards of living are rising; (b) more and more people have access to education and economic opportunities; and (c) climate-related risks are decreasing.

Insofar as they are right, this progress has not been made through one central planner like the czar. But it has also not been made solely through the workings of a laissez-faire market. Rather, it has been made through the efforts of multiple governments, agencies, and private-sector actors following the moral logic traced above in our discussions of ecomodernism. We can add to this the notion that per capita energy use seems to reach a "natural" saturation point without a czar's mandate to limit consumption. Recall the bottleneck metaphor—we may be going through an unsustainable period, but we will *breakthrough* to a more sustainable way of life if we just keep following the same moral logic that has created this temporary unsustainable situation. So, for the ecomodernists, things only appear irrational now, because we are in a temporary moment of overshoot. In one hundred years or so, we will be much further along to the Good Anthropocene, where standards of living are universally high and metrics of environmental sustainability (e.g., species loss, decarbonization, pollution) are all improved.

Of course, the ecomodernist project can be critiqued in the context of all three goals (see Briggle 2021). For the good life (a), one can argue that ecomodernists rely on shallow indicators of flourishing like GDP that fail to capture the pathologies of high-energy life (e.g., workaholism and lack of community) or the ways that life is stamped across the globe in a form of bio-cultural homogenization (see Rozzi 2013). For justice (b), one need only point to the massive distributional inequities that persist both within and between nations as well as shortcomings in recognition and participatory justice, say, for indigenous peoples. For sustainability (c), one can point out several things: the continuing failure to achieve general decoupling (Parrique et al. 2019), still growing carbon emissions, and all the other ecological problems in addition to carbon (such as species loss, plastics, ocean acidification, increased rates of zoonotic diseases, and "forever chemicals"). My point here is that this project—this ecomodernist moral logic—is nearly as powerful as the imagined energy czar. Indeed, at the heart of this project and its logic is what Pielke (2010) calls the "iron law" of climate policy. People's willingness to pay to achieve environmental objectives decays rapidly. This means, in aggregate, that any viable climate policy will have to be consistent with the objective of economic growth. Thus, green growth holds a tight ideological grip on the social actors and processes that drive our world.

In *The Imperative of Responsibility*—the book that arguably launched precautionary environmental politics—Hans Jonas (1984) develops a "heuristics of fear." Put in the present context, Jonas would say that people's willingness to pay for environmental objectives stems from an existential or aesthetic dilemma: environmental problems like climate change are often not immediately pressing. These "slow motion" disasters fail to elicit a strong emotional reaction like, say, the fear one feels when something goes bump in the night. This is the seed of the most central argument used by traditional environmentalists against ecomodernism. In brief, modern development allows us to build more layers of insulation around ourselves, further dampening our experience of the effects of our own actions. This only heightens the stakes of an eventual collapse, because we will fail to act until it is too late.

Our "naked" fear (at the bump in the night, say) comes to the aid of reason. We are acting irrationally, in Jonas's view, because we lack this fear to aid our reason when it comes to environmental issues. Thus, we need to develop and communicate publicly a more learned, even spiritual, fear. We have to learn to feel moved into action by the mere projections of future catastrophe. Doing so would supply urgency to overcome our low willingness to pay such that people would willingly sacrifice and reduce consumption.

Jonas's prescription of fear-inspired self-limitation has been the playbook for environmentalism ever since concerns about "limits to growth," "population bombs," and "silent springs." Yet as leading ecomodernists Ted Nordhaus and Michael Shellenberger (2011) argue, this strategy has been an abject failure. We won't solve global warming by "trying to scare the pants off the American public" or by "behavioral changes" such as voluntary reductions in consumption. As if following the commands of a czar, our only option is to press on with green growth.

REFERENCES

Arendt, Hannah. 1958. *The Human Condition*. Chicago: University of Chicago Press.

Baer, Paul, Sivan Kartha, Tom Athanasiou, and Eric Kemp-Benedict. 2009. "The Greenhouse Development Rights Framework: Drawing Attention to Inequality within Nations in the Global Climate Policy Debate." *Development and Change* 40, no. 6: 1121–38.

Aristotle. 1984. *Politics*. In *The Complete Works of Aristotle*. 2 vols, edited by Jonathan Barnes. Princeton, NJ: Princeton University Press.

Borgmann, Albert. 1984. *Technology and the Character of Contemporary Life*. Chicago: University of Chicago Press.

BP. 2019. *BP Statistical Review of World Energy, 68th Edition*. https://www.bp.com/content/dam/bp/business-sites/en/global/corporate/pdfs/energy-economics/statistical-review/bp-stats-review-2019-full-report.pdf.
Briggle, Adam. 2015. *A Field Philosopher's Guide to Fracking: How One Texas Town Stood up to Big Oil and Gas*. New York: Liveright.
———. 2021. *Thinking through Climate Change: A Philosophy of Energy in the Anthropocene*. New York: Palgrave Macmillan.
Drawdown. 2020. "The Drawdown Review: Climate Solutions for a New Decade." https://drawdown.org/drawdown-review.
Dworkin, Ronald. 1978. "Liberalism." In *Public and Private Morality*, edited by S. Hampshire, 113–43. Cambridge: Cambridge University Press.
Epstein, Alex. 2014. *The Moral Case for Fossil Fuels*. New York: Penguin.
Fuller, Steve. 2019. *Nietzschean Meditations: Untimely Thoughts at the Dawn of the Transhuman Era*. Basel: Schwabe Verlag.
Gates, Bill, and Melinda Gates. 2016. "Two Superpowers We Wish We Had." *GatesNotes*, February 22, 2016. https://www.gatesnotes.com/2016-Annual-Letter.
Ghosh, Amitav. 2017. *The Great Derangement: Climate Change and the Unthinkable*. Chicago: University of Chicago Press.
Hardin, Garrett. 1968. "The Tragedy of the Commons." *Science* 162 (3859): 1243–48.
Illich, Ivan. 1974. *Energy and Equity*. New York: Harper & Row.
Jochem, Eberhard, Daniel Favrat, Konrad Hungerbühler, Philipp Rudolph von Rohr, Daniel Spreng, Alexander Wokaun, and Mark Zimmermann. 2002. *Steps Towards a 2000 Watt Society—A White Paper on R&D of Energy-Efficient Technologies*. Zurich: ETH Zurich and Novatlantis.
Jonas, Hans. 1984. *The Imperative of Responsibility: In Search of an Ethics for the Technological Age*. Chicago: University of Chicago Press.
Karlsson, Rasmus. 2015. "Three Metaphors for Sustainability in the Anthropocene." *The Anthropocene Review* 3 (1): 21–32.
Latouche, Serge. 2010. *Farewell to Growth*. Translated by David Macey. Malden, MA: Polity.
Locke, John. [1689] 1994. *Two Treatises of Government*, edited by Peter Laslett. Cambridge: Cambridge University Press.
Lomborg, Bjorn. 2001. *The Skeptical Environmentalist: Measuring the Real State of the World*. Cambridge: Cambridge University Press.
Madrigal, Alexis. 2008. "MIT Class Calculates Carbon Footprint of 'the Man.'" *Wired* 30 (April). https://www.wired.com/2008/04/mit-class-calcu/.
Mill, John Stuart. [1861] 2001. *Utilitarianism*, edited by George Sher. New York: Hackett.
Mitcham, Carl, and Jessica Smith Rolston. 2013. "Energy Constraints." *Science and Engineering Ethics* 19, no. 2: 313–19.
Nordhaus, Ted, and Michael Shellenberger. 2011. "The Long Death of Environmentalism." https://thebreakthrough.org/issues/energy/the-long-death-of-environmentalism.

Okulicz-Kozaryn, Adam, and Micah Altman. 2020. "The Happiness-Energy Paradox: Energy Use Is Unrelated to Subjective Well-Being." *Applied Research in Quality of Life* 15, no. 4 (March): 1055–67.

Parrique, Timothé, Jonathan Barth, François Briens, Christian Kerschner, Alejo Kraus-Polk, Anna Kuokkanen, and Joachim H. Spangenberg. 2019. *Decoupling Debunked: Evidence and Arguments against Green Growth as a Sole Strategy for Sustainability*. European Environmental Bureau. https://eeb.org/library/decoupling-debunked/.

Pielke, Roger. 2010. *The Climate Fix: What Scientists and Politicians Won't Tell You about Global Warming*. New York: Basic Books.

Raworth, Kate. 2017. *Donut Economics: Seven Ways to Think like a 21st Century Economist*. London: Random House.

Rozzi, Ricardo. 2013. "Biocultural Ethics: From Biocultural Homogenization to Biocultural Conservation." In *Linking Ecology and Ethics for a Changing World*, edited by Ricardo Rozzi, S. T. A. Pickett, Clare Palmer, Juan J. Armesto, and J. Baird Callicott, 9–32. London: Springer.

Robinson, Kim Stanley. 2020. *The Ministry for the Future*. New York: Orbit.

Shellenberger, Michael. 2020. *Apocalypse Never: Why Environmental Alarmism Hurts Us All*. New York: Harper.

Shellenberger, Michael, and Ted Nordhaus. 2011. "Evolve." *Orion Magazine*, August 25, 2011. https://orionmagazine.org/article/evolve/.

Simon, Julian. 1981. *The Ultimate Resource*. Princeton, NJ: Princeton University Press.

Smil, Vaclav. 2010. "Science, Energy, Ethics, and Civilization." In *Visions of Discovery: New Light on Physics, Cosmology, and Consciousness*, edited by Raymond Y. Chiao, Marvin L. Cohen, Anthony J. Leggett, and Charles L. Harper Jr., 709–29. Cambridge: Cambridge University Press.

Smil, Vaclav. 2017. *Energy and Civilization: A History*. Cambridge, MA: MIT Press.

Strauss, Leo. 1953. *Natural Right and History*. Chicago: University of Chicago Press.

Symons, Jonathan. 2019. *Ecomodernism: Technology, Politics, and the Climate Crisis*. Cambridge, UK: Polity.

Tierney, Thomas. 1993. *The Value of Convenience: A Genealogy of Technical Culture*. Albany: State University of New York Press.

Tredgold, Thomas. 1828. "Civil Engineering," used in the Royal Charter of the Institution of Civil Engineers and published in *The Times*, London. Quoted in *Humanitarian Engineering* by Carl Mitcham and David Muñoz, 3, San Rafael, CA: Morgan & Claypool, 2010.

United Nations. 2020. "Progress towards the Sustainable Development Goals." https://undocs.org/en/E/2020/57.

Voosen, Paul. 2018. "The Realist." *Science* 359 (6382): 1320–24.

Chapter 24

Paradoxical Policy in Sub-Saharan Africa

Women's Farming, Oil, and Sustainable Development

Tricia Glazebrook and Gordon Akon-Yamga

This chapter aims to expose exacerbation of the growing threat from climate change on women's subsistence farming when African governments choose to reduce poverty using oil development. That is, women's responsibility for food security is paradoxically supported by governments that at the same time are contributing to the very causes of women's increasing struggle to meet food and nutritional needs. We analyze climate impacts on women farmers, childhood development, and humanitarian crises in hunger using an environmental justice approach, given impacts of environmental degradation on women, and a social justice approach, given the "feminization of poverty" and the impacts of hunger and malnutrition on childhood development and health. Because the basic principle of environmental justice is that environmental harms are also social and societal harms, our social justice approach is embedded in environmental justice. That is, in our approach, environmental issues are understood also to be social and societal issues, especially concerning women's agricultural production that can be for many in sub-Saharan Africa (SSA) the only barrier between food security and starvation.

Countries in SSA that discover oil can develop this "windfall resource" toward achieving the first Sustainable Development Goal (SDG1) of poverty alleviation (UN 2015). This is especially important for Africa, which was the only continent that, even after 4.5 percent economic growth during the 1990s, did not meet the 2015 deadline for the first of the Millennium Development

Goals (MDGs)—that is, MDG1 that combined eradication of extreme poverty and hunger in one Goal (FAO, IFAD, and WFP 2015). The World Bank anticipates that the world's poor will increasingly be African, even though the proportion of people living daily in Africa on less than the equivalent of $1.90 USD dropped from 56 percent in 1990 to 43 percent in 2012. Despite this decrease, population growth resulted in more people in poverty—that is, 284 million people in poverty in 1990 grew to 388 million in 2012—an increase of more than 100 million people (Beegle et al. 2016).

Though MDG1 combined poverty with hunger (World Bank n.d.), the SDGs separate them into SDG1 on poverty and SDG2 on hunger (UN 2015). Food security and poverty are inherently connected because, for instance, the poorer the farmer is, the less the farmer can afford resources to meet household needs or increase agricultural productivity. Separation of poverty from hunger in the SDGs is quite possibly creating an ideological partition between agriculture and economics in national policy planning, design, and implementation systems that fail to acknowledge women's role in food security. Offices responsible for policy decision-making may create policy dissociation by not regularly communicating with other offices.

The primary justices at issue in this chapter are distributive justice, climate justice, gender justice, food justice, and food sovereignty. Our arguments are not that women farmers in SSA are failing and need to be "saved" through interventions from the global North. Rather, the global North has an obligation to take responsibility for, and cut back substantially, its contribution to the atmospheric greenhouse gases (GHGs) driving climate change, and must deliver the funding promised in the Paris Agreement to rectify loss and damage that climate change has already caused and continues to cause in breach of distributive and other justices. Africa emitted only 3.8 percent of global GHG emissions in 2019 (CDP 2020), up by only 0.1 percent since 2009 (Canadell, Raupach, and Houghton 2009). Yet Africa is experiencing disproportionate climate-related damage (Lisk 2009). That is, Africa experiences high climate impacts while contributing very little to their source (see Glazebrook et al. 2020). The global North is thus not being called in this chapter to "help" Africa but to stop contributing to climate change that is driving mass crises in hunger already well under way in SSA.

Before we begin analysis, we here define key terms and provide an overview of this chapter's structure. *Food security* is understood in various ways. We use the UN Food and Agriculture Organization's definition: "when all people, at all times, have physical, social and economic access to sufficient, safe and nutritious food which meets their dietary needs and food preferences for an active and healthy life" (FAO 2003). Our focus is primarily on sufficiency and nutrition, though we also take the FAO's reference to "preferences" to include food sovereignty, also a disputed term (Patel 2009), following the 2007 Declaration of Nyéléni: *food sovereignty* is "the right of peoples to healthy and culturally appropriate food produced through ecologically sound and sustainable methods, and their right to define their own food and agriculture systems" (DoN 2007). The connection between food security and poverty is very much an issue in SSA, where women play a substantial role as subsistence agriculturalists (SOFA Team and Doss 2011) who grow the family's food with only a hand-hoe and fertilizer they make from animal dung (Glazebrook 2011). This connection is increasing for women subsistence farmers as climate impacts on crop production worsen.

Discussion focuses on Ghana in order to provide an account of women's on-the-ground lived experience, though we also draw on data concerning West Africa and SSA as a whole. *Oil* is the term used to indicate the broader range of petroleum resources, for example, crude oil, natural gas, ethane, butane, and propane. *Paradox* is intended etymologically from the Greek to indicate an "opinion" (*doxa*) that is contrary (*para*)—that is, an idea, way of thinking, or practice that works contrary to its own interests. A paradox is thus not unsolvable but must be identified and understood before it can be resolved. The paradox in SSA of oil development while climate change harms agriculture can be resolved by identifying and understanding the current situation of women farmers.

Finally, women farmers are vulnerable to poverty and hunger and are ultimately described in this chapter as *abject*. Rather than drawing from theoretical analyses of abjection, for example, Julia Kristeva's *Powers of Horror* (1982), our analysis is an etymological retrieval of the term to understand women's current lived experience as farmers and food providers in SSA. Our account of women's abjection does not mean that women are not strong, resilient, innovative agents of change, or weak and lacking in pride and dignity. On the contrary, "abject" is taken directly from the Latin *iacere*, meaning "to throw," and *ab*, meaning "away." That is, women subsistence farmers in SSA, though well recognized in the subsistence economy (Opoku and Glazebrook 2018), are largely invisible in local, regional, and international policy because they do not make direct economic contributions to their country's GDP or GNP. Women farmers are thus seemingly disposable, whether intended or not. We show that these women are crucial contributors to food security in SSA, yet their work is threatened by development policy that is causing paradoxical tension between SDG1 and SDG2.

Concerning structure, this chapter first lays out the long-standing debate about whether African oil resources are a "blessing or curse," with the conclusion that a "curse" is not inevitable but determined by governance and policy. The subsequent section describes women farmers' situation and role in food security in SSA, with specific attention to West Africa and Ghana. We show that women are relied upon as food providers yet economically invisible, and we detail challenges women farmers face, including climate impacts and adaptations. Finally, we extend existing arguments that *sustainable development* is an oxymoron (Sachs 2015) by arguing that the phrase functions globally to obscure the impossibility of sustaining constant growth. We find that international policy, including the SDGs, is committed to implementation of economic systems. The United Nations (UN), World Bank, International Monetary Fund (IMF), and other international governance and finance bodies are negative influences on SSA that undermine its capacity to achieve SDGs 1 and 2 concurrently. Using oil to reduce poverty in SSA contributes to climate impacts on food production and is paradoxically a threat to the most poor and vulnerable—that is, women subsistence farmers and their children.

OIL: BLESSING OR CURSE?

Historically, oil has long been a source of wealth in the global North. For example, oil took the state of Texas from a GDP of $119 million USD in the late nineteenth century to $29 billion USD by 1957 (Michaels 2010; Barr 2020). In 1947, in Alberta, Canada, Imperial Oil, after drilling 130 dry holes, discovered oil just outside Edmonton, thereby launching an "Oil Boom" in which the "very existence of active, wealthy cities in the Canadian prairies . . . reflects the potential offered to a region by nature's bounty of energy resources" (MacFadyen and Watkins 2014, 23). By the 1950s, Saudi Arabia and Venezuela had benefitted from oil wealth for decades (Ginsburg 1957), as have Chile, Australia, and Malaysia more recently (Amiri et al. 2018).

Oil wealth is not always beneficial, however. In 1957, Norton Ginsburg argued that resource endowment provides a path to economic growth, especially for poorer countries, by providing the initial capital needed to launch economic development, and later that fossil fuels are the best resource for that development (Ginsburg et al. 1986). Yet discovery of offshore oil in the Netherlands in 1959, anticipated to grow the national economy, had the opposite impact. The economic downturn showed that Ginsburg was right that poorer countries are likely to do better with resource exploitation than wealthier ones. The Netherlands was past early stages of economic development by 1959 and had a stable manufacturing and industry-based economy. The explanation given for the downturn was that investment in oil had pushed aside the manufacturing sector and driven up the value of the Dutch guilder, thereby making Dutch industries less competitive in international markets (Ebrahimzadeh 2020). This phenomenon became known as the "Dutch disease" (*Economist* 1977) and "the resource curse."

Poor countries in SSA do not always follow Ginsburg's pattern, however, and are prone to economic and political harms during natural resource development. A "windfall," especially of oil, can drive conflict and corruption. In the Sudan, for example, the Darfur genocide was funded by oil revenues, and foreign oil companies' resources, for example, fuel stations and runways, were used for refueling to attack villages of indigenous, traditional agriculturalists (Glazebrook and Story 2012). In the Niger Delta, oil development by corrupt governments in collaboration with multinational oil companies caused decades of damage to communities and ecosystems, upon which people depend for their livelihood and survival. Organized resistance led to the hanging of Ken Saro-Wiwa and the Ogoni Nine in 1995 after trial in an alleged "kangaroo court," with Shell Oil accused of complicity in unlawful arrest, detention, and execution (Amnesty International 2017), though the widows of four of the men recently lost their legal case against Shell (Holligan 2022). Armed "freedom fighters" still profiteer in the Delta by kidnapping foreigners for ransom, theft, and threatening communities with violence and extortion (Glazebrook and Kola-Olusanya 2011). These conflicts are obvious "curses," whereas the economic effects are unclear, with much debate about the unexpected consequences of natural resource wealth.

Gordon Akon-Yamga (2018, 36–73) documented factors identified in long-standing discussion of how natural resource wealth becomes a "curse," with skepticism that such negative consequences are inevitable. The "curse" has been explained in a variety of ways: the "Dutch disease" (Gelb 1988; Sachs and Warner 2001); a country's political system and level of industrialization (Auty 1994); low investment in human capital and poor education (Gylfason, Herbertsson, and Zoega 1999); declining terms of trade; unstable markets; dependency on multinationals; and policy failure to connect the natural resource sector with other sectors (Ross 1999). Discussions of why governments fail on policy include short-sighted "myopic sloth" or "myopic exuberance," empowerment of social groups that impede policy and growth, undermining of long-term development programs by "windfall" income, government ownership of the industry that would do better in the private sector, and government failure to enforce property rights (Neary and van Wjinbergen 1986; Ross 1999).

The curse is increasingly attributed to policy that allocates state resources poorly, and to the quality and role of institutions that largely manage the translation of politics into policy. Robinson, Torvik, and Verdier (2014) point to government over-expenditure and poor planning of public sector investment to raise employment and wages, and over-extraction that increases corruption, as politicians and their patrons seek to maximize revenues during their leadership and often divert rents from resource extraction infrastructure to the political elite rather than to the communities and population. Desire to retain "old privileges" can also result in policy corruptions instead of public benefit (Torik 2011). Ultimately, whether resources are a "blessing or curse" hinges on those governing and the policies they make.

Taking governance, policy, and corruption into account, the view that natural resources are advantageous for economic development began to hold sway among scholars. Papyrakis and Gerlagh (2004) and Davis and Tilton (2005) hold that natural resources would benefit economies absent other factors, for example, corruption, low investment, protectionism, deteriorating trade terms, and poor education. Bulte and Damania (2008) argue that the potential to bribe government officials can affect decision-making and make governance inefficient as entrepreneurial talent is pushed aside by those who accept bribes and neglect policies that could maximize growth. Brunnschweiler and Bulte (2008) argue that the "resource curse" is caused by the "red herring" (248) of assessing short-term fluctuations of "economic flows"—that is, the intensity of and dependence on oil in the market. They argue instead for tracking availability of oil resources over twenty- to twenty-five-year periods to better understand oil wealth over time. They also argue that "institutional quality," durable societal characteristics on the one hand and the flexibility of policy on the other, should be considered "constitutional variables." Policymakers can shape these variables by setting rules that establish new institutions that deter corruption, improve law enforcement, and promote investment.

Davis and Tilton (2005) also hold, however, that because every country uses a unique set of variables, no standardized approach or set of criteria can measure a country's development. We propose, however, that environmental justice functions as such a criterion in Ghana, where oil was found offshore in 2007 and first extracted in 2011 (Akon-Yamga 2018, 74). Our argument shifts the issue from economics alone to human experience, because development is not just measured as wealth. Amartya Sen's work and Mahbub al Haq's argument that data on GDP provides limited views of people's well-being led to the Human Development Index that began in 1990 to measure human development based on life expectancy, education, and gross national income per capita rather than simply GDP, in order "to emphasize that people and their capabilities should be the ultimate criteria for assessing the development of a country, not economic growth alone" (UNDP 2022). The idea that increased wealth drives development became weak in development theory more than two decades ago (Ranis, Stewart, and Ramirez 2000).

The "resource curse" debate described above does not consider the impact of oil wealth on agriculture. The focus was on economic impacts, and subsistence farming has little if any economic footprint. This oversight is exacerbated by the "enclave" nature of the petroleum industry in many SSA countries (Ferguson 2006; Ackah-Baidoo 2012), including Ghana, where it functions independently from the larger Ghanaian economy. As long as national economies in SSA rely on petroleum-based wealth, they will add to the greenhouse gases causing climate change. They thus harm food security by weakening agricultural production. The first and hardest hit in the humanitarian crises of hunger already underway in SSA are the poorest who have little alternative access to food, and meager resources to adapt to climate impacts.

This section has shown that decades of discussion of the "resource curse" have led to the insight that policy is crucial for causing or escaping the curse, but they have not led to understanding that agricultural and gender policy play a significant role in addressing the curse of climate-driven hunger. These discussions thus provide no basis for assessment of the sustainability of food security. This analytical lacuna is identified in the final section as an impediment to understanding how SSA risks SDG2 (hunger) in attempting to achieve SDG1 (wealth) by means of oil development. The next section shows that much of SSA's food security depends on women subsistence farmers who are extremely vulnerable to impacts of climate change on crop production.

WOMEN FARMERS, CLIMATE, AND HUNGER IN SUB-SAHARAN AFRICA

Globally, legal restrictions leave 2.7 billion women without the job options men have. Women's informal employment in family businesses is unrecognized—that is, they are not included in censuses as owners or employees (UN Women n.d.). When women do have paying jobs, the gender wage gap average is 23 percent. Women spend 2.5 times more time on domestic labor than men. Their rate of entrepreneurship is less than half that of men in at least forty countries. As of 2021, women make up just over 5 percent of Fortune 500 CEOs.

Similarly troubling statistics are found in agrarian areas. Close to half the global population—over 3 billion people—live in rural areas (FAO et al. 2021). About 1.4 billion women are farmers who contribute substantially to family food security (Glazebrook et al. 2020). Five hundred million of these women own no land and together have access to less than 5 percent of agricultural resources (FAO et al. 2021). Despite a third of women being employed in agriculture worldwide, less than 15 percent of landowners are women (UN Women 2021). Of the approximately 570 million farms in the world, some 51 million (9 percent) are in SSA (Lowder, Skoet, and Raney 2016), where roughly 41 million are family small-holdings (less than five acres) that together provide about 80 percent of what is eaten in SSA (FAO 2017, xi).

At the same time, across the planet, women are at 10 percent higher risk of hunger than men (UN Women 2019) due in part to the "feminization" of both poverty and agriculture. The "feminization of poverty" refers to the increasing female proportion of the global poor as well as women's lower pay, especially in agricultural employment. The "feminization of agriculture" manifests in women's small-scale, subsistence farming that can increase women's independence, responsibility, and decision-making power, and sometimes provide limited income, but also can increase women's vulnerability to poverty and food insecurity (Vaqué 2017).

Decades of evidence shows that women farmers suffer disproportionate economic and other harms caused by environmental degradation (Diamond 1990; Warren 2000). Philosophers and environmental ethicists have also made the direct connection between gender and environmental degradation to show that women's lives in particular worsen in direct proportion to local ecosystem damage (Shiva 1988; Glazebrook 2011). Given the feminization of poverty and agriculture, and the vulnerability of the poor to environmental hazards, it is no surprise that women are disproportionately threatened by degrading environments and the first affected (Eade and Williams 1994, 66).

Women farmers in SSA also face multiple inequities. They have weak land tenure and little control over the land they farm, which they might lose at any time. They face a large set of long-standing issues of persistent bias that affect access to agricultural supports, finance, and technologies. That is, they have less access to livestock, tools and equipment, markets, bank accounts, and credit (Glazebrook 2011; Vaqué 2017). These hardships are exacerbated by climate change that for at least two decades has increasingly impacted weather, water, and soil, and hence food security, food sovereignty, and nutrition.

A minority voice of feminist women lawyers in SSA, supported by international gender advocates and development experts, has brought attention to these problems. They argue for land tenure reform through state law and policy, despite retreat of state governance after failures in SSA's postcolonial states led to interventions by international bodies such as the World Bank that imposed structural adjustment programs (Whitehead and Tsikata 2003). The women maintained their position, despite lack of support. For example, the IMF rejected Ghana's inclusion of women farmers' needs in their 2003 Poverty Reduction Strategy (Opoku and Glazebrook 2018). In the meantime, women's capacity to provide food security is dropping.

Since at least 2007, SSA has experienced growing hunger crises. Food security has always been at risk in the dry regions because of the fragility of crop production. Only 4 percent of SSA is irrigated (Burney, Naylor, and Postel 2013), so farmers almost always rely on the rains. Climate is accordingly a major factor in SSA's agriculture. In West Africa, from 2019 to 2020, the proportion of people experiencing moderate food insecurity rose from 34.6 percent to 39.5 percent, and the proportion experiencing severe food insecurity rose from 19.6 percent to 28.8 percent (FAO et al. 2021, 19). This means that 54.2 percent of people in West Africa suffered either moderate or severe food insecurity in 2019, but just a year later, that proportion had risen to 68.3 percent. Put another way, the proportion of people experiencing no or minimal food insecurity in West Africa dropped from 45.8 percent in 2019 to 31.7 percent in 2020: *in one year, 14.1 percent of West Africa's population lost its food security.* Almost 159 million West Africans experienced moderate food insecurity in 2020, and almost 116 million were in severe food insecurity (FAO et al. 2021, 18). The farmers most vulnerable to climate changes are the poor with inadequate resources to compensate for crop shortage or failure. The poorest farmers are mostly women. Climate impacts thus lead to hunger and nutritional inadequacies that affect the most vulnerable—women farmers and their families.

Ghana provides a significant case study for understanding women farmers' current situation and experience of climate impacts. Northern Ghana is in the Sudan Savannah ecosystem that stretches from the west coast of SSA into East Africa. Ghana shares many of the cultural attitudes and traditions common in West Africa, such as crop selection, but also cultural bias against women and impediments to their farming. In northeast Ghana, women produce as much as 87 percent of what is eaten in the region (SWC 2010) and can be responsible for growing enough in the rainy season to feed as many as seventeen people for the year, using nothing but a hand hoe, saved seeds, and homemade animal dung fertilizer. The Ghana Environmental Protection Agency (2007, 28) thus acknowledged "the importance of the environment and particularly climate change in women's lives" as early as 2007.

Yet Ghanaian women continue to have limited access to machinery, credit, supporting labor, fertilizers, and agricultural extension services. Only 54 percent of Ghanaian women, versus 62 percent of men, have bank accounts (World Bank 2019a). These impediments have been recognized for decades (Baden et al. 1994) and will not be changed without federal policy and well-managed implementation at regional and local levels. Women's limited inclusion in agroforestry (IMF 2003), an industry shown to be beneficial to women, needs policy intervention to change common practice by providing supports and access to resources, as well as building strong women's groups (Kiptot and Franzel 2012). In 2012, men held 73 percent of small farms (less than five acres) and women 27 percent; men held 89 percent of larger farms and women only 11 percent (FAO 2012, 12). Women's lack of land tenure in Ghana (see Awanyo 2003) cannot be changed without policy reforms and other interventions for implementation, given global bias against women farming for profit (Glazebrook, Noll, and Opoku 2020).

Women and children in SSA have long and consistently been known to be extremely vulnerable to incremental climate impacts on food and agriculture, though they have the least political, economic, and social resources to recover (Bang 2008). Climate change has also been identified as a key development challenge (Kok et al. 2008; Gössling et al. 2009). Climate change is exacerbating women's situation as food growers and causing incapacity to meet food security needs in West Africa. Climate impacts include extreme weather events, such as heatwaves, drought, and flooding, as well as increased pest and disease events (IPCC 2014, 21).

The greatest challenge now facing Ghana's women farmers is sporadic and unpredictable rainfall. For generations, they had relied on consistent daily convectional rainfall in the rainy season, but now it is difficult for many farmers, especially corporate and industrial farmers, to know when to plant. Women often already do not know when they can plant because, with sparse resources, they are typically pushed into late planting because they do not always have the money to hire a man with bullocks to turn the soil. Women typically pay with a meal and promise of a share of the crop. The bullock drivers often support the local women farmers, though as businessmen dependent on their income, they first meet the needs of paying customers. Timing is not all that is affected, however: sporadic rains can have more of an impact for women subsistence farmers on *what to plant*—that is, crop selection—than on when to plant. A poor choice that creates a harvest shortfall can be catastrophic for food security.

The IPCC predicted in 2007, with "high confidence," that by 2020, climate impacts would include a 10 percent decline in millet (IPCC 2007). In 2020, Azare et al. (2020) found that millet yields were significantly down because of short, unreliable rainy seasons with frequent dry spells. Decline has been significant enough for women to move away from traditional food mainstays in favor of alternative crop selection. Millet is a mainstay because it is full of protein and calcium. Likewise, groundnut is traditionally eaten and is loaded with protein. Lack of protein stunts muscle and brain growth and can lead to Kwashiorkor, which causes an enlarged liver and fluid retention that brings swelling (see Healthline, "What Is Kwashiorkor?" https://www.healthline.com/health/kwashiorkor). Millet is especially good for pregnant and lactating mothers and their babies because the calcium supports milk production and bone health and growth. Millet is thus important for growing children. Sporadic rain hampers crop size for millet, however: if the rain is inadequate or there are substantial dry gaps, the crop may not suffice to the next growing season. Groundnut can also be harmed by sporadic rain because the plants need adequate rain in the early days after planting. If they do not get enough rain, the nuts do not grow, but because they grow underground, a farmer may waste space and resources by tending the plants all season but find no nuts when the plants are eventually pulled.

Farmers respond to these risks by growing more weather-adaptive crops, for example, drought-resistant rice that grows with the rain and, without rain, simply is dormant. Rice, however, has virtually no nutritional value. Losing both food sovereignty in crop selection and nutritional value in food consumption, women farmers are choosing poor nutrition over the risk of a crop shortfall that would lead to hunger and perhaps starvation. Given rapidly increasing food insecurity, the turn to rice may be a good choice now, though no help in the long term: every degree Centigrade increase in the planet's global mean surface temperature cuts rice productivity by 10 percent (Canadell, Raupach, and Houghton 2009). Women subsistence farmers are fighting a hopeless battle with little support.

It has been known for decades that women's labor is largely invisible because it does not figure in GDP and GNP market-based assessments (Waring 1988). The result is a paradoxical double-pronged bias of refusal to provide support and denial of the consequences of that refusal. Ghana is a country working against itself in policy—that is, governments are relying on what women grow while designing policy that does not support them to do so. This is not simply Ghana's doing, however. Ghana is caught in an international economic system within which SSA's women subsistence farmers (and presumably their counterparts everywhere) are at best invisible and at worst disposable.

The World Bank ranks countries on poverty and wealth. For a country to improve its status, the IMF completes an assessment and makes a recommendation to the World Bank. Between 2003 and 2012, Ghana submitted a series of Poverty Reduction Strategy (GPRS) papers and progress reports to the IMF that in return provided joint staff assessments as well as advisory notes. The first GPRS in 2003 included material collected in discussions with women's groups in Ghana's rural, northern areas, and for which access was made possible by, for example, use of local languages (Government of Ghana 2003, 7). One outcome was identification of gender equity as a poverty reduction strategy to achieve sustainable, equitable growth, accelerate poverty reduction, and protect the vulnerable and excluded (30). Gender was explicitly discussed, including the "gendering of poverty" and gender inequities in legal rights and protections, access and control of assets (e.g., land and credit), governance, education, literacy, health, workloads, and social participation (25–26). The need for action on these issues, integration of completion reports and target indicators, and attribution of ministry responsibility in poverty reduction policy matrices were laid out. Another section recognized constraints on women with respect to marketing, bargaining, and credit that hinder their added-value processing of, for example, shea butter, and expressed concern that improvements might pass women by (75). Rural areas were suggested as a catalyst for economic transformation through agriculture and "changing values . . . in relation to gender roles" (77).

The IMF response noted women's poverty and vulnerability (IMF 2003, ¶ 7), but then suggested that the GPRS understood women's challenges "in a narrow way" (¶ 25) based on new facilities rather than market demand—that is, based on spending money rather than making money. While the Ghanaians saw women farmers as agents playing a crucial role in food security, they were underperforming participants in the capital economy in the IMF's eyes. The remainder of the IMF document limits discussion of gender to education. Ghana's Progress Report to the IMF in 2004 followed suit and discussed gender almost exclusively in terms of education. In 2007, the year it discovered offshore oil, Ghana was promised advancement from a low-income to a middle-income country, a goal it had aspired to since its 1995 Vision 2020 plan (Government of Ghana 1995). On July 1, 2011, the first year that Ghana produced oil, its status change to middle-income was confirmed (World Bank 2011).

IMF intervention in Ghana's plans to integrate gender into agricultural policy appears to show that global economics does not value non-market production. This not only reinforces the lack of acknowledgment received by women subsistence farmers, making them invisible again, but also institutionalizes their invisibility. In the global conception of development, subsistence farmers should be replaced by corporate, industrial farming (which contributes to GDP), and they are seemingly treated as disposable because they do not contribute to the economy. The truth is that they do—the food they provide to the national food basket is an economic contribution even though it is not measured. Women subsistence farmers appear to be receiving little if any support, despite the Government of Ghana's ongoing efforts to eliminate gender inequities, for example, in their Poverty Reduction Strategies. The process of shifting from subsistence farming food systems to market systems in Ghana is a massive societal and economic change in the midst of climate and hunger crises. The invisibility of women's farming results in poor policy decisions in this transition, with the likely consequence of rendering women farmers abject through loss of access to land, lack of replacement livelihoods, and the hunger and suffering that accompanies these losses. While policy devalues women's non-market-based food provision, it has no problem measuring the promise of economic gain from oil development.

THE PARADOXES OF SUSTAINABLE DEVELOPMENT

The influential economist Jeffrey Sachs has argued that sustainable development brings in a "new era . . . in which economic progress is widespread; extreme poverty is eliminated; social trust is encouraged to strengthen the community; and the environment is protected from human-induced degradation" (Sachs 2015, 3). Not all agree. Brown (2015) calls Sachs's 2015 book on Sustainable Development "a bad book . . . deeply flawed from a scientific perspective and dangerously misleading from a policy perspective." The phrase "sustainable development" has been argued to be an oxymoron (Redclift 2005; Spaiser et al. 2017): growth cannot be endless on a finite planet, and, contrary to Sachs's promises, the phrase works against human and other interests by implying that it can. The consequence in SSA of the rhetoric of sustainable development—and its proclivity for economic interpretation—is the risk of failing to address climate-driven food security collapse that is underway in direct correlation with poverty. Women subsistence farmers have been shown in this chapter to be the most vulnerable to climate change and largely excluded in policy strategies to integrate SSA's countries successfully into global economics.

The idea of sustainable development emerged in economics debates in the 1970s. Intergenerational justice had become an issue out of concern that depletion of resources might deny future generations their fair share. This worry prompted quantitative assessments of resource management, and the concept of "conservation" emerged to protect reserves from depletion (Solow 1974; Hartwick 1977). Gro Harlem Brundtland was invited by the Secretary-General of the UN to assess the impacts of global development on ecosystems, and the ensuing 1987 report, *Our Common Future*, became known as the Brundtland Report. In this report, sustainable development was defined as development meeting present needs without compromising future ability of generations to meet their needs (Brundtland Report 1987). Three pillars—economic, environmental, and social—were identified as its foundational components, known informally as "people, planet, and profits."

A factor in this prioritizing was the two key concepts intended to function practically and fairly in global understanding and implementation of sustainable development—that is, "need" that prioritized the world's poor, and "natural limits" acknowledging that ecosystem capacity is finite. The Earth Summit in Rio de Janeiro in 1992 (UN 1992) complicated "need" by associating it with distributive justice and fairness in resource access at locations and in cultures. Because this justice approach aimed at human welfare and humanitarian advances, however, it reproduced anthropocentrism and the instrumentalist reduction of nature to resources for human use that accordingly prioritized human needs over natural limits. This approach fit conveniently with sustainable development policy because the idea of conservation was always already capital-based—that is, it depended on counting to determine the extent that resources could be sustainably exploited for global markets. The economic pillar thus took priority in decision-making at the expense of "people" and "planet." Despite the Brundtland Report's (1987, 42) close attention to global inequities and harm of natural systems, including climate, common understanding of "sustainability" favors intergenerational justice and rarely considers current harms beyond overuse, for example, climate change and its impacts.

There were other complications as the confluence of sustainability and resource exploitation was applied in local—that is, country-based—contexts. For example, in South Africa, policy allowed the removal of cultural groups from their long-standing homeland, allegedly to protect ecosystems in areas that were declared Parks intended "to conserve a wide range of biodiversity, landscapes, fauna and flora," according to the advertising intended to attract visitors ("Explore Parks," South African National Parks, Addo Elephant National Park, https://www.sanparks.org/parks/addo/, accessed 17 August 2022). It is paradoxical that something taken to be of great value needs to be protected from people who have lived there with that valued thing for generations, sometimes time immemorial, and further paradoxical that the "protectors" of this biodiversity, and so on, are embedded in market-based systems widely known to cause long-term damage to ecosystems for short-term profit, regardless of consumer innocence. "Conservation" in this context is, in truth, appropriation of land and animals as resources in the tourist industry. Wildlife watching accounts for 80 percent of South Africa's tourist travel revenues that generates $90 million USD annually from visitors to areas protected by removal of the human residents (UNWTO 2015, 25) who now live in a shanty town next to the Park's walls. These people are abject—thrown out of their homeland and left to their own resources. Sustainable development has thus failed to retain its originally intended priority and balance of people-planet-profits, and its principles of needs and natural limits have been lost to many people and other life and ecosystems on the planet but are still functional for profit.

Women farmers in SSA are also subject to this collapsed idea of sustainable development, so it seems unlikely that the SDGs will meet their needs. These women are crucial and depended-upon contributors to SSA's food supply—if their needs are not met, SDG2 will not be achieved. As noted above, less than one-third of Ghana's population currently has enough to eat, even though a decade ago it achieved middle-income status with the World Bank. That same year, Ghana first produced oil from its offshore reserves, thus demonstrating that it has the capacity to enter the global oil market and exchange. Oil is widely recognized as a primary source of the greenhouse gas emissions that drive climate change, and it is also widely understood that climate impacts are significantly damaging to agricultural production.

The World Bank's 2019 brief of Ghana's economic update on "enhancing financial inclusion" says that "Ghana can reach universal financial access across regions and key demographics using innovative technology," though financial access is still low, "particularly among women, poor and rural citizens" (World Bank 2019a). The World Bank Country Director for Ghana, Henry Kerali, said that universal financial access needed additional efforts in the medium term "to allow government's financial inclusion efforts to bear fruit," while the World Bank's senior financial sector economist, Carlos Vincente, says the Government of Ghana must lead implementation of financial inclusion and digital finance policy (World Bank 2019a). The call is thus for government policy to lead Ghana's economic development, and for patience as that process has impact.

The larger document of the Fourth Ghana Economic Update describes the demographics of those least financially included, as noted above, as "women, poor and rural citizens." The Update does not state that the phrase "poor and rural citizens" is gender-disaggregated to mean women, but, as shown in the section above, an overwhelming proportion of poor, rural women are farmers because of the double feminization of women and agriculture—women farmers do not necessarily choose to be farmers but learn from their mother at a young age (Glazebrook 2011) and continue because poverty makes it the only food security option in their household. "Poor, rural women" is an apt description of SSA's women subsistence farmers.

How will Ghana design policy to meet these rural women farmers' needs? The World Bank and IMF play different roles in international economics and finance—that is, broadly put, the World Bank addresses the economic development of poor countries that, once adequately developed, can be turned over to the IMF, which addresses the stability of functional economies. The lack of interest in women farmers in Ghana's 2003 Poverty Reduction Strategy indicates that the IMF did not see women's place in a stable and functional economy, while the World Bank did not see their necessity in development understood as economic. The women were abject because they were poor and had nothing to be seen to contribute to Ghana's economic growth. This is why the IMF, as noted above, called Ghana's perception of women farmers "narrow": the IMF thought that the Ghanaian government could not see that women farmers are an economic cost. The World Bank failed to acknowledge women farmers' contribution to supporting the monetary economy by operating outside that economy to provide a substantial portion of food to Ghanaians. The displacement of women farmers by acceding to the request that the gender policy defer to education caused women farmers to be displaced from the discourse and from the actions that discourse launched.

The awkward point here is what "poor" means. Vandana Shiva (1993) argues that "development" brings poverty because its model of "progress" is based on industrial economies of consumption in the global North that have only a culturally perceived understanding of poverty rather than real material poverty. This understanding causes inequality through ecologically disruptive economic policies that make for unequal access to resources—that is, government policy enables resource-intensive production to gain access to raw material, which impedes the access of the many people who depend on such resources for survival. She also identifies a gender bias in women's asymmetric participation in development, such as lack of access to land, technologies, and employment; invisibility of women's economic contributions; hidden costs women bear from ecological damage; denigration of their work respecting natural systems and destruction of those systems; and policy guidance and protection, including subsidies, of production and the industrial sector. Shiva's assessment is consistent with the gender bias discussed above on women farmers' experience in SSA. It may well be the case that governments in SSA cannot meet women farmers' needs because their policy approaches are too embedded in capital, market-based systems—in which, women subsistence farmers are abject in very much the ways Shiva describes, pushed into a created poverty.

Ghana's 8.1 percent economic growth in 2017 and 6.3 percent—slower but still impressive—in 2018 are attributed to mining, petroleum, agriculture, forestry, and logging. Growth of 7.6 percent was expected in 2019, "driven by both the oil and non-oil sectors," and the non-oil sector was expected grow as "policy interventions in agriculture and industry" revitalize them (World Bank 2019a). Given that the expectations of revitalization are described in the context of economic growth, they do not have subsistence farming in this vision. Though the Economic Update brief discussed above mentions low rates of financial inclusion—that is, access to money and markets—of women and the rural poor, none of its recommendations appear relevant to the needs of the rural poor who have little access to the technologies and services being recommended. Prior to climate impacts, women subsistence farmers were typically able to grow enough to feed the family until the next rains. Climate change means they often now cannot. Yet global economics provide no support and seem to be deterring countries from providing it.

In the meantime, the 2018 economy got little growth from oil production as investment in oil in 2016 could not be replicated in 2018 and the increased oil exports in 2017 were "one-off effects" (World Bank 2019b, 6). This data is deceptive, however, as it is based on the concept of growth that figures strongly in the economic and influential concept of "sustainable development" that emerged from the Brundtland Report. Oil has contributed significantly to Ghana's GDP, and was instrumental in Ghana's transition from a low-income to middle-income country in 2011, the first year it produced oil after its 2007 discovery. Between 2011 and 2020, Ghana collected the equivalent of $6.55 billion USD in oil revenues, according to its Public Interest and Accountability Committee (Clement Adzei Boye, "Ghana bags U.S. $6.550 billion in oil revenue . . . from 2011 to 2020," *Ghanaian Times*, September 22, 2021, https://www.ghanaiantimes.com.gh/ghana-bags-6-550-bn-in-oil-revenuefrom-2011-to-2020/). Yet many Ghanaians still live in poverty and collapsing food systems and have not seen the benefits of oil revenue. Ghana's middle-income status does not reflect the living conditions of all Ghanaians (Akon-Yamga 2018, 2).

CO_2 released from fossil fuel and industrial processes accounts for 65 percent of all greenhouse gas (GHG) emissions (IPCC 2014, 5). While agriculture is a source of methane, also a GHG more impactful than CO_2 but over less time, it is a necessary human activity. Fossil fuels are an energy source that could be displaced by renewable sources much more quickly than renewables are advancing now. For example, "conservative estimates put U.S. direct subsidies to the fossil fuel industry at roughly $20.5 billion per year" (Bertrand 2021). If globally such investment went instead to upscale and implement renewables, development and deployment of renewables could expand much faster than they are now. It is clear, however, that the oil industry is not yet ready to give up its energy provision and is supporting new countries to enter into the global oil economy. Ghana is one of these countries. Supporting renewables would be much better for the most vulnerable people of these countries who get little benefit from oil development and are losing their food base. Yet the UN Framework Convention on Climate Change (UNFCCC) is consistently in disagreement during annual Conferences of Parties in which the global South is increasingly experiencing loss and damage while the global North is reluctant to commit to, or provide promised funding to, the Loss and Damage Mechanism established at Warsaw in 2013 (UNFCCC 2022).

CONCLUSION

The information and discussion provided above indicate that both SSA countries and other sources are responsible for the damage climate change is doing to SSA's agricultural production, especially for women subsistence farmers who have sparse resources to adapt to climate impacts. These sources have been shown to be international organizations such as the World Bank and the IMF that do not value women's subsistence farming because it does not contribute measurably to the economy. Climate impacts are also exacerbated by the fossil fuel industry that promotes development of fossil fuel production in SSA and actively provides supports for development. Meanwhile, attempts by the UNFCCC to address damage in the global South are stymied by reluctance from the global North. These organizations and corporations are for the most part owned or controlled by the global North. Our analysis shows that the global North has an obligation to the global South to mitigate GHG production and provide finance to manage loss and damage because of the North's substantial role in creating the drivers of climate change.

We have shown above that oil development is overlooked as a curse on agriculture. The literature on the natural resource and oil "blessing or curse" debate over four decades shows that the determining factor depends on governance and policy choices and practices, which can be affected, in the case of the curse, by ineptitude or corruption. We have shown that West Africans suffer increasingly more intense hunger crises and that SSA's women subsistence farmers, who provide a substantially large contribution to food security, are receiving very little support from governments and their policies to eliminate persistent social and economic biases, even in the face of climate impacts that are obliging them to sacrifice their family's nutritional base. They are abject, disregarded in policy design, and rendered abject again by international development and economic organizations. We have provided evidence that in international policy "sustainable development" is a utopian promise because it is a capital-based system grounded on profits, not people, other life forms, and ecosystems.

Together, these issues indicate that counting on oil development to overcome poverty is paradoxical because oil is a substantial driver of the climate change that is accelerating damage to SSA's food security. Using oil wealth to eradicate poverty (SDG1) counteracts initiatives to address SDG2 on hunger in SSA, which is especially increasing in West Africa. If the 2019–20 decrease in food security continues, in less than three years, no-one in West Africa would have enough to eat. The wealthy will of course always eat, but West Africa is clearly in food crises that are already creating much suffering and harm. Ironically, in Ghana, the drive to meet SDG1 on poverty negatively affects the country's poorest. The women subsistence farmers of SSA are being pushed not just into poverty caused by impacts of postcolonialism but also hunger exacerbated by the impacts of climate change. Unless SSA's governments and international, market-based development and economic organizations develop policy directly addressing the situation of SSA's women subsistence farmers, including policy to reject oil-based wealth creation, there is no hope to address SSA's humanitarian hunger crises. Global entities such as the World Bank and IMF and oil-producing nation-states in SSA have already shown that they have little interest in supporting such policy and prefer instead to maintain the pie-in-the-sky of fossil fuel–based "sustainable development."

The failure of SSA governments to connect oil wealth with food security is an environmental justice issue in that environmental justice is grounded on the connection of environmental issues with social, economic, and governance issues in order to uncover injustices in which a vulnerable group suffers disproportionate harms from ecosystem deterioration. In this case, it connects economic policy and the impact of fossil fuels on the atmosphere with women farmers' capacity to meet family food security needs. This connection opens the question of how SDG1 on poverty is to be understood in the context of the UN and other international organizations that appear not to be committed to the full definition of "sustainable development" in the Brundtland Report but only to the initial discussion of leaving resources for the future. Replacing energy sources with renewables would support food security. Failure to focus on ecosystem health in preference for economic issues, however, does not support achievement of SDG2 on hunger. Addressing SDG1 on poverty by means of oil development policy contributes to climate change and is accordingly contrary to achieving SDG2. Relieving this policy tension and the environmental justice issues it creates would require improving the quality of institutions in SSA. Without strong institutions and thoughtful policies, women subsistence farmers remain abject in the face of SSA's oil wealth.

REFERENCES

Ackah-Baidoo, Abigail. 2012. "Enclave Development and 'Offshore Corporate Responsibility': Implications for Oil-Rich Sub-Saharan Africa." *Resources Policy* 37, no. 2: 152–59.

Akon-Yamga, Gordon. 2018. "Oil in Ghana: A Curse or Not? Examining Environmental Justice and the Social Process in Policy-Making." PhD diss., University of North Texas. https://digital.library.unt.edu/ark:/67531/metadc1157653/m2/1/high_res_d/AKONYAMGA-DISSERTATION-2018.pdf.

Amiri, Hossein Khalilpour, Farzaneh Samadian, Masoud Yahoo, and Seyed Jafar Jamali. 2018. "Natural Resource Abundance, Institutional Quality and Manufacturing Development: Evidence from Resource-Rich Countries." *Resources Policy* 62: 550–60.

Amnesty International. 2017. "Nigeria: Shell Complicit in the Arbitrary Executions of Ogoni Nine as Writ Served in Dutch Court." Accessed June 10, 2022. https://www.amnesty.org/en/latest/news/2017/06/shell-complicit-arbitrary-executions-ogoni-nine-writ-dutch-court/.

Auty, Richard. M. 1994. "Industrial Policy Reform in Six Large Newly Industrializing Countries: The Resource Curse Thesis." *World Development* 22, no. 1: 11–26.

Awanyo, Louis. 2003. "Land Tenure and Agricultural Development in Ghana: The Intersection of Class, Culture and Gender." In *Critical Perspectives on Politics and Socio-Economic Development in Ghana*, edited by Wisdom Tettey, Korbla Puplampu, and Bruce Berman, 273–304. Boston: Brill Academic.

Azare, I. M.; Dantata, I. J.; Abdullahi, M. S.; Adebayo, A. A.; Aliyu, M. 2020. "Effects of Climate Change on Pearl Millet (*Pennisetum glaucum* [L. R. Br.]) Production in Nigeria." *Journal of Applied Sciences and Environmental Management* 24, no. 1: 157. https://doi.org/10.4314/jasem.v24i1.23.

Baden, Sally, Cathy Green, Naana Otoo-Oyortey, and Tessa Peasgood. 1994. "Background Paper on Gender Issues in Ghana." Report prepared for the West and North Africa Department, Department for Overseas Development (DFID), UK. BRIDGE (development—gender). Accessed August 18, 2022. https://www.yumpu.com/en/document/view/5666095/background-paper-on-gender-issues-in-ghana-bridge-institute-of-.

Bang, Henry Ngenyam. 2008. "Social Vulnerability and Risk Perception to Natural Hazards in Cameroon Two Decades after the Lake Nyos Gas Disaster: What Future Prospect for the Displaced Disaster Victims?" Accessed June 10, 2022. http://citeseerx.ist.psu.edu/viewdoc/summary?doi=10.1.1.471.6902.

Barr, Alwyn. 2020. "Late Nineteenth-Century Texas." In *Handbook of Texas*. Austin, TX: State Historical Association. Accessed June 10, 2022. https://www.tshaonline.org/handbook/entries/late-nineteenth-century-texas.

Beegle, Kathleen, Luc Christiaensen, Andrew Dabalen, and Isis Gaddis. 2016. *Poverty in a Rising Africa*. Washington, DC: World Bank.

Bertrand, Savannah. 2021. "Fact Sheet: Proposals to Reduce Fossil Fuel Subsidies." Environmental and Energy Study Institute. Accessed 31 July 2022. https://www.eesi.org/papers/view/fact-sheet-proposals-to-reduce-fossil-fuel-subsidies-2021.

Brown, James H. 2015. "The Oxymoron of Sustainable Development." *Bioscience* 65, no. 10: 1027–29.

Brundtland Report. 1987. *Our Common Future*. Oxford: Oxford University Press.

Brunnschweiler, Christa N., and Erwin H. Bulte. 2008. "The Resource Curse Revisited and Revised: A Tale of Paradoxes and Red Herrings." *Journal of Environmental Economics and Management* 55, no. 3: 248–64.

Bulte, Erwin, and Richard Damania. 2008. "Resources for Sale: Corruption, Democracy and the Natural Resource Curse." *The B.E. Journal of Economic Analysis & Policy*, 8(1):1–37. http://doi.org/10.2202/1935-1682.1890.

Burney, Jennifer A., Rosamond L. Naylor, and Sandra L. Postel. 2013. "The Case for Distributed Irrigation as a Development Priority in Sub-Saharan Africa." *PNAS* 110(31):12513–17. https://doi.org/10.1073/pnas.1203597110.

Canadell, J. G.; M. R. Raupach, and R. A. Houghton. 2009. "Anthropogenic CO_2 emissions in Africa." *Biogeosciences* 6: 463–68.

CDP. 2020. "CDP Africa Report: Benchmarking Progress towards Climate Safe Cities, States, and Regions." London: CDP Worldwide. Accessed June 10, 2022. https://www.cdp.net/en/research/global-reports/africa-report.

Davis, Graham A., and John E. Tilton. 2005. "The Resource Curse." *Natural Resources Forum* 29, no. 3: 233–42.

Diamond, Irene. 1990. "Babies, Heroic Experts, and a Poisoned Earth." In *Reweaving the World: The Emergence of Ecofeminism*, edited by Irene Diamond, and Gloria F. Orenstein, 201–10. San Francisco, CA: Sierra Club Books.
DoN. 2007. *Declaration of Nyéléni*. Accessed June 10, 2022. https://nyeleni.org/spip.php?article290.
Eade, Deborah, and Suzanne Williams. 1994 "Emergencies and Development: Ageing with Wisdom and Dignity." *Focus on Gender* 2, no. 1:17–19.
Ebrahimzadeh, Christine. 2020. "Dutch Disease: Wealth Managed Unwisely." International Monetary Fund. Accessed June 10, 2022. https://www.imf.org/external/pubs/ft/fandd/basics/dutch.htm.
Economist, The. 1977. "The Dutch Disease." Business Brief. November 26, 1977: 82–83.
FAO (Food and Agriculture Organization). 2003. "Trade Reforms and Food Security: Conceptualizing the Linkages." Commodities and Trade Division. Rome. Accessed June 10, 2022. https://www.fao.org/3/y4671e/y4671e.pdf.
———. 2012. "Gender Inequalities in Rural Employment in Ghana: An Overview." Gender, Equity and Rural Employment Division. Rome. Accessed June 10, 2022. http://www.fao.org/docrep/016/ap090e/ap090e00.pdf.
———. 2017. "The State of Food and Agriculture: Leveraging Food Systems for Inclusive Rural Transformation." Rome. Accessed June 10, 2022. https://www.fao.org/policy-support/tools-and-publications/resources-details/en/c/1046886/.
FAO et al. (FAO, IFAD, UNICEF, WFP, WHO). 2021. "The State of Food Security and Nutrition in the World 2021." Accessed June 10, 2022. https://www.fao.org/publications/sofi/2021/en/.
FAO, IFAD, and WFP. 2015. "The State of Food Insecurity in the World 2015: Meeting the 2015 International Hunger Targets: Taking Stock of Uneven Progress." Rome: Food and Agriculture Organization. Accessed June 10, 2022. https://www.fao.org/policy-support/tools-and-publications/resources-details/en/c/469455/.
Ferguson, James. 2006. *Global Shadows: Africa in the Neoliberal World Order*. Durham, NC: Duke University Press.
Gelb. Alan H. 1988. *Oil Windfalls: Blessing or Curse?* New York: Oxford University Press.
Ghana Environmental Protection Agency. 2007. *Report on Sectoral Climate Change Impacts, Vulnerability and Adaptation Assessments in Ghana*. Technical Summary, 1–35. Accra: Environmental Protection Agency, Ghana and The Netherlands Climate Assistance Programme.
Ginsburg, Norton. 1957. "Natural Resources and Economic Development." *Annals of the Association of American Geographers* 47, no. 3: 197–212.
Ginsburg, Norton, James Osburn, and Grant Blank. 1986. *Geographic Perspectives on the Wealth of Nations*. University of Chicago Geography Research Papers. Chicago, IL: Committee on Geographical Studies, University of Chicago.
Glazebrook, Trish. 2011. "Women and Climate Change: A Case Study from Northeast Ghana." *Hypatia* 26: 762–82.

Glazebrook, Tricia, Samantha Noll, and Emmanuela Opoku. 2020. "Gender Matters: Climate Change, Gender Bias, and Women's Farming in the Global South and North." *Agriculture* 10, no. 7: 267–91.

Glazebrook, Trish, and Anthony Kola-Olusanya. 2011. "Justice, Conflict, Capital, and Care: Oil in the Niger Delta." *Environmental Ethics* 33, no. 2: 163–84.

Glazebrook, Trish, and Matt Story. 2012. "The Community Obligations of Canadian Oil Companies: A Case Study of Talisman in the Sudan." In *Corporate Social Irresponsibility: A Challenging Concept*, edited by Ralph Tench, William Sun, and Brian Jones, 231–61. Bingley, UK: Emerald Group Publishing.

Gössling, Stephan, C. Michael Hall, and Daniel Scott. 2009. "The Challenges of Tourism as a Development Strategy in an Era of Global Climate Change." In *Rethinking Development in a Carbon-Constrained World: Development Cooperation and Climate Change*, edited by Eija Palosuo, 100–119. Oy, Finland: Erweko Painotuote.

Government of Ghana. 1995. "Ghana—Vision 2020 (The First Step: 1996–2000)." Presidential Report on Co-ordinated Programme of Economic and Social Development Policies. Accessed June 10, 2022. https://www.ircwash.org/sites/default/files/Rawlings-1995-GhanaVision.pdf.

———. 2003. "Ghana Poverty Reduction Strategy: 2003–2005. An Agenda for Growth and Prosperity." Volume 1: Analysis and policy statement, February 19, 2003. Accessed January 9, 2022. https://www.ircwash.org/resources/ghana-poverty-reduction-strategy-2003-2005-agenda-growth-and-prosperity.

Gylfason, Thorvaldur, Tryggvi Thor Herbertsson, and Gylfi Zoega. 1999. "A Mixed Blessing: Natural Resources and Economic Growth." *Macroeconomics Dynamics* 3, no. 2: 204–25.

Hartwick, John M. 1977. "Intergenerational Equity and the Investing of Rents from Exhaustible Resources." *American Economic Review* 67: 972–74.

Holligan, Anna. 2022. "Ogoni Nine: Nigerian Widows Lose Case against Oil Giant Shell." BBC News, The Hague. 23 March 2022. Accessed August 18, 2022. https://www.bbc.com/news/world-europe-60851111.

IMF (International Monetary Fund). 2003. "Ghana: Joint Staff Assessment of the Poverty Reduction Strategy Paper." Country Report No. 03/12, May 2003. Washington, DC: IMF Publication Services. Accessed June 10, 2022. https://www.imf.org/en/Publications/CR/Issues/2016/12/30/Ghana-Joint-Staff-Assessment-of-the-Poverty-Reduction-Strategy-Paper-16554.

IPCC (Intergovernmental Panel on Climate Change). 2007. "AR4 Climate Change, 2007: Synthesis Report." In *Contribution of Working Groups I, II and III to the Fourth Assessment Report (AR4) of the Intergovernmental Panel on Climate Change*, edited by The Core Writing Team, Rajendra K. Pachauri, Andy Reisinger. Geneva, Switzerland: IPCC. Accessed June 10, 2022. https://www.ipcc.ch/report/ar4/syr/.

———. 2014. "Summary for Policymakers." In *AR5 Climate Change 2014: Impacts, Adaptation, and Vulnerability. Contribution of Working Group II to the Fifth Assessment Report of the Intergovernmental Panel on Climate Change*, edited by

The Core Writing Team, Rajendra K. Pachauri, and Leo Meyer, 1–32. Geneva, Switzerland: IPCC. Accessed June 10, 2022. https://www.ipcc.ch/report/ar5/syr/.

Kiptot, Evelyne, and Steven Franzel. 2012. "Gender and Agroforestry in Africa: A Review of Women's Participation." *Agroforestry Systems* 84: 35–58.

Kok, Marcel, Bert Metz, Jan Verhagen, and Sascha van Rooijen. 2008. "Integrating Development and Climate Policies: National and International Benefits." *Climate Policy* 8: 103–18.

Kristeva, Julia. 1982. *Powers of Horror: An Essay on Abjection*. Translated by Leon S. Roudiez. New York: Columbia University Press.

Lisk, Franklyn. 2009. "Overview: The Current Climate Change Situation in Africa." In *CIGI Special Report: Climate Change in Africa: Adaptation, Mitigation and Governance Challenges*, edited by Hany Besada and Nelson Sewankambo, 8–15. Waterloo, ON: Centre for International Governance Innovation.

Lowder, Sarah K., Jakob Skoet, and Terri Raney. 2016. "The Number, Size, and Distribution of Farms, Smallholder Farms, and Family Farms Worldwide." *World Development* 87: 16–29.

MacFadyen, Alan J. and G. Campbell Watkins. 2014. *Petropolitics: Petroleum Development, Markets and Regulations, Alberta as an Illustrative History*. Calgary, AB: University of Calgary Press.

Michaels, Guy. 2010. "The Long Term Consequences of Resource-Based Specialization." *The Economic Journal* 121, no. 551: 31–57.

Neary, J. Peter, and Sweder van Wjinbergen. 1986. *Natural Resources and the Macroeconomy*. Cambridge, MA: MIT Press.

Opoku, Emmanuela, and Trish Glazebrook. 2018. "Gender, Agriculture, and Climate Policy in Ghana." *Environmental Ethics* 40, no. 4: 365–80.

Papyrakis, Elissaios, and Reyer Gerlagh. 2004. "The Resource Curse Hypothesis and Its Transmission Channels." *Journal of Comparative Economics* 32, no. 1: 181–93.

Patel, Raj. 2009. "What Does Food Sovereignty Look Like?" *Journal of Peasant Studies* 36, no. 3, 663–706. https://doi.org/10.1080/03066150903143079.

Ranis, Gustav, Frances Stewart, and Alejandro Ramirez. 2000. "Economic Growth and Human Development." *World Development* 28, no. 2: 197–219.

Redclift, Michael. 2005. "Sustainable Development (1987–2005): An Oxymoron Comes of Age." *Sustainable Development* 13, no. 4: 212–27.

Robinson, James A., Ragnar Torvik, and Thierry Verdier, T. 2014. "Political Foundations of the Resource Curse: A Simplification and a Comment." *Journal of Development Economics* 106: 194–98.

Ross, Michael L. 1999. "The Political Economy of the Resource Curse." *World Politics* 51, no. 2: 297–322.

Sachs, Jeffrey D. 2015. *The Age of Sustainable Development*. New York: Columbia University Press.

Sachs, Jeffrey D., and Andrew M. Warner. 2001. "Natural Resources and Economic Development: The Curse of Natural Resources." *European Economic Review* 45, nos. 4–6: 827–38.

Shiva, Vandana. 1988. *Staying Alive: Women, Ecology, and Development*. London: Zed Books.

———. 1993. "The Impoverishment of the Environment: Women and Children Last." In *Ecofeminism*, edited by Maria Mies and Vandana Shiva, 70–90. Atlantic Highlands, NJ.: Zed Books.

SOFA Team and Cheryl Doss. 2011. "The Role of Women in Agriculture." Working paper no. 11–02, Agricultural Development Economics Division, FAO. Accessed June 10, 2022. https://www.fao.org/3/am307e/am307e00.pdf.

Solow, Robert M. 1974. "Intergenerational Equity and Exhaustible Resources." *The Review of Economic Studies* 41, no. 5: 29–45. Accessed June 10, 2022. https://doi.org/10.2307/2296370.

Spaiser, Viktoria, Shyam Ranganathan, Ranjula Bali Swain, and David J. T. Sumpter. 2017. "The Sustainable Development Oxymoron: Quantifying and Modelling the Incompatibility of Sustainable Development Goals." *International Journal of Sustainable Development & World Ecology* 24, no. 6: 457–70.

SWC. 2010. "Social Watch Coalition. National Reports—Ghana: MDGs Remain Elusive." Accessed June 10, 2022. http://www.socialwatch.org/node/12082.

Torik, Ragnar. 2011. "The Political Economy of Reform in Resource-Rich Countries." In *Beyond the Curse: Policies to Harness the Power of Natural Resources*, edited by Rabah Arezki, Thorvaldur Gylfason, and Amadou Sy, 237–55. Washington, DC: IMF.

UN. 1992. "Report of the United Nations Conference on Environment and Development." A/CONF.151/26/Rev. 1. Vol. 1. https://www.un.org/en/conferences/environment/rio1992.

———. 2015. "Transforming Our World: The 2030 Agenda for Sustainable Development." A/RES/70/1. https://sdgs.un.org/2030agenda.

UNDP. 2022. "Human Development Reports: Human Development Index." https://hdr.undp.org/data-center/human-development-index#/indicies/HDI. Accessed August 18, 2022.

UNFCCC (UN Framework Convention on Climate Change). 2022. "Warsaw International Mechanism for Loss and Damage associated with Climate Change Impacts." Accessed August 1, 2022. https://unfccc.int/topics/adaptation-and-resilience/workstreams/loss-and-damage/warsaw-international-mechanism.

UN Women. 2019. "Progress on the Sustainable Development Goals: The Gender Snapshot 2019." Department of Economic and Social Affairs. Accessed 10 June 2022. https://www.unwomen.org/en/digital-library/publications/2019/09/progress-on-the-sustainable-development-goals-the-gender-snapshot-2019.

———. 2021. "Learn the Facts: Rural Women and Girls." Accessed 10 June 2022. https://www.unwomen.org/en/digital-library/multimedia/2018/2/infographic-rural-women.

———. n.d. "Women Facts and Figures: Economic Empowerment." Accessed June 10, 2022. https://www.unwomen.org/en/what-we-do/economic-empowerment/facts-and-figures.

UNWTO (UN World Tourism Organization). 2015. "Towards Measuring the Economic Value of Wildlife Watching Tourism in Africa." Briefing paper, World Tourism Organization. Accessed June 10, 2022. https://sustainabledevelopment.un.org/content/documents/1882unwtowildlifepaper.pdf.

Vaqué, Jordi. 2017. "Rural Women: A Key Asset for Growth in Latin America and the Caribbean." *Agronoticias: Agricultural News from Latin America and the Caribbean.* Accessed June 10, 2022. http://www.fao.org/in-action/agronoticias/detail/en/c/501669/.

Waring, Marilyn. 1988. *If Women Counted: A New Feminist Economics.* New York: Harper & Row.

Warren, Karen. 2000. *Ecofeminist Philosophy: A Western Perspective on What It Is and Why It Matters.* Oxford, UK: Rowman & Littlefield.

Whitehead, Ann, and Dzodzi Tsikata. 2003. "Policy Discourses on Women's Land Rights in Sub-Saharan Africa: The Implications of the Re-Turn to the Customary." *Journal of Agrarian Change* 3, no. 1–2: 67–112.

World Bank. 2011. "Ghana Looks to Retool Its Economy as It Reaches Middle-Income Status." Accessed June 10, 2022. https://www.worldbank.org/en/news/feature/2011/07/18/ghana-looks-to-retool-its-economy-as-it-reaches-middle-income-status.

———. 2019a. "Ghana Economic Update: Enhancing Financial Inclusion. Media Brief." Accessed June 10, 2022. https://www.worldbank.org/en/country/ghana/publication/ghana-economic-update-enhancing-financial-inclusion.

———. 2019b. "Fourth Ghana Economic Update: Enhancing Financial Inclusion, Africa Region." June 2019. Accessed June 10, 2022. https://documents1.worldbank.org/curated/en/395721560318628665/pdf/Fourth-Ghana-Economic-Update-Enhancing-Financial-Inclusion-Africa-Region.pdf.

———. n.d. "Millennium Development Goals. Goal 1. Eradicate Extreme Poverty and Hunger by 2015." Accessed 10 June 2022. https://www5.worldbank.org/mdgs/poverty_hunger.html.

Chapter 25

The Pandemic and Clamor for Vaccines

Ethical-Legal Considerations for Intellectual Property Rights and Technology Sharing

Pamela Andanda

The COVID-19 pandemic has affected the world at an alarming rate, leading to widespread infections, overwhelmed health systems, and deaths at a rate that the world has not witnessed in a century. It also disrupted social, economic, and education systems. The response included the rapid development of diagnostics, therapeutics, and vaccines; even as these efforts continue, concerns have been raised regarding inequitable distribution. Although the main causes of inequitable distribution are still being debated, the ongoing uneven distribution of COVID-19 vaccines has shifted the spotlight to intellectual property rights (IPRs), which are barriers impeding the manufacturing and supply of the tools (MSF 2020; Erciyas and Üstün 2022). IPRs are blamed as enablers: "Vaccine developers have monopolized intellectual property, blocked technology transfers, and lobbied aggressively against measures that would expand the global manufacturing of these vaccines" (Amnesty International 2021, 4). In addition to concern about vaccines, some countries encountered similar intellectual property (IP) barriers when trying to procure the tools they needed during the peak of the pandemic. For example, South Africa faced a shortage of testing kits but could not develop their own kits since the diagnostic infrastructure required the use of proprietary test materials that were subject to trade secrets held by other manufacturers (Tomlinson 2020). And the Dutch competition authority and the European Commission had to force Roche to share its proprietary recipe with Dutch laboratories to enable the Netherlands to meet the demands of COVID-19 test materials (Van Ark and Strop 2020).

In view of the above challenges, South Africa and India initiated a call, supported by other low- and middle-income countries (LMICs), for temporary waiver from certain provisions of the WTO Agreement on Trade-Related Intellectual Property Rights (TRIPS) for the prevention, containment, and treatment of COVID-19 (WTO 2020). It is worth noting that, even before the pandemic, the two countries have been vocal advocates for a more favorable IP regime that more fairly balances IP protection with the protection of public health needs. During the negotiations of the TRIPS Agreement, India raised concerns that the proposed inclusion of pharmaceutical products in the IP regime could negatively affect its ability to manufacture these products for developing countries (Ganesan 2015).

South Africa broke from the global consensus when it amended the Medicines and Related Substances Act of 1997 to reduce the price of medicines in the country through compulsory licensing and parallel imports (*Pharmaceutical Manufacturers Association and others v President of the Republic of South Africa and others*, Case no 4183/98. Filed in the High Court of South Africa, Transvaal Provincial Division on February 18, 1998). The amendment inserted section 15C into the Act to empower the minister to "prescribe conditions for the supply of more affordable medicines in certain circumstances." Paragraph (a) of the section provides that "notwithstanding anything to the contrary contained in the Patents Act [the minister may] determine that the rights with regard to *any medicine* under a patent granted in the Republic shall not extend to acts in respect of such medicine which has been put onto the market by the owner of the medicine or with his or her consent" (emphasis added). This paragraph provides for compulsory licensing. Paragraph (b) in turn introduced parallel importation by empowering the minister to "prescribe the conditions on which any medicine which is identical in composition, meets the same quality standard and is intended to have the same proprietary name as that of another medicine already registered in the Republic but which is imported by a person other than the person who is the holder of the registration certificate of the medicine already registered and which originates from any site of manufacture of the original manufacturer as approved by the council in the prescribed manner, [to] be imported."

The pharmaceutical companies argued that the amendment was unconstitutional and in conflict with the provisions of Article 27 of the TRIPS Agreement since it discriminated against the enjoyment of patent rights in the pharmaceutical field (paragraph 2.4 of the Notice of Motion) but dropped the case in April 2001 due to international pressure.

The current call for a temporary waiver sought to encourage collaboration and ensure technology transfer to enable more producers to supply the market faster (Usher 2020). Additionally, independent experts jointly wrote letters calling on the WTO and other stakeholders to introduce a temporary waiver under TRIPS to eradicate barriers to the tools required to contain the pandemic (OHCHR 2021). A revised proposal was presented to the WTO's TRIPS Council in May 2021 seeking a waiver on patent enforcement and the disclosure of private information regarding vaccines and related health technologies for a period of three years from the date of the decision (WTO 2021).

The ongoing debate on access to COVID-19 vaccines and related technologies serves as a case study for this chapter, which proposes a new ethical-legal approach to IPRs and technology sharing for vaccines and related medical products in times of crisis such as the ongoing pandemic. It first highlights the issues, which led to calls by South Africa, India, and other countries for IP waiver over COVID-19 vaccines, and considers the various approaches that have been proposed to resolve lack of equitable access to COVID-19 vaccines. This is followed by an overview of the WTO Ministerial Conference's decision on the request for IP waiver. The last part proposes an ethical-legal approach to address future IPR issues, most immediately the WTO Ministerial Conference's reconsideration of the current waiver to cover the production and supply of COVID-19 diagnostics and therapeutics that will take place at the end of 2022 (six months from when this is written).

VACCINE ACCESS, INTELLECTUAL PROPERTY RIGHTS, AND HUMAN RIGHTS

The debate about IPRs and equitable access to medicines has been ongoing (Roffe et al. 2005; UNCTAD 2011). It has, however, become more heated with the emergence of the pandemic. High demands for vaccines require higher manufacturing capacity, which calls for sharing intellectual property or technical know-how (Forman et al. 2021, 557). The fact that the World Health Organization (WHO)-led COVID-19 Technology Access Pool (C-TAP), "[which] was established to pool intellectual property, data and manufacturing processes, licensing the production to other manufacturers and facilitating technology transfer" (WHO 2020), has not received patents or know-how from vaccine manufacturers has further intensified this debate (Amnesty International 2021, 5; Safi 2021). The situation was further compounded by the advance purchase of extra vaccine doses by high-income countries, which was dubbed "vaccine nationalism." For example, Canada ordered the largest number of doses, even though other countries had greater needs and would have expected better results if given the vaccines (Abbas 2021). Some of the advance purchase agreements between private companies and government agencies contained contractual terms that prohibited the sale or donation of surplus doses of the vaccine without the manufacturer's consent. An example is the Advance Purchase Agreement (APA) between CureVac and the European Union (European Commission 2021, Article I.10.2).

The failure of voluntary IP sharing through C-TAP and widespread vaccine nationalism triggered the South African and Indian request that the WTO grant a waiver of most obligations under TRIPS. The specific IPRs targeted by the proponents of the waiver are contained in sections 1, 4, 5, and 7 of Parts II and III of the TRIPS Agreement. These rights are exclusive patent rights (Articles 28(1); 31(a), (b), (f), and (h), and Article 31bis); undisclosed information (Article 39); copyright and related rights, trademarks and other rights provided for in Part II, and enforcement of IPRs (Part III of the TRIPS Agreement with respect to Sections 5 and 7 of Part II of the TRIPS Agreement).

Different camps emerged in response to this call. The European Union favored an approach that uses the existing flexibilities within TRIPS while others were opposed or remained non-committal (BBC News 2021b; European Union 2021). After initial opposition, the United States agreed to back negotiations at the WTO to secure a waiver while acknowledging that securing such a waiver will take time (BBC News 2021a).

The main concern about the proposed waiver was its scope: "The technologies may be applicable to the treatment of other diseases apart from COVID-19 therefore a waiver may have a knock-on effect" (Erciyas and Üstün 2022). Additionally, the technical knowledge to be shared may be protected through patents and trade secrets leading to debates on whether pharmaceutical companies should share their IP-protected technology (Gurgula and Hull 2021, 1244). The current situation is that technology owners continue holding onto their know-how and other IPRs while proponents of the waiver argue that since public funding supported much of the COVID-19 research, a waiver of all pertinent IP including trade secrets is warranted (Gostin, Karim, and Meier 2020; Sariola 2021). Waiver proponents argue that the wide variety of contributions from numerous stakeholders to a wide variety of initiatives that led to the rapid development of the vaccine makes it impossible to clearly distinguish proprietary IP from what should be in the public domain, as is usually done in non-emergency situations. Ultimately, the debate is about how IP such as trade secrets can be shared in ways that balance the needs of the public while ensuring fairness to trade secret owners at the same time (Gurgula and Hull 2021, 1244).

The effectiveness of the waiver is a second concern. The International Federation of Pharmaceutical Manufacturers and Associations (IFPMA) declares that it is "aligned with the goal to ensure COVID-19 vaccines are quickly and equitably shared around the world" but considered the proposed waiver of certain provisions of the WTO's TRIPS Agreement to be "the simple but the wrong answer to what is a complex problem" (IFPMA 2021). It argued that the proposed waiver is likely to lead to disruption and distraction from addressing the real challenges, "namely elimination of trade barriers, addressing bottlenecks in supply chains and scarcity of raw materials and ingredients in the supply chain, and a willingness by rich countries to start sharing doses with poor countries" (IFPMA 2021). Similarly, they argue that the circumvention of IPRs without consent of the rights holder will fail because "technology transfer goes far beyond the patent, is built on trust, know-how sharing and voluntary licensing" (IFPMA 2022). Moreover, they argue, at this point (July 2022), the problem of supply has been resolved.

One element of the EU response to the waiver request was to issue a counterproposal. It attributed inequitable COVID-19 vaccine access to insufficient manufacturing capacity that was unable to rapidly produce the required quantities of the vaccine (European Commission 2021). The EU also called on WTO members to agree on a global trade initiative for equitable access to COVID-19 vaccines and therapeutics consisting of the following three components: "(1) trade facilitation and disciplines on export restrictions; (2) expansion of production, including through pledges by vaccine producers and developers and; (3) clarification and facilitation of TRIPS Agreement flexibilities relating to compulsory licences" (European Union 2021).

The third component of the EU's counterproposal is especially relevant to the TRIPS waiver debate. The EU proposed that all WTO Members should agree on three points: first, that the pandemic is a national emergency thus warranting a waiver of the requirement to negotiate with the IPR holder as required under Article 31(b) of the TRIPS Agreement, which allows a compulsory license to be granted if "the proposed user has made efforts to obtain authorization from the right holder on reasonable commercial terms and conditions and that such efforts have been unsuccessful for a reasonable period of time." Secondly, affordable prices for vaccines that are produced under compulsory licenses can be ensured by allowing WTO members to "set the remuneration to the right holder at a level that reflects the price charged by the manufacturer of the vaccine or therapeutic under a compulsory licence" (European Union 2021, 11). This proposal varies the requirement under Article 31(h) of the TRIPS Agreement, which provides "that the right holder shall be paid *adequate remuneration in the circumstances of each case*, taking into account the economic value of the authorization" (emphasis added). Thirdly, the procedural aspect of Article 31bis and the Annex to the TRIPS Agreement should be simplified by allowing an exporting member utilizing compulsory licensing to "provide in one single notification a list of all countries to which vaccines and therapeutics are to be supplied directly or through the COVAX Facility" (European Union 2021, 12). This procedural change means that the eligible importing Members do not each have to notify the TRIPS Council. In addition to these three components, the EU issued a note on the points of convergence on the TRIPS issues for discussion. It indicated that Articles 28(1), 39, and Part III of the TRIPS Agreement should be considered outside the scope of the waiver since their inclusion is not justifiable or proportionate (Baker 2021; Love 2021).

Some have argued that compulsory licenses are sufficient. Erciyas and Üstün (2022) think it is unlikely that a patent owner would refrain from granting a license since Article 30 of the TRIPS Agreement allows WTO members to provide limited exceptions to patent rights that respect the legitimate interests of the patent holder and third parties. Additionally, Article 31 allows the use of the invention by a government or third party authorized by the government without seeking the consent of the patent owner "in the case of a national emergency or other circumstances of extreme urgency or in cases of public non-commercial use," and Article 31bis sets out conditions in which products developed through compulsory licensing authorized in one country can be sent to other developing countries.

Available evidence, however, contradicts this view. First, the Bolivia-Biolyse compulsory licensing case, which is yet to be resolved, illustrates a situation where an IP owner refused to grant a license (Schouten 2021). This case resulted from Johnson & Johnson (J&J) refusing to grant a voluntary license to Biolyse, a Canadian generic company that could manufacture and supply the vaccines to LMICs. After J&J's refusal, the attempt to obtain a compulsory license through the Canadian Access to Medicines Regime has stalled due to red tape and bureaucratic procedures (Schouten 2021).

Moreover, IP waivers, which will ensure that countries that decide to use flexibilities do not face trade retaliation or legal action, are important enablers that must be combined with knowledge sharing and technology transfer in order to be effective. Evidently, technology owners have not been willing to share knowledge or transfer technology, fearing that disclosing their trade secrets when contributing to initiatives such as C-TAP entails too great a loss (Gurgula and Hull 2021, 1243). Accordingly, relying on flexibilities in TRIPS can be challenging since the vaccines may be subject to other types of IPRs such as trade secrets that are not subject to compulsory licensing mechanisms, thus making it impossible to rely on compulsory licensing (1245).

The EU's counterproposal has been called a delaying tactic since it only addresses the compulsory licensing of patents rather than the problem at hand. Compulsory licensing has proven to be complex, time consuming, and not an ideal solution to the public health emergency, which the proposed waiver seeks to address (Marans 2021). The counterproposal also excludes relevant IPRs such as the mandatory disclosure of proprietary information and other demands that were part of the waiver. Additionally, the flexibilities that are provided for in the TRIPS Agreement are insufficient to ensure equitable access to COVID-19 vaccines. The Committee on Economic, Social and Cultural Rights (CESCR) has therefore recommended that "all mechanisms, including voluntary licensing, technology pools, use of TRIPS flexibilities and waivers of certain intellectual property provisions or market exclusivities should be explored carefully and utilized" (2021, 13).

Those attempting to provide more equitable distribution under TRIPS have proposed several approaches beyond what the EU is willing to do. First, compulsory licensing of trade secrets has been proposed (Gurgula and Hull 2021, 1250; Levine 2020, 850) as a possible solution. This entails requiring disclosure of the secret or transfer coupled with a strict "obligation of confidentiality" for those with access to limit damage to the IPR holder, an obligation that is usually not part of the established public interest defense in law (Gurgula and Hull 2021, 1250). Additionally, Article 39 of the TRIPS Agreement only explicitly prohibits disclosure of undisclosed information or trade secrets when it will lead to unfair competition, but says nothing when the concern is public interest (WHO, WIPO, and WTO 2020, 81; Andanda 2013, 150). In practice, undisclosed information related to clinical trials has been subordinated to public health priorities in Europe (European Medicines Agency 2019), and countries such as Argentina, Brazil, Japan, Israel, and the United States authorize the use of undisclosed information by regulatory authorities for abridged regulatory approval of generic medical products without any harm to the right holders (Andanda 2013, 146). Opinions and national practices differ on the requirements of this Article as there is no WTO jurisprudence or authoritative WTO guidance on it, although "regulatory agencies may, however, disclose the data when disclosure is necessary to protect the public or where steps are taken to ensure that there is no unfair commercial use of the data concerned" (WHO, WIPO and WTO 2020, 81). In the case of COVID-19 vaccines, diagnostics, and treatments, the public health priorities of "speed, adequacy of supply, and affordability" are sufficient to force disclosure (Levine 2020, 849).

A second approach entails using contract law by encouraging patent owners to cooperate and provide other manufactures or companies with the necessary licenses. Such licenses can then be used to support international initiatives like COVAX to ensure equitable access to vaccines (Erciyas and Üstün 2022). The presence of such licenses is not enough to ensure that this approach leads to equitable access to the products generated through the IP, even in cases where public funds have been used to generate such IP: absence of political or moral will has led states to contribute to exclusive ownership of IP in ways that result in inequitable access to products. An excerpt from the contract between the Advance Purchase Agreement between CureVac and the

European Union, which was funding some of the research, illustrates a typical exclusive ownership of IPRs:

> The Parties acknowledge and agree that the contractor shall be the sole owner of all intellectual property rights generated during the development, manufacture, and supply of the Product, including all know-how (collectively, the "Product IP Rights"). The contractor shall be entitled to exclusively exploit any such Product IP Rights. Except as expressly set forth in this APA, the contractor does not grant to the Commission and/or the participating Member States by implication, estoppel or otherwise, any right, title, license or interest in the Product IP Rights. All rights not expressly granted by the contractor hereunder are reserved by the contractor. (European Commission 2020c, Article 1.20.1)

The EU contributed 2.7 billion euros under the Emergency Support Instrument and provided loans from the European Investment Bank (European Commission 2020a). By entering into these contractual terms, which led to the manufacturer exclusively owning the IP, the EU backtracked from its earlier decision that the Commission will promote COVID-19 vaccine as a global public good:

> In the negotiations with the pharmaceutical industry under the present Agreement, the Commission will promote a COVID-19 vaccine as a global public good. This promotion will include access for low and middle income countries to these vaccines in sufficient quantity and at low prices. The Commission will seek to promote related questions with the pharmaceutical industry regarding intellectual property sharing, especially when such IP has been developed with public support. . . . Any vaccines available for purchase under the APAs concluded but not needed and purchased by Participating Member States can be made available to the global solidarity effort. (European Commission 2020b)

Ultimately, signing contractual terms that deviated from the above objective put the EU in a weaker position when negotiating favorable prices (Boschiero 2021). The contract also contained terms prohibiting the EU and participating states from reselling, exporting, distributing, or donating the products to another country outside the EU, to NGOs, or to the WHO without the contractor's consent (Article 1.10.2). The ongoing crisis relating to inequitable access to vaccines can therefore also be blamed on states, which have left "decisions around availability, accessibility and affordability in the hands of businesses" (Amnesty International 2021, 8). States have an obligation to ensure access to essential medicines by preventing unreasonably high prices so they cannot be exonerated from contributing to the ongoing crisis (CESCR 2020b).

A more radical approach is to reconsider IPRs in light of human rights. States are required to exercise their human rights obligation of protecting the right to health (United Nations, International Covenant on Economic, Social and Cultural Rights 1966, Article 12(2)). The Covenant also provides for the responsibilities of other non-state actors such as the private business sector in realizing the right to health (General comment 14, paragraph 42). In fact, the TRIPS Agreement hints at human rights provisions by stipulating that IPRs should be protected and enforced "in a manner conducive to social and economic welfare, and to a balance of rights and obligations" (Article 7). The Agreement also recognizes that WTO Members "may adopt measures necessary to protect public health" (Article 8.1).

A number of United Nations committees have also issued statements emphasizing the need to utilize IPRs in ways that are consistent with respect for human rights during the ongoing pandemic. CESCR, which is responsible for monitoring the implementation of the International Covenant on Economic, Social and Cultural Rights by member states, has issued a statement urging private actors to "refrain from invoking intellectual property rights in a manner that is inconsistent with the right of every person to access a safe and effective vaccine against COVID-19 or the right of States to exercise the flexibilities of the TRIPS Agreement" (2021, 8). Accordingly, the way the ongoing debate on equitable access to COVID-19 vaccines is handled by private actors such as IPR holders can adversely affect the obligation of states to protect the right to health by ensuring the "most advanced, up-to-date, and verifiable science available" (CESCR 2020a).

To avoid this eventuality, political and moral will are required to waive IPR protection during the ongoing pandemic. Erfani et al. (2021) correctly argue that a waiver is required to remove IP barriers to "enable future hubs, engage a greater number of manufacturers, and ultimately yield more doses faster." Forman et al. (2021, 554) have framed the issue of ensuring equitable vaccine access globally as a policy challenge. They blame inequitable access on "vaccine nationalism" and the refusal to distribute vaccines across national borders (556). Political and moral will can be achieved through the proposed ethical-legal framework that is discussed in this chapter.

THE WTO'S DECISION ON THE WAIVER OF COVID-19 PATENTS

The EU, United States, India, and South Africa eventually brokered a compromised deal which influenced the WTO Ministerial Council's decision that was adopted on June 17, 2022. The deal entailed authorizing "eligible WTO members to use patented ingredients and processes for the production and supply of COVID-19 vaccines without the consent of the right holder for a limited period" (WTO 2022b). The deal was criticized for deviating from the initial broad waiver that had been proposed by India and South Africa (Correa and Syam 2022; Zarocostas 2022). It was limited to clarifying aspects of Articles 31 and 39.3 of the TRIPS Agreement without addressing other obligations related to other intellectual property rights, therapeutics, diagnostics and other technologies (Correa and Syam 2022). All these unaddressed issues were proposed to be addressed by Members at a later stage.

After long deliberations, the WTO's Ministerial Conference decision mirrored the contents of the above deal (WTO 2022a). The conference decided that eligible Members may limit patent rights "by authorizing the use of the subject matter of a patent required for the production and supply of COVID-19 vaccines without the consent of the right holder to the extent *necessary* to address the COVID-19 pandemic, in accordance with the provisions of Article 31 of the Agreement" (WTO 2022a, ¶1, emphasis added). All developing countries qualify as eligible Members in terms of this decision and can apply the decision until five years from the date it was made (WTO 2022a, ¶6). The decision may be extended taking into consideration the exceptional circumstances of the COVID-19 pandemic. Notably, developing country Members with existing capacity to manufacture COVID-19 vaccines are encouraged by the WTO to make a binding commitment not to exercise the rights afforded by this decision.

The inclusion of the word *necessary* in ¶1 of the decision implies that eligible Members will have to meet the necessity test to rely on this decision. The challenge with this requirement is that the current jurisprudence at the WTO shows that the "necessity test" is interpreted narrowly such that eligible members will have to prove that limiting patent rights is necessary and that there are no other less restrictive means of achieving their public health objectives (Correa and Syam 2022). Notably, narrow interpretation of necessity can result in a hypothetically available alternative defeating a claim of necessity (Kapterian 2010, 103).

Some procedural issues under the TRIPS Agreement were also clarified in the decision: eligible Members are not required to make efforts to obtain an authorization from the right holder as required by Article 31(b); the products developed under the waiver are not limited to the supply of the domestic market as required by Article 31(f) but may be exported to eligible Members; the "determination of adequate remuneration under Article 31(h) may take account of the humanitarian and not-for-profit purpose of specific vaccine distribution programs aimed at providing equitable access to COVID-19 vaccines in order to support manufacturers in eligible Members to produce and supply these vaccines at affordable prices for eligible Members" (WTO 2022a, ¶3).

Two positive outcomes from this decision are worth highlighting. First, eligible Members do not need to have national legislation that provides for compulsory licensing to utilize the decision. Where no such legislation is available, Members can rely on executive orders, emergency decrees, and judicial or administrative orders (¶2). The second outcome, an assurance that eligible Members who use the decision will not face legal challenges, is contained in ¶7 of the decision. The paragraph provides that "members shall not challenge any measures taken in conformity with this Decision under subparagraphs 1(b) and 1(c) of Article XXIII of the GATT 1994."

The decision is limited, however, to COVID-19 vaccine patents, and Members will decide after six months whether to extend it to the production and supply of COVID-19 diagnostics and therapeutics (¶8). It mentions, however, that "Article 39.3 of the Agreement does not prevent an eligible Member from enabling the rapid approval for use of a COVID-19 vaccine produced under this Decision" (¶4). This paragraph essentially extends the decision beyond patents and endorses the use of abridged regulatory approval through the use undisclosed information such as trade secrets by authorities as discussed in the previous section. However, this slight extension from an exclusive focus on patents does not sufficiently address the concerns that warranted the request for a waiver. An attempt to resolve issues of inequitable access to medical products and find models of biotechnology intellectual property that work properly can lead to an erroneous conflation of different intellectual property regimes (Gold et al. 2002). This error results from "lack of an integrated, transdisciplinary methodology to understand and analyze the social, ethical and economic impact of intellectual property rights" (327): intellectual property protections should be treated as a transdisciplinary issue, not as a protected silo of legal and regulatory interest (328). This argument supports the proposal for an ethical-legal consideration in dealing with the ongoing debates on intellectual property rights and inequitable access to vaccines and other medical products related to COVID-19 in times of crisis as discussed in the next section.

AN ETHICAL-LEGAL PARADIGM SHIFT IN MANAGING INTELLECTUAL PROPERTY RIGHTS

Noting that the TRIPS Agreement was not designed for handling situations such as pandemics, UNESCO has underscored the need for new global approaches and mechanisms that can ensure efficient development and production of vaccines (2021, 3). Such global approaches and mechanisms require a coordinated approach that combines legal and ethical principles to ensure that IPR holders manage their rights in a way that balances their proprietary rights with concerns for ethical principles and human rights obligations. The fact that even before the pandemic, developing and least developed countries faced constraints in using TRIPS flexibilities to ensure access to essential medicines testifies to the need for this coordinated approach. According to the World Intellectual Property Organization's Standing Committee on the Law of Patents, these constraints arise from the practical complexity of implementing the flexibilities (WIPO 2017, 23); the need for clarity on the scope of the flexibilities (24); "insufficient local legal and technical expertise to incorporate and implement the TRIPS flexibilities into the national law and policy" (26); lack of coordination among various government departments and ministries (30); and external political and economic pressure that prevents governments from issuing compulsory licenses (31). Moreover, vaccine production is a complex process, and the shortage has been attributed to lack of raw materials, production capacity constraints, and the complex nature of the production of drugs (Erciyas and Üstün 2022). Consequently, attempts to resolve the issue of inequality by only targeting vaccine patents, even if export restrictions are changed, will likely yield suboptimal outcomes.

This realization raises a more fundamental question: has the debate over COVID-19 correctly shifted the spotlight to protection of IPRs, or should it shine on fundamental problems with the business models and government policies that lead to inequitable access to vaccines? I propose that a new ethical-legal framework is needed to address longstanding physical and legal barriers that will yield favorable outcomes that can benefit all stakeholders with minimum legal and ethical risks. In developing this framework, I utilize Carl Mitcham's idea of assessing the ethics of technology by moving "From Thinking Big to Small—and Big Again" (2020). To use Mitcham's apt metaphor, the ongoing attention given to IPRs as a barrier to equitable access to vaccines is the tip of the iceberg while the icy mass beneath, which deserves focus, consists of multifactorial barriers such as a lack of transparency on the costs of production, factors that determine the prices that are charged, and what leads to failure to share technology with other stakeholders.

Erciyas and Üstün have correctly observed that IPRs should not be impacted negatively "without making sure that the IP right waiver is indeed the ultimate solution to vaccine inequity" (2022). In this regard, Amnesty International has observed that "waiving of intellectual property rights could lift legal and bureaucratic deterrents to manufacturing COVID-19 vaccines . . . [but this] alone would not, however, automatically accelerate manufacturing and thus increase availability of the COVID-19 vaccine" (2021, 23). Zoltán Kis and Zain Rizvi (2021) have shown that other avenues such as resource and knowledge sharing and repurposing of existing facilities to produce vaccines can be used to meet the demand for vaccines in one year.

Ensuring resource and knowledge sharing can only be realized if IPR holders "increase efforts to license, transfer technology and expand manufacturing" (Mercurio 2021, 986). IPR holders must also "support efforts to enable prompt and effective use of existing flexibilities in the TRIPS Agreement and concerted and coordinated efforts involving governments and the private sector to ensure all qualified generic producers willing and capable of manufacturing vaccines are doing so" (987). For IPR holders to act in this way, a paradigm shift is urgently needed. This shift should entail an ethical management of IPRs by stakeholders in a way that prioritizes human rights and life instead of insisting on their exclusive IPRs and refusing to grant requests for licenses and technology transfer. States that contract with IPR holders or vaccine manufacturers must also honor their human rights obligations by not leaving important decisions that affect equitable distribution of vaccines and other products to businesses. For example, in relation to the EU's APA with CureVac, Boschiero correctly observes that "the EU has shamefully guaranteed a de facto veto right to the private company to provide universal, fair access of vaccines through the current channel of global solidarity [since] none of the EU APAs have clauses that will ensure global non-exclusive licenses to third parties, which would be able to guarantee the accessibility/affordability of COVID-19 health technologies" (2021, 28).

It is in the spirit of ensuring equitable access that Amnesty International has insisted that vaccine manufacturers "share their knowledge and technology and train qualified manufacturers committed to contributing to the ramp-up of the production of COVID-19 vaccines" (2021, 9). Their call resonates with the human rights obligations of pharmaceutical companies stipulated in the UN Guiding Principles, which require these companies to "respect the rights of countries to use, to the full, the provisions in the Agreement on Trade-Related Aspects of Intellectual Property Rights (TRIPS) . . . which allow flexibility for the purpose of promoting access to medicines" (OHCHR 2008, 26). The companies are also required to ensure that the pricing of their medicines is affordable (33).

One justification for a new ethical-legal approach is the fact that inequitable access to vaccines will leave parts of populations in countries that lack access to vaccines unvaccinated, thus creating opportunities for more variants of concern to develop, which increases the risk of vaccine escape (Williams and Burgers 2021). UNESCO's ethics commission has also underscored the need for an ethical approach to the ongoing debate by proposing that vaccines be considered global common goods and that the current regulation of patenting and ownership rights should be considered as an ethical concern (2021, 3). They have concluded that "a utilitarian approach based on the benefit of the greatest number of people is not acceptable as the sole criterion from an ethical perspective" (4). Accordingly, a utilitarian approach of protecting IPRs to incentivize innovation cannot be justified in the context of the pandemic where much of the research leading to the development of the vaccines and other products has been supported through public funds (Katz, Weintraub, Bekker, and Brandt 2021). The "public good" nature of these products cannot be overlooked. Consequently, deontological approaches that vest IPRs in pharmaceutical companies to reward them for their labor and justify returns on investment fall short due to public investment in the funding and research. Jecker and Atuire have correctly concluded in this regard that these products equally belong to the public due to public investment in their production (2021, 596).

My proposed paradigm to ensure equitable access to vaccines is a synthesis of three ethical-legal principles already established in the literature: justice, fairness, and global solidarity. These principles should guide stakeholders in managing IPRs in ways that are beneficial to the society and ensuring that the rights are not only managed from a purely legal perspective. As discussed below, these principles are ethical-legal in nature.

Ethical-Legal Principle 1: Justice

Justice should be viewed, in the context of this chapter, from an ethical-legal perspective due to the subject matter under discussion, namely IPRs. This perspective is based on Ghosh's characterization of intellectual property law as a system that defines and regulates creative activity (2008, 106). This is conceptualized as regulatory justice. In this regard, IPRs have a social dimension in the sense "that they are the means to a social end beyond the protection of individual self-interest" (114). Ignoring this social end is what leads to structural injustice that has been witnessed during the pandemic where IPRs are not shared to ensure the realization of human rights and the protection of public health.

The global system of managing property rights such as IPRs has been identified as a structural injustice that deprives certain parts of the global community of the means that are needed to ensure equitable access to the tools that are necessary to combat the ongoing pandemic (Jecker 2022). In the context of this chapter, the relevant system refers to multilateral trade rules that were established through the TRIPS Agreement. These rules tend to intrude into the domestic policy space of WTO members (Ganesan 2015, 212). This intrusion has indeed played out in the ongoing debates on TRIPS waiver as had been feared by developing countries during the negotiations of the TRIPS Agreement, which led to the inclusion of the enforcement of IPRs as trade issues. Developing countries were not keen on this inclusion since they feared that including IPRs in trade issues would limit their ability to provide affordable health care in their countries (213).

A duty of justice arises in this regard due to the IPR holders having benefited from the structural injustice. Such structural injustice allows IPR holders to monopolize rights that are derived from public funds, data, samples and expertise from other stakeholders while limiting the rights of competitors to gain access to protected technology that can enable them to increase manufacturing capacity and supply developing countries with much needed products, even as global vaccine justice could be achieved through "a rapid shift in trade regulations and contract transparency that streamlines IP sharing and technology transfers" (Harman et al. 2021).

Ethical-Legal Principle 2: Fairness

Fairness is used in IP law context in two senses: to restrict abusive and anti-competitive conduct by holders and users of IPRs and to protect inventors and creators of IP (Gervais 2020, 1). Gervais (2020) correctly notes that the principle confers the protection of IPRs with a degree of proportionality and is one of the justificatory theories of IP. Questions of fairness, therefore, arise from the way IPRs are used or managed. In practical terms, fairness dictates that IPRs are protected and enforced in ways that also respect human rights and ensure the intended social value of IP.

The principle thus obliges IPR holders not to insist on enforcing their rights and instead transfer and share the technology that can help ramp up the production of vaccines and other products that can contain the global pandemic. Additionally, pharmaceutical companies should change from adopting a business-as-usual attitude during the pandemic and fairly distribute the vaccines (Emanuel et al. 2021). This principle thus makes the restrictive approach that was used in the Advance Purchase Agreement between CureVac and the European Union unethical.

To ensure fairness, there is a need to adopt solutions that address the root causes of the problem of inequity by challenging unethical intellectual property regimes, which are currently being avoided through short-term solutions such as charitable donations of surplus vaccines. Fairness requires sharing COVID-19 technologies broadly and quickly (Harman et al. 2021).

Ethical-Legal Principle 3: Global Solidarity

Article 13 of the Universal Declaration on Bioethics and Human Rights (UNESCO 2005, 77) encourages solidarity and international cooperation among human beings. Specific aspects of solidarity are outlined in Article 15, which states that "benefits resulting from any scientific research and its applications should be shared with society as a whole and within the international community." Notably, sharing of benefits is also recognized as a human right in international human rights instruments such as Article 27 of the Universal Declaration of Human Rights, which states that "everyone has the right freely to participate in the cultural life of the community, to enjoy the arts and to share in scientific advancement and its benefits." Article 15 (1)(b) of the International Covenant on Economic, Social and Cultural Rights equally requires States to recognize the right of everyone to enjoy the benefits of scientific progress and its applications.

The Special Rapporteur in the field of cultural rights issued a report highlighting the potential of intellectual property regimes to obstruct new technological solutions to critical human problems such as food, water, health, chemical safety, energy, and climate change (United Nations General Assembly 2012, ¶56). In this regard, the Special Rapporteur proposed "the adoption of a public good approach to knowledge innovation and diffusion and suggest[ed] reconsidering the current maximalist intellectual property approach to explore the virtues of a minimalist approach to IP protection" (¶65).

The above ethical-legal principle accordingly confirms that the principle of global solidarity is a moral rights-based approach, which can be used to ensure equitable distribution of resources to contain the pandemic by recognizing the pandemic as a common threat to all (Afilalo and Burns 2021, 5). Framing global solidarity in this way is aligned with the view that "global collaboration requires a collective response to shared risks and fundamental rights, where all states have mutual responsibilities" (Gostin et al. 2010, 719). The link between solidarity and human rights has become more evident during the pandemic as people felt more interconnected in the face of global vulnerability, and the link calls for a paradigm shift that categorizes COVID-19 vaccine as a common good to be protected from exclusionary market models (Borges and Dos Santos 2021).

A starting point in the application of the principle is ensuring that pandemic-related research is free of patent restrictions and research participants are assured of access to the vaccines that prove effective since their participation in research demonstrates an act of solidarity (Torres et al. 2021). Additionally, research leading to the development of vaccines and other products often benefit from public funds. Accordingly, a commitment to global solidarity requires all countries to ethically and efficiently manage the distribution of vaccines (Binagwaho et al. 2022, 101). If all countries embraced global solidarity, the issue of "vaccine nationalism" would not have emerged and the WHO-led C-TAP would not have experienced a breakdown of solidarity that is evident from the ongoing lack of support. Disregard for international mechanisms, for sharing of vaccines and other tools, which were established at the beginning of the pandemic, ultimately undermine global solidarity (Torreele and Amon 2021, 276).

A clear demonstration of global solidarity can be glimpsed from the way researchers freely shared their research articles and findings to ensure global containment of the pandemic (Chatfield and Schroeder 2020) and high levels of transparency in sharing genomic data (Berditchevskaia and Peach 2020). These good examples should be formalized through an internationally coordinated initiative to facilitate technology transfer and work collectively to ensure sufficient vaccine supply and contain the pandemic (Figueroa et al. 2021). Voluntary licensing, which should be led by public-private partnerships as recommended by the WTO can work effectively if an internationally coordinated initiative is in place (Zarocostas 2020).

CONCLUSION

The paradigm shift proposed in this chapter should guide IPR holders and other stakeholders in managing these rights in an ethical way based on the principles of justice, fairness, and solidarity. Applying these principles will encourage stakeholders to share technology with other manufactures to improve the capacity for the production of COVID-19 vaccines and ensure equitable access across the globe. Other avenues that are mentioned in this chapter should also be explored so that the solution to the ongoing public health crisis is not limited to the interim decision of the WTO on the TRIPS waiver, which has only focused on vaccine patents and adopted a phased approach in addressing the remaining issues that were identified in the initial proposal for a waiver.

REFERENCES

Abbas, Zaheer Muhammad. 2021. "Canada's Political Choices Restrain Vaccine Equity: The Bolivia-Biolyse Case." *South Centre Research Paper* 136, September 2021.

Afilalo, Ari, and Brittany Burns. 2021. "Global Pandemic and International Crisis." *Rutgers Business Law Review* 17, no. 1 (Fall): 2–16.

Amnesty International. 2021. "A Double Dose of Inequality: Pharma Companies and the COVID-19 Vaccines Crisis." September 22, 2021. https://www.amnesty.org/en/wp-content/uploads/2021/09/POL4046212021ENGLISH.pdf.

Andanda, Pamela. 2013. "Managing Intellectual Property Rights over Clinical Trial Data to Promote Access and Benefit Sharing in Public Health." *The International Review of Intellectual Property and Competition Law* 44, no. 2: 140–77.

Baker, Brook K. 2021. "EU's Proposal on Convergence on WTO TRIPS Waiver Only Addresses Compulsory Licensing on Patents, Ignores Trade Secrets, and Is Wholly

Inadequate to Solve Inequitable Access." infojustice, October 14, 2021. http://infojustice.org/archives/43712.

BBC News. 2021a. "Covid: Germany Rejects US-backed Proposal to Waive Vaccine Patents." *BBC News*, May 6, 2021. https://www.bbc.com/news/world-europe-57013096.

———. 2021b. "Covid: US Backs Waiver on Vaccine Patents to Boost Supply." *BBC News*, May 6, 2021. https://www.bbc.com/news/world-us-canada-57004302.

Berditchevskaia, Aleks, and Kathy Peach. 2020. "Coronavirus: Seven Ways Collective Intelligence Is Tackling the Pandemic." *The Conversation*, March 12.

Binagwaho, Agnes, Mathewos Kedest, and Davis Sheila. 2022. "Equitable and Effective Distribution of the COVID-19 Vaccines: A Scientific and Moral Obligation." *International Journal of Health Policy and Management* 11, no. 2: 100–102. https://doi.org/10.34172/ijhpm.2021.49.

Borges, Gustavo Silveira, and Benício Fagner dos Santos. 2021. "COVID-19 Vaccine as a Common Good." *Journal of Global Health*, 11:03109.

Boschiero, Nerina. 2021. "COVID-19 Vaccines as Global Common Goods: An Integrated Approach of Ethical, Economic Policy and Intellectual Property Management." *Global Jurist* 2021: 20210042. https://doi.org/10.1515/gj-2021-0042.

CESCR (UN Committee on Economic, Social and Cultural Rights). 2020a. "Statement on the Coronavirus Disease (COVID-19) Pandemic and Economic, Social and Cultural Rights." April 6, 2020. E/C.12/2020/1.

———. 2020b. "General comment No. 25 (2020) on Science and Economic, Social, and Cultural Rights (Article 15 (1) (b), (2), (3) and (4) of the International Covenant on Economic, Social and Cultural Rights)." April 30, 2020. E/C.12/GC/25.

———. 2021. "Statement on Universal Affordable Vaccination against Coronavirus Disease (COVID-19), International Cooperation and Intellectual Property." April 23, 2021. E/C.12/2021/1.

Chatfield, Kate, and Doris Schroeder. 2020. "Ethical Research in the COVID-19 Era Demands Care, Solidarity, and Trustworthiness." *Research Ethics* 16, nos. 3–4: 1–4.

Correa, Maria Carlos, and Nirmalya Syam. 2022. "Analysis of the Outcome Text of the Informal Quadrilateral Discussions on the TRIPS COVID-19 Waiver." South Centre, Policy Brief 110, May 5, 2022.

Emanuel, Ezekiel J., Allen Buchanan, Chan, Shuk Ying Chan, Cécile Fabre, et al. 2021. "What Are the Obligations of Pharmaceutical Companies in a Global Health Emergency?" *Lancet* 398, no. 10304: 1015–20.

Erciyas, Selin Sinem, and Zeynep Çağla Üstün. 2022. "Vaccine Patent Rights—A Scapegoat for Inequality?" *IAM,* January 12, 2022. https://www.iam-media.com/article/vaccine-patent-rights-scapegoat-inequality.

Erfani, Parsa, Agnes Binagwaho, Mohammed Juldeh Jalloh, Muhammad Yunus, Paul Farmer, and Vanessa Kerry. 2021. "Intellectual Property Waiver for COVID-19 Vaccines Will Advance Global Health Equity." *BMJ* 374, no. 1837. http://dx.doi.org/10.1136/bmj.n1837.

European Commission. 2020a. "Communication from the Commission to the European Parliament, the European Council, the Council and the European Investment Bank: EU Strategy for COVID-19 Vaccines." Brussels, June 17, 2020, COM (2020) 245 final.

———. 2020b. "Annex to the Commission Decision on Approving the Agreement with Member States on Procuring COVID-19 Vaccines on behalf of the Member States and Related Procedures." Brussels, June 18, 2020, C (2020) 4192 final.

———. 2020c. "Advance Purchase Agreement for the Development, Production, Advance Purchase and Supply of a COVID-19 Vaccine for EU Member States." Number Sante/2020/C3/049.

———. 2021. "EU Proposes a Strong Multilateral Trade Response to the COVID-19 Pandemic." June 4, 2021. https://ec.europa.eu/commission/presscorner/api/files/document/print/en/ip_21_2801/IP_21_2801_EN.pdf.

European Medicines Agency. 2019. Policy on Publication of Clinical Data for Medicinal Products for Human Use." EMA/144064/2019, March 21, 2019. https://www.ema.europa.eu/en/documents/other/european-medicines-agency-policy-publication-clinical-data-medicinal-products-human-use_en.pdf.

European Union. 2021. "Communication from the European Union to the Council for TRIPS Urgent Trade Policy Responses to the COVID-19 Crisis: Intellectual Property." Brussels, June 4, 2021. https://trade.ec.europa.eu/doclib/docs/2021/june/tradoc_159606.pdf.

Figueroa, J. Peter, Peter J. Hotez, Carolina Batista, Yanis Ben Amor, Onder Ergonul, Sarah Gilbert, et al. 2021. "Achieving Global Equity for COVID-19 Vaccines: Stronger International Partnerships and Greater Advocacy and Solidarity Are Needed." *PLoS Med* 18, no. 9: e1003772. doi:10.1371/journal.pmed.1003772.

Forman, Rebecca, Soleil Shah, Patrick Jeurissen, Mark Jit, and Elias Mossialos. 2021. "COVID-19 Vaccine Challenges: What Have We Learned So Far and What Remains to Be Done?" *Health Policy* 125, no. 5: 553–67.

Ganesan, Venkatachalam Arumugamangalam. 2015. "Negotiating for India." In *The Making of the TRIPS Agreement: Personal Insights from the Uruguay Round Negotiations*, edited by Jayashree Watal and Antony Taubman. Geneva: World Trade Organization.

Gervais, Daniel J., ed. 2020. *Fairness, Morality and Ordre Public in Intellectual Property*. Northampton: Edward Elgar.

Ghosh, Shubha. 2008. "When Property Is Something Else: Understanding Intellectual Property through the Lens of Regulatory Justice." In *Intellectual Property and Theories of Justice*, edited by Axel Gosseries, Alain Marciano, and Alain Strowel, 106–21. London: Palgrave Macmillan.

Gold, E. Richard, David Castle, L. Martin Cloutier, Abdallah S. Daar and Pamela J. Smith. 2002. "Needed: Models of Biotechnology Intellectual Property." *Trends in Biotechnology* 20, no. 8: 327–29.

Gostin, Lawrence O., Safura Abdool Karim, and Benjamin Mason Meier. 2020. "Facilitating Access to a COVID-19 Vaccine through Global Health Law." *Journal of Law, Medicine and Ethics* 48, no. 3: 622–26. doi:10.1177/1073110520958892.

Gostin, Lawrence O., Mark Heywood, Gorik Ooms, Anand Grover, John-Arne Røttingen, and Wang Chenguang. 2010. "National and Global Responsibilities for Health." *Bulletin of the World Health Organization* 88, no. 10: 719–719A. doi:10.2471/BLT.10.082636.

Gurgula, Olga, and John Hull. 2021. "Compulsory Licensing of Trade Secrets: Ensuring Access to COVID-19 Vaccines via Involuntary Technology Transfer." *Journal of Intellectual Property Law & Practice* 16, no. 11: 1242–61.

Harman, Sophie, Parsa Erfani, Tinashe Goronga, Jason Hickel, Michelle Morse, and Eugene T. Richardson. 2021. "Global Vaccine Equity Demands Reparative Justice—Not Charity." *BMJ Global Health* 6: e006504. doi:10.1136/bmjgh-2021-006504.

IFPMA (International Federation of Pharmaceutical Manufacturers & Associations). 2021. "IFPMA Statement on WTO TRIPS Intellectual Property Waiver." https://www.ifpma.org/resource-centre/ifpma-statement-on-wto-trips-intellectual-property-waiver/.

———. 2022. "IFPMA statement on TRIPS discussion document." https://www.ifpma.org/resource-centre/statement-ifpma-trips-discussion-document/.

Jecker, Nancy S. 2022. "Global Sharing of COVID-19 Vaccines: A Duty of Justice, Not Charity." *Developing World Bioethics*. https://doi.org/10.1111/dewb.12342.

Jecker, Nancy S., and Caesar A. Atuire. 2021. "What's Yours Is Ours: Waiving Intellectual Property Protections for COVID-19 Vaccines." *Journal of Medical Ethics* 47: 595–98.

Kapterian, Gisele. 2010. "A Critique of the WTO Jurisprudence on 'Necessity.'" *International & Comparative Law Quarterly* 59, no. 1: 89–127.

Katz, Ingrid T., Rebecca Weintraub, Linda-Gail Bekker, and Allan M. Brandt 2021. "From Vaccine Nationalism to Vaccine Equity—Finding a Path Forward." *New England Journal Medicine* 384, no. 14: 1281–83.

Kis, Zoltán, and Zain Rizvi. 2021. "How to Make Enough Vaccine for the World in One Year." *Public Citizen*. Accessed August 20, 2022. https://www.citizen.org/wp-content/uploads/mRNA-vaccine-roadmap-May-26-final.pdf.

Levine, David S. 2020. "Trade Secrets and the Battle against Covid." *Journal of Intellectual Property Law & Practice* 15, no. 11: 849–50.

Love, James. 2021. "KEI Comments on the Recent EU TRIPS Waiver Papers." *Knowledge Ecology International*, October 15, 2021. https://www.keionline.org/36768.

Marans, Daniel. 2021. "New European Vaccine Proposal Offers Limited Help to Developing Countries." *HuffPost* (USA), October 13, 2021. https://www.huffpost.com/entry/european-union-covid-vaccine-intellectual-property-proposal_n_61664498e4b0f26084edbbff.

Mercurio, Bryan. 2021. "The IP Waiver for COVID-19: Bad Policy, Bad Precedent." *International Review of Intellectual Property and Competition Law* 52: 983–88.

Mitcham, Carl. 2020. "The Ethics of Technology: From Thinking Big to Small—and Big Again." *Axiomathes* 30, 589–96.

MSF (Medecins Sans Frontieres). 2020. "WTO COVID-19 TRIPS Waiver Proposal: Myths, Realities and an Opportunity for Governments to Protect Access to

Lifesaving Medical Tools in a Pandemic." https://msfaccess.org/sites/default/files/2020-12/MSF-AC_COVID_IP_TRIPSWaiverMythsRealities_Dec2020.pdf.

OHCHR (Office of the High Commissioner for Human Rights). 2021. "Information Note: Experts Send Pharma Companies, States, EU and WTO Letters Calling for Urgent Action on COVID-19 Vaccines." https://www.ohchr.org/en/press-releases/2021/10/information-note-experts-send-pharma-companies-states-eu-and-wto-letters.

———. 2008. Human Rights Guidelines for Pharmaceutical Companies in Relation to Access to Medicines, Commentary to Guidelines 6–8, A/63/263, 11 August 2008, https://digitallibrary.un.org/record/637422/files/A_63_263-EN.pdf.

Roffe, Pedro, Christoph Spennemann, and Johanna von Braun. 2005. "From Paris to Doha: The WTO Doha Declaration on the TRIPS Agreement and Public Health." In *Negotiating Health: Intellectual Property and Access to Medicines*, edited by Pedro Roffe, Geoff Tansey, and David Vivas-Eugui, 9–26. London: Routledge. https://doi.org/10.4324/9781849772082.

Safi, Michael. 2021. "WHO Platform for Pharmaceutical Firms Unused since Pandemic Began." *Guardian*, January 22, 2021. https://www.theguardian.com/world/2021/jan/22/who-platform-for-pharmaceutical-firmsunused-since-pandemic-began.

Sariola, Salla. 2021. "Intellectual Property Rights Need to be Subverted to Ensure Global Vaccine Access." *BMJ Global Health* 6:e005656. https://doi.org/10.1136/bmjgh-2021-005656.

Schouten, Arianna. 2021. "Canada-based Biolyse Pharma Seeks to Manufacture COVID-19 Vaccines for Low Income Countries, May Test Canada's Compulsory Licensing for Export Law." *KEI*, 12 March 2021. http://www.keionline.org/35587.

Tomlinson, Catherine. 2020. "COVID-19: Behind SA's Shortages of Test Materials." *Spotlight*, May 5, 2020. Accessed August 20, 2022. https://www.spotlightnsp.co.za/2020/05/05/COVID-19-behind-sas-shortages-of-test-materials/.

Torreele, Els, and Amon J. Joseph. 2021. "Virtual Round Table: Equitable COVID-19 Vaccine Access." *Health and Human Rights Journal* 23, no. 1: 273–88.

Torres, Irene, Daniel Lopez-Cevallos, Osvaldo Artaza, Barbara Profeta, JaHyun Kang, and Cristiani Vieira Machado. 2021. "Vaccine Scarcity in LMICs Is a Failure of Global Solidarity and Multilateral Instruments." *The Lancet* 397, no. 10287: 1804.

UN (General Assembly). 1966. "International Covenant on Economic, Social, and Cultural Rights." *Treaty Series* 999 (December): 171.

———. 2012. "Report of the Special Rapporteur in the Field of Cultural Rights, Farida Shaheed: The Right to Enjoy the Benefits of Scientific Progress and its Applications." A/HRC/20/26 (May 14, 2012).

UNCTAD (UN Conference on Trade and Development). 2011. *Using Intellectual Property Rights to Stimulate Pharmaceutical Production in Developing Countries: A Reference Guide*. New York: United Nations. https://unctad.org/system/files/official-document/diaepcb2009d19_en.pdf.

UNESCO (UN Educational, Scientific, and Cultural Organization). 2005. Records of the General Conference, Paris, October 2005. 33C/Res. 15. http://unesdoc.unesco.org/images/0014/001428/142825e.pdf#page=80.

———. 2021. "UNESCO's Ethics Commissions' Call for Global Vaccines Equity and Solidarity." Joint Statement by the UNESCO International Bioethics Committee (IBC) and the UNESCO World Commission on the Ethics of Scientific Knowledge and Technology (COMEST). February 24, 2021. https://unesdoc.unesco.org/ark:/48223/pf0000375608/PDF/375608eng.pdf.multi.

Usher, Danaiya Ann. 2020. "South Africa and India Push for COVID-19 Patents Ban." *The Lancet* 396: 1790–91.

Van Ark, Eelke, and Strop Jan-Hein. 2020. "Roche Releases Recipe after European Commission Considers Intervention due to Lack of Coronavirus Tests 2020." Accessed August 24, 2022. https://www.ftm.eu/articles/roche-releases-recipe-after-public-pressure-while-european-commission-considers-intervention-due-to-coronavirus-test.

WHO (World Health Organization). 2020. "C-TAP: A Concept Paper." Accessed August 28, 2022. https://www.who.int/publications/m/item/c-tap-a-concept-paper.

WHO, WIPO, and WTO (World Health Organization, World International Property Organization, and World Trade Organization). 2020. *Promoting Access to Medical Technologies and Innovation: Intersections between Public Health, Intellectual Property and Trade*. Second edition. Geneva: WHO. WIPO, WTO.

Williams, Thomas C., and Wendy A Burgers. 2021. "SARS-CoV-2 Evolution and Vaccines: Cause for Concern?" *Lancet Respiratory Medicine* 9, no. 4: 333–35.

WIPO (World Intellectual Property Organization) Standing Committee on the Law of Patents. 2017. "Constraints Faced by Developing Countries and Least Developed Countries (LDCs) in Making Full Use of Patent Flexibilities and Their Impacts on Access to Affordable Especially Essential Medicines for Public Health Purposes in Those Countries." June 2, 2017. SCP/26/5.

WTO (World Trade Organization). 2020. "Waiver from Certain Provisions of the TRIPS Agreement for the Prevention, Containment and Treatment of COVID-19." Communication from India and South Africa. IP/C/W/669 (October 2, 2020).

———. 2021. "Waiver from Certain Provisions of the TRIPS Agreement for the Prevention, Containment and Treatment of COVID-19, Revised Decision Text." Communication from the African Group et al. IP/C/W/669/Rev.1 (May 25, 2021).

———. 2022a. "Ministerial Decision on the TRIPS Agreement." WT/MIN (22)/30 WT/L/1141 (June 22, 2022).

———. 2022b. "Communication from the Chairperson." IP/C/W/688, May 3, 2022.

Zarocostas, John. 2020. "New WTO Leader Faces COVID-19 Challenges." *The Lancet* 397, no. 10276: 782.

———. 2022. "Mixed Response to COVID-19 Intellectual Property Waiver." *The Lancet* 399, no. 10332: 1292–93.

Chapter 26

An Effective History of the Basic-Applied Distinction in "Science" Policy

J. Britt Holbrook

Today, we speak not only of science policy but also of technology policy.[1] We often run the two together and speak of science and technology (S&T) policy. Our understandings of science, of technology, of the relation between the two, of how we ought to develop policies for science and technology, and of how science and technology ought to influence policy are changing. In truth, they have been changing for some time.

Post–Cold War S&T policy has been searching for new directions for over three decades. The late 1990s, especially, were marked by a proliferation of new ideas and new policies aimed at replacing, or at least modifying, the foundational philosophy behind post-World War II science policy, expressed most fully in Vannevar Bush's ([1945] 2020) *Science, the Endless Frontier*. The year 2020 marked its seventy-fifth anniversary, along with the seventieth anniversary of the US National Science Foundation (NSF), which has identified more closely than any other S&T funding agency with the policy for science outlined by Bush. Indeed, NSF published a "75th Anniversary Edition" of *Science, the Endless Frontier* that includes a foreword by France A. Córdova, NSF's fourteenth director.

This chapter focuses on NSF and its philosophical underpinnings. It does so in part because the seventy-fifth anniversary of *Science, the Endless Frontier* sparked a great deal of reflection in the science policy community.[2] Focusing on the philosophical underpinnings of NSF also allows us to see recent moves in S&T policy in stark relief. As is usually the case when one writes about S&T policy, things are in flux. As of May 2022, the US House of Representatives and the US Senate were resolving differences on the United States Innovation and Competition Act of 2021 (H.R. 4521), a bill that could change the face of NSF if it is enacted.[3] In something of a preemptive move, in 2022, NSF established its first new directorate in thirty years—the Directorate for Technology, Innovation and Partnerships (TIP), which closely resembles the description of a new directorate of the same name in President Biden's proposed FY2022 NSF budget.[4] According to NSF's website, the point of the TIP Directorate is "maximizing NSF's impact."

> NSF's TIP Directorate doubles down on the agency's commitment to support use-inspired research and the translation of research results to the market and society. In doing so, the new directorate strengthens the intense interplay between foundational and use-inspired work, enhancing the full cycle of discovery and innovation.[5]

Anyone familiar with *Science, the Endless Frontier* must be tempted to see the TIP Directorate as a fundamental shift in how NSF does business. Bush argued for an agency that would fund what he called "basic research," research performed without any consideration of its practical use, aiming instead at fundamental understanding. This chapter explores, from a philosophical perspective, the roots of the idea of basic research and what it means for NSF to pursue an S&T policy that runs science and technology together in pursuit of the idea of use-inspired basic research.

A BRIEF EFFECTIVE HISTORY OF "SCIENCE" POLICY

Donald E. Stokes (1997) devotes the second chapter of his book to tracing the history of the idea that basic research is conceptually distinct from applied research and the corollary that basic research ought to be kept separate from applied research in practice. He identifies an ideal of "pure inquiry" set forth by the ancient Greeks as the source of this dual prejudice. To paraphrase his argument, the Greeks invented science by moving beyond the mere applications of knowledge—Stokes mentions the ancient Egyptians and Babylonians as peoples who had enough knowledge to apply in the prediction of floods and astronomical events, respectively, but lacked any ultimate understanding of these events as natural phenomena—and focusing on knowledge for its own sake. Although Stokes describes the Greeks, especially Plato and Aristotle, as valuing pure inquiry over the applications of knowledge, he bypasses some important details about the arguments they offered in support of their judgment. Because these details have important policy ramifications, I discuss them briefly here.

Stokes spends some time discussing whether the Greeks' "pure inquiry" was actually science. Unfortunately, this discussion obscures the connection between the policy import of the Greeks' views on knowledge and the recommendations Bush laid out in *Science, the Endless Frontier*. Although the activity that Plato and Aristotle engaged in—which I would call philosophy, rather than science—was clearly not the same as what we today call science, it is equally clear that both were pursuing knowledge. Insofar as today's scientists are engaged in the activity they believe will lead to more—and more secure—knowledge, they are involved in the same activity that engaged Plato, Aristotle, and the ancient Greeks. To put the point more philosophically, today's scientists are engaged in an activity that bears a strong family resemblance to that of the ancient philosophers, especially when we view these activities in terms of the ends they pursue.

Take the question of whether Aristotle was engaging in the same sort of physics as Isaac Newton, then, as the same sort of question as whether Newton was engaging in the same sort of physics as Albert Einstein. Although that might be interesting in a Kuhnian way, it does not help in understanding the policy ramifications of Aristotelean views of knowledge, to which I now turn.

Aristotle divides the pursuit of knowledge into three types: (1) theoretical, (2) practical, and (3) productive (see, for example, *Metaphysics* Books VI and XI). Aristotelean physics (and, for that matter, Newtonian and Einsteinian physics) would be included under the rubric of theoretical knowledge, as would any of today's exact sciences. Aristotle also includes metaphysics as a theoretical science, indeed, as the theoretical science that deals with the most certain knowledge. For Aristotle, ethics and politics constituted the practical sciences. Many of today's social sciences would be included here. Book I of *Nicomachean Ethics* clearly states that the practical sciences are not exact sciences, which means that we are dealing with likelihoods rather than certainties when we reason about ethics and politics. Productive science, which for Aristotle includes the arts and which would include much of today's engineering, aims at producing a product, some thing that is beautiful or useful to attain some other end. This brings us to the key point. For the Greeks, and this is most clearly articulated by Aristotle in the opening paragraphs of *Nicomachean Ethics*, there exists a hierarchy of the ends we pursue. Today we speak of the contrast between intrinsic and instrumental value. Ends pursued for their own sake possess intrinsic value, while ends pursued for the sake of something else possess (merely) instrumental value. Anything pursued for its own sake is more valuable than anything pursued for the sake of something else. It follows that knowledge pursued for its own sake is more valuable than knowledge pursued for the sake of something else.

Clearly, then, the Greeks valued theoretical science over productive science. For productive science, by definition, is pursued for the sake of producing something else. Now comes the astonishing finesse with policy ramifications. Practical science, whether ethics on the level of the individual or politics on the level of the state, also aims for something else—*eudaimonia*, or human flourishing. Since the highest good for human beings, according to Aristotle, is what he calls "the life according to intellect" or "a contemplative activity," practical science should aim, as far as possible, to bring about the practical conditions for the possibility that human beings will be able to engage in such contemplative activity (see *Nicomachean Ethics*, Book X, especially 1178a5–8 and 1178b8). In other words, the practical science of politics should be geared toward making the life devoted to the pursuit of theoretical science both possible and actual.[6] To put the point in Bushian terms, it is the proper role of government to support basic science.

Stokes (1997) interprets the Greeks as evidencing what he calls a "bias against practical use" (32).[7] Yet nothing could be more practical than organizing all of society to support human flourishing. Of course, this is a difficult task, even when dealing with city-states. Once states have become countries (or empires), things become even more complicated. But there is an appealing elegance to the argument that the state should support human flourishing. How we understand human flourishing has changed; but the idea that the state should support human flourishing has remarkable resilience, as we shall see.

By the nineteenth century, the state had delegated the task of ensuring human flourishing in terms of knowledge to universities. Just as Bush articulated the foundational philosophy of post–World War II S&T policy, Wilhelm von Humboldt's (1809–10) "On the Spirit and the Organisational Framework of Intellectual Institutions in Berlin" lays the philosophical foundation for the modern research university.[8] Remarkably, Humboldt suggests that the details of the organizational framework are relatively unimportant, as long as everyone—especially the state—respects the spirit of the university. I interpret what Humboldt refers to as the spirit of the university as a set of philosophical foundations.

First and foremost, universities must be designed for the pursuit of *Wissenschaft*. It is worth noting here that the German word *Wissenschaft* is usually translated into English as "science and scholarship" to include the liberal arts and humanities, as well as what we in the Anglophone world normally consider sciences. It is certainly clear that Humboldt, himself a linguist, would include a broad range of today's disciplines under the rubric of *Wissenschaft*. For Humboldt (1809–10), the pursuit of *Wissenschaft* brings about the fulfillment of what he calls "intellectual and moral culture," the cultivation of the individual who combines "objective scientific and scholarly knowledge with the development of the person" (243). In other words, the universities provide the conditions for the cultivation of citizens with *Bildung*.

Second, universities and the individuals who make them up must regard *Wissenschaft* "as dealing with ultimately inexhaustible tasks: this means that they are engaged in an unceasing process of inquiry" (243). The individuals at universities engage in this endless process of inquiry collaboratively, which means that the intellectual achievements of one excite the intellectual passions of others, such that "what was at first expressed only by one individual becomes a common intellectual possession instead of fading away in isolation" (243).

Third, this process of collaboration entails the idea that professors and students collaborate in endless inquiry in characteristic ways. Professors know more, in most cases, than the students. Yet the professors need the students both to communicate their own knowledge and to continue to learn (cultivate their own *Bildung*). For Humboldt, teaching and research are integrated in the person of the professor. The same individual teaches their discoveries to students, and in the process of doing so makes new discoveries. The students keep the professors fresh. Meanwhile, the professors turn the students onto the path of their own *Bildung*.

Finally, the state should provide the resources needed in order to establish and maintain the university, but then leave to the university and its denizens the endless pursuit of *Wissenschaft*.

> The state must always remain conscious of the fact that it never has and in principle never can, by its own action, bring about the fruitfulness of intellectual activity. It must indeed be aware that it can only have a prejudicial influence if it intervenes. The state must understand that intellectual work will go on infinitely better if it does not intrude. The state's legitimate sphere of action must be adapted to the following circumstances: in view of the fact that in the real world an organizational framework and resources are needed for any widely practised activity, the state must supply the organisational framework and the resources necessary for the practice of science and scholarship. (244)

In other words, universities must remain autonomous from the state, lest the pure pursuit of knowledge be ruined. In return for its support for the autonomous, endless pursuit of knowledge for its own sake, the state will be populated by citizens who possess *Bildung*.

Humboldt gives the state a modicum of control over the universities, namely in the form of hiring the faculty. Once that is accomplished, however, the state must leave the faculty to their own devices. For Humboldt, the successful pursuit of higher education "depends on strict adherence to the principle that science and scholarship [*Wissenschaft*] do not consist of closed bodies of permanently settled truths; effective intellectual accomplishment is to be sought in ceaseless effort" ([1809–10] 1970, 244). Humboldt allows the state full control over the lower education institutions, the schools, which provide students with the proper preparation for university by teaching them the requisite knowledge and skills.[9]

The similarities between the ideas Bush expressed in *Science, the Endless Frontier* and Humboldt's ideas is astounding. Whereas Wolfgang Rohe (2017) picks up on the frontier metaphor in Bush's title to explore the notion of scientist as pioneering hero, what is most striking to me in light of reading Humboldt is Bush's evocation of the idea that science is an endless pursuit. Of course, Bush's conception of science was limited to what could be called the hard sciences. But Bush was above all concerned to guarantee the autonomy of science from state control. For Humboldt, the fact that the pursuit of *Wissenschaft* is an endless task is precisely what guarantees, if the state will simply recognize this fact, the autonomy of the university. Humboldt gives the state control over the lower schools, because comparatively little harm can come from the state teaching all students reading, writing, and arithmetic. If the state were to treat universities as if they were merely instruments for the communication of permanently settled truths, however, it would immediately ruin the pursuit of *Wissenschaft*.

Bush is similarly intent on preserving freedom of inquiry. Although Daniel Sarewitz (2016) criticizes Bush for invoking "the free play of free intellects," the idea has an impressive lineage. Aristotle was a student at Plato's Academy, yet he developed his own approach to knowledge, one that is critical of Plato in many respects. There is little indication that Plato was irked by this, and quite a bit of evidence that Plato went to great lengths to avoid anyone simply accepting his views as settled truths. Aristotle consistently begins his own thinking with reference to the opinions of others. The interaction Humboldt describes between professors and students is in the same spirit—a spirit of free and necessarily ceaseless inquiry, rather than indoctrination into settled, dogmatic truths.

Whereas Aristotle argues that the state should provide the conditions necessary for this life of contemplation, Humboldt argues that the state should provide the financial support and organizational structure that would best allow universities to pursue *Wissenschaft* free from the state's other concerns. Coming out of the Second World War, Bush ([1945] 2020), too, is concerned to limit the power of the state over scientific inquiry.

> Many of the lessons learned in the war-time application of science under Government can be profitably applied in peace. The Government is peculiarly fitted to perform certain functions, such as the coordination and support of broad programs on problems of great national importance. But we must proceed with caution in carrying over the methods which work in wartime to the very different conditions of peace. We must remove the rigid controls which we have had

to impose, and recover freedom of inquiry and that healthy competitive scientific spirit so necessary for expansion of the frontiers of scientific knowledge.

Scientific progress on a broad front results from the free play of free intellects, working on subjects of their own choice, in the manner dictated by their curiosity for exploration of the unknown. Freedom of inquiry must be preserved under any plan for Government support of science. (9–10)

Like Humboldt, Bush identifies universities as the proper place for such inquiry. The publicly and privately supported colleges, universities, and research institutes are the centers of basic research. They are the wellsprings of knowledge and understanding. As long as they are vigorous and healthy and their scientists are free to pursue the truth wherever it may lead, there will be a flow of new scientific knowledge to those who can apply it to practical problems in government, in industry, or elsewhere ([1945] 2020, 9).

Where the Greeks, Humboldt, and Bush evidence remarkable agreement about the notion of inquiry and the need for the state to support it, they offer different views on what the state should expect in return. For the Greeks, since the point of the state is to ensure human flourishing, and since human flourishing consists in the life of contemplation, it is obvious that a well-functioning state should support the life of contemplation. There is no exchange at stake here, nothing the state should expect in return for supporting the life of contemplation.

For Humboldt, the state can expect citizens with *Bildung* in return for its support of *Wissenschaft*. Indeed, rather than expecting anything from the university to satisfy its immediate needs, the state "should instead adhere to a deep conviction that if the universities attain their highest ends, they will also realise the state's ends too, and these on a far higher plane" (1809–10, 246). This bargain presumes that the state, once it has taken care of its citizens in the lower schools, turn them over to the university, recognizing that "the transition from school to university constitute[s] a stage in the life of a young person which—when it is successful—brings him to a point where physically, morally and intellectually he can be entrusted with freedom and with the right to act autonomously" (246). In the sense that there is some sort of exchange proposed here, Humboldt takes a step beyond the Greeks. However, he still supposes that the ends of the state and the ends of the university are the same, and these amount to the same ends proposed by the Greeks.[10]

As Stokes observes, in formulating *Science, the Endless Frontier*, "In the broadest terms, the task Bush and his advisers set for themselves was to find a way to continue federal support of basic science while drastically curtailing the government's control of the performance of research" (1997, 52). This task is essentially the same as Humboldt's. However, Bush ([1945] 2020) differs most clearly from either the Greeks or Humboldt in terms of what he offers the state in return for its support of the free play of free intellects.

> Progress in the war against disease depends upon a flow of new scientific knowledge. New products, new industries, and more jobs require continuous additions to knowledge of the laws of nature, and the application of that knowledge to practical purposes. Similarly, our defense against aggression demands new knowledge so that we can develop new and improved weapons. This essential, new knowledge can be obtained only through basic scientific research.
>
> Science can be effective in the national welfare only as a member of a team, whether the conditions be peace or war. But without scientific progress no amount of achievement in other directions can insure our health, prosperity, and security as a nation in the modern world. (1)

In return for government support for basic scientific research, the state can expect an endless flow of new knowledge. The new knowledge is necessary to achieve the listed set of national needs; and the new knowledge can only be guaranteed by the free play of free intellects.

Stokes (1997) remarks on the "paradox" that, although Bush's blueprint for the National Research Foundation failed, the rhetoric of and ideas underlying *Science, the Endless Frontier* triumphed, becoming the dominant discourse of S&T policy. Stokes explains the paradoxical success of Bush's ideas as a combination of intellectual and institutional factors. The idea of the value of the free play of free intellects has been part of the intellectual tradition of the West since the ancient Greeks. And when NSF was finally created in 1950, it saw the idea that it was supporting basic research at universities to ensure the flow of knowledge that would drive technological progress as attractive, in part because of the sort of autonomy that vision entails. Sarewitz (2016) also suggests that Bush's vision is so successful, despite its initial lack of policy bite, because it jives with a collective prejudice about the value of unfettered scientific research. Benoît Godin (2003; 2006) argues that it was because of numbers, specifically statistics used to measure basic research, rather than rhetoric that the Bush ideas gained currency and have persisted for so long. Of course, measuring something has its own rhetorical force (cf. Stone 1988). But recognizing the rhetorical force of numbers is no argument against Godin's view, which strikes me as more than plausible.

Sarewitz also argues that, in Bush's framing, science is unassailable, suggesting, "Science can get the credit, but not the blame" (2016, 17). Sarewitz claims that by characterizing basic science as necessary but not sufficient to meet national needs, Bush succeeds in immunizing basic science from critique. Yet, the obvious reply to Bush's characterization of basic science as necessary for us to meet national needs is to argue that this claim is false. This is the force of Project Hindsight, which demonstrated that the vast majority of advances in defense technology stemmed from incremental changes to existing technologies in response to reports from the field, rather than from advances in basic research (Stokes 1997). In other words, in a vast number of cases, basic research is not at all necessary for technological advancement. Indeed, as Stokes (1997) discusses, quoting Kuhn, most countries enjoy success in either science or technology, but not both simultaneously. Thus, it is possible to make technological advances that meet societal needs in countries that do not invest huge sums of money into basic research. Today's most salient example is China, which brings us back to NSF's new TIP Directorate.

BEYOND BASIC AND APPLIED?

Although China is working hard to close the gap on the United States in terms of spending on research and development, as of 2021, the United States maintains the number one spot, as determined by the Organization for Economic Cooperation and Development (OECD).[11] Despite the US lead, China increasingly concerns US policymakers.[12] Were it simply a matter of spending more on basic research to up the pace of technological advance, it seems the United States would possess more than enough political will to make it happen. For the most part, the policymakers' emphasis on the commercial and military threat posed by China to the US research enterprise focuses not on outpacing China in funding and producing basic research, but instead on technology.

Outpacing China's technological development is perhaps the main impetus behind the various plans to create a new "technology directorate" within NSF. Enter the aforementioned TIP Directorate. If the launch of Sputnik spurred the United States to unprecedented investments in both basic and applied research and technology, today's China is doing something similar—but with policymakers, and perhaps many members of the scientific community, now less settled on the basic-applied distinction outlined by Bush.

In her foreword to the seventy-fifth anniversary edition of *Science, the Endless Frontier*, then NSF director France Córdova looks back at Bush's vision and remarks on what remains relevant today and what has changed at NSF. Among the changes, she notes that engineering has become much more prominent.

> While Bush's original document would have focused the agency on basic research alone and left more applied research to industry, NSF has gone on to fund many programs with technology transfer in mind, including the Engineering Research Centers and the more recent Innovation Corps. These

efforts build faster tracks to tangible outcomes, have produced many startups, and have earned approval from the Administration and Congress. (2020, vi)

Córdova also mentions the social sciences, which were added to the NSF's domain in 1957 at the urging of then–Vice President Richard Nixon and became a stand-alone directorate in 1992, and the role of NSF's Merit Review process in ensuring that NSF funds excellent research in the national interest (see Holbrook 2018). She concludes her foreword with a section titled "The Heart of the Matter."

> The dividing line today between basic and applied research is not as clear as it was when Bush wrote his report. Many university investigators are engaged in the full spectrum of research, from fundamental discovery to application, and are blind to a division between these phases of research development. Today, use-inspired basic research and curiosity-driven basic research motivate many young scientists and engineers. There is an increased emphasis in all sectors on technology transfer and return on investment, which . . . has been adopted by academia as well as the private sector. New "accelerators" funded by both private entities and the government strive to bring research and new ideas to fruition faster. (Córdova 2020, ix–x)

Córdova here anticipates much of what the new TIP Directorate seems designed to accomplish. The actual NSF, not Bush's vision for it, does not rely on the distinction between basic and applied research to guarantee the autonomy of inquiry. Instead, to guarantee a degree of autonomy balanced with attention to social responsibility, NSF relies on its Merit Review Process, which incorporates not only questions of "intellectual merit" (is the proposed activity good science?) but also of "broader impacts" (how will the proposed activity benefit society?) (see Holbrook 2012). NSF itself has never been as wedded to the primacy of basic research as Bush would have wanted. But it is now continuing to move in directions that connect the research it funds to societal needs while maintaining a degree of autonomy from the state.

Comparing Córdova's foreword to Rush D. Holt's introductory essay (2021) of the Princeton reissue of *Science, the Endless Frontier* also yields some interesting results. Where Córdova understandably focuses on NSF and how it resembles and differs from Bush's vision, Holt focuses instead on Bush's vision of science. Holt is sharply critical of both Bush and other scientists who seek to separate science from society. Because Holt believes that science is essentially a way of asking questions that yields the most reliable knowledge we have of the way things are, he argues that the public can and should think more like scientists. For Holt, a well-funded scientific elite cannot replace a well-informed, engaged public.

Carl Mitcham (2021) criticizes Holt's idea that the public needs to think more like scientists, writing, "Although the scientific method can serve as a glue for the scientific community, it is doubtful that it can do the same for a larger public" (3). Indeed, Mitcham sees science as a destabilizing force in society, rather than as a capable social bond.

> It is difficult to wonder whether Holt himself has fully recognized the challenge of historical and social science evidence against what may be his own wishful thinking. The complexities of science and of democracy suggest the inadequacy of any American science policy default to either the Vannevar Bush's original or to Rush Holt's emendation. (4)

Perhaps an emendation to Holt's emendation would be more palatable. Earlier in this chapter, I suggested that today's scientists are involved in the same activity that engaged Plato, Aristotle, and the ancient Greeks just insofar as they are engaged in the activity they believe will lead to more—and more secure—knowledge. When Holt describes science, he suggests that science is essentially a way of asking questions that yields the most reliable knowledge we can have. The question is whether we can stretch our imaginations enough to see that science is separable from "the scientific method," whatever that is. Humboldt is instructive here. *Wissenschaft* is not science in the narrow sense we employ today, to which both Holt and Mitcham fall prey. It is not the scientific method the public needs to adopt, but *Bildung* it needs to acquire. Only if all of us, including us philosophers and scientists, cultivate our intellectual and moral characters will we acquire the ability to live autonomously.

THINKING THROUGH "SCIENCE" POLICY

This chapter presents an effective history of the basic-applied distinction, in contrast to a traditional history (see Foucault 1977). Whereas traditional history is often seen as an end in itself (a bona fide discipline within the academy), effective history is a means to some other end. Where a traditional history is often seen as the attempt to secure the objective truth about the development of a set of facts or an idea, an effective history attempts to provoke us to think through an idea, often with the intent of disrupting the sorts of continuities—or facts—traditional history establishes. In the case of this chapter, the point is not to disrupt a traditional historical approach to science policy, but rather to propose a sort of continuity of ideas that a traditional historical approach would dispel as a myth.

If effective history is the means, what is the end? In this chapter, there are several. The overarching end is to encourage us to rethink S&T policy. On the way to achieving that end, I argue we need to rethink "science." As long as we continue to ask (and answer) the question "What is science?" we continue the thinking of the Greeks, which seems so obviously operative in today's, as well as yesterday's, "science" policy. I was tempted to replace "science" with "knowledge," but justifying that move would have taken more space than I have here. Instead, I invoked philosophy (including metaphysics, ethics, and politics), *Wissenschaft*, the basic-applied distinction, and technology all as "science" or in relation to "science." To what end? Unless we rethink what we mean by "science" to be more inclusive, including especially assumed non-sciences such as liberal arts and humanities, as well as philosophical reflection on engineering and technical disciplines, we run the risk that whatever intellectual activity we argue the state should support will be reduced to a mere means to satisfy the desires of the state. There is no surer way to guarantee an absolutist state than to give in to the idea that all "science" should support the needs of the state as defined by the state.

That last claim, if true, has existential implications for the future of "science" policy. We should avoid, for instance, the pursuit of technological supremacy as the end of "science" policy. We should be more careful about endorsing a policy for "science" that reduces easily to a transactional relationship. Humboldt provides a middle ground that we should explore between the Greeks' assumption that society should serve "science" in order to create the conditions for contemplation and Bush's proposal that "science" should serve society by providing tidbits of knowledge leading to technological widgets to meet national needs. Yes, Bush also tried to isolate "science" from societal control, but there has always been something questionable—both for scientists and for society—about the bargain he proposed (see Sarewitz 2020). We need to rethink, in short, the value of knowledge and knowledge production.

Of course, these concluding musings hold water only if we believe, and are committed to the belief, that limits on state power should be maintained. If that requires us to rethink "science," knowledge, technology, and society, so much the better.

NOTES

1. In the spirit of brevity, I will hereafter often refer to "policy" rather than "science and technology policy." All references to "policy" mean "science and technology policy" or "knowledge policy" unless otherwise noted.

2. Sarewitz (2020) announced a yearlong series of articles in *Issues in Science and Technology* devoted to *Science, the Endless Frontier*, and Princeton University Press published another reissue of the document in 2021. The Princeton edition features an introductory essay by Rush Holt, former chief executive officer of the American

Association for the Advancement of Science (AAAS). For a commentary on Holt's essay, see Mitcham (2021). This chapter is inspired by Mitcham's essay, as well as by his career-long effort to bring philosophers into closer engagement with S&T policy.

3. Despite the name of the bill, H.R. 4521 has not been enacted into law as of May 2022. The latest text of the bill is available here: https://www.congress.gov/bill/117th-congress/house-bill/4521/text. Accessed May 7, 2022.

4. Available for download here: https://www.nsf.gov/about/budget/fy2022/toc.jsp. Last accessed May 9, 2022.

5. https://beta.nsf.gov/tip/latest. Accessed May 6, 2022.

6. This is also why Plato argued that philosophers should be kings, namely to bring about the just polis, one in which a philosopher may live a life devoted to philosophy without fear.

7. It would be interesting to pursue the idea of a prejudice against practical use even further, but space will not allow me to do so here. But reading H. A. Rowland's (1883) "A Plea for Pure Science," especially in light of David A. Hounshell's (1980) remarkable article on "Edison and the Pure Science Ideal in 19th-Century America" reveals not simply a prejudice, but full-blown resentment of the honors bestowed on what Rowland called "mere tinkerers."

8. By no means do I wish to suggest that Humboldt was the actual architect of university reforms in Germany. As Mitchell G. Ash (2006) argues, Humboldt's actual role in reforming German higher education was quite limited. But a similar point can be made about Bush with regard to the founding of NSF (see Blanpied 1988; Godin 2003). Just as Ash speaks of a "Humboldt myth" that grew up to provide the rhetorical underpinnings for higher education policy, so to we can identify a "Bush myth" that underlies postwar S&T policy. I am not, therefore, making a traditional historical argument, but rather an effective historical argument (Foucault 1977). The ideas I am tracing from the Greek, through Humboldt, to Bush are there, whether Bush ever read Humboldt or Humboldt ever read Aristotle, or whether anyone ever actually reformed higher education or S&T policy on the basis of *their* writings. I am willing to bite the bullet of contributing to the Bush myth. Cf. Rohe (2017), who also notes the close kinship between Humboldt's and Bush's ideas.

9. Humboldt explicitly mentions mathematics as important, but it is hardly a stretch to include also reading and writing along with arithmetic.

10. Here I am treating *Bildung* as the German analog for the Greek *eudaimonia*.

11. As determined by querying the OECD search tool highlighting both US and Chinese investments in R&D as a percentage of gross domestic product from 2000 to 2020. https://issues.org/necessary-but-not-sufficient/. Accessed May 9, 2022.

12. Searching the latest available text of H.R.4521 reveals 663 occurrences of the word *China*, and an additional 190 occurrence of the word *Chinese*. By comparison, note the number of occurrences of the following terms: *Russia* = 9, *Russian* = 18; *Korea* = 30; *Iran* = 42. Policymakers routinely invoke China in their discussions of H.R. 4521 and related bills.

REFERENCES

Ash, Mitchell G. 2006. "Bachelor of What, Master of Whom? The Humboldt Myth and Historical Transformations of Higher Education in German-Speaking Europe and the US." *European Journal of Education* 41, no. 2: 245–67.

Aristotle. 1984. *Metaphysics*. In *The Complete Works of Aristotle: The Revised Oxford Translation*. 2 vols. Translated by Jonathan Barnes. Princeton, NJ: Princeton University Press.

———. 1984. *Nicomachean Ethics*. In *The Complete Works of Aristotle: The Revised Oxford Translation*. 2 vols. Translated by Jonathan Barnes. Princeton, NJ: Princeton University Press.

Blanpied, William A. 1998. "Inventing US Science Policy." *Physics Today* 51, no. 2: 34–40.

Bush, Vannevar. (1945) 2020. *Science, The Endless Frontier*. 75th Anniversary Edition. Washington, DC: National Science Foundation.

———. (1945) 2021. *Science, The Endless Frontier*. Princeton: Princeton University Press.

Córdova, France A. 2020. Foreword to Bush (1945) 2020, iii–x.

Foucault, Michel. 1977. "Nietzsche, Genealogy, History." In *Language, Counter-Memory, Practice*. Ithaca, NY: Cornell University Press: 139–64.

Godin, Benoît. 2003. "Measuring Science: Is There 'Basic Research' without Statistics?" *Social Science Information* 42, no. 1: 57–90.

———. 2006. "The Linear Model of Innovation: The Historical Construction of an Analytical Framework." *Science, Technology, & Human Values* 31, no. 6: 639–67.

Holbrook, J. Britt. 2012. "Re-assessing the Science–Society Relation: The Case of the US National Science Foundation's Broader Impacts Merit Review Criterion (1997–2011)." In *Peer Review, Research Integrity, and the Governance of Science–Practice, Theory, and Current Discussions*, 328–62. Dalian, China: People's Publishing House and Dalian University of Technology.

———. 2018. "Philosopher's Corner: What Is Science in the National Interest?" *Issues in Science and Technology* 34, no. 4: 27–29. https://www.jstor.org/stable/26597986.

Holt, Rush D. "The Science Bargain." In Bush (1945) 2021, 1–42.

Hounshell, David A. 1980. "Edison and the Pure Science Ideal in 19th-Century America." *Science* 207, no. 4431: 612–17.

Humboldt, Wilhelm von. (1809–10) 1970. "On the Spirit and Organisational Framework of Intellectual Institutions in Berlin." *Minerva* 8, no. 2: 242–50.

Mitcham, Carl. 2021. "Science Policy and Democracy." *Technology in Society* 67 (November): 1–4. https://doi.org/10.1016/j.techsoc.2021.101783.

Rohe, Wolfgang. 2017. "The Contract between Society and Science: Changes and Challenges." *Social Research: An International Quarterly* 84, no. 3: 739–57.

Rowland, Henry Augustus. 1883. "A Plea for Pure Science." *Science* 2, no. 29: 242–50.

Sarewitz, Daniel. 2016. "Saving Science." *The New Atlantis* 49 (Spring/Summer): 4–40.

———. 2020. "Necessary but not Sufficient?" *Issues in Science and Technology* 36, no. 2 (Winter): 17–18.
Stokes, Donald E. 1997. *Pasteur's Quadrant: Basic Science and Technological Innovation*. Washington, DC: Brookings Institution Press.
Stone, Deborah A. 1988. *Policy Paradox and Political Reason*. Glenview, IL: Scott Foresman.

Chapter 27

Technological Risks, Institutional Wariness, and the Dynamics of Trust

José A. López Cerezo

Two related phenomena have conditioned the political evolution of our societies during the last decades. One is a growing social concern about the risks and negative effects derived from technological development. Global warming, nuclear threats, new carcinogens, and many other consequences of an intense technology-based industrial development are today objects of social and institutional concern. The second is a growing political activism on the part of a diversity of publics and advocacy groups that demand more participation and accountability. Public policies are now subject to greater scrutiny and stronger demands for citizen involvement in all areas of social interest, including science and technology.[1] It is a state of affairs that is visible especially in the most industrialized countries but has a global reach given the transnational characteristics of the so-called risk society and the growing internationalization of trade and communication networks.

In this framework of concern and involvement, several technology-related crises of the last decades have produced a gradual deterioration of institutional trust. While expressed globally, it is manifested with great intensity in Western Europe. Public opinion is increasingly suspicious of messages from industry and regulatory authorities regarding the safety of technological products and systems. Risk society seems to become a "post-trust society": we live surrounded by artifacts, yet they often arouse suspicion. In this chapter, by highlighting the strong parallels of the notion of trust with regard to the notion of risk, I examine the nature and social dynamics of trust in modern Western society. The erosion of trust results not only in the necessity of enhanced surveillance mechanisms but also, through the phenomenon of trust-trust tradeoffs, in evolving reconfigurations of the mapping of trust location in social agents, individual and collective. On this basis, I defend the value of critical awareness and well-grounded distrustful attitudes as essential for mature citizenship that will also facilitate governance in today's world.

THE EROSION OF TRUST

The media visibility of several food, health, and industrial crises during the last decades has produced a gradual deterioration of trust in the world that has arisen after the conclusion of the Cold War. It is not difficult to make a brief list of well-known scandals produced by lies, negligence, information gaps, or simple greed in the recent history of technological development: the Spanish toxic oil syndrome in the early 1980s, the mad cow crisis in Europe since the late 1980s, blood contaminated with AIDS in France in the mid-1990s, the bird flu of the late 1990s, the anti-inflammatory Vioxx in the USA in early 2000 and many other drugs that have followed in the wake of Thalidomide,[2] Iraq's nonexistent weapons of mass destruction as justification of the 2003 war, contaminated children's milk in China in 2008, Volkswagen's diesel deception in 2015, Cambridge Analytica data traffic in 2018, early anti-COVID campaigns by various governments during 2020,[3] etc.

These scandals occur in populations increasingly better educated and more aware of the dependence of science on the interests of industry,[4] growing demands concerning institutional accountability, an adversarial political climate, the tendency to amplify risks and create polarization by the communication media, and an increasing aversion to risk produced by higher standards of life. The logical result is growing citizen suspicion regarding the social agents that make the decisions that affect our lives (see Bauer, Shukla, and Allum 2012; Weinstock 2013; Kasperson and Stallen 1991; and Lofstedt 2009, 2013). It is what the French political scientist Pierre Rosanvallon (2008) has called the "age of distrust," what the Swedish-American sociologist Ragnar Lofstedt (2009) calls "post-trust society." But what is at stake in all this? What social function does trust play in today's technological society?

THE NOTION OF TRUST

The notion of trust is a complex one. As an intellectual virtue and moral virtue, it includes cognitive and affective capacities that cannot be described by an algorithm or a single definition (see Zagzebski 1996, 46; Simpson 2012, 553–54). At first glance, an elementary basis of trust is necessary for the very life of individuals: we must trust our teachers for the acquisition of knowledge, we must trust engineers and doctors to take planes and pills, and so on. It is precisely the ubiquity of trust that makes it invisible (Stern 2017, 274).

According to Ortwin Renn (2008), in the field of public life, the term *trust* refers to judgment of the degree to which an organization meets the expectations of social actors regarding the fulfillment of its institutional mission, if it does so with professionalism, honesty, competence, and so on. We can understand credibility as a variety of trust referring only to communication, defining it as the degree to which the communicative efforts of an organization satisfy social expectations in terms of honesty, transparency, and professionalism (Walaski 2011).

Concerning interpersonal trust in communication, as Aristotle pointed out in the *Rhetoric* (1378a) when talking about persuasion: "For the orator to produce conviction three qualities are necessary . . . good sense [*phrónesis*], virtue [*areté*], and goodwill [*eunoia*]."[5] The trust or credibility given to an interlocutor thus depends not only on the demonstration of good sense or practical wisdom (*phrónesis*),[6] but also on the manifestation of excellence of character, as shown by moral virtues such as honesty or equity, as well as goodwill or benevolence. The speaker would need to combine them in what he does to be understood as someone worthy of credit.

As a future-oriented attitude, there are certain aspects that need to be highlighted in the notion of trust, concerning expectation, vulnerability, and risk (see Hupcey et al. 2001, 290; McCraw 2015, 416–18). Trust can be understood in terms of what we expect with respect to other persons or institutions, and, in this sense, it involves the expectation of the realization of some action or state of affairs (the target object of trust). The expectations involved in trust may also have a normative dimension, in addition to a predictive dimension, providing thus a basis for moral critique when trust is betrayed (Dormandy 2020, 4–7). Besides, the agent who trusts, as dependent on another agent or institution for the realization of what is expected, is in a vulnerable position with respect to failure or noncompliance. The concept of trust is nevertheless more restricted than that of dependence: trust implies dependence but not the other way around, because we may depend on other people (or social systems) even if we distrust them (see Zagzebski 1996, 160; Hawley 2017, 231; Davidson and Satta 2021, 131ff). Trust also implies reliance, which, as trust, implies choice—although the latter concept is richer and goes beyond that of reliance regarding the above mentioned moral dimension: misplaced reliance on people or inanimate objects drives one to disappointment, and misplaced trust drives one to anger and often to feeling betrayed (Baier 1986: 234–35).[7] Trust also implies risk since it requires considering courses of action and the choice between alternatives (trusting, distrusting, or simply not trusting),[8] placing oneself before a possible frustrated expectation.[9]

In his 1979 pioneering contribution *Trust and Power*, Niklas Luhmann understands trust as a mechanism for reducing complexity, underlining the relationship between the notion of trust and those of risk and expectation. To him, trust constitutes a risky bet of anticipation of the future. In his words:

> The problem of trust therefore consists in the fact that the future contains far more possibilities than could ever be realized in the present and thus be transferred into the past. The uncertainty about what will happen is simply a consequence of the very elementary fact that not all futures can become the present and hence become the past. The future places an excessive burden on a person's ability to represent things to himself. People have to live in the present along with this everlasting over-complex future. They must, therefore, prune the future so as to measure up with the present, that is, to reduce complexity. (2017, 49)

In this way, the attribution of trust "sets" the future behavior of other actors as a presupposition in order to decide our own behavior in the present, reducing uncertainty and making social action possible. But in what sense do we trust when we claim to trust someone or some institution?

According to the specialized literature, despite specific debates in empirical studies,[10] with respect to the trustee, trust has three main dimensions,[11] and therefore three ways of understanding it or weighing its factors:

1. Epistemic: perceived competence (degree of expertise, including efficiency)
2. Moral: perceived independence and integrity (lack of bias, objectivity, consistency, sincerity)
3. Affective: perceived empathy (similarity of values or worldviews, benevolence)

As pointed out by Mayer, Davis, and Schoorman (1995, 717), these dimensions roughly coincide with the factors that make someone worthy of credit according to Aristotle's *Rhetoric*. The differentiation ultimately depends on the kind of expectations implied by the exercise of trust. In the sociological terminology, at stake in the first case are instrumental expectations related to technical or epistemic aspects, in the second one, axiological expectations related to moral values, and, in the third one, fiduciary expectations related to setting others' interests above one's own (Sztompka 2000, 53ff).

Considering trust as a four-way relationship (A trusts B to do C within context X) (e.g., Bauer and Freitag 2018, 16), the attribution of credibility or the exercise of trust as an expectation may frequently combine different modalities (competence, independence, empathy) with different intensities along a bandwidth (Rousseau et al. 1998, 394–95).

POST-TRUST SOCIETY

There are certain features of contemporary society that give particular relevance to trust. According to Ulrich Beck (1992) and his vision of the contemporary world as a "risk society," we have gone from a society based on destiny to one modeled by intentional human action, with growing power to modify social life and the environment. The technological and industrial development has produced new forms of danger and has pushed risk to the center of social and personal life, generating uncertainty and vulnerability. Moreover, as pointed out by Luhmann (1993), many of the damages that in the past were attributed to nature, destiny, or supernatural beings, and were seen as unavoidable dangers, today are usually imputed to human actions and decisions, and, therefore, they are given the form of risks. Hazards thus become a political issue and a springboard for the mobilization of social agents in the public arena.

Such new conditions are part of a broader phenomenon of increasing artificialization of the environment and even of interpersonal relationships. The public administration and the market, through material or bureaucratic devices, have an increasing presence in our life by enabling, conditioning, or regulating the various areas of human behavior. A growing range of goods and services, which were traditionally free and whose provision depended on family, friends, or simple nature, now depend on administrations or industries: the preservation of health, the care of the elderly, advice and emotional support, leisure and sports, small pleasures of daily life, and even clean water and air.

We must also take into account, according to Piotr Stzompka (2000), the high degree of functional differentiation, interdependence, and globalization that we have achieved in our risk society. The increasing complexity of the institutions and technological systems, their interdependence and the global reach of their operation, has made large segments of the social world opaque for their members. And, in turn, the opacity of organizations and institutions has produced the growing anonymity and depersonalization from those on which our welfare depends. We do not know the agents that are behind the financial system, the scientists who elaborate new active principles in medicines, or the engineers responsible for the design of airplanes. Social trust is a bet about the contingent actions of opaque institutional agents, a mechanism to manage an uncertain and uncontrollable future in the midst of an artificial world (Weber and Carter 2003, 5ff).

In this context, trust is often considered as a critical asset for governance in the risk society (e.g., Renn 2008). The bad news is that empirical studies on risk perception show today a continuous erosion of trust, which is associated with decreasing levels of risk tolerance regarding involuntary hazards among the population (e.g., Slovic 2000, 2013). From the traditional point of view, trust generates efficient policies through the production of social capital.[12] However, trust is only one way to ground an efficient risk policy and management. This is an important point. When trust in other people or organizations fails, we are still willing to depend on them as long as we have appropriate institutional structures for control (Rousseau et al. 1998, 399), such as laws that are enforced, judicial independence, transparency, accountability, freedom of information, and opportunities for participation. These are control mechanisms designed to eliminate the need for trust or at least to reduce costs in the case that trust was disappointed (Lofstedt 2009, xi). Expressed briefly: if the "handshake" fails then what we need is a contract. It is important to note that, unlike deontic attitudes or contracts, trust neither implies nor presupposes an authority or obligation that may compel the trustee to act as expected, so that it acquires a firm legal or moral responsibility as depositary of the trust (Darwall 2017, 40).[13] In fact, distrust (not lack of trust) exercises a similar function of reducing the complexity of the system (Luhmann 2017, ch. 10). The key question now is: Does trust in post-trust society really tend to disappear?

THE DYNAMICS OF TRUST

This complex question cannot be adequately answered by a simple "yes" or "no." Post-trust society is not a society of distrust. This point is also important. It is not merely the dissipation of trust and its replacement by regulatory institutional structures that guarantees governance. Rather, trust seems to behave similarly to risk in the situations of risk-risk tradeoff—that is, the elimination or reduction of a target risk for a population entails the rising of a countervailing risk (of the same or a different type) for another population or else the appearance of a new risk for the original population (of the same or a different type).[14] For example, the removal of asbestos in old buildings reduces the exposure to this carcinogen by users of such facilities, but in turn exposes operators who perform extraction to a much higher exposure. Avoiding aspirin because of possible stomach troubles often results in the use of another type of painkiller with different side effects. Like energy, risk does not disappear (although it can be created) but it moves and transforms itself.

Risk-risk tradeoffs can be classified into four types, taking into account two dimensions: the affected population and the type of countervailing risk, as shown in table 27.1.

In the case of trust, more than just disappearing (or simply turning into distrust), it seems to migrate and sometimes change in nature. Institutional

Table 27.1. Target and Countervailing Risk Tradeoffs

		The target and countervailing risk are:	
		The same type	Of different type
The target and countervailing risk affect:	The same population	Risk offset	Risk substitution
	A different population	Risk transfer	Risk transformation

Source: Adapted from Graham and Wiener (1995a, 22).

mechanisms of surveillance, sometimes seen as alternatives to trust, are in reality forms of trust-trust tradeoffs to other social agents (Weinstock 2013, 213).[15] Opting for a contract when distrusting another person entails assuming expectations, and establishing a dependency relationship, with respect to the individuals or institutions responsible for ensuring compliance—the law and law enforcement officials, in this case. The same individuals and social groups that previously placed their trust in government and industry increasingly tend now to do so in the judiciary, interest groups such as consumer or patient associations, a broad diversity of NGOs, or the academic world. By trusting we become vulnerable to risk; by shifting trust we change our inclination (willingness) to take risks from one source to another (Mayer et al. 1995, 712).

But displaced trust may also change in nature, emphasizing the dimension of empathy at the expense of the dimension of independence (in the case of interest groups), the dimension of competence at the expense of the dimension of empathy (in the case of the academic world), or the dimension of independence at the cost of the dimension of empathy (case of the judiciary). Withdrawing trust in the government due to the opaque and controversial management of an environmental threat can lead us to place trust in environmental organizations that we consider more independent of the industry and therefore more credible (trust transfer: same type of trust, different type of social agent); withdrawing trust in doctors for their bad news about a disease we suffer and their lack of sensitivity to patients can dangerously redirect a new type of trust towards homeopaths or healers who are able to give us some hope and talk to us about our worries in a comprehensible language (trust transformation: different kind of trust, different type of social agent), as shown in table 27.2.

This phenomenon allows us to understand the enormous power that those interest groups, scientific organizations, independent NGOs, and the judiciary, have achieved in recent decades with respect to their ability to influence public administration (Löfstedt 2009, xv). While trust in public administration and financial power has been declining between 2007 and 2015, trust in

Table 27.2. Target and Countervailing Trust Tradeoffs

		The target and countervailing trust are:	
		Of the same type	Of different type
The target and countervailing trust are addressed to:	The same type of social agent	Trust offset	Trust substitution
	A different type of social agent	Trust transfer	Trust transformation

the judiciary has increased in OECD countries, according to the Gallup World Survey (OECD 2017, 19).

It is a dynamic where citizens tend to weaken their dependency links with respect to traditional institutions like public administration or the corporate world, somehow gaining autonomy and assuming a leading role in redirecting their expectations towards other social actors. The good news is that the weakening of institutional trust seems to have a healthy effect on democracy. But the empowerment of personal autonomy to the detriment of institutional dependence in contemporary society is not free of risks. Citizens are willing to make their own decisions about risk acceptability. However, these are decisions not informed by regulatory authorities but by the social agents that have the most efficient communication strategies at any moment. It is a scenario where unfounded but well-communicated alerts can easily cause considerable public alarm, as it has been shown with HPV vaccine, although the same could be said of the management by regulatory authorities of recent health threats such as the H1N1 virus.[16] These are risks inherent in the democratic game.

The formula usually invoked to try to overcome institutional distrust is to open decision-making and communication processes to participation of social agents (i.e., trusted spokespersons representing relevant interest groups), and particularly to adopt deliberative risk management strategies implementing mechanisms that involve affected or interested social agents, especially in contexts with low public trust where regulators or industry are seen as biased (see, e.g., Löfstedt 2009, 10, 21, 129, 132).

CRITICAL TRUST, RESPONSIBLE DISTRUST

A common message in the media and specialized literature is that trust is necessary for the proper functioning of a society (see, e.g., OECD 2017, 2018; Slovic 2000, 2013). Trust is certainly a central element of social capital in contemporary society, but it is easily reduced to vertical or *institutional* trust and perhaps tends to be overrated in detriment of public scrutiny and accountability (see Poortinga and Pidgeon 2003; Weinstock 2013). Horizontal or interpersonal trust as necessary for cooperation and fluid social functioning is beyond the present argument, but this is not the case for trust in institutions (particularly regulatory authorities and the industry). With or without risk amplification, mad cows, dioxins in food, and other recent crises that have particularly affected European countries, together with the intensification of citizen activism and the action of advocacy groups, have generated a new setting in which it is not possible or perhaps even desirable to recover the traditional role of institutional trust as the main support for efficient social functioning.[17]

The erosion of institutional trust in today's society is not necessarily bad news. Between visceral rejection and naive acceptance, there is a broad territory for what may be called "critical trust" or "responsible distrust." In any case, trust does not disappear: it is rather qualified through caution and criticism regarding traditional actors (government, industry) as well as occasionally redirected as an asset to new actors (NGOs, universities, or others), often changing in nature (competence-independence-empathy). Cultivated distrust (in opaque institutional agents) plus retargeted trust jointly contribute to the reduction of complexity so as to make social action and democratic governance possible.

An example of such "critical trust" can be found in the early COVID-19 vaccination campaign in Spain. In autumn 2020, there was very strong reluctance in the Spanish population regarding the willingness to receive the new COVID-19 vaccines, which were being developed in record time by pharmaceutical companies with the support of public funds. Taking into account accumulated data between September and December 2020, approximately 40 percent of the Spanish population were not willing to be immunized with the new vaccines, while another 40 percent were inclined to do it, with a high percentage of people doubting (around 10 percent).[18] This data was very concerning. Public health authorities launched vigorous communication campaigns that tried to change these figures through the transmission of information: the key was trying to foster willingness by correcting a deficit of public knowledge (about the safety and efficacy of the new vaccines). However, the key was somewhere else: among the reticent, a very high percentage were well-educated people.[19] The issue at stake was not so much a knowledge deficit (in educated publics who were already well informed) but rather a trust deficit in public authorities and private pharmaceutical companies—a deficit that was effectively corrected, not with technical information, but with the testimonies offered by the mass media when vaccination programs began in Western countries in December 2020, showing that the new vaccines did not produce serious side effects. In January 2021, the percentage of reticent people fell to 16 percent, and a significant part of the statistical weight of this drop corresponded to people who were not ignorant but cultivated cautious persons.[20] As also shown by statistical data,[21] the reticent supported the health system and vaccines in general but tended to hold a wary attitude regarding politicians and pharmaceutical corporations. Put simply, in autumn 2020, facing a very serious health crisis, many citizens did not trust the simple technical competence and experts at the service of the government or private firms. While this critical attitude did not eventually undermine the vaccination campaign, it was seriously misdiagnosed by public authorities as a knowledge-deficit problem (Goldenberg 2021), thus creating obstacles for the management of uncertainty among distrusting citizens.

Demographically, science-informed "critical trust" or "responsible distrust," as shown by the former example, is not an imaginary chimera. It is the type of attitude which can be found through cluster analysis in the "distrustful engagers" of the British PAS survey of 2014 or the "many-benefits/many-risks" population of the Ibero-American Survey of 2007.[22] They are well-educated citizens very interested in science and informed about it, holding qualified and differentiated opinions regarding different fields of science and technology instead of a naive, global pro-science enthusiasm. They think that science is generally beneficial for society although they are aware and cautious about the risks that accompany its development in particular applications, especially with regard to the growing field of commercial science. Usually cultivated urbanites, these are people who view science with familiarity and realism, who have a high political value due to their high consumption of scientific information and their inclination to participate in science and technology–related affairs of social interest. It is the population segment which is called "loyal skeptics" by Martin Bauer and collaborators (2012): a population located on the right extreme of their well-known inverted U, in the association between attitude and knowledge that is especially characteristic of postindustrial societies and large cities.[23]

It is not something new to highlight the value of criticism for the formation of sound opinions and decision-making. The roots of this idea can be found in the enlightened thought of the eighteenth century (Phillips 2015, 25), for example in the writings of Thomas Jefferson and other pioneers of the American and French revolutions, as well as in later key contributions to political thought, such as John Stuart Mill's *On Liberty* (1859). These were thinkers committed to rationality and progress but who also claimed a critical stance toward human affairs.[24] Actually, science itself exemplifies the value of evidence-based criticism and skeptical attitudes for the advancement of knowledge and social progress.

A long tradition of philosophical reflection on technology also supports a cultivated distrust standpoint: Jacques Ellul's ethics of non-power before technological determinism, Ivan Illich's conviviality society as alternative to the techno-bureaucratic fascism, Herbert Marcuse's rational rebellion against the tyranny of things, and so on. It is the path also followed by contemporary STS thinkers such as Carl Mitcham, Jerome Ravetz, Kristin Shrader-Frechette, Langdon Winner, and Brian Wynne, who promote a critical, reflective and well-informed attitude toward our artificial world while recognizing the value of science and engineering. Certainly, we cannot escape the artificial and risky world created by new technologies, but criticism and reflection by STS scholarship and "STS citizens" can perhaps facilitate a better coexistence with the modern Prometheus, echoing the famous Gothic novel by Mary Shelley.

As a final reflection, we must unfortunately assume that Aristotle's practical wisdom, moral virtue, and goodwill cannot always be presupposed in those on whose decisions our well-being depends. Facilitating governance in the risk society requires citizens who do not restrain themselves from the great issues and daily threats. It also requires critical and well-informed citizens willing to act as a democratic counterweight to traditional institutional social agents, as well as some worrying newcomers who use Big Data and social networks to gain credibility and influence.

In a world undergoing an accelerated technological transformation, with an overdose of information and misinformation coming from a diversity of actors, and with a growing dependence of the springs of development on opaque organizations and interests, critical awareness and a certain institutional distrust are not a problem but a very valuable resource.

ACKNOWLEDGMENTS

I would like to thank the volume editors for their helpful comments regarding the improvement of this contribution. For economic support, I am also indebted to the research project "Post-normal cultures of science and technology" (PID2021–123454NB-C41) of the Spanish Ministry of Science.

NOTES

1. This trend can be seen in the increasing social orientation of the framework science programs of the European Union and the salience of citizen engagement in a broad diversity of science policies or organizations. As to the first point, notice the

evolution, since the early 2000s, of names for specific key sub-programs from "science and society," through "science with and for society," to "responsible research and innovation." Concerning the second point, see, for example, the emphasis the American Association for the Advancement of Science (AAAS) and the European Union give "public engagement," available at https://www.aaas.org/focus-areas/public-engagement and https://op.europa.eu/en/publication-detail/-/publication/2d7d42ad-d69e-46ab-94bd-035b068ae676/language-en (accessed July 17, 2022).

2. Lipobay, Trasylol, Yasmin, and others since 2001.

3. Particularly Jair Bolsonaro's government in Brazil, and to some extent the Trump administration in the United States and the initial reactions of Boris Johnson's government in the United Kingdom, deliberately downsizing the pandemic threat in early 2020. See, in general, Claessens (2021).

4. Together with a growing public awareness about how experts frequently misrepresent and systematically underestimate technological risks (Wynne 1983; Shrader-Frechette 1991; Mitcham 2021).

5. See the commentaries of Rapp (2010) and Edward M. Cope (Cope and Sandys 2010, 5) on this passage.

6. According to Book VI of *Nicomachean Ethics*, practical wisdom or prudence (*phrónesis*) is an intellectual virtue regarding the ability to define ends and the means to achieve them in human action. It must be distinguished from theoretical wisdom (*sophia*) because in *phrónesis* rational thought is directed toward the particular and contingent (rather than to universal truths), as well as from technique (*techné*) since *phrónesis* is directed toward human action (not toward the production of objects). Its roots are experience and responsibility, and it involves not only the ability to decide how to reach a certain goal but also to determine the good ends (consistent with the full life—*eudaimonia*).

7. On the relationship between the concepts of trust and reliance, see, e.g., the discussions by Goldberg (2020) and McLeod (2020).

8. *Trust* and *distrust* are contrary terms, not contradictories. They cannot both be true, but they can both be false (see, e.g., D'Cruz, 2020).

9. The concept of trust should also be differentiated from that of confidence, although both imply vulnerability. In the first the risk must be recognized and assumed, something that does not occur in the concept of confidence. Möllering (2013, 56) discusses this concept as a weaker, one-sided form of trust based on predictability.

10. For instance, the classical model about perceived trustworthiness by Mayer et al. (1995) differentiates the dimensions of ability, integrity, and benevolence. See, in general, Pidgeon, Poortinga, and Walls (2007, 119 ff.).

11. Regarding the trustor, the main factor would be the propensity to trust depending on the different types of personality, experiences, political climates, and so on.

12. A common way to define the concept of "social capital" is in terms of a number of factors that permit cooperation and explain the effective functioning of groups. These include trust as well as a shared identity, interpersonal relationships, shared norms and values, and others (e.g., OECD 2017).

13. As mentioned above, trust might include notwithstanding a normative dimension, besides its predictive dimension related to expectations, providing a basis for reproach or moral critique when trust fails (Dormandy 2020, 4–7).

14. See Graham and Wiener (1995b) and Löfstedt and Schlag (2017).

15. These would be a case of second-order institutional trust in the approach of Warren (2018)—first-order institutional trust would instead refer to the agencies and services provided from government in order to provide common goods such as health care, statistics, energy, clean air, transportation, and so on (including risk regulation). As exemplified by this same author (Warren 2018), a fairly common place in the literature is that the basic form of trust is social or interpersonal trust, so that institutional trust in their various layers (or other derived types of trust such as "brand trust"—see OECD 2017) would be higher-order forms of interpersonal trust with respect to the agents, anonymous or no, who stand for the institutions. Still, this is an open question which deserves further discussion elsewhere.

16. Just the opposite of what seems to be the case in the origin of the COVID-19 pandemic during January–February 2020: a well-founded but badly communicated alert by public health authorities.

17. Institutional trust might be overrated, but this is not to claim that vertical trust is impossible or even unnecessary, for the defense of this latter claim, following for example the path of Hardin (2013), would require the assumption of a too narrow, rationalistic concept of trust (Möllering 2013, 53). Concerning this discussion, see Cook, Hardin, and Levi (2005) and Hardin (2013), as well as the critiques by Yang (2008) and Möllering (2013). A step beyond in this direction, see also the discussion about the supposedly fictitious character of the concept "trust" by Williamson (1993) and Karpik (2014).

18. Data from the Spanish "Centro de Investigaciones Sociológicas" (CIS) (Center for Sociological Research). Accessed May 1, 2021. http://www.cis.es/cis/opencm/ES/11_barometros/depositados.jsp.

19. These individuals consumed quite a bit of scientific information and had a higher than average inclination to participate in public discourse, but also an inclination to skepticism regarding the myths of science (such as the one conveyed by the phrase "science can solve any problem"). While recognizing that people with a good educational level are not automatically "critical trusters," in this context it seems a relevant proxy. Perhaps a novelty worthwhile mentioning was the high incidence of women among them in this particular case.

20. Between November 2020 and January 2021, the percentage of people with higher education not willing to be vaccinated fell from 54.6 percent to 15.3 percent, according to CIS data. It was the largest decline among population segments analyzed by education level.

21. See the CIS webpage (https://www.cis.es/cis/opencm/ES/11_barometros/index.jsp, accessed July 21, 2022), where statistical data and analyses are available, as well as the last demoscopic study of the Spanish Foundation for Science and Technology (FECYT) in Lobera Serrano and Cabrera Álvarez (2021)—available at https://www.fecyt.es/es/publicacion/evolucion-de-la-percepcion-social-de-aspectos-cientificos-de-la-covid-19, accessed July 21, 2022.

22. See, respectively, Ipsos MORI (2014) and Cámara Hurtado and López Cerezo (2012). See also Cámara Hurtado et al. (2018) and López Cerezo and Laspra (2018).

23. Surveys regularly show a significant positive association between level of scientific knowledge and pro-science attitude, but, in certain contexts (e.g., postindustrial societies of Northern Europe) this association typically inverts its sign at a certain level of knowledge, revealing the existence of an educated population aware of the great potential of science and technology, familiar with them, but also cautious about the risks and possible negative effects of technology-based industrial development (see Bauer et al. 2012).

24. In fact, the term *criticism* comes from the Greek *kritikos*, which refers to a person with good judgment.

REFERENCES

Aristotle. 1926. *The 'Art' of Rhetoric*. Second edition. Translated by J. H. Freese. London: William Heinemann, G.P. Putnam's Sons.

———. 2009. *Nicomachean Ethics*. Translated by David Ross. New York: Oxford University Press.

Baier, Annette. 1986. "Trust and Antitrust." *Ethics* 96, no. 2: 231–60.

Bauer, Paul C., and Markus Freitag. 2018. "Measuring Trust." In *The Oxford Handbook of Social and Political Trust*, edited by Eric M. Uslaner, 15–36. Oxford: Oxford University Press.

Bauer, Martin W., Rajesh Shukla, and Nick Allum, eds. 2012. *The Culture of Science: How the Public Relates to Science Across the Globe*. New York: Routledge.

Beck, Ulrich. 1992. *Risk Society: Towards a New Modernity*. London: Sage.

Cámara Hurtado, Montaña, and José A. López Cerezo. 2012. "Political Dimensions of Scientific Culture: Highlights from the Ibero-American Survey on the Social Perception of Science and Scientific Culture." *Public Understanding of Science* 21, no. 3: 369–84.

Cámara Hurtado, Montaña, Ana Muñoz van den Eynde, and José A. López Cerezo. 2018. "Attitudes towards Science among Spanish Citizens: The Case of Critical Engagers." *Public Understanding of Science* 27, no. 6: 690–707.

Claessens, Michel. 2021. *The Science and Politics of COVID-19: How Scientists Should Tackle Global Crises*. Cham, Switzerland: Springer.

Cook, Karen S., Russell Hardin, and Margaret Levi. 2005. *Cooperation without Trust*. New York: Russell Sage Foundation.

Cope, Edward Meredith, and John Edwin Sandys, eds. 2010. *Aristotle: Rhetoric*. Vol. 2. Cambridge: Cambridge University Press.

Darwall, Stephen. 2017. "Trust as a Second-Personal Attitude (of the Heart)." In *The Philosophy of Trust*, edited by Paul Faulkner and Thomas Simpson, 35–50. Oxford: Oxford University Press.

Davidson, Lacey J., and Mark Satta. 2021. "Justified Social Distrust." In *Social Trust*, edited by Kevin Vallier and Michael Weber, 122–47. New York: Routledge.

D'Cruz, Jason. 2020. "Trust and Distrust." In *The Routledge Handbook of Trust and Philosophy*, edited by Judith Simon, 41–51. New York: Routledge.

Dormandy, Katherine. 2020. "Introduction: An Overview of Trust and Some Key Epistemological Applications." In *Trust in Epistemology*, edited by Katherine Dormandy, 1–40. New York: Routledge.

Goldenberg, Maya J. 2021. *Vaccine Hesitancy: Public Trust, Expertise, and the War on Science*. Pittsburgh, PA: University of Pittsburgh Press.

Goldberg, Sanford C. 2020. "Trust and Reliance." In *The Routledge Handbook of Trust and Philosophy*, edited by Judith Simon, 97–108. New York: Routledge.

Graham, John D., and Jonathan B. Wiener. 1995a. "Confronting Risk Tradeoffs." In Graham and Wiener 1995b, 1–41.

———., eds. 1995b. *Risk versus Risk: Tradeoffs in Protecting Health and the Environment*. Cambridge, MA: Harvard University Press.

Hardin, Russell. 2013. "Government without Trust." *Journal of Trust Research* 3, no. 1: 32–52.

Hawley, Katherine. 2017. "Trustworthy Groups and Organizations." In *The Philosophy of Trust*, edited by Paul Faulkner and Thomas Simpson, 230–50. Oxford: Oxford University Press.

Hupcey, Judith E., Janice Penrod, Janice M. Morse, and Carl Mitcham. 2001. "An Exploration and Advancement of the Concept of Trust." *Journal of Advanced Nursing* 36, no. 2: 282–93.

Ipsos MORI. 2014. *Public Attitudes to Science 2014*. Accessed July 18, 2021. https://www.ipsos.com/ipsos-mori/en-uk/public-attitudes-science-2014.

Karpik, Lucien. 2014. "Trust: Reality or Illusion? A Critical Examination of Williamson." *Journal of Trust Research* 4, no. 1: 22–33.

Kasperson, Roger E., and Pieter Jan M. Stallen, eds. 1991. *Communicating Risks to the Public: International Perspectives.* Dordrecht: Kluwer.

Lobera Serrano, Josep, and Pablo Cabrera Álvarez. 2021. *Evolución de la percepción social de los aspectos científicos de la COVID-19 (julio 2020–enero 2021)* [Evolution of the social perception of scientific aspects of COVID-19, July 2020–January 2021]. Madrid: FECYT.

Löfstedt, Ragnar. 2009. *Risk Management in Post-Trust Societies*. London: Earthscan.

———. 2013. "Communicating Food Risks in an Era of Growing Public Distrust: Three Case Studies." *Risk Analysis* 33, no. 2: 192–202.

Löfstedt, Ragnar, and Anne Schlag. 2017. "Risk-Risk Tradeoffs: What Should We Do in Europe?" *Journal of Risk Research* 20, no. 8: 963–83.

López Cerezo, José A., and Belén Laspra. 2018. "The Culture of Risk: STS Citizens Facing the Challenge of Engagement." In *Spanish Philosophy of Technology: Contemporary Work from the Spanish Speaking Community*, edited by Belén Laspra and José A. López Cerezo, 87–100. Dordrecht: Springer.

Luhmann, Niklas. 1993. *Risk: A Sociological Theory.* Berlin: De Gruyter.

———. 2017. *Trust and Power*. Cambridge: Polity Press.

Mayer, Roger C., James H. Davis, and F. David Schoorman. 1995. "An Integrative Model of Organizational Trust." *Academy of Management Review* 20, no. 3: 709–34.

McCraw, Benjamin W. 2015. "The Nature of Epistemic Trust." *Social Epistemology* 29, no. 4: 413–30.
McLeod, Carolyn. 2020. "Trust." In *The Stanford Encyclopedia of Philosophy*, Fall 2020 edition, edited by Edward N. Zalta. https://plato.stanford.edu/archives/fall2020/entries/trust/.
Mill, John Stuart. [1859] 2001. *On Liberty*. Ontario: Batoche Books.
Mitcham, Carl. 2021 (July). "Engineering Existential Risks." Paper presented at 2021 ASEE Virtual Annual Conference Content Access, Virtual Conference. https://peer.asee.org/37060. Accessed January 13, 2021. 8 pp.
Möllering, Guido. 2013. "Trust without Knowledge? Comment on Hardin, 'Government without Trust.'" *Journal of Trust Research* 3, no. 1: 53–58.
OECD (Organization for Economic Co-Operation and Development). 2017. *Trust and Public Policy: How Better Governance Can Help Rebuild Public Trust*. Paris: OECD Publishing.
———. 2018. *Trust and Its Determinants*, OECD Working Paper 89, Paris: OECD Publishing.
Pidgeon, Nick, Wouter Poortinga, and John Walls. 2007. "Scepticism, Reliance and Risk Managing Institutions: Towards a Conceptual Model of 'Critical Trust.'" In *Trust in Risk Management: Uncertainty and Scepticism in the Public Mind*, edited by Timothy C. Earle, Michael Siegrist, and Heinz Gutscher, 117–42. London: Routledge.
Rapp, Christof. 2010. "Aristotle's Rhetoric." In *The Stanford Encyclopedia of Philosophy*, edited by Edward N. Zalta. https://plato.stanford.edu/archives/spr2010/entries/aristotle-rhetoric/.
Renn, Ortwin. 2008. *Risk Governance: Coping with Uncertainty in a Complex World*. London: Routledge.
Rosanvallon, Pierre. 2008. *Counter Democracy: Politics in the Age of Distrust*, translated by Arthur Goldhammer. Cambridge: Cambridge University Press.
Rousseau, Denise M., Sim B. Sitkin, Ronald S. Burt, and Colin Camerer. 1998. "Not So Different after All: A Cross-Discipline View of Trust." *The Academy of Management Review* 22, no. 3: 393–404.
Phillips, Angela. 2015. *Journalism in Context: Practice and Theory for the Digital Age*. London: Routledge.
Poortinga, Wouter, and Nick F. Pidgeon. 2003. "Exploring the Dimensionality of Trust in Risk Regulation." *Risk Analysis* 23, no. 5: 961–72.
Shrader-Frechette, Kristin. 1991. *Risk and Rationality: Philosophical Foundations for Populist Reforms*. Berkeley: University of California Press.
Simpson, Thomas W. 2012. "What Is Trust?" *Pacific Philosophical Quarterly* 93: 550–69.
Slovic, Paul. 2000. *The Perception of Risk*. London: Earthscan.
———. 2013. "Perceived Risk, Trust, and Democracy." In *Social Trust and the Management of Risk*, edited by George Cvetkovich and Ragnar Löfstedt, 42–52. London: Earthscan.

Stern, Robert. 2017. "'Trust Is Basic': Løgstrup on the Priority of Trust." In *The Philosophy of Trust*, edited by Paul Faulkner and Thomas Simpson, 272–93. Oxford: Oxford University Press.

Sztompka, Piotr. 2000. *Trust: A Sociological Theory*. Cambridge: Cambridge University Press.

Walaski, Pamela (Ferrante). 2011. *Risks and Crisis Communication: Methods and Messages*. Hoboken, NJ: Wiley.

Warren, Mark. 2018. "Trust and Democracy." In *The Oxford Handbook of Social and Political Trust*, edited by Eric M. Uslaner, 75–94. Oxford: Oxford University Press.

Weber, Linda, and Allison I. Carter. 2003. *The Social Construction of Trust*. New York: Springer.

Weinstock, Daniel. 2013. "Trust in Institutions." In *Reading Onora O'Neill*, edited by David Archard, Monique Deveaux, Neil Manson, and Daniel Weinstock, 199–218. London: Routledge.

Williamson, Oliver E. 1993. "Calculativeness, Trust, and Economic Organization." *Journal of Law and Economics* 36, no. 1: 453–86.

Wynne, Brian. 1983. "Redefining the Issues of Risk and Public Acceptance: The Social Viability of Technology." *Futures* 15, no. 1 (February): 13–32.

Yang, Kaifeng. 2008. "Cooperation without Trust? A review of Cook, Hardin, and Levi's Book." *Public Administration Review* 68, no. 6: 1164–66.

Zagzebski, Linda Trinkaus. 1996. *Virtues of the Mind: An Inquiry into the Nature of Virtue and the Ethical Foundations of Knowledge*. Cambridge: Cambridge University Press.

Index

AAAS. *See* American Association for the Advancement of Science (AAAS)
AAC. *See* augmentative and alternative communication (AAC)
Abimelech, 223n7
abstraction and engineering, 374
Abt Suger von Saint-Denis, 250–51
Academica (publisher), 172
accessibility, 14, 468, 473
Achterhuis, Hans, 29
"Actions, Reasons, and Causes" (Davidson), xii
activists and activism, 290, 347, 348, 381, 393, 499, 507
activity, technology as, 57
Adderall, debate over use of, 311
administration of things, 44, 46
Adorno, Theodor, 23
Advance Purchase Agreement (APA) between EU and CureVac, 464, 467–68, 473, 475
affectivity and moral sensitivity, 143–44, 145, 148, 150, 155, 159, 160
affordances, 132
Africa: policy paradoxes in sub-Saharan Africa, 8, 435–59. *See also individual African countries*
Agamben, Giorgio, 6, 256, 267, 269–71
Agazzi, Evandro, 324, 331

Agreement on Trade-Related Intellectual Property Rights (WTO). *See* TRIPS
Agricultural University, 350
agriculture, feminization of, 442, 449
AI. *See* artificial intelligence (AI)
Akon-Yamga, Gordon, 8, 435–59
Alexa (as voice-activated technology), 114
Alexander, Jennifer Karns, 6, 225–41
alienation, 172, 273, 289, 295, 380; alienation/bridging (as used by Mitcham), 369, 374–78, 383; Odell on, 372, 373, 384–85
allegory, 21, 22, 97
Alonso Puelles, Andoni, 6, 277–96
Amazon (company), 255, 429
American Association for the Advancement of Science (AAAS), 494n2, 509–10n1
American Association of Industrial Management, 44
American Home Mission Society, 227
American Symposium on Philosophy of Engineering, 175
Amish communities and technology, 123–24, 282, 287
Amnesty International, 472, 473
Ancient Greeks and "pure inquiry," 484–86, 489, 493, 494, 495n9

Andanda, Pamela, 8, 461–82
Anders, Günther, 95–97, 280
Andreas, Joel, 199
Angel, Hayyim, 299, 300
angelism, 48–50
angels and angelic structure, 249, 251, 256, 259–60, 261, 263, 273n5, 274n10. *See also* hierarchical order
Anthropocene, 3, 378, 430
anthropocentrism, 82, 447
Antigone (Sophocles), 17
Antonsen, Trine, 136
Apple, engineers working for, 158
applied science, 8; demarcation between basic (pure) science and applied science, 320, 321, 323, 324, 325, 331; in "science" policy, 483–97
applied-science: view of engineering and technology, 151, 152–53
Arendt, Hannah, 35, 73, 95, 420
"Are Technological Artifacts Mere Tools?" (Peterson), 134
Aristophanes, 73, 77
Aristotelian thinking, 14, 23, 72, 247, 264, 298, 301, 305, 309
Aristotle, 76, 222n4, 253, 269, 288, 312, 493, 495n8, 509; on ethics, 300, 485; and the golden mean, 422; Heidegger on, 17; on intellect and contemplation, 486, 488; on levels of the mind, 309; on politics and political science, 75, 290, 485; on the practice of philosophy, xii, 76; and "pure inquiry," 484–85; and the pursuit of happiness, 307; on retail trade, 422, 424; on sages, 321, 322; as a student at Plato's Academy, 488; on tools and technology, 264–65; on trust, 501, 502
Arndt, H. W., 357
Arnhart, Larry, 7, 297–316
art distinguished from technical artifacts, 18–19, 21
artifacts, 2, 24, 124, 133, 153, 158, 170, 197, 375; Aristotle on, 17;

"convivial" as a feature of artifacts, 244; design of, 21, 121–22, 128, 198; the embodied or embedded values inserted in, 121–22, 132; embodying personal intentions, 243, 247; health care artifacts, 154–55; Heidegger on, 14–15; intrinsic final values, 130; material artifacts, 31; as morally value-laden, 127, 130; *qi* of an artifact, 198; second-order responsibility of, 133; social artifact, 215, 217; "systemic pseudo-tools," 245; technology as artifact, 55, 57, 61; value embodiment within, 120, 121–22, 128–29, 130, 132; value neutrality of, 153. *See also* technical artifacts; technological artifacts; tools
artificial intelligence (AI), 59, 60, 61, 64, 394, 399, 405, 429
artisan economy, 5, 114–15
Asian Development Bank, 356
"Asia's Needless Hunger" (Goodstadt), 354, 355
"assemblage," 64
assurance game, 99–100, 111n4
atheistic religiosity of Leon Kass, 298, 301, 303, 304
Athens: Athens/Jerusalem: reason vs. revelation, 4, 74, 81, 83–84, 87–89, 298–304, 305; intellectual understanding representing Athens, 304; as a technoscientific Athens, 83
attention. *See* speed/attention (as used by Mitcham)
Atuire, Caesar A., 474
augmentative and alternative communication (AAC), 154, 155, 159
Augustine of Hippo, (Saint Augustine), 211, 263
automation, 46, 379, 405
automobiles, 20, 22, 121, 123–24
autonomy, 34, 74, 122, 157, 401; autonomy of inquiry, 492; autonomy of technology debate, 126–27, 135;

and neutrality, 124–28; personal autonomy, 37, 310, 506; as a principle of bioethics, 310–11; and technology, 15, 16; of the university, 488, 490

backward induction reasoning, 99, 108, 111n7
Bacon, Francis, 82, 286, 396
Bacon, Roger, 288
Bai, Shuying, 200
Baltimore, David, 326
Baoshan Iron and Steel project, 173
Barbour, Ian, 87, 282
Barker, Tim, 55–56, 58, 59, 62, 64
Barth, Karl, 231
Barthes, Roland, 20, 21–22, 23
basic (pure science): demarcation between basic (pure) science and applied science, 320, 321, 323, 324, 325, 331; in "science" policy, 483–97
Bates, M. Searl, 231, 232–33
Baudrillard, Jean, 21, 22–23
Bauer, Martin, 508
Beck, Ulrich, 503
becoming, process of, 55, 57–58, 64
The Beginning of Wisdom: Reading Genesis (Kass), 298–99
behavioral economics, 135–36, 137
Being and Time (Heidegger), 14–15, 17
Being Human: Readings from the President's Council on Bioethics (President's Council on Bioethics), 311–12
Beiyang West Learning College in Tianjin, China, 196
belief, 84, 96, 146, 200, 337, 395, 408, 494; and the Amish, 124; belief-desire model, 101; concerning technology, 264–65, 266, 282; false beliefs, 84, 111n1, 280; human beliefs, 77, 395; moral beliefs, 345; positivism as a belief, 75; psuedo-logic, 279; and *qi*, 198; rejecting truth in favor of belief, 336; religious beliefs, 277, 279, 280, 281, 282–83, 287, 288, 292n2, 304; and revelation, 84, 87; scientific beliefs, 7, 282; unbelief, 87; and the World Council of Churches, 227, 229
Benedict XVI (pope) (Joseph Ratzinger), 223n8
beneficence, 70, 310, 311
Benjamin, Walter, 75
Bennett, Frank, 230
Bennett, J. C., 236
Bentham, Jeremy, 23
Berdyaev, Nikolai, 4, 33–34, 37, 50
Berger, Gaston, 111n6
Bergson, Henri, 37, 55, 62, 80, 96; and the possible, 93–94
Berle, A., 45
Berlin, Isaiah, 397
Bestand, 25
Beyond Therapy: Biotechnology and the Pursuit of Happiness (President's Council on Bioethics), 304–5, 306–7, 308, 311, 312, 314
Bezos, Jeff, 429
bias against practical use, 486, 495n7
Bible, 76, 107, 230; authority of, 227; Biblical revelation, 301; describing the political foundations of Israel, 222n5; and Greek philosophy, 81–82, 88; Hebrew Bible not having a word for "nature," 300–301; Kass's books on, 298–300, 301, 304–5, 306; modern biblical criticism, 84; questioning of the Bible, 84–85, 87; source of morality, 325; and typology, 211; understanding the Israelite monarchy, 212–17
Biblical citations from the Old Testament: Genesis 1:27, 209, 302–3; Genesis 9:2 and 9:9–17, 222n1; Genesis 49, 216; Exodus 19:6, 215; Exodus 29:38–42 and 29:43–46, 302; Judges 8:22–23, 214; Judges 9:19–20, 223n7; Judges 17:6, 21, 25, and 18:1 and 19:1, 213; Judges

19, 223n11; 1 Samuel 8:4–18, 215; 1 Samuel 8:7–9, 214; 1 Kings 4, 215–16; 1 Kings 5:6 and 11:13, 216; 2 Kings 17, 18, and 25, 216; Proverbs 3:18, 82
Biblical citations from the New Testament: Matthew 16:18 and 18:17, 219; Luke 4:16–30, 219; John 15:7, 16, 250–51; Acts 11:26, 222n3; Romans 5:14, 210; Hebrews 1:1–2, 211
biblical cosmology, 274n13
Biden, Joe, 341, 484
BIES. *See Bulletin of Indonesian Economic Studies* (BIES)
Bildung, 486–87, 489, 493, 495n10
BIMAS program in Indonesia, 351, 352, 357
bioethics, 297–316; biotechnology debate over thick and thin bioethics, 310–13; natural biological ethics, 306; public bioethics, 314; Universal Declaration on Bioethics and Human Rights (UNESCO), 475
Biolyse (company), 466
"bioregionalism," 373
biotechnology debate, 304–7; natural limits of biotechnology, 313–14; over thick and thin bioethics, 310–13
Bjerknes, Mari Skancke, 146–47
Bjørk, Ida Torunn, 146–47
Bliss, Kathleen, 231, 232, 235
Block, Fred, 410n10
Bloom, Harold, 283–84, 286–87, 293n9
Blue Origin (company), 429
Boethius, 247
Bolivia-Biolyse compulsory licensing case, 466
Bolsonaro, Jair, 510n3
Bordeaux School, 32, 235, 285, 294n16
Borgmann, Albert, 5, 18, 113–16, 136, 423
Boschiero, Nerina, 473
Bostrom, Nick, 60–61
Boylan, Michael, 322

BP (company), 422
Breakthrough Energy initiative, 427
breakthroughs, 23, 393, 428, 430
Brew to Bikes (Heying), 115
Brey, Philip, 29
bridging. *See* alienation/bridging (as used by Mitcham)
Briggle, Adam, 8, 326–27, 332, 402, 417–33
British PAS survey of 2014, 508
Broad and Narrow Interpretations of Philosophy of Technology (Durbin, ed.), 167
Brodie, Bernard, 107–8, 109
Broome, Taft H., Jr., 168, 175
Brown, James H., 447
Brundtland, Gro Harlem, 447
Brundtland Report, 447, 448, 451, 453
Brunnschweiler, Christa N., 440
Buber, Martin, 223n7
Bucciarelli, Louis L., 148–49, 156, 171–72
Buddhism, 282, 283, 289–90, 291, 293n12
Bukharin, Nikolai, 398
Bulletin of Indonesian Economic Studies (BIES), 353, 360, 361, 362, 363; reporting on the Green Revolution, 357–59
BULOG (Badan Urusan Logistik) Indonesian food logistics organization, 357, 359–60
Bulte, Erwin, 440
Bundy, McGeorge, 106
Bunge, Mario, 30, 323, 324
bureaucracy, 6, 16, 39, 41, 43, 44, 243, 255–75, 285, 351, 503; bureaucratic depersonalization, 36; Facebook as an exemplary instance of, 272–73; in health care, 160n3, 466, 472; in Indonesia, 351, 352, 355; techno-bureaucratic fascism, 509
Burnham, James, 37, 41, 45–46
Bush, George W., 340, 380

Bush, Vannevar, 8, 485, 493, 494; on basic and applied science, 490–92; Bush myth, 495n8; compared to Humboldt, 486, 488–90; on National Science Foundation, 484, 492, 495n9; relationship between science policy and technology policy, 483; "science" should serve society, 494

CAE. *See* Chinese Academy of Engineering (CAE)
Cahill, Joseph P., 211
California Council of Churches, 238
Canada: Access to Medicines Regime, 466; and oil resources, 438; and vaccines, 463, 466
"Can Technological Artefacts Be Moral Agents?" (Peterson and Spahn), 133
Capital (Marx), 396
capitalism, 37, 232, 374, 384, 400, 403, 404, 419, 421; capitalist economy, 393, 396, 404, 407; and churches, 234, 236–37; cognitive capitalism, 404; digital capitalism, 404; market capitalism, 393, 407; Marx on, 396; as a political religion, 237; in Soviet Union, 45; "turbocapitalism," 373
carbon currency, 421
care ethics in engineering education, 5, 142, 157, 159, 160; role of emotions and attitudes, 161n9. *See also* generalized care; particularized care; universalized care
Carnegie, Andrew, 262
Carson, Rachel, xii, 338
catastrophes, to prevent must believe in possibility of, 95, 111n1
Catechism of the Catholic Church, 212
Cather, Willa, 312
Catholic International Humanum Foundation, 225
Catholicism and Catholic Church, 211–12, 265, 268; on accepting or rejecting technology, 282, 287; Catholic Church as basis for contemporary bureaucratic order, 256; Catholic Church not participating in World Council of Churches formation, 228
Catroga, Fernando, 397
causae mediae, 263
causa instrumentalis, 6, 125, 246, 247–48, 249, 252, 256, 265
causality, 4, 100, 106, 209, 281, 300, 313, 331; causal mechanisms, 130, 136; causal paradoxes, 99, 111n3; causal relations, 14, 15, 16; circular causality, 38, 46, 48–49; quadri-causality, 248, 249, 253; reactions to the past causally bring about future, 103; and technology, 17, 24; uni-causality, 248, 254
Cayley, David, 223n10
Ceccarelli, Leah, 367–68
Cech, Erin, 160n6
celestial hierarchy, 6, 259–60, 271, 273n5, 274n7, 401
The Celestial Hierarchy (Pseudo-Dionysius), 273n4
Cérézuelle, Daniel, 4, 29–53
CESCR. *See* Committee on Economic, Social and Cultural Rights (CESCR) (UN)
Challenger disaster (as poor engineering), 370, 385n2
Charbonneau, Bernard, 4, 32–33, 50, 50n5; concerns about technical action, 43, 46–47; on organization, 36–40, 41; and social totalization, 34–36
charisma (*charism*), 267, 268
Chavez, Cesar, 238
CHEN, Changshu, 192–93, 194
CHEN, Fan, 193
CHENG, Sumei, 166
Chicago Ecumenical Study Group, 239n7
Chicago Tribune (newspaper), 297
China, 165–89, 191–205, 232, 255–56; Emerging Engineering Education

(3E) initiative in, 198, 202; mentions of China in H.R. 4521 (United States Innovation and Competition Act of 2021), 495n12; technological development in, 491, 495n11
Chinese Academy of Engineering (CAE), 182, 194, 198, 201; Division of Engineering Management (DEM), 174, 178–79, 181, 184
Chinese Academy of Sciences, Graduate University of (GUCAS), 173–74, 178
Chinese Academy of Sciences, University of (UCAS), 173–74, 178
Chinese Conference on Philosophy of Engineering, 174, 184
Chinese Institute of Engineer code of ethics, 199
Chinese Ministry of Education, 198
Chinese People's Political Consultative Conference, 200
Chinese Society for Philosophy of Engineering (CSPE), 174, 184
Chinese Society for the Dialectics of Nature (CSDN), 174, 178, 184, 193
Christianity, 6, 209–24, 225–41, 288, 293n12; and anarchism, 284; and bureaucratic order, 256, 262–63, 271; Christianism, 289, 292n6, 293n12; the Church and charisma, 267; the Church and power of divine redemption, 249; the Church and the sacraments, 249, 252–53, 256, 259–60, 264–65, 266, 274n10; the Church as a unique institution, 268; comparing Greco-Roman cosmology with Jewish-Christian theology, 274n13; and Eastern Orthodox Christianity, 229; emphasizing God as "creator of creators," 410n5; Foucault on the Church as a unique institution, 268–69; how Christians respond to God, 6, 219; and ideology, 246; and liturgical order, 262–63; Medieval Christian theology leading to contemporary technology, 286; Mitcham on how Christianity relates to technology, 288–89; and modern technology, 6, 39, 74, 220, 278, 285, 285–86, 287, 293n17; as pro-environment, 289; relationship of the Christian God to the world, 269; as revealed theology, 89; on revelation, 78; and the System, 246; theological paradigms of modern hierarchy, 255–75; Thomas Aquinas on the Church, 263, 264–65, 266; and the Trinitarian argument, 265, 269–71; typological understanding of the Church, 209–24; varieties of religious and technological experiences, 286–91. See also Bible; Catholicism and Catholic Church; Judeo-Christian thought; religion; theology
Christian Newsletter, 231, 232, 235
Churchill, Winston, 80
Ciba (company), 351–52, 354, 357
Cicero, 77
Citroën DS (car), design of, 21–22
city, Leo Strauss on, 70, 74, 76, 79, 87–88
civil engineering, definition of, 424
"The Classical Solution" ("What Is Political Philosophy?") part two (Strauss), 77–78
climate change, 113, 409, 476; causes and consequences, 244, 339, 340, 370; climate change deniers, 341, 421; climate fiction novel about, 421; climate justice, 436; fossil fuel and the energy dilemma, 377, 381, 417, 425, 427, 430; Framework Convention on Climate Change of the UN (UNFCC), 451, 452; and the "iron law," 426, 430; "iron law" of climate change, 430; as a "slow motion" disaster, 431; in sub-Saharan Africa, 8, 435–59. See

also energy ethics; greenhouse gas (GHG) emissions
Clinton, Bill, 105
code of conduct for engineers, 151
Coeckelbergh, Mark, 4, 55–67
cognitive capitalism, 404
Cold War and nuclear deterrence, 104–8
Collier, William L., 352
Colorado School of Mines, 175, 177; Daniels Fund Program in Professional Ethics Education, xvi
commercial innovation, 394, 398
Committee on Economic, Social and Cultural Rights (CESCR) (UN), 467, 469
commodities and commodification, 5, 18, 40, 115, 272, 394, 405, 406–7, 423, 425; commodification, 113–14, 405, 408; "fictitious commodities," 392, 402, 406, 407; making labor and land into commodities, 400; technological commodities, 398, 407
Communist Party of the USSR, 398
compartmentalization, 374–75
compatibilism, 97, 99; "incompatibilitst" thesis, 98, 102
competence, 44, 356, 359, 501, 503, 507, 508; and trust, 502
computerization, Charbonneau on risks involved in, 40
computers, doctors hating, 47, 50n8
Confucianism, 6, 197; and Marxism, 197–99
Congress of US, 483–84
"consilience," 312
consumption, 22, 403, 431, 445, 450, 508; comfort and consumption, 114; energy consumption, 8, 376–77, 418–21, 425–27, 428, 429, 430, 431; and "fictitious commodities," 407; forces of consumption dominating media, 115; and green technologies, 428; in a market society, success means consumption, 409n3; production and consumption, 23, 82, 114, 115, 289; "pseudo-innovation" increasing consumption, 406; world destined to be consumed, 222n1
convenience and energy ethics, 423–25
conviviality and convivial tools, 244, 371, 372, 373, 385, 386n3, 509
"Convivial Software" (Mitcham), 371, 372
convocatio, 220, 223n12
Córdova, France A., 483, 491–92
corpora coelestia, 249
cosmology, 76, 77, 199, 249, 320–21; Greco-Roman cosmology compared to biblical cosmology, 274n13
Council of Nicaea (in 325), 269–71
counterfactual determination of the past by the future, 99–100, 111n3
"counterproductivity," 245, 386n3; counterproductive rebound effect, 429
COVAX, 466, 467
COVID-19, 8, 122, 268, 381, 461–82; handling of the pandemic, 341, 461, 511n16; vaccination campaign in Spain, 507–8
COVID-19 Technology Access Pool (C-TAP), 463, 464, 466, 476
creation, the ethos of, 209
credibility as a variety of trust, 501, 502–3
"The Crisis of Our Time" (Strauss), 80
Critical Perspectives on Nonacademic Science and Engineering (Durbin), 166–67, 168–69, 183
critical theorist quadrant in Feenberg model, 124, 125
critical trust and responsible distrust, 506–9, 511n19
criticism, Greek origin of word, 512n24
Cropsey, Joseph, 71
CSDN. *See* Chinese Society for the Dialectics of Nature (CSDN)
CSPE. *See* Chinese Society for Philosophy of Engineering (CSPE)

C-TAP. *See* COVID-19 Technology Access Pool (C-TAP)
cultural processes. *See* games, narratives, and cultural processes
cultural socialization, 401
CureVac (company), 464, 467–68, 473, 475
Cutcliffe, Stephen H., 167
cybernetics, 40, 44, 46, 48
Cybernetics or Control and Communication in the Animal and the Machine (Wiener), 46

Damania, Richard, 440
D'Ancona, Matthew, 335
dangerous knowledge, 324, 327–31
Dao, 198–99
Daqing oilfield, construction of, 173
Darfur genocide, 439
Darwin, Charles, 308, 309
Darwinian science and liberal education, 6, 298, 301, 305, 308, 309, 310, 313, 314
Dasein, Heidegger's concept of, 13–14, 15, 16, 19, 20, 23
David (king), 215
Davidson, Donald, xii
Davis, Graham A., 440
Davis, James H., 502
De Anima (Aristotle), 309
Declaration of Nyéléni, 437
dehumanization, 36, 78, 226, 231, 233, 298, 306, 313
De institutione musica (Boethius), 247
Deleuze, Gilles, 56, 62, 64
Delft, University of Technology, 151, 153, 161n8, 175
Delft University Press, 171
DEM. *See* Division of Engineering Management (DEM)
democracy, 78, 84, 259, 493, 506; liberal democracy, 422–23
"Demonstrasi Massal" (mass demonstration) in Indonesia, 350–51
Department of Defense (US), 258

Department of Energy (US), 343n5, 422
Department of State (US), 392
depersonalization, 4, 32–43, 46–47, 49–50, 268, 503; of charisma (*charism*), 267, 268
Descartes, René, 82, 170, 193
The Descent of Man (Darwin), 308
descriptive vs. normative statements, 153
design, 23, 36, 172, 180; of a bridge, 375–76, 377; and causal relations, 15, 16, 26n5; design of the Citroën DS (car), 22; and ethics, 307, 314, 371; future designs, 25; God's design, 232; incorporating assumptions and biases, 154; and maintenance, 143, 148, 157, 169; of the Manhattan Project, 374–75; of modern technologies, 65, 288–89; and moral sensitivity, 160; and movement, 61; non-neutrality of, 137; policy design, 436, 445, 449, 452; process of, 4, 20–21, 149–50, 155, 159, 178, 288; of *qi*, 198; redesign, 154, 394; shaping and reshaping of, 26; of technical artifacts, 16–17, 85, 152; value-sensitivity and design, 57, 121–22, 128, 130, 131, 132, 153; and *Wissenschaft*, 486
desires. *See* natural human desires
determinism, 101, 136, 314, 379, 397, 408, 509; determinist quadrant in Feenberg model, 125; seeing technology as neutral and autonomous, 125; "soft determinism," 97–100
deterrent intention, 104–5, 106, 111n11
Devanandan, Paul, 231
Dewey, John, 57, 119
Dharmakīrti, 325
dialectical materialism, 103
digital capitalism, 404
digital hermeneutics, 63

digital media/technology, 56, 57, 59–60, 63, 65, 283, 406. *See also* social media
Dihle, Albrecht, 274n13
Dilthey, Wilhelm, 396–97
Diodorus Kronos, 101, 102, 108
Directives pour un manifeste personnaliste (Charbonneau and Ellul), 36–37, 38, 41
Directorate for Technology, Innovation and Partnerships (TIP) (NSF), 484, 491–92
discernment, 214, 226, 237, 367–89
discouraged knowledge, 319, 323, 324, 327–31
disembeddedness, 7, 391–92, 400, 401, 402; disembedding of *causa instrumentalis*, 247; "first great disembeddedness, 398–403; "second great disembeddedness," 403–9. *See also* embeddedness
disembodiment, 33, 61, 245, 314
disenchanted modern life (*Entzauberung*), 277
distributive justice, 428, 436, 447
distrust, 8, 70, 212, 500, 501–2, 504, 505–6, 510n8; institutional distrust, 506; responsible distrust and critical trust, 506–9, 511n19. *See also* trust
di-symmetry, 223n10
dividing loyalties, 143–45, 159, 160
divine knowledge, 260, 261
Division of Engineering Management (DEM) of Chinese Academy of Engineering (CAE), 174, 178–79, 181, 184
DNA, recombinant, 326, 331
domination, 38, 209, 217, 220, 221, 290, 408
Doorn, Neelke, 5, 141–63, 178, 185
Downey, Gary L., 174
Draper, John William, 282, 288
DU, Cheng, 174–75
Duany, Andres, 114
Dulles, John Foster, 231

Du mode d'existence des objets techniques (*On the Mode of Existence of Technical Objects*) (Simondon), 31, 47, 60
Dunlap, John R., 45
Dupuy, Jean-Pierre, 4–5, 93–112
Durbin, Paul, 166–67, 168, 183
"Dutch disease," 438, 439
Dworkin, Ronald, 422
dynamic momentum, 380
dynamos, 378–79, 380, 384
dystopianism, 125

Earth Summit in Rio de Janeiro in 1992, 447
Eastern Orthodox Christianity, 229
Eastland, James A., 327
Eccles, John, 320
ecclesiastical hierarchies, 259, 260, 261, 271
The Ecclesiastical Hierarchy (Pseudo-Dionysius), 273n4
ecomodern energy ethics, 425–27, 428, 429–30, 431
economics and economic policy, 391–413; artisan economy, 5, 114–15; behavioral economics, 135–36, 137; "donut economy," 421; ecologically disruptive policies, 450; ecomodern energy ethics, 425–27, 428; economic commodification, 113; economic determinism, 400; economic efficiency, 36, 357, 358; economic growth, 393, 409, 426, 427, 431, 435, 438, 440, 449, 450; economic invisibility of women, 446, 450; economic technique, 41; in economic theory economy is autonomous, 410n10; favoring innovation as an economic instrument, 393; market economy, 195, 255, 398–99, 401–2, 403, 404, 405, 407, 409, 410n10; sustainable development in economic debates, 447–48; techno-economic sphere,

391, 408, 409. *See also* commodities and commodification; markets
Economist (journal), 353, 354, 355, 359, 362, 363
"Edison and the Pure Science Ideal in 19th-Century America" (Hounshell), 495n7
efficiency, 3, 42, 45, 46, 126, 284, 375, 429, 502; of counterproductivity, 245; decreased efficiency, 377; economic efficiency, 36, 357, 358; Ellul on, 286, 293n13; energy efficiency, 421, 425; and good governance, 31; and inefficiency, 50n7; and nuclear deterrence, 108, 110
Ehrlich, Paul, 353, 354
Eindhoven, University of Technology, 153
Einstein, Albert, 102, 321, 322, 336, 375, 485
Eisenhower, Dwight D., 337–38, 341
élan vital (life force), 55
Eliot, George, 312
Ellul, Jacques, 4, 6, 50, 235, 293n10; concerns about technical action, 43, 46–47; as a critic of technology, 226, 278, 283, 284–85, 288; on the end of era of the System, 245–46; ethics of non-power, 508–9; and Illich, 244–45, 278, 283, 284–85, 287, 288, 293n14; Mitcham on Ellul's writings, 293n12; naming the post-instrumental age: Erewhon, 244; on organizational technology and techniques of organization, 29, 32–33, 34, 36–37, 40–43, 46–47, 286, 293n13; political study of technology based on revelation, 74; on pre-technological culture, 289; as a substantivist, 126; on technological system, 30, 49, 246–47, 287, 293n13; on the totalitarian State, 36; and the World Council of Churches 1948 meeting, 226, 235

Elohim. *See* Yahweh
Elsevier (publisher), 177
embeddedness, 355, 369, 370, 374, 391–92, 400–401, 410n10. *See also* disembeddedness
embodiment, 127, 370; Church embodying doctrine and practice, 210; embodied value of neutrality, 135; Klenk on, 132; Miller on, 129–30; Pitt on, 129, 130; value embodiment, 128, 129, 130, 131, 132, 134
emergence, Mitcham on concept of, 369, 378, 379, 381, 385
"The Emergence of Chinese Discussions of Engineering, Philosophy, and Ethics" (Nan WANG), 166
Emerging Engineering Education (3E) initiative in China, 198, 202
Emmerson, Harrington, 45
empathy, 20, 144, 155, 502, 503, 507
empirical turn or "thing turn," 4, 29, 31, 32, 48–50, 55
ends and means. *See* means and ends
energy and energy consumption, 376–77, 418, 419; energy and threats to the environment, 8, 417–18, 421, 425, 426, 427, 430–31; energy poverty, 425–26, 428; renewable energy, 114, 379, 425, 451, 453. *See also* oil (petroleum resources)
energy ethics, 417–33; ecomodern energy ethics, 425–27, 428, 429–30; and sufficiency, 419, 420, 421, 422, 423, 424, 425–26, 428; "Type II energy ethics," 418, 428. *See also* climate change
enframing, concept of, 16, 17, 19, 20, 23; Barthes and Baudrillard on, 21; as *Gestell*, 32, 287
engineering, 7, 69, 119–40, 148, 157, 160n2, 161n10, 165–89, 191–205, 290, 292n1, 367–89; abstraction and engineering, 374; applied science view of engineering and technology,

151, 152–53; Center for Research of Engineering and Society, 173–74; distinguishing between engineering and technology, 31; engineer as social actor, 151; Engineering and Public Policy in China, 199; engineering as a technical space without social or political issues, 160n6; engineering as philosophy of *zaowu*, 198; engineering as productive activity of making, 169; engineering epistemology, 179, 180, 181; engineering innovation, 197, 382; engineering methodology, 179–80, 181; engineering philosophy of technology (EPT), 55, 194; engineers and maintenance, 161n10; engineers thinking about their role, 64; etymology of engineering, 195–96; history of, 5; "The History of Engineering in Modern China" (research project), 196; how engineers relate to "The Other," 142, 148–50, 155, 160; ideological mechanisms and world of engineering, 150–56; linguistic philosophical approach to engineering, 63; Mitcham's use of as a frame of reference for development of the philosophy of engineering in China, 191–205; myth and engineering, 63; philosophy of engineering, 166, 168, 173, 177; tripartite relationship between science, technology and engineering, 197; as value laden, 195; ways students are socialized into world or engineering, 156; world of nursing and world of engineering, 145–50, 155. *See also* engineering ethics

"Engineering, Philosophy and Ethics: Emerging Chinese Discussions on Technology in Society" (Nan WANG), 166

engineering ethics, 119–40, 141–63, 169, 182, 186, 201–2, 346, 347, 373, 382; care ethics in engineering education, 5, 142, 156–57, 159, 160, 161n9; engineering and technology ethics, 346; ethical ideologies of engineers in China, 199; Mitcham on, 199, 201, 202, 373, 382; Mitcham on engineering ethics, 382; neutrality of engineering, 156. *See also* generalized care; universalized care; Chinese Institute of Engineer code of ethics, 199

"Engineering Meets Philosophy and Philosophy Meets Engineering" (in Delft, Netherlands), 175

"The Engineering Method" (Koen), 169–70

Engineering Philosophy (Bucciarelli), 171, 184

engineering philosophy of technology (EPT). *See* EPT (engineering philosophy of technology)

Engineering Studies (journal of INES), 174, 184, 185

English persons, pedestrian mentality of, 281, 292n4

Entzauberung (disenchanted modern life), 277

environment: artificialization of, 503; Christianity as pro-environment, 289; and energy, 7, 417–18, 421, 425, 426, 427, 430–31; environmental justice, 377, 430, 435, 440, 453; environmental movement, 23, 25; environmental sustainability, 120, 430; and farming in sub-Saharan Africa, 435, 440, 442, 447, 453; Heidegger on, 13–14, 15, 17, 19; politics and the environment, 339, 340–41, 431; pollution of, xii, 376–77, 381, 421, 426, 430; socio-environmental impacts, 8; technological environment, 18, 136, 159; threats and problems for, 47,

113, 258, 341, 405, 408, 420, 425, 431, 506; trusting environmental organizations, 505; virtual environments, 369. *See also* urban environment
Environmental Protection Agency (US), creation of, 343n5
EPT (engineering philosophy of technology), 55, 194
Erciyas, Selin Sinem, 466, 472
Erdös, Paul, 88
Erewhon, 243–54; post-instrumental age named by Ellul and Illich, 244
Erfani, Parsa, 469
Ethical Issues in Social Science Research (Beauchamp, et al), 328
ethics, 319–34, 461–82; Aristotle on ethics, 300, 485; biological ethics, 305–6; Chinese Institute of Engineer code of ethics, 199; ethical aspects of maintenance, 65; ethical naturalism, 298, 301, 305–6, 307–8, 309; ethical sensitivity, 48, 50, 141, 145, 160n1; ethical values, 154, 325; ethics of non-power before technological determinism, 508–9; ethics-politics, 7, 422; and political science, 7; technology ethics, 65, 72, 346, 472. *See also* bioethics; engineering ethics; moral sensitivity
Ethics and Science (Briggle and Mitcham), 326–27
"Ethics into Design" (Mitcham), 371
etymology of engineering, 169, 195–96
EU. *See* European Union
Eucharist, 250–51, 253, 274n10
Euclid, 102, 247
eudaimonia, 486, 495n10, 510n6
Eudemian Ethics (Aristotle), 264
Euler, Leonard, 88
European Commission, 461
European Investment Bank, 468
European Union, 394, 468; Advance Purchase Agreement (APA) between EU and CureVac, 464, 467–68, 473, 475; brokering a compromise on waiver of COVID-19 patents, 469–70; counter proposal on waiver of TRIPS for COVID-19 vaccines, 465–67; desire for green and sustainable fuels, 381; Emergency Support Instrument, 468; framework science programs of the EU, 509–10n1; website for stimulating innovation, 392–93
Evangelical Alliance, 227
Evans, John, 310
evolutionary continuity, 308
evolutionary neuroscience, 309
evolutionary science of human nature, 305
evolutionary science of life, 298
evolutionary science of religion, 304
existential deterrence, 106–8
"exposome," 340
extropianism, 278, 283, 292n7

Fabini, Tibor, 211
Facebook, 60, 158, 272–73, 372
fairness: and energy policy, 418; as an ethical-legal principle, 8, 474, 477; and intellectual property rights, 8, 464, 475; and resource allocation, 447
Far Eastern Economic Review (publication), 353, 354, 355, 363
farmers: during Indonesia's Green Revolution, 7, 345–66; women farmers in sub-Saharan Africa, 8, 435–59
fascism, 236–37, 403, 509
fatalism, 97, 102, 103, 395
fate, 75, 233, 284, 375; and the chance of nuclear disaster, 107–8, 109
Fayol, Henri, 273n3
fear, Jonas's heuristics of, 89, 431
Federal Reserve Bank (US), 257
Feenberg, Andrew, 4, 13–27, 73, 124–27, 130

"feminization": of agriculture, 442, 449; of poverty, 435, 442, 449
fertilizer, 115, 233; use of in Indonesia, 346, 349, 350, 351–52, 355, 357, 358, 361; use of in sub-Saharan Africa, 437, 443
Feynman, Richard, 312
"fictitious commodities," 392, 402, 406, 407
final values, 128, 130–31, 132
First Chinese Conference on Philosophy of Engineering, 174, 184
Floridi, Luciano, 55
The Fog of War (movie), 104, 111n9
food, 39, 115, 255, 350, 392, 421; and climate change, 8, 350, 438, 442–43, 444, 447, 452, 476; commodified food, 114; food crises, 8, 357, 452, 500; food insecurity, 230, 354, 442, 443, 445; food justice, 436; food production, 115, 257, 353, 437, 438; food security, 8, 354, 435–36, 437–38, 441, 442–43, 444, 446, 449, 452–53; food sovereignty, 436–37, 442, 445; genetically modified, 339, 405–6; and greenhouse gases, 441; and the Green Revolution in Indonesia, 345–66; in sub-Saharan Africa, 435–59; total energy used from food calories, 417. *See also* hunger, problem of
Forbidden Knowledge and Other Essays on the Philosophy of Cognition (Rescher), 329
"forbidden" knowledge or truth, 319, 323, 324–31
Forman, Rebecca, 469
forms of life, 4, 63
Fortuna (Roman goddess), 109
"Forum on Philosophy, Engineering, and Technology" (fPET), 166, 175, 177
fossil fuel industry. *See* oil (petroleum resources)
Foucault, Michel, 6, 256, 268–69, 495n8

Founding God's Nation: Reading Exodus (Kass), 299
fPET. *See* "Forum on Philosophy, Engineering, and Technology" (fPET)
fracking boom, 379–81
Framework Convention on Climate Change of the UN (UNFCC), 451, 452
Francis (pope), 225
Francis of Assisi, Saint, 287, 288, 289
Frankfurt School, 16
Franssen, Maarten, 161n8
Frazer, James George, 279, 281, 292n4
freedom, 271, 397, 417, 418; of action, 34; balancing order and freedom, 235; Charbonneau on freedom, 34–35, 38, 39, 43, 50; destructive potential of human freedom, 236; and economic theory of *laissez-faire*, 235; Ellul on freedom, 42, 43, 50, 244; farmers in Indonesia needing, 357, 358, 359; freedom and technological society, 33–34, 38, 42, 244, 287, 408; and God (Yahweh), 218, 233, 237, 303; Illich on freedom, 223n10; of information, 504; and innovation, 391; of inquiry, 326, 329, 488–89; Jack Sparrow (fictional character) on freedom, 121, 122; Kass on freedom, 303, 305; loss of freedom, 48, 50; National Rifle Association on freedom, 122; personal freedom, 84, 86; preserving freedom, 4; and responsibility, 218; Simondon on, 48; Strauss on freedom, 72, 79, 81, 82, 84; technologies promising freedom, 287; tools as compatible with freedom, 244; unlimited freedom, 325
free will, 97, 101, 270
French Philosophy of Technology (Loeve, Guchet and Bensaude-Vincent), 29

French Reformed Church, 230
Fried, Charles, 329, 332
"From Thinking Big to Small—and Big Again" (Mitcham), 472
Frye, Northrop, 211
Fryer, John, 196
functionality, 21, 120, 371, 388n3
The Fundamental Concepts of Metaphysics (Heidegger), 14, 19–20
future, 2, 5, 7, 29, 86, 135, 219, 220, 297, 340, 349, 395; and artificial intelligence, 59; determined by human actions, 396, 397; and dialectical materialism, 103; digital information influencing the future, 59–60; economic future, 293; and energy, 8, 380, 418, 422, 426, 427–29, 431; environmental future, 384; faith in a perfect future, 286; and future catastrophes, 431; future effecting the present, 94, 111n8; future prospect of philosophy of engineering in China and the West, 166, 170, 186; Heidegger on, 21; Illich on, 245; impact of engineering on, 60; the indeterminacy of the future, 110–11; and industrial and technical society, 36, 44, 396; intellectual property rights (IPR) and the future, 463, 469; and intergenerational justice, 447; Janus face looking toward, 26; leaving resources for the future, 453; of literature, 94; posthuman future, 313; and potentialities, 349; and predetermined future, 101, 108, 110; predictability of, 217; in projected time the future is necessary, 108, 110; and the prophecy of doom, 96–97, 106; relation of past and future, 99–100, 101, 103, 308; of "science" policy, 494; Simondon on, 48; and temporality, 100, 101; trust and the future, 501–2, 504; uncertainty of the future, 110, 217,
502; use of future perfect tense, 95; of work, 405. *See also* "projected time" metaphysics; temporality
Futurism, 410n6
"futurology," 111n6

Galbraith, J. K., 37
Gallup World Survey (OECD), 506
Galston, William, 71
games, narratives, and cultural processes, 56, 62–64
game theory, 111n4
Garcia, José Luis, 7, 391–413, 410n7
Gates, Bill, 427–28
Gellner, Ernest, 397
gender, 445–46, 449; and environmental degradation, 442; gender bias, 450; gender inequity, 441, 445, 446; gender justice, 436; gender policy, 441, 450. *See also* "feminization"; women farmers in sub-Saharan Africa
generalized care, 5, 142, 143, 155–56, 158
genetic engineering, 42, 310, 314, 326, 406, 408
Gerlagh, Reyer, 440
Gestell, 32, 249, 287
Ghana, women farmers in, 435–59
Ghana Environmental Protection Agency, 443
Ghanaian Times (newspaper), 451
Ghana Poverty Reduction Strategy papers (GPRS), 445–46, 449
GHG. *See* greenhouse gas (GHG) emissions
Ghosh, Amitav, 428
Ghosh, Shubha, 474
Gideon, 214, 223n7
Gilson, Étienne, xii
Gingrich, Newt, 293n9
Ginsburg, Norton, 438, 439
Ginsburg, Simona, 309
Girard, René, 107
Glazebrook, Tricia, 8, 435–59

global solidarity, 8, 468, 473, 474; as an ethical-legal principle, 475–77
global warming. *See* climate change
Gnosticism, 284, 293n9; neo-Gnosticism, 286–87
God, 82, 211, 269, 274nn7,8,10, 395; and divine knowledge, 260, 261; emphasizing God as "creator of creators," 410n5; and freedom, 218, 233, 237, 303; God's judgment, 244; how Christian's respond to God, 6, 39, 74; Kass's thoughts on God, 297–316; kingdom of God, 219, 220, 251; love for God, 214; as mystery, 261, 270; portrait of God, 292n5; and the prophet Jonah, 111n8; responsibility to God, 226, 234, 239n6; revelation from God, 84, 305; right relationships with God, 229, 232, 234, 239n6; Thomas Aquinas on God, 263–66; unity of God, 265, 269–71; will of God, 226, 234, 237, 266, 300; word of God, 83, 212, 259, 269; working through secondary or instrumental causes, 267, 274n10; World Council of Churches on, 226–27, 229, 232, 233–34, 237, 239n6. *See also* angels and angelic structure; Bible; Christianity; hierarchical order; Trinitarian argument; Yahweh
Godin, Benoît, 395, 409n4, 490
Golden Bough (Frazer), 279
Goldman, Steven L., 167–68, 170, 171, 183, 184
gong, ministry of, 196
gongcheng, meaning of Chinese word, 192, 196–97
Gongcheng fangfa lun (The theory of engineering methodology) (YIN Ruiyu, WANG Yinluo and Enjie LUAN), 179, 180
Gongcheng lunli (*Engineering Ethics*) (Zhengfeng LI, Hangqing CONG, and Qian WANG), 201

Gongcheng yanhua lun (The theory of engineering evolution) (YIN Ruiyu, WANG Yingluo, and LI Bocong), 179
Gongcheng yanjiu: Kua xueke shiye zhong de gongcheng (Engineering studies: Engineering in interdisciplinary perspectives) (journal), 174–75
Gongcheng zhexue (Philosophy of Engineering) (YIN Ruiyu, WANG Yingluo, and LI Bocong), 170, 172, 173, 178, 179, 180, 181, 201
Gongcheng zhexue yinlun (An introduction to the philosophy of engineering) (LI Bocong), 169, 170, 171–72, 174, 181, 193
Gongcheng zhishilun (The theory of engineering knowledge) (YIN Ruiyu, WANG Yingluo, and LI Bocong), 179
Gongcheng zuofa zeli (Patterns of gongcheng), 196–97
Gordon, Robert J., 114
gotong-royong program in Indonesia, 351
GPRS. *See* Ghana Poverty Reduction Strategy papers
Graduate University of Chinese Academy of Sciences (GUCAS). *See* Chinese Academy of Sciences, University of UCAS)
"grandfather paradox," 111n3
Gray, Asa, 309
"Great Moulting," 34
Great Tradition of Proportionality, 247
The Great Transformation (Polanyi), 392, 399, 400, 403
Greece. *See* Ancient Greeks and "pure inquiry"
Greek philosophy and the Bible, 81–82
green growth and energy ethics, 426, 428, 431
Greenhouse Development Rights, 418

greenhouse gas (GHG) emissions, 103, 436, 441, 449, 451. *See also* climate change
Green Revolution and moral narratives of technological change, 7, 345–66
Grote, Jim, xv
groundnuts, 444–45
Guardini, Romano, 32, 50n1
GUCAS (Graduate University of Chinese Academy of Sciences). *See* Chinese Academy of Sciences, University of (UCAS)
guns as a value-laden topic, 122–23, 264

H1N1 virus, 506
Habermas, Jürgen, 73
Halbertal, Moshe, 213, 222n6
Haldeman, H. R., 109
Handbook of the Philosophy of Science: Philosophy of Technology and Engineering Sciences (Meijers), 177
happiness, 305, 306–7, 417, 418, 419, 420
Haq, Mahbub al, 440
Hardin, Garrett, 421
Hardin, Russell, 511n17
Harvard Business School, 42, 45
Hawking, Stephen, 320
Hecht, Gabrielle, 377
Hegel, George Wilhelm Friedrich, 24, 87, 215, 396
Heidegger, Martin, 26n1, 29, 95, 277–78, 280; concept of "world" (*Dasein*), 4, 13–14, 15, 16, 19, 20, 23; on the disenchantment of the world, 286; Ellul not quoting, 293n10; and essence of technology, 284, 285; Feenberg on Heidegger's philosophy of technology, 13–23; on *Gestell* (*Ge-Stell*), 21, 32, 249, 287; on nature of organization, 50n2; on pre-technological culture, 289; and *techné*, 17–19, 24; valuation of modern technology, 293n11

Heisenberg, Werner, 110
Hempel, Carl, 322–23, 331
Heraclitus, 55
Herf, Jeffrey, 410n9
hermeneutics: Bible from a hermeneutical point of view, 211; digital hermeneutics, 63; (material) hermeneutics, 64; of the Pauline and Johannine Theology of the New Testament, 223n9; process-oriented phenomenology and hermeneutics of technology, 4, 55–67
Heron, Nicholas, 263
Heying, Charles, 115
hierarchical order, 6, 43, 256, 258, 259–60, 259–61, 262, 267, 271–72. *See also* angels and angelic structure
Hippocrates, 312
Hiroshima, bombing of in 1945, 7, 232, 336
historicism, 397; marketological historicism, 391; and positivism, 75, 77; and rationality, 396; techno-liberal historicism, 392, 399, 407, 410n7; technological historicism, 7, 395–98, 410nn7,9
history and philosophy, 345–46
History and Technology (journal), 225
"The History of Engineering in Modern China" (research project), 196
Hobbes, Thomas, 99, 105, 312, 422
Ho Chi Minh, 109
Hoinacki, Lee, 248, 249
Holbrook, J. Britt, 8, 483–97
Holmes, Stephen, 213, 222n6
Holt, Rush D., 492–93, 494n2
Homer, 300, 312
Horkheimer, Max, 23
Hottois, Gilbert, 48, 49, 51n9
Hounshell, David A., 495n7
How to Do Nothing: Resisting the Attention Economy (Odell), 372, 384
HPT (humanities philosophy of technology), 194
HPV vaccines, 506

H.R. 4521. *See* United States Innovation and Competition Act of 2021 (H.R. 4521)
Hromadka, Joseph, 230
Hughes, Thomas, 232
Hugh of Saint Victor, 287, 288
Hui, Yuk, 60, 199
human capabilities, 2; differences in kind from other animals, 308–9, 314
Human Development Index, 440
humanities, 55, 177, 195, 298, 314, 383, 405, 486, 494; humanities philosophy of technology (HPT), 194
human rights, 472, 473, 474, 475, 476; and Intellectual Property Rights, 463–69
Humboldt, Wilhelm von, 486–88, 493, 494; compared to Vannevar Bush, 489–90; "Humboldt myth," 495n8; importance of mathematics, 495n9
Hume, Edward, 23, 121
hunger, problem of: global hunger, 454; in Indonesia, 346, 350, 353, 355, 357, 359, 361, 362–63; reduction of hunger, 8, 346, 362; in sub-Saharan Africa, 8, 435–36, 437, 441–46, 452, 453. *See also* food
The Hungry Soul (Kass), 306
Husserl, Edmund, 50n3, 58
hydroxychloroquine, 341

Ibero-American Survey of 2007, 508
I Ching (The book of change), 198
ideology, 292n8; and Christianity, 246, 285, 287; of depoliticization, 160n6; of domination, 220–21; ideological mechanisms and world of engineering, 150–56; of innovation, 86; of networks, 44; of neutrality, 142, 144, 150, 151–52, 160, 161n9; of progress, 244, 287
IFPMA. *See* International Federation of Pharmaceutical Manufacturers and Associations (IFPMA)

Ihde, Don, 57–58, 59, 61, 126; hermeneutics and technology, 63–65; and phenomenology, 56, 127, 133
Ikea (company), 115
Iliad (Homer), 300
Illich, Ivan, xi, 6, 29, 265, 293n14, 509; on bureaucratic order, 256, 273; and concept of *causa instrumentalis*, 6, 249; on convivial tools, 244, 371, 372, 373, 509; on "counterproductivity," 245, 386n3; as a critic of technology, 278, 283, 284–86, 288; description of the present, xi; and Ellul, 244–45, 278, 283, 284, 285, 286, 287, 288, 293n14; idea of di-symmetry, 223n10; influence of on Mitcham, 293n12, 371, 372–73, 380; on instrumental cause, 266; on instrumentality, 246–47, 249; on proportionality, 247, 290; sharing ideas with Ellul and Wittgenstein, 287; on the System, 246–47; technology as a degradation of Christian faith, 278; on the tool and the user, 265
Illies, Christian, 125–26, 133–34, 135
IMF. *See* International Monetary Fund (IMF)
"Immediate Tasks of the Power of the Soviets" (Lenin), 45
The Imperative of Responsibility (Jonas), 431
"The Importance of Philosophy to Engineering" (Mitcham), 169, 183, 194, 378
Incarnated Verb, 253–54
India, 255, 256, 350; asking for temporary waiver of Trade-Related Intellectual Property Rights, 462–63, 464; brokering a compromise on waiver of COVID-19 patents, 469–70; and World Council of Churches, 228, 231, 232, 235
Indonesia and the Green Revolution, 7, 345–66

industrialism, 285
industrial machinism, 37
Industrial Mission Movement, 230, 238
Industrial Revolution, 43, 113, 152, 182, 395, 398, 402, 403, 429; Third Industrial Revolution, 394
INES. *See* International Network for Engineering Studies (INES)
information technology, 40, 42, 46, 48, 368, 399, 405
innovation, 391–413; characteristics of an innovative paradigm, 165; commercial innovation, 394, 398; and the development of the plow, 251–52, 253; "disembedded innovation," 406; distinguished from invention, 409n3; engineering innovation, 197, 382; European Union's website for stimulating innovation, 392–93; hope for salvation through innovation, 71; ideology of innovation, 86; mobilization of innovation, 392–94; and models, 409n4; National Research Foundation for innovation and R&D, 393; processes and products of, 158; "pseudo-innovation," 406, 407; and rationality, 397; techno-scientific innovation, 32; theological innovation, 6; and value-neutrality thesis, 153–54. *See also* technological innovation
Innovation Policy document of US Department of State, 392
Instagram, engineers working for, 158
instrumentalist quadrant in Feenberg model, 125
instrumentality, 17, 246, 254; framed by proportionality, 247–48; and liturgical order, 262–67, 274n10; religious and cosmogonical origins of, 249; theological expression of, 252–54

instrumentalization, 253–54, 273; of "ends," 258
instrumental period in history, 243, 244
instrumental rationality, 16, 420; of bureaucracy, 256, 257–58
instrumental reality, 30
instrumental value, 120, 122–23, 131, 485
instrumenta separata, 264, 265
instruments. *See* tools
intellectual property rights (IPRs), 8, 461–82
intellectual understanding. *See* Athens
International Asilomar Conference, 326
International Committee for Nanking Safety Zone, 232–33
International Covenant on Economic, Social and Cultural Rights (UN), 469, 476
International Energy Agency, 422
International Federation of Pharmaceutical Manufacturers and Associations (IFPMA), 464–65
International Missionary Council, 232
International Monetary Fund (IMF), 438, 446, 451, 452, 467; Ghana's Progress Report to, 442, 446, 449
International Network for Engineering Studies (INES), 174, 184
International Research Conference (in Denmark), 175, 184
International Rice Research Institute (IRRI), 350
internet, 59, 244, 342, 405, 406; Internet of Things, 114
interruption. *See* momentum/interruption (as used by Mitcham)
intrinsic final values, 130
Introduction to Metaphysics (Heidegger), 17
intstrumentum separatum, 247
invention, 21, 36, 114, 152, 172, 391, 393, 404, 405, 466; invention distinguished from innovation,

409n3; Machiavelli on, 70; mechanical inventions, 48, 402
IPRs. *See* intellectual property rights (IPRs)
IR8 and IR5 rice varieties, 350, 352
Iran, 495n12
"iron law" of climate policy, 426, 430
irrationality, 8, 89, 154, 409; of miracles, 84; of revelation, 82
IRRI. *See* International Rice Research Institute (IRRI)
"is" and "ought," 4, 23–25
Islam, 74, 225, 282, 288, 291
Issues in Science and Technology (journal), 494n2

Jablonka, Eva, 309
Jacobs, Alan, 298, 299
James, William, 93, 279, 289
Jecker, Nancy, 474
Jefferson, Thomas, 307, 508
Jerónimo, Helena Mateus, xi, xv–xvi, 1–9
Jerusalem/Athens: revelation vs. reason, 4, 74, 81, 83–84, 87–89, 298–304, 305; religious understanding representing Jerusalem, 304
Jesus, 210, 211–12, 219–20, 221, 251, 261, 269, 270; believers as disciples of, 222n3; World Council of Churches on, 226, 229, 239n6
Jevon's Paradox, 421
"Jishu kexue zhong de fangfalun wenti (Methodological issues in technical science)" (Qian), 192
Jishu zhexue yinglun (An introduction to the philosophy of technology) (CHEN Changshu), 194
Johannine Theology and the New Testament, 223n9
John Hus Faculty, 230
Johnson, Boris, 510n3
Johnson & Johnson (company), 466
Jonah (prophet), 111n8
Jonas, Hans, 89, 95, 103, 320–21, 431

Jones-Imhotep, Edward, 345
Jotham, 215, 223n7
Jouvenel, Bertrand de, 111n6
Judaism, 74, 79, 225, 283, 288, 291; Church's view of relations to its Jewish origins, 211; and Divine Law, 89; Kass and Judaism, 299–300; understanding the Israelite monarchy, 212–18
Judeo-Christian thought, 209, 222n1, 283; compared to Greco-Roman cosmology, 274n13
Jünger Brothers, 32, 50n1
jury cases studied in University of Chicago Jury Project, 327–28
justice, 82, 328, 425, 426–27, 477; climate justice, 436; distributive justice, 428, 436, 447, 448; environmental justice, 377, 430, 435, 440, 453; ethical-legal principles and IPRs, 474–75; food justice, 436; gender justice, 436, 441–42; intergenerational justice, 447; Kass on, 306, 310, 311, 314; Niebuhr on, 230, 236, 237; as a principle of bioethics, 310, 311; reason and justice, 104; social justice, 225, 238, 435; Socrates on, 77; vaccine justice, 475

Kahn, Pinchas, 213
Kahneman, Daniel, 136
Kant, Immanuel, 23, 94, 396
Kass, Leon, 71; critical assessment of Kass's thought, 7, 297–316
Kavka, Gregory, 111n11
Kelly, Kevin, 292n5
Kerali, Henry, 449
Kierkegaard, Søren, 289
Kim Jong Un, 109
kinesis, 61
kingdom of God, 219–20, 221, 251
Kis, Zoltán, 472
Klenk, Michael, 131, 132

knowledge: adopting industrial objectives producing goods rather than knowledge, 404; dangerous and discouraged knowledge, 319, 323, 324, 327–31; engineering-relevant knowledge, 151, 153–54; "forbidden" knowledge or truth, 319, 323, 324–31; ideal of total knowledge, 322–23, 331; knowledge deficit and trust deficit, 507–8, 511n19; and the knowledge of reality, 62; knowledge sharing, 406, 466, 472–73; light and divine knowledge, 260, 261; and meaning, 62; positive association between level of scientific knowledge and pro-scientific attitudes, 512n23; "pure" knowledge, 320, 320–21, 323–24, 324, 329, 331; revelation as knowledge (for Christians), 89; right knowledge, 325; search for scientific knowledge, 319–34
Koen, Billy Vaughn, 168–69
Kojève, Alexandre, 70
Koran, 292n6
Korea, 495n12
Koselleck, Reinhart, 396, 397
Kroes, Peter, 29, 128, 130–32
Kuchenbuch, Ludolf, 253
Kuhn, Thomas, 165, 200, 491
Kurzweil, Ray, 60–61

labor: artifacts as products of design and labor, 198; as an element of industry, 402; exploitation of, 233, 407; and functionality, 23; labor and value, 423–24, 426; labor costs, 403, 405; people vis-à-vis labor, 236, 289
La bureaucratisation du monde (Rizzi), 45
LaDuke, Winona, 377
Lakatos, Imre, 179
land: land ownership, 441, 442, 443; land redistribution policies, 355; land use planning techniques, 47

Langsdorf, Lenore, 58
language, 22, 74, 223n10, 367–68; Adamic language, 283–84; biblical language, 82; "Emergence" language, 378; Heideggerian language, 60; Hempel on, 322–23; language as the marker of a rational soul, 309; language differences at World Council of Churches, 229, 231; language games, 63, 279, 291, 292n3; language of intention, 105; language of proportionality, 250; language of research, 394; language of science, 281, 339; limited to humans, 308–9; philosophy of language, 63, 170, 277; symbolic language, 15; Wittgenstein on, 63, 277, 278, 279, 280, 291, 292nn1–3
La technique ou l'enjeu du siècle (*The Technological Society*) (Ellul), 40, 41
Latour, Bruno, 26, 30, 49, 64, 347–48
Laudato Si (papal encyclical), 225
L'avenir de la science (*The Future of Science*) (Renan), 34, 50n5
The Laws (Plato), 78
Lee, Steven P., 104
Lefebvre, Henri, 397
legal considerations for intellectual property rights and technology sharing, 461–82
Lehigh University Press, 166–67
Leibniz, Gottfried Wilhelm, 396
Lenin, Vladimir, 45
Les Demoiselles d'Avignon (painting by Picasso), 95
Les deux sources de la morale et de la religion (*The Two Sources of Morals and Religion*) (Bergson), 37
Le Systèm des Objets (Baudrillard), 22
Le système et le chaos (Charbonneau), 38–39
Le système technicien (*The Technological System*) (Elllul), 40, 42
L'état (Charbonneau), 35, 37–38

Levite of Ephraim in Gibeah, 223n11
Lewis, C. S., 313
Lewis, David K., 97, 98, 99, 107
LI, Bocong, 5, 165–89, 193, 201; compared to Louis Bucciarelli, 172; compared with Steven Goldman, 170–71; engineering as philosophy of *zaowu*, 198; on ethics, 202; initiating Center for Research of Engineering and Society, 173; leading a research project "The History of Engineering in Modern China," 196; starting an engineering journal, 174; on tripartite relationship between science, technology and engineering, 197; use of Confucian concepts, 198–99
LI, Tong, 5, 191–205
liangdan yixing (Two Bombs and One Satellite) (Chinese national initiative), 192
lianzheng gongcheng (Government Integrity Project), 197
liberal democracy, 422–23
liberal education, 6, 311–12, 314; Darwinian liberal education, 298, 310, 313, 314
Life, Liberty, and the Defense of Dignity (Kass), 306
L'illusion politique (Ellul), 40–41
liturgical order and instrumentality, 262–68, 274n10
LIU, Dachun, 194, 202
LIU, Yongmou, 5, 191–205
"living instrument" (*instrumentum animatum*), 265, 266, 267, 271, 273; as instrument with intelligence, 274n11
Locke, John, 307, 423–24, 425, 426
Logos, 221, 247
London Missionary Society, 227
López Cerezo, José Antonio, 8, 499–515
L'organisateur (Saint Simon), 43, 46

loss and damage, 436, 452; Loss and Damage Mechanism established in 2013, 451
loyalties, dividing, 143–45, 159, 160
Lucena, Juan C., 174
Lucretius, 312
Luhmann, Niklas, 502, 503
Lukács, György, 16, 21
Łukasiewicz, Jan, 321–22
Lumen Gentium, 211
Lundestad, Eric, 136
Lutheran Church (Missouri Synod), 228
Lycurgus, 84

Machiavelli, Niccolo, 70, 82, 422
MacIntryre, Alasdair, 328, 332
Mackey, Robert, xv, 63, 292n1
Maclaurin, Rupert W., 394
MAD (mutually assured destruction approach to nuclear deterrence), 4, 104–6, 108
"Madman Theory" of Richard Nixon, 109–10
maintainers, engineers as, 156–60
maintenance, 2, 143; budget spent on, 161n10; digital maintenance, 57; ethical aspects of, 65; as part of technology, 56, 57, 157
management techniques, 41, 46, 47, 48; computerized management and hating computers, 47, 50n8; from organization to total management, 43–46; principles of management, 259, 273n3; professionalization of, 45; in research and development, 32, 49
managerial revolution, 44, 45
The Managerial Revolution (Burnham), 41, 45
managers, 37, 45, 258, 272
Man and Machine (Berdyaev), 33
Manhattan Project, 379, 381; as an example of compartmentalization, 374–75
manifest destiny, 384–85

Manifesto of the Communist Party (Marx and Engels), 169
Marcuse, Herbert, 20, 509
Marin, Lavinia, 5, 141–63
markets, 391–413; commercial innovation, 394, 398; construction of national and international markets, 399; and dependence on oil, 440; disconnecting economic production from social contexts, 392; global markets, 392, 448; market capitalism, 393, 407; market economy, 195, 255, 398–99, 401–2, 403, 404, 405, 407, 409, 410n10; market liberalism doctrine, 392; marketological historicism, 391; market society, 409n3; market system, 391, 399, 400, 401, 446; and public administration, 503; as self-regulated, 401. *See also* economics and economic policy
Mar Thoma Syrian Church, 231
Martins, Hermínio, 398, 410n5
Marx, Karl, 198, 271, 289, 396, 398
Marxism and Marxist philosophy, 6, 103, 192, 197–99, 374, 397, 398
mass-energy equivalence, 336
materialization of science, technology as, 277
material techniques, 31–32, 38, 40, 41–43, 46, 47, 49
Matthew (apostle), 219
Mayer, Roger C., 502, 510n10
McAllister, Brian J., 378
McCawley, Peter, 357
McLuhan, Marshall, 2
McNamara, Robert, 104
MDGs. *See* Millennium Development Goals of the UN
meaning, 16, 21, 23, 63; art revealing meaning, 17, 21; computational relations structures meaning, 60; and culture, 57, 63; eternal meaning, 292n6; Heidegger on, 14, 15, 16, 17, 18, 19, 20–21, 26n1; hidden meanings, 20; imaginary meaning, 21; and knowledge, 62; loss of freedom and meaning, 48, 50; "meaning" of life, 25, 219, 279, 288; meaning of words as a matter of language use, 63; the moment of time giving meaning, 223n13; perceived meaning, 19, 217; political meaning, 4, 192, 399; relationship between meaningfulness and means, 243; and religion, 279; and technology, 23, 25, 57, 59, 64, 280, 349, 362

Means, G., 45
means and ends, 243–54; in bioethics, 310–11; of a bureaucracy, 258; evolution of natural kinds and natural ends, 307–10
mechanical production techniques, 41
mechanism and mysticism, 80
"mechanology" of Simondon, 48
media, reporting on Green Revolution in Indonesia: done by *Bulletin of Indonesian Economic Studies*, 357–59; done within Indonesia, 359–61; in the international Anglophe media, 353–56
mediation, 2, 19, 49, 133, 136, 246; postphenomenological mediation theory, 58; through which subjects and objects become, 56, 57–58
Medicines and Related Substances Act of 1997 (South Africa), 462
medieval period. *See* Middle Ages
Meijers, Anthonie W. M., 29, 125–26, 133–34, 135, 176
memory, 308, 510n10; collective memory, 2; technology exteriorizing memory, 60
metacriticism, 7, 267, 367, 368
metaphysics, 30, 131, 168, 271, 283, 494; analytical metaphysics, 97, 99; the multiplicity of metaphysics, 101–3; of nuclear deterrence, 103–10; and "political theology," 222n6; process metaphysics, 58; "projected time"

metaphysics, 4, 110; of the prophecy of doom, 93–97, 100; substance metaphysics, 58; of temporality, 101; as theoretical science, 485
Metaphysics (Aristotle), 485
Michelfelder, Diane P., 176, 178, 185
"Midas problem" of biotechnology, 307
Middle Ages, 248, 253, 282, 286, 378, 395; medieval churches and theology, 6, 249, 250, 251, 286; Medieval Koran, 292n6
Mill, John Stuart, 421, 508
Millennium Development Goals of the UN (MDGs), 435–36
Miller, Boaz, 129–30, 131, 132
Miller, Glen, xi, xii, xiii, xv–xvi, 1–9
millet, importance of in food supply, 444
miniature making, 375, 382, 383. *See also* models and modeling
The Ministry for the Future (Robinson), 421
"Miracle, Maybe" (*Economist*), 354
miracles, 84–85, 89; Wittgenstein on, 280
Mislow, Kurt, 326
missio divina, 209
Mitcham, Carl, xi–xiii, xv–xvi, 55, 64–65, 183, 185, 283, 294m17, 385, 385n1, 494n2, 509; on activism, 290; on advanced forms of energy production, 380; on *causa instrumentalis*, 256; on Christianity and technology, 288–89; on Christians' understanding of "sacred technique" and the sacraments, 274n10; on concept of emergence, 378; on convivial tools, 371–72, 373, 386n3; cultural context of philosophy of technology, 64; cultural influence of engineering, 377; on discernment, 369; on disembedding, 402; on division of technological activity, 170; on Ellul's writings, 293n12; on energy consumption, 418, 428; on engineering, 290; engineering as a dynamo, 379; engineering destroying as it builds, 384; on engineering ethics, 373, 382; on ethical concerns in science, 326–27; on ethics of technology, 472; on evolution of philosophy of technology, 293n10; as "International Distinguished Professor of Philosophy of Technology," 177, 194; on John Dewey's view of technology, 119; on LI Bocong, 181; linguistic philosophical approach to engineering, 63; metaphors used by, 369–82; and moral sensitivity, 143–45, 159; on need for bridges in the philosophy of engineering, 375–76; on need to pay attention, 383; on need to understand from different cultural perspectives, 291; on "nodal events," 381; on nursing ethics and moral perception applied to engineering, 141, 145–46, 148, 154, 155; on philosophy of engineering, 166, 168, 169, 173, 176, 177, 181, 292n1, 374; on process of technical development, use and maintenance, 56; productive activity of making as engineering, 169; questioning whether there should be limits on science research, 332; on scientific method, 493; serving as a frame of reference for development of the philosophy of engineering in China, 5–6, 191–205; on Strauss and the political philosophy of technology, 4, 69–91; on technological design, 122; "technology as activity," 57; throughlines in Mitcham's works on the philosophy of engineering, 7, 367–89; on transcendence, 294n17
mobility, 61, 155
models and modeling, 35, 90, 238, 246, 340, 374, 384, 393, 397, 503; belief-desire model, 101;

business models, 393–94, 472, 476; cybernetic model, 46; in engineering, 177; explaining theologic concepts, 265, 268, 269; Feenberg's model and neutrality, 124–27; for looking at concept of the world, 14, 15; machine model, 45; mathematical models, 340; models in innovation literature, 394, 409n4; of moral sensitivity, 143; natural science models, 75, 471; for perceived trustworthiness, 510n10; philosophical model, 31; for rationality, 171; for social unity, 43; technological models, 8, 44, 287. *See also* miniature making

The Modern Corporation and Private Property (Berle and Means), 45

Mohamed, Goenawan, 360

momentum/interruption (as used by Mitcham), 369, 378–82, 383, 384

monarchy: comparing the Church to, 220; Israelite monarchy, 210, 212–19; typological image of, 221

Moon, Suzanne, 7, 345–66

morality, 77, 82, 197, 212, 347, 424; and Chinese engineers, 201; Christianity and morality, 210, 221; "constitutive political morality," 422; "expressive-collaborative" theory of, 349, 363; grounded in human nature, 309–10; moral commodification, 113; moral dimension of trust, 502; morality and research, 326, 328; morality and the Bible, 301, 305; moral perception, 143–44, 145, 148, 150, 155, 159; moral relevance, 126, 127, 128, 133–34, 135; and neutrality, 422; no morally prohibited truths, 331; political morality, 348, 422; search for truth and moral objections, 323–24; social morality, 348, 421; and technology, 137, 141–63, 348; truth as immunization to moral objections, 323–24

Morality, Prudence, and Nuclear Weapons (Lee), 104

moral sensitivity: in engineering ethics education, 141–63; and ethical sensitivity, 48, 50, 141, 145, 160n1; including perception affectivity and dividing loyalties, 143–45; and nursing ethics, 5, 143, 144, 145–48; toward well-being of others, 152–53

Morris, Errol, 104, 111n9

Morrow, David, 135–36

Morse, Janice, 148, 155

Moses, 84, 212–13, 216, 329

Mounk, Yascha, 97

multiple temporalities, 56, 59–61

"multistability," 58, 59

Mumford, Lewis, 29, 232

Musso, Pierre, 44, 45

mutually assured destruction (MAD) approach to nuclear deterrence, 4, 104–6, 108

My Journey at the Nuclear Brink (Perry), 104

mysticism and mechanism, 80

Mythologies (Barthes), 22

myths and mythology, 63, 109, 280, 336, 493; biblical stories as blend of myth and interpretive history, 212, 213, 325; "Bush myth," 495n8; demythologizing truth, 342; "Humboldt myth," 495n8; myth and engineering, 63, 374; myths of science, 511n19; as pseudo-science, 279, 281; technological mythology, 19–23

NAE. *See* National Academy of Engineering (NAE) (in U.S.)

narrative, 4, 65, 210, 345–66, 378, 381, 382; games, narratives, and cultural processes, 56, 62–64; progress narratives, 369, 378; technological narratives, 60–61, 63–64

National Academy of Engineering (NAE) (in U.S.), 123, 175, 182, 184

National Academy of Sciences (US), 326
National Conference on Philosophy of Engineering, 181
National Health Service, 255
National Oceanic and Atmospheric Administration (US), creation of, 343n5
National Research Foundation, 393, 490
National Rifle Association, 122
National Science Foundation (NSF) (in U.S.), 8, 483–97; created in 1950, 490; Directorate for Technology, Innovation and Partnerships (TIP), 484, 491–92; Merit Review process, 492; Vannevar Bush providing blueprint for, 483, 490, 495n8
natural biological ethics, 306
natural human desires, 298, 304–7, 305–6, 312–13
natural resources, 338, 439, 440, 452
"near-hits" and "near-misses," 104, 106, 111n10
"The Need for a Philosophy of R&D" (Durbin), 167
neo-Gnosticism, 286–87
neolatry, 395, 410n6
neoliberalism, 7, 255, 358, 391, 392, 407, 409
neo-positivism, 278, 279
Netflix, 255, 421
Netherlands, 175; and COVID-19, 462; discovery of offshore oil and impact on economy, 438
neutrality, 119–40; Feenberg's model, 124–26; ideology of neutrality, 142, 144, 150, 151, 155, 160, 161n9; meaning of, 127, 128, 136; and modern energy, 423; and morality, 422; neutrality thesis of technology, 151–52, 153; normative neutrality, 136; recent literature on neutrality/value-ladenness, 5, 128–37; technological neutrality debate,

5, 119, 120, 122, 123. *See also* value-ladenness; value-neutrality
"the new" (neolatry), 395, 410n6
Newberry, Byron, 5, 119–39, 161n7
"The New Biology: What Price Relieving Man's Estate?" (Kass), 71
New Deal in the United States, 401
The New Industrial State (Galbraith), 37
New Order Green Revolution programs in Indonesia, 350
New Testament, 223n9, 263. *See also* Bible
Newton, Isaac, 1, 282, 322, 485
New York Times (newspaper), 97, 353, 356, 359, 362, 363
Nicomachean Ethics (Aristotle), 300, 485, 510n6
Niebuhr, H. Richard, 231, 236–37
Niebuhr, Reinhold, 225, 230, 231, 234, 235, 236, 237
Niger Delta, 439
Nine, Ogoni, 439
Niniveh, prophesy on fall of, 111n8
Nixon, Richard, 109, 492
Noah and the flood parable told by Anders, 95–96
Nobel Prizes: in Chemistry, 336; in Economic Sciences, 410n10; Peace Prize, 336; in Physiology, 320
Noble, David F., 286
nonmaleficence as a principle of bioethics, 310–11
(non)neutrality, 119–40
Nordhaus, Ted, 427, 431
normative neutrality, 136
NSF. *See* National Science Foundation (NSF) (in U.S.)
nuclear power, 39, 339, 379
nuclear weapons, 93–112; nuclear deterrence, 106, 111n11; nuclear disarmament debate in 1958, 336–37, 338, 339, 341, 342; nuclear threats, 105, 499; stockpiling of, 336
"Nuli xiang jingji zhexue he gongcheng zhexue lingyu kaituo: Jian lun

21shiji zhexue de zhuanxiang" (Working hard to initiate philosophy of engineering and of economy: On the turn of philosophy of the 21st century) LI Bocong), 170
nursing and caring, 5, 141–42, 143, 144, 145–48, 150–51, 159; comparing world of nursing and world of engineering, 145–50; moral perception and affectivity affecting who a nurse listens to, 150, 160n3
Nyāya-Bindu (*A Short Treatise of Logic*) (Dharmakīrti), 325
Nygren, Anders, 87

Obama, Barack, 225, 340
Oberdiek, Hans, 125, 127
"objective spirit," 24
obsolescence, programmed, 409
"Occupational *Bildung* and Philosophy of Engineering" (in Denmark), 175, 184
occurring time, 101, 102, 104, 108, 110
Odell, Jenny, 368, 372–73, 382, 384, 385
OECD. *See* Organization for Economic Cooperation and Development (OECD)
Oedipus, 109
oikonomia, 269–71
oil (petroleum resources), 417–33, 435–59; exacerbating climate change, 451–52. *See also* energy consumption; energy ethics
Oldham, J. H., 231, 232, 235–36
Old Testament, 210, 211; authority of, 212; understanding the Israelite monarchy, 212–18. *See also* Bible
Olmstead, Gracy, 122–23
Omega Point, 278
Omens of the Millennium (Bloom), 283
one-dimensionality, 18, 20
On Liberty (Mill), 508
On the Mode of Existence of Technical Objects (*Du mode d'existence des objets techniques*) (Simondon), 31, 47, 60
"On the Spirit and the Organisational Framework of Intellectual Institutions in Berlin" (Humboldt), 486
organic farming, 5, 114, 115
organizational techniques, 29–53, 50n2, 235–36, 255–75, 286
Organization for Economic Cooperation and Development (OECD), 394, 491, 495n11, 506
organon, 247, 264–65
"The Origin of the Work of Art" (Heidegger), 17–18
Ortega y Gasset, José, 284, 287, 289, 293n10
"The Other": characterized by proximity or distance, 142, 148, 155; chart comparing of nurse's and engineering's "Other," 149; in the world of engineering, 141–42, 143, 146, 148–50, 154, 155–56, 157, 159–60; in the world of nursing, 142, 145, 146–48, 158
Ottinger, Gwen, 377
"ought" and "is," 4, 23–26
Our Common Future (Brundtland Report). *See* Brundtland Report
Overton Window, 420
Oxford Dictionaries, 335
Oxford English Dictionary, 1

pandemic, 8, 268, 339, 341, 461–82, 510n3, 511n16. *See also* COVID-19
Papyrakis, Elissaios, 440
paradoxes, 3, 9, 31, 82, 107, 109, 254, 382, 437, 490; grandfather paradox, 111n3; Jevon's paradox, 422; of preventing a catastrophe, 95; and projected time, 102; of the prophecy of doom, 96–97; of reason and revelation, 82; of sustainable development, 447–51
parallel axiom, 102

Paris Agreement on climate change, 436
Parthenon, building of, 18, 20
particularized care, 5, 141, 142, 147, 148, 155, 157, 159
"Part Moon, Part Travelling Salesman: Complementarity in the Thought of Ivan Illich" (Cayley), 223n10
Parviainen, Jaana, 61
Pascal, Blaise, 396
pastoral order and universality, 268–71
Patagonia (company), 258
patents, 394, 406, 465, 466, 467, 473; making pandemic-related research free of patent restrictions, 476; Patent Act (South Africa), 462; patent owners encouraged to cooperate with others, 467; for vaccines, 463, 464, 469–71, 472, 477
Pauck, Wilhelm, 237, 239n7
Paul, Saint (the apostle), 87
Pauline Theology of the New Testament, 223n9
Pauling, Linus (Pauling-Teller debate on nuclear disarmament in 1958), 336–37, 338, 339, 341, 342
perception, moral, 143–44, 145, 148, 150, 155, 159
Perelman, Grigori, 88
performance processes, 56, 61–62
Perry, William, 104
Persecution and the Art of Writing (Strauss), 89
Person, H. S., 45
personnalité juridique, 210
pesticides, 115, 338, 346, 349, 350, 351–52, 355, 357
Peters, Tom, 258
Peterson, Martin, 127–28, 133–34, 135
pharmaceutical companies, 464, 468, 473, 474, 475, 507–8; on South Africa amending the Medicines and Related Substances Act of 1997, 462
phenomenology, 14, 75, 127, 133, 141–42, 146, 155, 246; postphenomenology, 55, 56, 57, 58, 60, 61, 62, 64; process-oriented phenomenology, 4, 64–65
Philo of Alexandria, 274n10
Philosophy and Engineering: An Emerging Agenda (van de Poel and Goldberg), 175, 176
Philosophy and Engineering: Historical-Philosophical and Critical Perspectives (Mitcham, ed.), 193–94
Philosophy and Engineering: Reflections on Practice, Principles and Process (essays from WPE-2008), 175, 184
Philosophy in Engineering (Christensen, Meganck, and Delahousse), 172, 173, 184
Philosophy of Engineering, East and West (Mitcham, et al, eds.), 166, 177, 202
"Philosophy of Engineering and Technology" (Springer series), 175, 176, 185
philosophy of engineering in China and the West, parallel steps toward, 165–89
Philosophy of Engineering: The Proceedings of a Series of Seminars Held at the Royal Academy of Engineering, 175
philosophy of language, 63, 170, 277
Philosophy Today (journal), 71
phrónesis (practical wisdom/good sense), 369, 501, 510n6
physical techniques, 30, 31
Picasso, Pablo, 95
Pielke, Roger, 426, 430
Pirates of the Caribbean—Curse of the Black Pearl (movie), 121
Pitt, Joe, 120, 128–29, 130, 131, 134
Plantinga, Alvin, 99, 282
Plater-Zyberg, Elizabeth, 114
Plato, 78, 82, 170, 171, 248, 269, 300, 312, 325, 493; Aristotle a student of, 488; philosopher king, 420, 495n6; and "pure inquiry," 484–85

"Playing Technology in Religious-
Philosophical Perspective: A
Dialogue among Traditions"
(Mitcham), 288–89
"A Plea for Pure Science"
(Rowland), 495n7
plow, development of, 251–52, 253
POET. *See* "Philosophy of Engineering
and Technology" (Springer series)
Poincaré, Henri, 102
Polanyi, Karl, 200, 391–92, 399–400,
401, 402–3, 404, 406, 410n10
policy, 391–413, 483–97; risk-related
policy debates, 339. *See also*
economics and economic policy;
science and technology policy;
technology policy
*Political Philosophy, Facts, Fiction and
Vision* (Bunge), 30
politics and political science, 4, 69–91,
335–44, 347, 423; beginning
of politics, 213; "constitutive
political morality," 422; and
energy ethics, 419, 420; ideology
of depoliticization, 160n6;
"perfectionist" political regimes,
422; and political activism
demanding accountability, 499;
political economics, 338, 339, 342;
political evolution of society, 499;
political meaning, 4, 192, 399;
political morality, 348, 422; political
theology, 73, 74, 83–84, 86, 88, 213,
222n6; politicization of science, 7, 8,
339, 384; politics of technology, 21;
power politics, 77, 284; symbiosis of
the political and the technical, 35
pollution, 376–77, 381, 421, 426, 430
Popper, Karl, 30, 322, 331, 397
The Population Bomb
(Ehrlich), 353, 354
population growth and food inadequacy.
See food; hunger, problem of

positivism, 44, 75, 77; neo-positivism,
278, 279, 281; positivist social
science, 79
possibilities: belief in necessary to
prevent catastrophes, 95, 111n1;
definition of possible worlds, 111n2;
unrealized possibilities, 25
"The Possible and the Real"
(Bergson), 94
post-instrumental age, 243–54; Ellul and
Illich on, 244
Postman, Neil, 123
postphenomenology, 55, 56, 57, 58, 60,
61, 62, 64
"post-trust society," 8, 499, 500, 504
post-truth era, 7, 335–44
*Post-Truth: The New War on Truth and
How to Fight Back* (D'Ancona), 335
potentia absoluta (absolute potency),
398, 410n8
potentiality, 4, 24–25, 26, 266
poverty: development as a source of,
450; energy poverty, 425–26, 428;
"feminization of poverty," 435, 442,
449; and food security, 435, 437,
449; gendering of poverty, 445;
Ghana Poverty Reduction Strategy
papers (GPRS), 442, 445–46, 449;
impact of Green Revolution in
Indonesia, 346; poverty alleviation,
435–36; Poverty Reduction Strategy
of IMF, 442; in sub-Saharan Africa,
8, 435–59; supporting children
in poverty, 197; and sustainable
development, 447; use of oil to
reduce poverty, 438; World Bank
ranking countries on poverty and
wealth, 445–46; World Council of
Churches on, 227, 231, 233, 234, 236
power, 222n1, 273n5, 292n8, 323, 401,
505; of bureaucratic order, 256, 272;
concentration of power, 235, 236;
counterfactual power, 99–100, 102,
111n3; and energy ethics, 417–33; of
engineering, 195, 198, 201; ethics of

renunciation to power, 244; freedom and power, 81, 223n10; God's power, 226, 234, 249, 263, 274n13, 303; human power, 13, 26, 39, 77, 80, 230, 237, 503; political power, 347, 381; power techniques, 35, 49, 287; relationship of Christianity to power, 209–24, 218–22, 268–71, 293n14; sanctification of power, 259–61; of science, 319, 325, 327, 335, 336, 339; technological power, 6, 13, 16, 26, 80, 84, 218–22, 234–35, 237, 277, 403, 408
practical wisdom (*phrónesis*), 501, 509, 510n6
pragmatism, 57, 64
predeterminism, 97–100
prejudice, 167, 215, 484, 490, 495n7
President's Council on Bioethics (US), 297, 304, 306–7, 310–13, 314
Preston, Christopher, 113
principles of management, 259, 273n3
Principles of Scientific Management (Taylor), 45
Prisma (journal), 353, 359, 363
privatization, 404
"The Problem of Political Philosophy" (Strauss), 71–72, 73
"Problems of Consent in Sex Research: Legal and Ethical Considerations" (Fried), 329
"Problems of Forbidden and Discouraged Knowledge: Intrinsic and Extrinsic Constraints" (Rudner), 324–25
process, 55–67; ascending process of freedom and rationalization, 391; design process, 20, 21, 178, 288; games, narratives, and cultural processes, 56, 61–62; Heidegger on, 20, 21, 23, 24; industrial processes, 374, 407, 451; intangible process, 33, 37; mediation process, 56, 57–58; Mitcham on, 80, 169, 198, 290, 371–72, 386n3; performance processes, 56, 61–62; practices and process, 30; processes and products, 158, 407; process of making time and being made by time, 56, 59–61; process of organization, 34, 40, 41, 44, 47; process-oriented conceptualizations, 56; scientific processes, 1; technical development process, 120; technological development process, 31, 56–57; toward a process-oriented phenomenology, 4, 64–65
productivity, 40–41, 179, 372, 393, 407, 423; agricultural productivity, 436, 445; counterproductivity, 245, 386n3
programmed obsolescence, 409
Program to Combat Racism (World Council of Churches), 238
progress, 37, 281, 396, 450; "costs of progress," 38, 39; and "domain codes," 21; economic progress, 353, 447; engine of progress, 379, 381; Feenberg on, 125; future progress, 397; ideology of progress, 244, 277, 287; intellectual progress, 79; openness to new ideas, 398–99; in philosophy of engineering, 178, 181; progress narratives, 369, 378; progress of material power, 43, 48, 49; progress of technique, 38, 42, 48–49; scientific progress, 279, 313, 476, 489, 490; social progress, 42, 79, 508; Strauss on, 79, 80–81, 89; technical/technological progress, 29, 35, 41, 44, 287, 291, 372, 410n9, 490; and technological singularity, 60–61
"Progress or Return? The Contemporary Crisis in Western Civilization" (Strauss), 79, 81, 83, 87, 90n1
"projected time," 5, 90n1, 101, 103, 104; chart illustrating, 100; nuclear deterrence in projected time, 108–9; and occurring time, 102; uncertainty of the future in projected time, 109–10

Project Hindsight, 491
promised land, 222n5
prophecy of doom, 93–112; quandary of the Prophet of Doom, 95–97
prophets of the Bible, 102–3, 211; Jonah's refusal to prophesy effecting the future, 111n8
proportionality, 244, 245, 249, 285, 290, 475; Great Tradition of Proportionality, 247; instrumentality framed by proportionality, 247–48; theological expression of, 250–54
"prospective," 100, 111n6
Pseudo-Dionysius the Areopagite, 256, 259–60, 261, 273nn4–5, 274n6
"pseudo-innovation," 406, 407
"public good," 376, 468, 474, 476
"pure inquiry," 484–85
"pure" knowledge, 320, 324, 329; and the search for truth, 320–21, 323–24, 331
"pure science," 320, 321, 323, 324, 331, 495n7
Putin, Vladimir, 105

qi, Confucian concept of, 198–99
Qian, Xuesen, 192, 198, 200
Qu'est-ce que la technologie? (What is technology?) (Raynaud), 31
The Question Concerning Technology and Other Essays (Heidegger), 18
The Question Concerning Technology in China (Hui), 199
Quietism, 287

Rappin, Baptiste, 32, 46, 47, 50
rational choice theory, 99
rationalistic view of engineering-relevant knowledge, 151, 153–54
rationality, 105, 396, 421, 422, 424, 508; and cybermanagement, 46; and emotion, 153–54; formal rationality, 310, 311; *Ge-Stell*, 20; and innovation, 397; instrumental rationality, 16, 256, 257–58, 420; linear development of, 407; of means-to-ends relationships, 243; of modern technology, 15–16; necessity-based model of, 171; and nuclear deterrence, 105, 106–7; rationality of engineering, 167; role of in modern societies, 13; scientific rationality, 36, 152; subjecting man to technical rationality, 50n1; substantive rationality, 310
rationalization, 38, 47, 277, 391, 410n9; of production, 23, 45
rational science, 50n5
rational soul, 309
Ratzinger, Joseph (Pope Benedict XVI), 223n8
Ravetz, Jerome, 509
Rawls, John, 73
Raworth, Kate, 421
Raynaud, Dominique, 31
reactionary modernists, 410n9
Reaganism, 392
reality, 20, 26, 47, 55, 62, 101, 113–16, 209, 210, 219, 246, 291, 308, 335; anomalies of, 220; creating reality, 34, 94, 217; instrumental reality, 30; intelligibility of, 395; multifaceted reality, 280; new reality, 34; social reality, 215, 330, 401; technical/technological reality, 30, 48, 50
reason: reason and justice, 104; reason vs. revelation, 4, 74, 81, 83–84, 87–89, 298–304, 305
"rebound effect," 421; counterproductive rebound effect, 429
recognition/concealment (as used by Mitcham), 370
recombinant DNA, debate on research of, 326, 331
red engineers in China, 199–200, 201
redistribution, 355, 419
Reform and Opening Up policy in China, 182, 193, 195, 196, 200, 202
Reijers, Wessel, 63

religion, 6–7, 209–24, 225–41, 243–54, 255–75, 277–96, 297–316; difference between religion and superstition, 280–81, 288; efforts to exclude religion from the public sphere, 86; efforts to harmonize the Bible and Greek philosophy, 81–82; evolutionary science of religion, 304; how religion relates to technology, 281–83; inefficiency as sin against Holy Ghost, 50n7; natural human desire for religious belief, 304; religious and cosmogonical origins of instrumentality, 249; religious beliefs, 227, 277, 279, 280, 281, 282–83, 287, 288, 292n2, 304. *See also* Bible; God; science, religion, and technology; *specific religions*; theology

The Religion of Technology (Noble), 286

Renaissance, 395, 410n5

Renan, Ernest, 34, 50n5

renewable energy, 114, 379, 425, 451, 453

renewal of reality, 5, 113–16

Rengong lun tigang (*An outline for the theory of human mak*ing) (LI Bocong), 169

Renmin University of China, 177, 194

Renn, Ortwin, 501

reporting on the Green Revolution: done by *Bulletin of Indonesian Economic Studies*, 357–59; done within Indonesia, 359–61; in the international Anglophone media, 353–56

Republic (Plato), 300, 325

Rescher, Nicholas, 329, 332

research, 4, 7, 152, 165–89, 191–205, 319–34, 337–38, 483–97; academic research, 13, 177, 487, 489, 490; adopting industrial objectives producing goods rather than knowledge, 404; federal non-defense research expenditures, 1958 to 2021, 342n3; foundation for a modern research university, 486; government support for basic scientific research, 490; growth of, 339, 342n4; historical research, 348; integration of multiple research areas, 405; and intellectual property rights, 464, 468, 474, 475–76; investing in research, 393; marginalization of research areas less susceptible to innovation, 404–5; operational research, 30; research and development (R&D), 32, 49, 114, 167, 255, 342n3, 393, 394, 491, 495n11; research ethics, 509–10n1; social research, 115; subjects needing research, 63, 169, 339, 362; technoscientific research, 391, 407; US investments in scientific research starting in the 1950s, 338

Research in Technology Studies (Lehigh University), 166–67

Research with Recombinant DNA (National Academy of Sciences), 326

resource curse, 438–41

Rest, James R., 143, 146

restoration, Bloom on, 286

revelation, 89, 220; divine revelation, 74, 78, 82, 84, 300, 301; interpretation of, 89, 217; relationship of to power, 218; revelation vs. reason, 4, 74, 81, 83–84, 87–89, 298–304, 305

Rhetoric (Aristotle), 501, 502

Rice, Condoleezza, 105

rice growing: in Ghana, 445; in Indonesia during the Green Revolution, 346, 349–52, 354, 359, 360

Ricoeur, Paul, 56, 63, 64

Riemannian manifold, 102

right relationships with God, 229, 232, 234, 239n6

risks: vulnerability to, 505

risks, technological, 499–515; risk-risk tradeoff, 504; target and countervailing risk tradeoffs (table 27.1), 505
"risk society," 338, 340, 509; risk society as "post-trust society," 499–500, 503–4
Ritalin, debate over use of, 311
Rizvi, Zain, 472
Rizzi, Bruno, 37, 45
Robert, Jean, 6, 243–54
Robinson, James, 439
Robinson, Kim Stanley, 421
Roche (company), 462
Rockefeller, John D., Jr., 228
Roeser, Sabine, 5, 141–63
Rohe, Wolfgang, 488, 495n8
Rolston, Jessica Smith, 418, 428
Roman Catholic Church. *See* Catholicism
Roman Empire, 262, 417
Routledge (publisher), 177; *The Routledge Handbook of the Philosophy of Engineering* (Michelfelder and Doorn, eds.), 178, 185
Rowland, H. A., 495n7
Royal Academy of Engineering, 175, 182, 184
Rudner, Richard, 324–25, 330–31, 332
Russell, Andrew, 156–57, 159, 161n10
Russia/Soviet Union, 45, 103, 105, 173, 229, 495n12; Soviet Union's nuclear stockpile, 336

S&T. *See* science and technology (S&T) policy
Sachs, Jeffrey, 447
sacred texts, 209, 213, 292n6. *See also* Bible; Koran
safety, 145, 151, 156, 157, 348, 359, 476, 499, 507; as a final value, 131
Saint-Simon, Henri de, 43–44, 46, 396
Saint-Simonians, 34, 44
Sajogyo, 352

Samuel (prophet), 213, 214–16, 218, 219, 223n7
Samuel, Sajay, 6, 255–75
Sarewitz, Daniel, 7, 335–44, 488, 490, 494n2
Saro-Wiwa, Ken, 439
Saul (first king of Israel), 213
scandals involving science and industry, 500, 510n4
scenario method (prospective), 111n6
Schelling, Thomas, 109
Schmitt, Carl, 70, 222n6
Schneider, Jen, 7, 367–89
Schoorman, F. David, 502
Schopenhauer, Arthur, 289–90
Schumpeter, Joseph, 393–94, 409n3
Schwartz, Melvin, 321
science, religion, and technology, 209–24, 225–41, 243–54, 255–75, 277–96, 297–315
science, technology, and human values (STHV), 168
Science, Technology and Public Policy in China (STPP), 199
Science, the Endless Frontier (Bush), 8, 483, 484, 485, 488, 489, 490, 491, 492, 494n2
science and technology policy, 7, 8, 417–33, 435–59, 461–82, 483–97, 499–515. *See also* policy
science and technology studies, 319–34, 335–44, 345–66, 367–89, 391–413; connection between science and truth, 7; extrinsic constraints on science, 325, 330, 332; science as part of K–12 education system, 1. *See also* STS (science, technology studies)
scientific humanism, 237
"Scientific Management as a Philosophy and a Technique of Progressive Industrial Stabilization" (Person), 45
scientific method, 1, 44, 169, 192, 325, 330, 493; engineering methodology not a variant of, 180

scientific organization of society, 4, 44
scientific research. *See* research
scientific truth. *See* truth
scientism, 201, 287, 292n8
SDG1 and SDG2. *See* Sustainable Development Goals of UN
Second Treatise (Locke), 423
self-restraint, 421
Self-Strengthening Movement in China, 192, 196, 199
Selzer, Richard, 312
Sen, Amartya, 440
Shakespeare, William, 312
Shellenberger, Michael, 427, 431
Shelley, Mary, 509
Shell Oil (company), 439
Sherlock, Richard, 298, 299
Shiva, Vandana, 450
Shrader-Frechette, Kristin, 509
Silent Spring (Carson), xii, 338
Simon, Julian, 426
Simondon, Gilbert, 29, 31–32, 47–48, 49, 51n9, 58, 60
"Sino-American Seminar on STS," 193
Sinsheimer, Robert L., 328–29, 332
skepticism: about natural resource wealth, 439; farmer skepticism in Indonesia, 352, 357; "loyal skeptics," 508; regarding myths of science, 511n19
"sleeping policeman," 348
Smil, Vaclav, 418, 425–26, 428
Smolin, Lee, 88
social capital, 504, 506–7, 510n12
"The Social Captivity of Engineering" (Goldman), 167–68
social change, 70, 79, 290, 379
social engineering, 30, 200
socialist technological historicism, 398
socialization: cultural socialization, 401; of engineering students into the world of engineering, 156
social justice, 225, 238, 435
social media, 59, 61, 62, 63, 152, 342, 373, 381

social morality, 421
social philosophy, 74, 88
social progress, 42, 79, 508
"social technology" and social techniques, 30–31
social trust, 447, 503–4
Society for Philosophy and Technology (SPT), 167, 175
Socrates, xiii, 2, 73, 76–77, 78, 325
"soft determinism," 97–100
Solomon (king), 215–16
Sophocles, 17
South Africa, 233, 461; amending the Medicines and Related Substances Act of 1997, 462; asking for temporary waiver of Trade-Related Intellectual Property Rights, 462–63, 464, 470; brokering a compromise on waiver of COVID-19 patents, 469; and sustainable development, 448
Southern Baptist Convention, 228
Soviet Union. *See* Russia
Spahn, Andreas, 127–28, 133–34, 135
Spain, COVID-19 vaccination campaign in, 507–8
Sparrow, Jack (fictional character), 121, 122, 123
Speck, Jeff, 114
speed/attention (as used by Mitcham), 369, 370–74, 383
"speed bump," 153, 348
"speed trap," 372, 381
Spinoza, Benedict, 15, 82, 84–86, 87, 88, 89
Spinoza's Critique of Religion (Strauss), 85
Springer (publisher), 171, 175, 176, 177, 181
SPT. *See* Society for Philosophy and Technology (SPT)
SSA. *See* Sub-Saharan Africa, paradoxical problems in
Stalin, Joseph, 398
Stalnaker, Robert, 97, 99

standard of living and energy ethics, 417, 419, 421, 427
Standing Committee on the Law of Patents (WIPO), 472
Stanford University, xii
State Council of China, 182
Steps toward a Philosophy of Engineering: Historico-Philosophical and Critical Essays (Mitcham), 176–77, 185, 382, 385n1
STHV (science, technology, and human values), 168
Stich, Stephen P., 326
Stiegler, Bernard, 60
Stiglitz, Joseph, 410n10
Stoics and Stoicism, 269
Stokes, Donald E., 484–86, 489, 490, 491
Stone, Taylor, 5, 141–63
STPP. *See* Science, Technology and Public Policy in China (STPP)
The Strategy of Conflict (Schelling), 109
Strauss, Leo, 4, 397; impact of on political philosophy of technology, 69–91
STS (science, technology studies), xii, 7, 62, 192, 193, 199, 345, 347, 378, 509
Stzompka, Piotr, 503
Sub-Saharan Africa, paradoxical problems in, 8, 435–59
subsistence: of the Church, 211–12, 219; subsistence farming in sub-Saharan Africa, 8, 435, 437, 438, 440, 441–46, 447, 449, 450, 451–52, 453; "*subsistit in*," Aristotle on, 222n4
substantive rationality, 310
substantivist quadrant in Feenberg model, 125, 126
Suburban Nation: The Rise of Sprawl and the Decline of the American Dream (Duany, Plater-Zyberg, and Speck), 114
Sudan, 439, 443

sufficiency: and energy ethics, 419, 420, 421, 422, 423, 424, 425–26, 428; and nutrition, 436; self-sufficiency, 232, 350
Suger, Abbot of Saint-Denis, 250–51
Suharto, 350, 359, 360
suitcase bomb, example of, 134–35
Sukarno, 350
Summa Theologica (Thomas Aquinas), 252
Sunderland, Mary, 160
superstition, 95, 280–81, 287; difference between religion and superstition, 288; removal of, 82; Spinoza on, 84; technological superstition, 278, 291
supplément d'âme, 80
Survivre et Vivre, 292n8
sustainability: and energy, 418, 420, 426, 429, 453; environmental sustainability, 23, 120, 381, 426, 428, 430; in sub-Saharan Africa, 435, 437, 438, 441, 445, 447–51, 453
Sustainable Development Goals of UN, 418, 424, 435, 436, 437, 438, 441, 448, 452, 453
Suzhou, Jiangsu Province, China, 181
The Synthetic Age (Preston), 113
the System, Ellul on, 244–46
System 1 thinking (taking mental shortcuts), 136
System 2 thinking (making considered choices), 136

Tabernacle, 301–2
Taylor, Frederick and the Taylor Society, 45
techné and technology, 2, 17–19, 24, 76, 77, 247, 510n6. *See also* technics; techniques
technical artifacts, 29; Baudrillard on, 22–23; embodying values, 128; Heidegger on, 15–16, 17, 18–19; value-neutrality of, 130–31, 156. *See also* artifacts; technological artifacts; tools

technical development, 38, 236; processes of, 56–57
technical ontology and the role of time, 25
Technical University Delft. *See* Delft, University of Technology
technics, 76, 199, 232, 233, 235, 283, 287, 289. *See also techné* and technology
Technics and Civilization (Mumford), 232
techniques, 4, 25, 29–53, 126, 237, 246, 277–78, 282, 285, 308, 402; difference between technique and technology, 290; intangible techniques and technologies, 4, 30, 31–32, 42, 47, 48, 49–50; intellectual techniques, 37, 41; modern technique, 33, 41, 235, 285; objectifying power of, 49; organizational techniques, 6, 235–36, 286; physical techniques, 30, 31; sacred technique, 274n10; techniques of man, 41–42; thought techniques, 30. *See also techné* and technology
technocratic view of the engineer as social actor, 151, 154
"techno-liberal historicism," 392, 398, 399, 407, 410n7
Technological Age, 244, 247, 248
technological artifacts, 6, 18–19, 20, 56, 65, 150; care ethics in evaluating technological artifacts, 157; considering whether they are moral agents, 127, 133; as more than tools, 57–58; neutrality theory of technological artifacts, 127, 152; process of technical development, 56–57; treating the Earth as a technological artifact, 85; valuing of, 134; workshop studying technological artifacts, 158–59. *See also* artifacts; technical artifacts; tools

technological change, 70, 345–66, 396, 398, 405, 406; speed of, 370
"technological historicism," 7, 398, 410nn7,9
technological idolatry, 7, 278, 283–86
technological innovation, 71, 154, 404, 407, 423; for commercialization, 392, 398; deregulation of, 409; expanding meaning of the term, 395; first use of the term, 394; as technological historicism, 395–98
technological mythology, 19–23
technology, 13–27, 29–53, 55–67, 69–91, 113–16, 119–40, 209–24, 225–41, 277–96, 461–82; classified with ordinary tools, 15–16; difference between technique and technology, 290; hermeneutics of technology, 4, 63, 64–65; as materialization of science, 277; moral and ethical dimensions of technological change in the Green Revolution, 7, 345–66; needing a user, 137; pervasiveness of, 1–3; as realm of neutral tools, 264; techno-economic sphere, 391, 408, 409; technological neutrality debate, 5; technological neutrality-value-ladenness of, 124–28; technology dependency, 158–59; technology sharing and intellectual property rights, 461–82; transformation of, 1–2. *See also* conviviality and convivial tools; philosophy and technology; science, religion, and technology; science and technology studies; STS (science, technology studies); *techné* and technology
technology and philosophy of engineering, 177, 191
Technology and the Character of Contemporary Life (Borgmann), 18
"Technology as Applied Science" (Bunge), 30

"Technology: Autonomous or Neutral?" (Oberdiek), 125
technology games, 63
Technology in Society (journal), 166
technology policy, 8, 417–33, 435–59, 461–82, 483–97, 499–515. *See also* science and technology policy
Teilhard de Chardin, Pierre, 49
teleology, 4, 24, 308, 309, 398
Teller, Edward, 336–37; Pauling-Teller debate on nuclear disarmament in 1958, 336–37, 338, 339, 341, 342
Tempo (magazine), 353, 359, 360, 361, 363
temporality, 250, 349, 396; metaphysics of, 101–2; multiple temporalities, 56, 59–61, 62, 63; temporal chain, 98, 99–100. *See also* time
"terministic screens," 7, 367, 370, 374, 380, 384
Tertullian, 87, 269
Thales, 312
Thatcherism, 392
theology, 225–41, 249, 255–75; analytical theology, 97; Calvinist theology, 99; and *causa instrumentalis*, 6, 249; Christian theology, 74, 256, 269, 286, 287; comparing Greco-Roman cosmology with Jewish-Christian theology, 274n13; Ellul's theological appraisal of modern technology, 74; political theology, 73, 74, 83–84, 86, 88, 213, 222n6; revealed theology, 89; and Strauss, 73–74, 83–84, 86, 87, 88, 89; technicist theology, 49; theological expression of instrumentality, 252; theological-scientific-technological parallels, 410n5; Trinitarian theology, 271. *See also* Christianity; religion
thick and thin bioethics, 310–13
"The Thing" (Heidegger), 18
"thing turn." *See* empirical turn or "thing turn"

Thinking through Technology: The Path between Engineering and Philosophy (Mitcham), xv–xvi, 119, 169, 183, 193, 195
Third Industrial Revolution, 394
Third International Missionary Council, 232
"The 35,000 Villages That Know That Growth Works" (*Economist*), 355
Thomas, M. M., 231, 232–33
Thomas Aquinas, Saint, 87, 252–53, 263–66
Thompson, Paul, 57
Thorne, Kip, 88
thought techniques, 30
Three Gorges Dam, construction of, 173, 182
Tianjin University, 198
Tierney, Thomas, 424
Tiles, Mary, 127
Tilton, John E., 440
Timaeus (Plato), 248
time, 56; counterfactual determination of the past by the future, 99–100; end-time, 211; entry of the Logos into human time and the world of time, 221, 223n13; indeterminacy of future in conception of tim making future necessary, 110; liturgical time, 250; objective time vs. experience time, 55; occurring time, 101, 102, 104, 108, 110; philosophical views on concept of time, 395–97; process of making time and being made by time, 56, 59–61; projected time, 5, 101, 102, 103, 104, 108–10; real time, 46, 48; role of in technical ontology, 25; time travel paradoxes, 111n3. *See also* future; temporality
Time (magazine), 353
TIP. *See* Directorate for Technology, Innovation and Partnerships (TIP) (NSF)
Tollon, Fabio, 132
Tolstoy, Leo, 289, 312

tools, 33, 243–54, 287, 288, 307, 373; embedding of, 158; equitable access to tools, 474; as "generalized hardware," 180; Heidegger on, 14–15, 26n2; instrument as separate from the user, 264–65, 267; intellectual tools, 36; knowledge as a tool, 325; and the knowledge of reality, 62; multifunctional tools, 196; and the pandemic, 461, 463; perfecting of, 37; philosophical tools, 128; for research, 338; strategic tools, 109; technology differing from tools, 15–16, 17; truth as a tool, 335; ways of producing tools, 290. *See also* artifacts; technical artifacts; technological artifacts

Tools for Conviviality (Illich), 244, 386n3

Torah, 82, 84, 88, 216, 299, 300, 302

Torvik, Ragnar, 439

totalitarianism, 34–36, 40, 236

Toward a More Natural Science (Kass), 301, 305, 307–8

Tractatus (Wittgenstein), 278, 280–81, 288, 290, 291

Tractatus Politicus (Spinoza), 84

Tractatus Theologico-Politicus(Spinoza), 84

Trade-Related Intellectual Property Rights (TRIPS) (WTO), 462–63, 464–71, 472, 473, 474, 477

transcendence, 274n8, 287, 294n17, 301; transcendent Deity, 260, 261, 269

transgenic seeds, 406

Treasury Department (US), 257

Tredgold, Thomas, 424

Trinitarian argument, 265, 269–71

TRIPS. *See* Trade-Related Intellectual Property Rights (TRIPS) (WTO)

Troeltsch, Ernst, 396–97

"The True Grand Challenge for Engineering: Self Knowledge" (Mitcham), 385

Trump, Donald, 97, 109, 341, 510n3

trust, 499–515; complex nature of, 500–503; and confidence, 510n9; credibility as a variety of trust, 501, 502–3; critical trust and responsible distrust, 506–9; dimensions of, 502; displaced trust, 505; and distrust, 510n8; dynamics of trust, 504–6; erosion of trust, 8, 500; between farmers and students in Indonesia, 350–51; institutional trust, 499, 506, 506–7, 511nn15,17; "post-trust society," 8, 499, 500, 503–4; social trust, 447, 503–4; target and countervailing trust tradeoffs (table 27.2), 506; trust-building, 351; trust deficit, 507; trustor and trustee, 510n11; trust-trust tradeoffs, 505; trustworthiness, 510n10; vertical trust, 511n17. *See also* distrust

Trust and Power (Luhmann), 502

truth, 7, 171, 291, 306, 319–34, 335–44; connection between science and truth, 7; criterion for testing, 198; demythologizing truth, 342; dogmatic truths, 488; "forbidden" knowledge or truth, 319, 323, 324–31; heavenly truth, 221; Noah "clothed in the habit of truth," 95; no morally prohibited truths, 331; objective truth, 493; post-truth era, 7, 335–44; pursuit of truth, 489; quest for truth, 72; of revelation, 301, 304; scientific truth, 336, 339–40, 341; self-evident truth, 303; settled truths, 304; truth of facts, 281; truths of religion, 277, 279, 281, 299, 303, 304; truth value of a proposition, 95; universal truths, 322, 510n6; utilized by technology, 3; varieties of, 20

Tsinghua University, 193, 201

TU Delft. *See* Delft, University of Technology

Tversky, Amos, 136

Twelve Principles of Efficiency (Emmerson), 45
The Two Sources of Morality and Religion (Bergson), 37, 93
"Type II energy ethics," 418, 428
"Types of Harm in Social Research" (Warwick), 328
typewriter as an example of convivial technology, 371–72, 386n3
typological understanding of the Church, 209–24

UCAS. *See* Chinese Academy of Sciences, University of (UCAS)
Ullate, José Antonio, 6, 209–24
Ulrich, Laurel Thatcher, 113
Unbestimmtheitsrelation (principle of uncertainty), 110
uncertainty, 217, 401, 503, 508; uncertainty of the future, 109–10, 502
UNESCO, 471, 473; Universal Declaration on Bioethics and Human Rights, 475–76
UNFCCC. *See* Framework Convention on Climate Change of the UN (UNFCC)
unintended consequences, 377, 427
Union Theological Seminary, 225, 230, 236
United Christian Missionary Society, 232–33
United Farm Workers, 238
United Kingdom, reaction to COVID-19, 500, 510n3
United Nations (UN), 228, 438, 453; Bruntland Report, 447, 448, 451, 453; Committee on Economic, Social and Cultural Rights (CESCR), 466, 469; Food and Agriculture Organization, 436; Framework Convention on Climate Change (UNFCC), 451, 452; Guiding Principles, 473; International Covenant on Economic, Social and Cultural Rights, 469, 476; Millennium Development Goals of the UN, 435–36; Special Rapporteur in field of cultural rights, 476; Sustainable Development Goal of UN, 418, 424, 435
United States: brokering a compromise on waiver of COVID-19 patents, 469; and COVID-19, 341, 510n3. *See also individual agencies or departments*
United States Innovation and Competition Act of 2021 (H.R. 4521), 483–84, 495n3; searches for China in, 495n12
Universal Declaration on Bioethics and Human Rights (UNESCO), 475–76
universality and the pastoral order, 268–71
universalized care, 5, 142, 143, 156–60
universal truths, 322, 510n6
universities, importance of for inquiry and research, 487–89, 495n9
University of Chicago Jury Project, 327–28
University of Nanking, 232
University of Twente (Netherlands), 175
University of Wisconsin, 50n8
urban environment, 5, 20, 36, 40, 61, 114, 233, 508; urban planning, 41, 47
urbanism, 5, 114
Üstün, Zeynep Çağla, 461, 466, 472

vaccines: COVID-19 vaccines, 8, 255, 341, 461–82, 507–8, 511n20; HPV vaccines, 506; "vaccine nationalism," 463, 464, 469, 476
value-ladenness, 119–20, 121–24, 168, 199; fuzziness of neutrality/value-ladenness, 124–28; normative value-laden activities, 142; *qi* as value-laden, 198; recent literature on neutrality/value-ladenness, 5, 128–37
value-neutrality, 119, 120, 122, 124, 130, 133; value-neutral attitudes and

viewpoints, 5, 119, 122, 123, 130, 156, 159; value-neutrality thesis, 153–54, 158, 161n7; value-neutral technical artifacts, 152, 156. *See also* neutrality
values, 119–40, 339, 368, 407, 410n6, 446; community values, 123–24; conflicting values, 328, 337; conflicts of values, 195, 200; embedded value, 121–22, 129; embodied value, 121–22, 130; engineering not value free, 195; ethical values, 154, 325; extrinsic final values, 128, 130–31, 132; human values, 3, 120, 125, 126, 168, 331–32; instrumental value, 120, 122–23, 131, 485; intrinsic values, 120, 123, 129, 130, 131, 485; moral relevance, 127; moral values, 362, 502; norms and values, 177, 510n12; societal values, 123–24, 327, 332, 400; truth value of a proposition, 95; value embodiment, 128, 129, 130, 131, 132, 134, 177; values and facts, 23, 25, 75, 80; value sensitivity design, 57, 128
Van de Poel, Ibo, 128, 130–32, 176
Van Dusen, Henry, 230
Van Gogh, Vincent, 18
Van Grunsven, Janna, 5, 141–63
Vantage Technology Consulting Group, 50n8
Verbeek, Peter-Paul, 55, 58, 61–62, 127, 133
Verdier, Thierry, 439
Verdugo-Serna, Carlos, 7, 319–34
Vermaas, Pieter, 152, 156, 176, 185
Vincente, Carlos, 449
Vincenti, Walter, 180
Vinsel, Lee, 156–57, 159, 161n10
Visser 't Hooft, W. A., 229–30
Voegelin, Eric, 237
voluntarism, 102, 103, 398
Vuillemin, Jules, 101

Walker, Margaret Urban, 7, 347, 349, 363
WANG, Nan, 5, 165–89
WANG, Yingluo, 184, 185, 200, 201
Warren, Mark, 511n15
Warwick, Donald P., 328
Warwick, Kevin, 287
"Washington consensus," 392
Watanabe, Takeshi, 356
Watson, James, 312
Weaver, Kathryn, 141, 144–45, 148, 154, 155; comparing nursing and engineering, 145–46
Weber, Max, 255, 256–58, 259, 267, 268, 277
Weinberg, Steven, 88
Welfare State in Europe, 401
the West and China having parallel steps toward philosophy of engineering, 165–89
Western States Communication Association, 367
Wetmore, Jameson, 345, 348
Wharton Business School, 45
What Engineers Know and How They Know It (Vincenti), 180
What Is Political Philosophy? (Strauss), 71–72, 77, 78; originally given as a lecture at Hebrew University, 87–88
Whelchel, Robert, 120
"When Technologies Makes Good People Do Bad Things" (Morrow), 135
Whitehead, Alfred North, 55, 56, 58, 62
Whitman, Walt, 312
"Why We Need a Philosophy of Engineering: A Work in Progress" (Goldman), 171
Wichita case in 1954 as example when research can cause harm, 327–28, 332
Wiener, Norbert, 46
Wieseltier, Leon, 106
Wildavsky, Aaron, 212
Wilhelmina (queen), 228

Wilson, Edward O., 312–13
Winner, Langdon, 73, 123, 152, 161n7, 509
WIPO. *See* World Intellectual Property Organization (WIPO)
Wissenschaft ("science and scholarship"), 486–87, 488, 489, 493, 494
Wittgenstein, Ludwig, 6, 56, 63, 292nn1–4; a Wittgensteinian approach to religion and technology, 277–96
women farmers in sub-Saharan Africa, 435–59
word processor. *See* typewriter as an example of convivial technology
Workshop on Philosophy of Engineering (WPE) (in the Netherlands and in London), 175, 184
World Bank, 353, 436, 438, 442, 445, 448, 451, 452; brief on Ghana's economic update, 449, 450
World Council of Churches (1948), 225–41; statement of unity as "the basis," 229
World Health Organization (WHO), 467, 468; COVID-19 Technology Access Pool (C-TAP), 463, 466, 476
World Intellectual Property Organization (WIPO), 467, 472
World Missionary Conference in Edinburgh in 1910, 228
World Student Christian Federation, 231
World Trade Organization. *See* WTO (World Trade Organization)
"Wo zaowu gu wozai: Jian lun gongcheng shizai lun (I create, therefore I am: On engineering realism)" (LI Bocong), 170, 183, 193
WPE. *See* Workshop on Philosophy of Engineering (WPE)

WTO (World Trade Organization): Agreement on Trade-Related Intellectual Property Rights (TRIPS), 462, 464–71, 474, 477; efforts to revise Agreement on Trade-Related Intellectual Property Rights, 463; Ministerial Conference, 463, 469, 470; on public-private partnerships, 477; US backing negotiations to waive TRIPS, 464
WU, Qidi, 196
Wynne, Brian, 509

Xenophon, xiii, 77
xiwang gongcheng (I Hope Project), 197
XU Kuangdi, 174, 178

Y2K, fears of a universal computer collapse, 111n1
Yahweh, 213, 214–15, 216, 217–18, 263
YIN, Ruiyu, 174, 184, 185, 200, 201, 202
YIN, Wenjuan, 197
YMCA, 228
Young, Mark Thomas, 57
YouTube, 421

Zain, Harun, 351
zaowu, 193, 198
ZHONG, Denghua, 198
Zhongguo gongchengshi shi (The history of Chinese engineers) (Qidi WU), 196
"Zhongguo gongchengshi xintiao (The code of ethics of the Chinese Institute of Engineer) from 1933 to 1996" (Junbin SU and CAO Nanyan), 199
ZHOU, Qiu, 200
ZHU, Qin, xi, xiii, xv–xvi, 1–9, 177

About the Contributors

Gordon Akon-Yamga is a research scientist at the CSIR-Science and Technology Policy Research Institute of Ghana in Accra, Ghana. He obtained his PhD in philosophy from the University of North Texas. His dissertation examined and gave critical analysis of environmental justice in Ghana's oil and gas sector. His areas of research also include science, technology, and innovation (STI) policymaking, STI policy evaluation, resources recovery from waste, and gender in STI.

Jennifer Karns Alexander is a historian of technology specializing in technology and religion. She is director of Graduate Studies in History of Science, Technology and Medicine at the University of Minnesota, where she is also associate professor in the Department of Mechanical Engineering. Her book *The Mantra of Efficiency* (2008) was awarded the Edelstein Prize from the Society for the History of Technology. She is especially interested in the intersection of technology and orthodoxy, in its many different forms.

Andoni Alonso Puelles is professor in the Department of Philosophy and Society at Universidad Complutense de Madrid. He has published widely in philosophy of technology and STS, often focusing on computer and communication technologies. Areas of interest include the free software and communitarian technological movements; diasporas and the use of telecommunications; and *luddismo*, its contemporary versions, and criticisms of progress. He has written *La quinta columna digital: Antitratado comunal de hiperpolitica* (2005) and coedited (with Pedro Oiarzabal) *Diasporas in the New Media Age: Identity, Politics, and Community* (2010).

Pamela Andanda is professor of law at the University of the Witwatersrand, Johannesburg. She previously practiced as a commercial litigation advocate and was a policy research consultant before joining academia. She has acted as an expert for the World Intellectual Property Organization's Standing

Committee on the Law of Patents (WIPO/SCP) and was an external expert reviewer of the World Health Organization's guidance document on the *Ethics and Governance of Artificial Intelligence for Health* (2021). Her research interests are in the areas of intellectual property, privacy and data protection, research integrity and ethics, biotechnology, health law, and regulation of emerging technologies.

Larry Arnhart is professor emeritus of political science at Northern Illinois University. He studies the history of political philosophy and the application of biological science to issues in political philosophy. He has published four books and many articles. He was an associate editor of the *Encyclopedia of Science, Technology, and Ethics* (2005), working under Carl Mitcham.

Albert Borgmann is regents professor emeritus of philosophy at the University of Montana, Missoula, where he has taught since 1970. His special area is the philosophy of society and culture. Among his publications are *Technology and the Character of Contemporary Life* (1984), *Crossing the Postmodern Divide* (1992), *Holding On to Reality: The Nature of Information at the Turn of the Millennium* (1999), *Power Failure: Christianity in the Culture of Technology* (2003), and *Real American Ethics* (2006).

Adam Briggle is associate professor and the director of graduate studies in the Philosophy and Religion Department at the University of North Texas. His teaching and research focus on the philosophy and politics of science and technology. He is particularly interested in environmental ethics and climate change. He is the author of *A Field Philosopher's Guide to Fracking* (2015) and *A Rich Bioethics* (2010). He has also served as a longtime Governing Board member for the Public Philosophy Network.

Daniel Cérézuelle was born in Bordeaux in 1948. He studied philosophy and social sciences with Jean Brun, Hans Jonas, and Jacques Ellul. As a philosopher he has taught philosophy of technology in France and in the United States and since 1991 served on the board of the Société Francophone pour la Philosophie de la Technique. His main field of research is the technological imaginary. As a sociologist he is investigating the social importance of non-monetary economy in a modern society and serves as scientific director of the Programme Autoproduction et Développement Social (PADES). Among his publications is *La technique et la chair* (2011).

Mark Coeckelbergh is professor of philosophy of media and technology at the University of Vienna, where he leads a philosophy of technology research group, and the former president of the Society for Philosophy and

Technology. He is the author of numerous publications including *Introduction to Philosophy of Technology* (2020), *AI Ethics* (2020), and *The Political Philosophy of AI* (2022).

Neelke Doorn is professor of ethics of water engineering at TU Delft. She is director of education of the Faculty of Technology, Policy and Management. Her research concentrates on moral and methodological issues in engineering and technology. She is coeditor of *The Routledge Handbook Philosophy of Engineering* (2021).

Jean-Pierre Dupuy is professor emeritus of social and political philosophy at Ecole Polytechnique, Paris, and professor of political science and of science, technology, and society at Stanford University. He is a member of the French Academy of Technology, a spinoff of the Academy of Sciences, and an honorary member of the Conseil Général des Mines, the French High Magistracy that oversees and regulates industry, energy, and the environment. He chairs the Ethics Committee of the French High Authority on Nuclear Safety and Security. He is the director of the research program of Imitatio, a San Francisco foundation devoted to the dissemination and discussion of René Girard's mimetic theory. He is author of numerous major works, including, in English, *The Mechanization of the Mind* (2000); *On the Origins of Cognitive Science* (2009); *The Mark of the Sacred* (2013); *Economy and the Future: A Crisis of Faith* (2014); *A Short Treatise on the Metaphysics of Tsunamis* (2015); and *How to Think about Catastrophe: Toward a Theory of Enlightened Doomsaying* (2022).

Andrew Feenberg teaches in the School of Communication, Simon Fraser University, where he directs the Applied Communication and Technology Lab. He served as directeur de programme at the Collège International de Philosophie in Paris from 2013 to 2019. His books include *Questioning Technology* (1999), *Transforming Technology* (2002), *Heidegger and Marcuse* (2005), *Between Reason and Experience* (2010), and *The Philosophy of Praxis* (2014). His most recent book is *Technosystem: The Social Life of Reason* (2017).

José Luís Garcia is senior researcher at the Institute of Social Sciences, University of Lisbon. Garcia has held visiting positions and lectureships at various universities in Portugal, Spain, France, Italy, Argentina, Brazil, and the United States. He was a member of the National Ethics Committee for Clinical Research, president of the Ethics Council of the Portuguese Sociological Association and of the Portuguese Observatory of Cultural Activities, and member of the board of the Society for Philosophy and

Technology. He has written numerous major essays in the fields of social and critical theory, philosophy of technology, social studies of science and technology, and communication and media studies. In English, he is editor or coeditor of *Media and Portuguese Empire* (2017), *Pierre Musso and the Network Society: From Saint-Simonianism to the Internet* (2016), and *Jacques Ellul and the Technological Society in 21st Century* (2013).

Tricia Glazebrook is professor of philosophy in the School of Politics, Philosophy and Public Affairs at Washington State University. She publishes on Heidegger and science, the history and philosophy of science and technology, ecofeminism, international development, gender and climate change, military ethics, and the ethics of "big data." Current research projects investigate the capacity of climate finance to meet the needs of women subsistence farmers in Ghana and Ghana's oil development strategies to address poverty. She also is funded by the United States Department of Agriculture to investigate gene editing in dairy cows to prevent horn growth.

J. Britt Holbrook has a doctorate in philosophy and is associate professor in the Department of Humanities at New Jersey Institute of Technology. His research combines topics relevant to contemporary science policy (interdisciplinarity, open science, open access, altmetrics, broader impacts requirements for grants) and of perennial concern (ethics education, peer review, academic freedom, the role of the university in society). Holbrook served on the AAAS Committee on Scientific Freedom and Responsibility from 2012 to 2018 and as a member of the European Commission Expert Group on Indicators for Open Science in 2019. He is the editor in chief of *Ethics, Science, Technology, and Engineering: A Global Resource* (2015).

LI Bocong is professor of philosophy at the University of Chinese Academy of Sciences. His main areas of research are philosophy of engineering, sociology of engineering, and history of engineering. He has published several books including *An Introduction to Philosophy of Engineering* (2002), *An Introduction to Sociology of Engineering: Research on Engineering Community* (2010), and *A Brief History of Modern Engineering in China* (2017). He is the founder of the *Journal of Engineering Studies in Chinese* (2009).

Tong LI is assistant professor at the School of Medical Humanities, Capital Medical University, and member of the Society for Natural Dialectics of Beijing, China. Her teaching and research are about the historical, social, and philosophical issues in biology and medicine. Another line of her research concerns philosophy of technology and engineering. She is currently carrying on a project focused on the potential philosophical problems of cyborgs.

Yongmou LIU is professor in the School of Philosophy at the Renmin University of China, Beijing. He is the managing director of the Chinese Society of the Dialectics of Nature and director of the Society for the Dialectics of Nature of Beijing. His work focuses on the philosophy of science and technology, science, technology and society (STS), and science, technology and public policy (STPP).

José A. López Cerezo is professor of logic and philosophy of science and director of the Research Group on Social Studies of Science at the University of Oviedo. His main academic interest is the study of social and philosophical aspects related to science and technology (STS), particularly topics linked to scientific culture and citizen participation. His last publication in English within this field, coedited with Belén Laspra, is *Spanish Philosophy of Technology: Contemporary Work from the Spanish Speaking Community* (2018).

Lavinia Marin is assistant professor in the ethics and philosophy of technology section, TU Delft. Her current research investigates the conditions of possibility for epistemic agency for users of social networking platforms. Another line of research concerns educational development in ethics education for engineering programs. Together with colleagues from the Comet project, she is currently building a framework toward theorizing an experiential approach in ethics education.

Carl Mitcham is international distinguished professor of philosophy of technology at Renmin University of China and professor emeritus of humanities, arts, and social sciences at Colorado School of Mines (USA). His publications include *Thinking through Technology: The Path between Engineering and Philosophy* (1994) and *Steps toward a Philosophy of Engineering: Historico-Philosophical and Critical Essays* (2019).

Suzanne Moon is associate professor in the history of science at the University of Oklahoma. Her research focuses on the culture, history, and politics of technology in Southeast Asia, especially Indonesia. Her recent work explores Indonesia's postcolonial industrialization (1950–2010), paying special attention to the moral concerns and debates that informed Indonesia's emerging sociotechnical order. Between 2010 and 2020, she was editor-in-chief of *Technology and Culture*, the flagship journal in the history of technology.

Byron Newberry is professor of mechanical engineering at Baylor University. He holds BS and MS degrees in Aerospace Engineering and a PhD in Engineering Mechanics. His technical background is in aerospace materials and structures, and in ultrasonic non-destructive evaluation; his non-technical research topics include engineering ethics, philosophy of engineering and technology, and higher education research. He primarily teaches engineering design, engineering and technology ethics, and sustainable engineering. Newberry has served as an associate editor for the journal *Science and Engineering Ethics* and serves as editor for the Springer book series "Philosophy of Engineering and Technology."

Jean Robert was a Swiss architect who migrated to Mexico in 1972, which became his home for the next fifty years and (he says) where his intellectual biography began. "The contact with a culture in which so much can be achieved with so little radically questioned my professional certainties." The first person who helped put his culture shock into words was John Turner, whom he met at Cuernavaca, where he lived most of the time since. Under the guidance of Ivan Illich, Robert began to examine how what people can best do by themselves (walking or biking) was being replaced by what technologies can do for them (trains, automobiles, airplanes) and the ways "increased speed in transport dissolves the particularities of a landscape into fleeting images, apparently confirming the coordinate space of mathematical physics." He has authored *Le temps qu'on nous vole: Contre la société chronophage* (1980), coauthored (with Jean-Pierre Dupuy) *La trahison de l'opulence* (1976), and contributed to numerous publications (e.g., Wolfgang Sach's *The Development Dictionary*, 1991; Lee Hoinacki and Carl Mitcham's *The Challenges of Ivan Illich*, 2002) but is most well known in Mexico as the designer of a flush-less toilet and as an adviser to the Zapatistas.

Sabine Roeser is head of the Ethics and Philosophy of Technology Section at TU Delft. She teaches and publishes on emotions and ethics of risk, also in the context of nuclear energy, climate change, and public health issues. Her latest book is *Risk, Technology, and Moral Emotions* (2018).

Sajay Samuel has a PhD in business administration and is clinical professor of accounting at Penn State University. His research aims at clarifying some of the foundational assumptions of management thought and practice, as well as parallel research on political philosophy and history, especially as they apply to technology. In addition to many publications on management and accounting, he is also the book series editor of *Ivan Illich: 21st Century Perspectives*.

Daniel Sarewitz is professor of science and society, and codirector and cofounder of the Consortium for Science, Policy, and Outcomes (CSPO), at Arizona State University. He is interested in relationships among knowledge, technology, uncertainty, disagreement, policy, and social outcomes. He is editor-in-chief of *Issues in Science and Technology* and was a regular columnist for *Nature* from 2009 to 2017.

Jen Schneider is professor of public policy and administration and interim associate dean of the School of Public Service at Boise State University. Her research addresses the communication of science and technology controversies, ranging from energy transitions to rapid urban growth. Among her publications is *Under Pressure: Coal Industry Rhetoric and Neoliberalism* (2016). She teaches courses in US energy policy, qualitative methods, and engagement and empathy in public service.

Taylor Stone is a postdoctoral researcher in the Sustainable AI Lab at Bonn University's Institute for Science and Ethics. He received a PhD in ethics of technology from TU Delft in 2019, and has since held postdoctoral and lecturer positions in industrial design and ethics of technology at TU Delft. Originally from Canada, he received a BA from the University of Toronto in architectural design and a master's in environmental studies from York University, and he previously worked in urban policy and community development as a project manager in the nonprofit sector. His research explores the convergence of philosophy and design, with a specialization in urban lighting.

José Antonio Ullate is an independent researcher and publisher. He collaborates with the University of the Earth (Oaxaca). His initial training is as a jurist, and his field of research for more than twenty years has been the classical doctrines of *ius naturale*. He obtained a PhD in philosophy from the National University of Distance Education in Spain, with a thesis on the philosophical and theological foundations of the doctrine of private property in the teaching of the Church. He is currently conducting research on mutability and permanence in human nature.

Janna van Grunsven is assistant professor in the Ethics and Philosophy of Technology Section at TU Delft. Under the auspices of the 4TU.Center for Engineering Education, she does collaborative research on best practices for ethics education in engineering and design curricula. Additionally, she works on human embodiment and social interaction and the ways in which these are affected by different clinical and communication technologies.

Carlos Verdugo-Serna is emeritus professor of philosophy at the Universidad de Valparaiso, Chile. He has been a Fulbright and British Council scholar. His research areas include philosophy of science, philosophy of technology, bioethics, and the philosophy of Karl Popper. He was the founder and director of the Center for the Study of Science, Technology and Society at the Universidad de Valparaiso. He was an international consulting editor of *Ethics, Science, Technology, and Engineering: A Global Resource* (2015). He is the author of many papers on ethical problems in science and technology, on bioethics, and on Popper.

Nan WANG is associate professor and the associate director of the Philosophy Department at the University of Chinese Academy of Sciences. Her teaching and research focus on the philosophy of engineering and technology, and the sociology of engineering. She is particularly interested in engineering knowledge, and deviances in engineering and engineering ethics.

About the Editors

Glen Miller is instructional associate professor in philosophy at Texas A&M University (USA). He has a bachelor's degree in chemical engineering and a master's and doctorate in philosophy. He coedited (with Ashley Shew) *Reinventing Philosophy and Technology, Reinventing Ihde* (2020) and is an associate editor of the journal *Science and Engineering Ethics* and an executive editor of *Philosophy & Technology*. He is secretary for the Society of Philosophy and Technology and a member of the Executive Committee of the Society for Ethics Across the Curriculum. His current research focuses on philosophy and engineering, environmental philosophy, philosophy of technology, and cyber and AI ethics.

Helena Mateus Jerónimo is a sociologist with a PhD from the University of Cambridge, UK, and assistant professor at ISEG, Lisbon School of Economics and Management, Universidade de Lisboa, Portugal. Her research focuses on a philosophical and sociological study of science and technology, with a particular emphasis on the topics of risk and uncertainty, scientific expertise and decision-making, and environmental and technological calamities. Her books in English include *Jacques Ellul and the Technological Society in the 21st Century* (coeditor, 2013) and *Portuguese Philosophy of Technology: Legacies and Contemporary Work from the Portuguese-Speaking Community* (editor, 2023). She has been a member of the Executive Board of the Society for Philosophy and Technology (SPT) and is currently a member of the UNESCO World Commission on the Ethics of Science and Technology (COMEST).

Qin Zhu is associate professor of engineering education at Virginia Tech (USA). He is an associate editor for *Science and Engineering Ethics* and *Studies in Engineering Education* and an Executive Committee member of the Society for Ethics Across the Curriculum (SEAC). His major research interests include the cultural foundations of engineering ethics, global engineering education, and the ethics and policy of robotics and computing technologies. His most recent work examines the ethical values communicated in the "hidden curriculum" in engineering and the design of robots that contribute to a flourishing "moral ecology" of human–robot interaction.

www.ingramcontent.com/pod-product-compliance
Lightning Source LLC
Chambersburg PA
CBHW022132300426
44115CB00006B/149